ANNUAL REVIEW OF
FLUID MECHANICS

ANNUAL REVIEW OF FLUID MECHANICS

VOLUME 15, 1983

MILTON VAN DYKE, *Co-Editor*
Stanford University

J. V. WEHAUSEN, *Co-Editor*
University of California, Berkeley

JOHN L. LUMLEY, *Associate Editor*
Cornell University

ANNUAL REVIEWS INC.   4139 EL CAMINO WAY   PALO ALTO, CALIFORNIA 94306 USA

ANNUAL REVIEWS INC.
Palo Alto, California, USA

International Standard Serial Number: 0066-4189
International Standard Book Number: 0-8243-0715-1
Library of Congress Catalog Card Number: 74-80866

PREFACE

A significant component in the planning of these volumes is topics suggested by readers. Although the members of the Editorial Committee are chosen to embrace as much as possible of the wide world of fluid mechanics, interested readers can discern subjects ripe for review that we would miss. Experience has shown that a suggestion is ordinarily most useful when the proposed author is someone other than the suggester himself. Ideally, in fact, we want a list of the three people in the world best qualified to survey the subject. We have by now settled on meeting the Saturday before Thanksgiving Day to plan the next volume. Any Editor or member of the Editorial Committee will therefore particularly appreciate a suggestion received shortly before mid-November.

We are indebted to Raymond J. Emrich of Lehigh University for arranging the prefatory article. It began with Emrich's visit in 1967 to the Ernst-Mach-Institut in Freiburg, when he learned that most of Mach's scientific correspondence is kept in the institute library. On a later visit in 1980, he saw a recently acquired box containing photographic plates that had been provided by Karma Mach, the wife of Ludwig Mach. Amazingly modern shadowgraphic pictures and interferograms of flows containing shock waves, many probably taken by Ernst Mach himself, are preserved in excellent condition. It seemed to Emrich that physicists and historians of science and technology would be interested in knowing more about a milieu where a man could think deeply about the nature of things, and at the same time take the first spark pictures of a projectile in flight. He proposed to Steven L. Goldman, who was organizing a series of lectures at Lehigh on "Science, Technology and Society," that Heinz Reichenbach, director of the Ernst-Mach-Institut since 1972, be invited to discuss Mach's contributions to experimental physics. Our article is the text of the lecture that Reichenbach delivered at Lehigh on 14 April 1981, subsequently lightly edited by Emrich.

Our thanks go to John F. Kennedy for his service as a member of the Editorial Committee in planning five consecutive volumes, of which this is the last.

THE EDITORS

SOME RELATED ARTICLES IN OTHER *ANNUAL REVIEWS*

From the *Annual Review of Astronomy and Astrophysics*, Volume 20 (1982):

   Interaction Between a Magnetized Plasma Flow and a Strongly Magnetized Celestial Body with an Ionized Atmosphere: Energetics of the Magnetosphere, S.-I. Akasofu
   Planetary Magnetospheres, D. P. Stern and N. F. Ness
   Gas in the Galactic Halo, D. G. York

From the *Annual Review of Biophysics and Bioengineering*, Volume 11 (1982)

   Solid and Liquid Behavior of Red Cell Membrane, R. M. Hochmuth

From the *Annual Review of Earth and Planetary Sciences*, Volume 10 (1982)

   Poleward Heat Transport by the Ocean: Observations and Models, Kirk Bryan
   The Transport of Contaminants in the Great Lakes, Wilbert Lick
   Heat and Mass Circulation in Geothermal Systems, Ian G. Donaldson
   Magma Migration, D. L. Turcotte
   The Heat Flow from the Continents, Henry N. Pollack

Special Announcement: Volume I of the *Annual Review of Immunology* (Editors: W. E. Paul, C. Garrison Fathman, and Henry Metzger) will be published in April, 1983.

Annual Review of Fluid Mechanics
Volume 15, 1983

CONTENTS

ANNUAL REVIEWS INC. is a nonprofit scientific publisher established to promote the advancement of the sciences. Beginning in 1932 with the *Annual Review of Biochemistry*, the Company has pursued as its principal function the publication of high quality, reasonably priced *Annual Review* volumes. The volumes are organized by Editors and Editorial Committees who invite qualified authors to contribute critical articles reviewing significant developments within each major discipline. The Editor-in-Chief invites those interested in serving as future Editorial Committee members to communicate directly with him. Annual Reviews Inc. is administered by a Board of Directors, whose members serve without compensation.

For the convenience of readers, a detachable order form/envelope is bound into the back of this volume.

ca. 1910

Ernst Mach

Ann. Rev. Fluid Mech. 1983. 15:1–28
Copyright © 1983 by Annual Reviews Inc. All rights reserved

CONTRIBUTIONS OF ERNST MACH TO FLUID MECHANICS[1]

H. Reichenbach

Ernst-Mach-Institut, 7800 Freiburg i. Br., West Germany

Often I am asked why the Ernst-Mach-Institut bears the name of a famous philosopher of the 19th century. The question is certainly justified, because my institute deals with research in experimental gas dynamics, ballistics, high-speed photography, and cinemaphotography. We are not in any way doing philosophical studies. In this article, I describe the contributions of Ernst Mach to experimental physics, especially to gas dynamics and ballistics, and thus not only explain our institute's name but also serve a great physicist's reputation for noteworthy discoveries in natural science.

1. ERNST MACH THE UNIVERSALIST

The wideness of Ernst Mach's intellectual world enabled him to influence many branches of physics and medicine, and, as we shall see, to relate his findings to philosophy and to strongly influence that branch of philosophy known as positivism.

One of the fields, for instance, in which he was especially interested, was the analysis of the human senses. By analyzing the sensations of bodily movement, he was the first to explain correctly the function of the labyrinth of the inner ear. In the course of his studies, he constructed a chair that could rotate about two perpendicular axes (see Figure 1), which is noticeably similar to modern devices for training astronauts.

During his investigations of the physiology of the human eye, he discovered a phenomenon of visual contrast that is today called "Mach bands."

[1] The author gratefully acknowledges the assistance of Prof. Raymond J. Emrich, Lehigh University, Bethlehem, Pa., in preparing and editing this article.

1

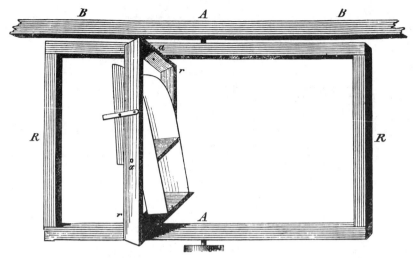

Figure 1   Mach's chair to study human sensations of movement (Mach 1875).

Mach was even more interested in epistemology than in the physiology of the senses, however, because he believed that things only existed if they could be perceived by the senses. The interaction between observer and observed object fascinated him especially. This interest also formed the basis of his critical analyses of the foundations of science.

It is well known that Mach's criticism of Newtonian mechanics and his proposal for relativistic mechanics provided the basis of Einstein's theory of gravitation and general relativity. In a letter of 1901, Einstein called himself a student of Mach. In his commemorative address on Ernst Mach, Einstein surmised that Mach could have stated the special theory of relativity fifty years earlier if physicists had taken more seriously the constancy of the speed of light experimentally evident at that time. It was also Einstein who introduced the name "Mach's principle" for the hypothesis that it is not Newton's *absolute space* but distant masses in the galaxies that form the origin of Coriolis and centrifugal forces.

Mach's epistemology has had enormous influence on science. His theory of knowledge originates in the principle of economy of thought. Following this principle, science must keep away from all metaphysical constructions. Natural laws are, for Mach, only useful abbreviations that make a repetition of measurements or calculations unnecessary.

2.   MACH NUMBER

Today in the era of supersonic flight and space flight, the term "Mach number" has become a household word. Every schoolboy and schoolgirl indulges in the romance of flying at Mach 1 or Mach 2, or whatever his or her counting ability permits. It was, however, J. Ackeret (1929), in his thesis submitted for the certificate of *Habilitation*, who suggested calling the ratio of flow speed to local sound speed the Mach number:

$M = v/a$.

If one looks at a modern textbook on aerodynamics or gas dynamics, the name Mach appears also in "Mach angle," "Mach wave," and "Mach reflection." Ernst Mach carried out the fundamental research explaining and defining these terms. He can properly be considered the originator of supersonic aerodynamics.

He was also innovative in his pioneering work in ballistics and in ballistic measurement technique. He detected and visualized the bow shock wave in front of a projectile moving with speed higher than the speed of sound.

It was more than a hundred years ago that Ernst Mach found the essential difference between subsonic and supersonic flight speeds. He was first to recognize the real nature of shock waves in air and to study their reflection from boundaries and from other shock waves. He was also first to recognize what the various gas elements are doing as they emerge from a supersonic nozzle and encounter the shock fronts that are needed to accommodate the surrounding gas to the jet. To carry out these studies, Mach had to invent new experimental techniques such as trigger events, delay circuits, and time-interval measurements at a time when even the electron tube had not yet been invented.

Mach's experimental work should not be considered in isolation from his philosophy and his whole life's work. His experimental methods exemplify his philosophy, for many of the phenomena he investigated are not visible with the naked eye. By means of new experimental techniques, he made physical processes perceptible to the human senses, and found, in many cases, a correct physical interpretation. He practiced what he preached.

3.   A SKETCH OF MACH'S LIFE

Ernst Mach was born on 18 February 1838 in Chirlitz-Turas, Moravia, not far from Vienna, Austria. His father, Johann Mach, was tutor for the family of Baron Breton. He had studied philosophy at Prague and was interested in animal psychology and in agriculture. One of his achievements was starting silkworm culture in Europe. Ernst inherited his stubborn personality from his father.

Josephine Mach (*nee* Langhans), Ernst's mother, was artistic, having a talent for music, drawing, and poetry. Both parents were idealistic and solitary people.

Mach described himself as a "weak pitiful child who developed very slowly." At the age of ten, he entered the lowest class of a classical secondary school (*Gymnasium*) directed by Benedictine monks, but his intellect and temperament were not well suited to such an education. He had not the slightest taste for Latin and Greek. The clerical teachers found the boy to be "very much without talent" and "unfit for study." They advised his father to let him learn a trade or business. Like many original thinkers, Ernst Mach found no purpose in the meaningless parroting of facts. He preferred to try to discover for himself the origins and causes of things.

Disappointed with his son's failure, Mach's father resumed teaching him Latin, Greek, history, algebra, and geometry at home.

After the suppression of the revolution of 1848, Austria entered a reactionary clerical period. Mach's liberal family thought for a time of emigrating to America, and Ernst asked permission to learn a trade to be able to earn money in the New World. For two years he was apprenticed to a cabinet maker in a neighboring village. This was a pleasant period in his life, since he always enjoyed making things with his own hands. But in the end, the family remained in Austria.

In 1853, when he was fifteen, Mach entered the sixth class of the *Gymnasium* at Kremsier in Moravia, also directed by monks. He finished his studies two years later, and wrote afterwards, "I passed the examination only by pure chance."

In the same year, at the age of seventeen, he went to the university in Vienna to study mathematics and physics. Mathematics and natural sciences were rather poorly represented there, but Mach found the philosophy, philology, and history teaching to his liking. One wonders if this, combined with the inspiration of reading Kant's *Prolegomena to Any Further Metaphysics* and the early education by his father, is the reason for the antimetaphysical direction of Mach's thoughts and his special interest in physics, psychophysiology, and epistemology.

At the age of twenty-two, Mach took the degree Doctor of Philosophy with a thesis titled "On Electrical Discharge and Induction." In 1861, he qualified as *Privatdozent* in physics at the University of Vienna. His first lectures were "Physics for Medical Students" and "Advanced Physiological Physics."

In 1864, at the age of twenty-six, Mach was appointed to the chair of mathematics at the University of Graz. Incidentally he was considering at the same time accepting the chair in surgery at the University of

Salzburg. In 1867, Mach became Professor of Experimental Physics at the German University of Prague. Here he spent twenty-eight prolific years.

Important critical and historical books and papers were written in this period: *History and Root of the Principle of the Conservation of Energy*; *The Science of Mechanics*; *The Analysis of Sensations and the Relation of the Physical to the Psychical*; *Studies of Flying Projectiles*.

In 1895 Mach was appointed to a chair in Vienna created for him in "philosophy, especially the history and theory of the inductive sciences." In Vienna he published "Principles of Heat," a critical account of the science of heat.

In 1898 Mach suffered a paralytic stroke. His right side was paralyzed, but fortunately he remained mentally alert. Even after his illness he finished the book *Perception and Error*.

In 1901 he finished his active career. In the same year, Emperor Franz Joseph appointed Mach a member of the Austrian *Herrenhaus*, but Mach refused to be ennobled. For him only science was nobility.

In 1913 Mach moved to Vaterstetten near Munich to live with his son Ludwig. Together they worked on his last major book, *The Principles of Physical Optics*. However, he could not finish this work. The volume was first published in 1921.

Mach died one day after his seventy-eight birthday, on 19 February 1916.

By this short sketch of his life, we see that Ernst Mach belonged to an era where universality and broad education were a matter of course for a scientist. He was among the great scholars who had the privilege of actively searching for and assessing the wealth of scientific knowledge appearing in many fields in his time.

Many books and articles have been written about the personality of Ernst Mach. Most of these have analyzed and criticized the importance of his contribution to the fundamentals of science and the philosophical value of his work. Unfortunately most publications, with some exceptions (Merzkirch 1966, 1970, Kutterer 1966), barely mention his research in gas dynamics and other parts of experimental physics.

Let us now pick out some examples of his experimental contributions to physics, especially to gas dynamics and ballistics.

4. DOPPLER EFFECT

When he was only a twenty-two-year-old student, Mach contributed to the understanding of the Doppler effect. Christian Doppler (1803–1853),

Assistent at the Polytechnic Institute in Vienna, observed in 1841 that at the approach of a whistling train a stationary observer hears a higher frequency sound than the natural tone of the whistle, and then hears a lower frequency after the train has passed.

Doppler's discovery, however, was not appreciated at the time, as he had included some untenable ideas about the color of stars. In particular, the respected Viennese mathematician Petzval, famous for his work in photographic optics, did not accept Doppler's discovery and explanation. Together with the astronomer Mädler, he quarreled with Doppler for years (Herrmann 1966). Ernst Mach was successful in producing the Doppler effect in the laboratory by means of a simple piece of apparatus and in demonstrating the correctness of Doppler's formula for the dependence of the sound-wave frequency on the movement of the source. This report was published in 1860 under the title "On the Change of Color and Tone by Movement" (Mach 1860). Here he acknowledged the correctness of Doppler's interpretation and disproved Petzval's theory (Mach 1861). Obviously, Mach, even in his youth, was more respectful of facts than of authority.

In order to make the change in the received frequency audible, Mach constructed the device shown in Figure 2. A 1.8-m rod A-A' turns in bearings C-C' about axis B-B'. A hole F bored in A and connecting through B' and C' to bellows H supplies air to whistle G of known frequency f. The arm A-A' is rotated by a cord passing over pulley D. An observer standing in the plane of rotation of the whistle hears a frequency f' alternately higher and lower than f as the whistle approaches and recedes. Mach reported that the amount of the frequency change clearly depends on the whistle speed.

Some years later in Prague, Mach initiated further experiments to test the correctness of the Doppler formulas (Mach 1878a). Our institute has two hand-written documents in which it is mentioned that Ernst Mach was present, with some colleagues, at an experiment conducted with two trolley cars passing on adjoining rail tracks. On one car was a whistle, and on the other were the observers. In a second experiment, the observers stood near the track and listened as the trolley with the whistle passed. Those present certified by signature on the documents that in the second case they perceived a smaller Doppler frequency shift than in the first.

Concerning the astronomical consequences of the Doppler effect, Mach proposed to analyze the spectrum of a star and to determine the speed of the star from the displacement of the spectral lines (Mach 1878a). Mach informed Kirchhoff of this idea. In a reply letter of October 1860, of which we have the original, Kirchhoff shares Mach's

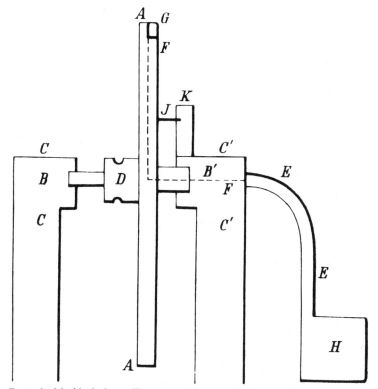

Figure 2   Mach's device to illustrate and explain the Doppler effect (Mach 1860).

opinion that by study of the spectrum one can determine the motion of the source. Kirchhoff points out that the spectra of sodium and magnesium have characteristic lines that could be easily identified in a shifted recording. No doubt Mach was one of the pioneers in spectral astronomy.

5. WAVES GENERATED BY ELECTRIC-SPARK DISCHARGES

One sound source Mach used for his physiological and physical studies in acoustics was the electric spark (Mach & Fischer 1873). He was quite interested to learn from his assistant Dvorak of a publication by K. Antolik (1875) that dealt with acoustic phenomena of spark discharges. Inspired by the work of Antolik, a Hungarian high school teacher, a number of experimental studies (Mach & Wosyka 1875, Mach & Sommer 1877, Mach et al. 1878, Mach 1878b, Mach & Grüss 1878, Mach & von

Weltrubsky 1878, Mach & Simonides 1879, Mach & Wentzel 1885) were made during the years 1875 to 1885 at the Prague Physical Institute. In addition to the development of ingenious experimental methods, two significant advances in shock wave physics were made, namely (*a*) the elucidation of the properties of propagating shock waves in air, and (*b*) the discovery of the irregular reflection of shocks.

The experimental procedure was to detect the air motion induced by shock waves reflecting from a sooty glass plate. A thin layer of soot is deposited on a glass plate by holding it over a smoky flame, such as a candle. The glass plate is laid on a table, soot side up, and above it two small spark discharges are produced simultaneously. Figure 3 is a drawing of successive positions of the shock fronts as they reflect from the

Mach V

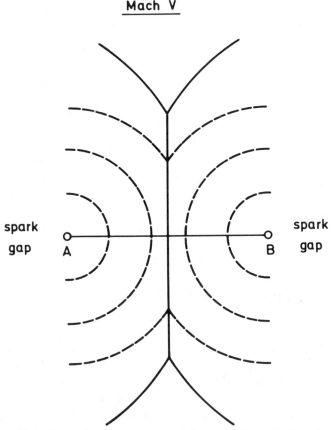

Figure 3   Schematic drawing of shock fronts and the trace of their reflection left on a sooty plate (Mach 1878b).

plate surface (*dashed lines*) and the trace of their mutual reflection from each other (*solid lines*). As the separate shocks reflect from the glass surface, the air next to the surface is put into motion and the soot is blown off; however, when two equal shocks moving oppositely mutually reflect from each other and the surface, the gas next to the surface remains at rest and the soot is not removed.

This explains the straight line marking the plane of symmetry of the two explosions. But beyond a certain point, the soot line diverges into two branches and forms a "V" at each end, the so-called Mach-V. This unexpected result, which Mach called irregular reflection, is today called Mach reflection.

Recently I duplicated this experiment. A photograph of the sooted plate after the two shock waves met and reflected from it appears in Figure 4.

In addition to various spark gaps, exploding wires were used as sources of shock waves by Mach. A straight fine wire absorbs the electrical energy stored in a capacitor, raising the wire's temperature and pressure in a very short time, so that a finite pressure wave propagates out from a line. Figure 5*a* illustrates an arrangement with a wire forming an angle so that two cylindrical shocks start out from the exploded wire and their intersection can be studied. The capacitor—a Leyden jar—was charged in Mach's experiments by a hand-operated influence machine until the spark gap fired, thus closing the circuit. Most of the stored electrical

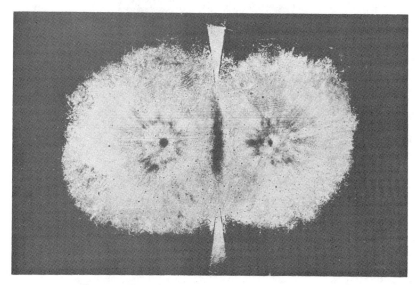

Figure 4 Photograph of 1981 sooted plate with Mach-V.

energy was delivered to the wire. A sooted glass plate below the wire revealed the intersection of the shock waves produced and the Mach-V into which the line of intersection bifurcated. This can be seen in Figure 5*b*, which is an original photograph by Ernst Mach.

Antolik was convinced one could learn from such soot patterns something about the features of the electric discharge. Mach realized very

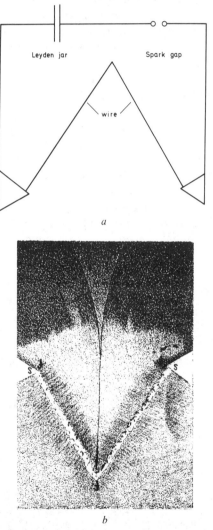

Figure 5 Study of two cylindrical shock waves meeting at an angle. (*a*) Exploding wire circuit; (*b*) Mach's photograph of the sooted plate beneath the exploded angle wire (Mach & Wosyka 1875).

early that it was not electrical phenomena but mechanical-acoustical phenomena that were revealed by the soot patterns. Two small, simultaneous chemical explosions at points A and B (see Figure 3) give the same soot pattern as the discharge of two spark gaps.

Mach and his colleagues were successful in gathering speed-distance data on the wave-front progression from a spark discharge by use of a rotating disk chronograph (Mach et al. 1878). Two channels, ab and ac, were bored in a wooden block P, down which pulses were started simultaneously by discharge of a spark at a, as shown in Figure 6. The spark was produced between electrodes EE and confined with cover plate D. At the outlet of the channels b and c, a rotating, circular sooted plate detected the arrivals of the pulses. The speed of rotation of the sooted disk was adjusted until it was synchronized audibly with a 63.05-Hz tuning fork. The difference in the arrival times at the sooted disk was determined by the angular displacement of the marks relative to calibration marks impressed on the disk at rest.

By extending the lengths of the channels in thicker blocks, successive additional travel times were determined, so that a composite speed-distance curve of the wave-front propagation could be constructed. Figure 7 shows the results of these measurements. Near the source, a blast wave such as is produced by a spark differs considerably in its propagation speed from an acoustic wave, i.e. a small-amplitude wave. Mach concluded that the spark-produced wave is a compression wave of finite amplitude. He writes (Mach & Sommer 1877):

> It does not contradict the theory [of sound] to assume that the speed of sound is independent of the amplitude. But this is not valid for oscillations of finite amplitude

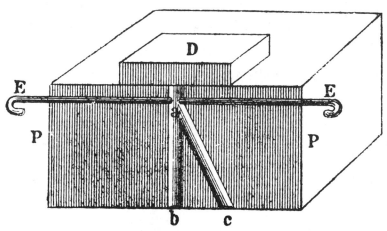

Figure 6   Apparatus for measuring travel times of a blast wave (Mach et al. 1878).

as has been shown by Riemann, 1860. Speed of sound has, for finite amplitudes, a quite different meaning; it is different at different places in the wave and alters during the wave motion. It appears that we deal in our experiments with such waves as are described by Riemann.

Although Mach clearly recognized that the spark-produced wave is different from a sound wave, he did not employ the expression "shock wave" (*Stosswelle*).

Mach obtained other quantitative results on spark-produced blast waves while photographing them with a Jamin interferometer. A reproduction of his results published in 1878 can be seen in Figure 8. The relative density change is about 50 times the change that Toepler, Boltzmann, and Mach himself had measured in the sound wave from a whistle.

We may remark at this point that Ernst Mach later developed, with his son Ludwig, a modification of the Jamin interferometer that has been widely used for aerodynamic studies. As seen in Figure 9, the Mach-

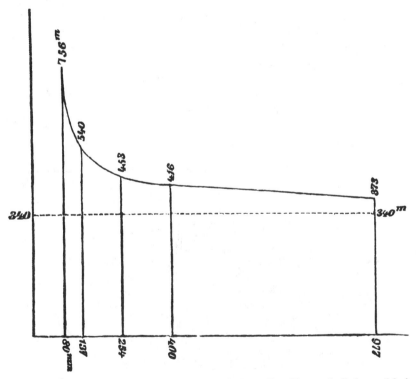

Figure 7 Distance dependence of speed of wavefront produced by spark discharge (Mach et al. 1878).

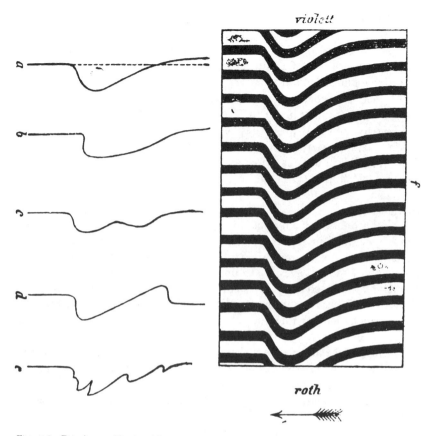

Figure 8 Density profiles in a blast wave as obtained from a Jamin interferogram (Mach & von Weltrubsky 1878).

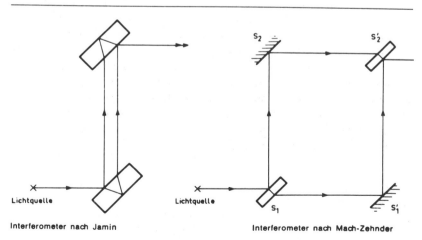

Figure 9 Jamin and Mach-Zehnder interferometers.

Zehnder modification, with four optically flat glass plates instead of two, permits an arbitrarily great separation of the coherent beams, thus furnishing a large field of view.

As well as recognizing the greater speed of propagation of shock waves, Mach's studies in the 1870s with sooted plates gave other information on the physical properties of shock waves. When he mounted a second glass plate over and parallel to the sooted plate, Mach observed that a cylindrical wave is converted into a plane wave as it propagates in the channel formed by the two plates (Mach & Grüss 1878). The arrangement has features of the present-day shock tube with electric driver.[2]

By interpretation of a number of experiments such as the ones described, Mach succeeded in giving the correct gas-dynamic interpretation of the Mach-V and thus of the criteria for the occurrence of irregular Mach reflection of shock waves. By examination of the reflected light from the sooted plate, Mach surmised that in the flow after the Mach-V the pressure is higher than in the flow behind the regular reflection. But how could Mach demonstrate the existence of a more intense wave after irregular reflection?

Mach used the arrangement shown in Figure 10. A straight wire source *i-g* and an angle wire source *b-c-d* are initiated by discharge of separate capacitors. A time delay between the explosions is varied in successive "exposures" of sooted plates, giving the intersection patterns traced in Figure 10.

These intersection paths give a lot of information about the propagation of the colliding cylindrical waves emanating from the two arms of the angle wire. The successive patterns, recorded when the counter-moving waves intersect, depict in a rough way successive shapes in time of the colliding cylindrical waves as they propagate away from the angle wire. It is almost as if we have snapshots of the wave front reflecting obliquely from a rigid boundary represented by the line of symmetry leaving *c* and bisecting *i-g* in Figure 10.[3] If we think of the incident wave reflecting from a rigid boundary, after the Mach-V forms we can pick out the incident shock and the reflected shock meeting at the triple point, and a new wave appears that is today called a "Mach stem." Furthermore, the trace of the triple point appears in Figure 10 and coincides

---

[2] The modern shock tube is described in a number of books and articles. See, for example, Bershader (1981).

[3] The traces sketched in Figure 10 are not truly snapshots of the wave front(s), because the counter-moving wave from *i-g* does not intersect with the wave system under study at all places at the same instant.

with the edge of the Mach-V, which appears so clearly in the original soot-plate experiments. (See Figures 4 and 5.)

Nowadays, by using modern apparatus, particularly the shock tube (Bleakney et al. 1949), the phenomenon of Mach reflection can be even more directly demonstrated photographically. Figure 11 is an example. It was only in 1944 that R. J. Seeger (Keenan & Seeger 1944) named the irregular oblique reflection of shock waves "Mach reflection."

How did Mach interpret the kind of reflection he discovered? He knew that waves of very small amplitude, i.e. sound waves, can pass through one another without being influenced by the meeting. When waves of finite amplitude intersect, however, the intersection becomes the starting point of a new wave, as illustrated in Figure 12. As the primary waves propagate, the angle α between their normals grows smaller and the new wave's speed must become larger in order to keep up with the point of intersection. When the speed w must be greater than $V/(\cos\alpha/2)$, the reflection is no longer regular, and irregular reflection with a Mach stem must occur. More than 100 years ago, Mach stated this criterion that allowed a computation of the limiting angle α_g. Below this limiting angle, regular reflection should not be possible. α_g is a function of the shock strength, because the speeds w and V depend on the shock strength. Actually the criterion for the start of Mach reflection is even today not

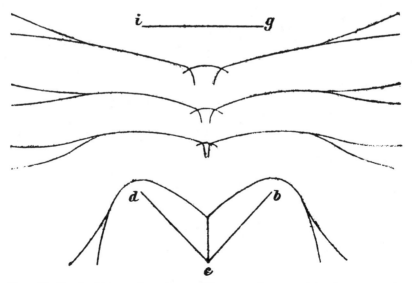

Figure 10   Sketch of intersection patterns of blast waves from a straight wire source and an angle wire source exploded at varying time delays (Mach 1878b). See text for explanation of symbols.

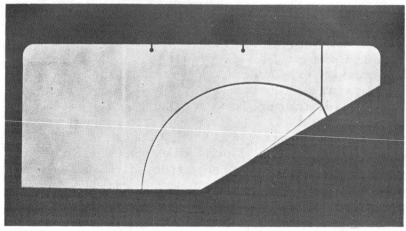

Figure 11 Snapshot with schlieren optics through shock tube windows of Mach reflection. A plane shock traveling to the right has met a sloping floor. The reflected shock is curved and intersects the incident shock at the triple point. The Mach stem extends from the triple point to the wall. The fourth line emanating from the triple point is not a shock but a contact surface separating gases with the same pressure but different entropies and densities.

Condition for Mach reflection

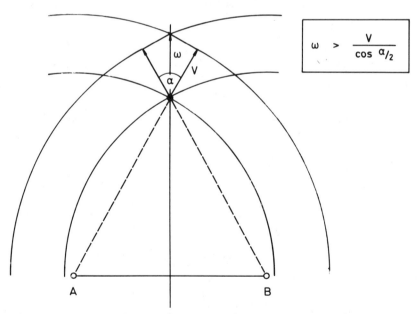

$$\omega \quad > \quad \frac{V}{\cos \alpha/2}$$

Figure 12 Cross-sectional drawing of two intersecting cylindrical waves from sources *A* and *B*. The wavefronts are shown at two positions reached a small time apart. A new wave starts and spreads continually from the intersection with speed *w*. The criterion for the start of Mach reflection is given by the formula (Mach 1878b).

completely clear in detail, and research is continuing (Whitham 1957, Ben Dor et al. 1980).

It is remarkable that Mach, without the help of apparatus that could give him pictures such as Figure 11, discovered the irregular reflection of shock waves merely by interpretation of soot patterns. It seems curious, on the other hand, that he was concurrently engaged in the first optical recording of blast waves using the Toepler schlieren method. The experimental problem consisted in producing the exact time delay between the spark discharge generating the blast wave and the spark discharge photographing the front of the blast wave. Electronic circuits did not exist; the electron tube had not yet been invented.

6.   ELECTRIC DELAY CIRCUIT

Mach experimented with different arrangements of Leyden jars. One of these circuits (Mach & Grüss 1878), which worked well as an adjustable delay circuit, is shown in Figure 13. In Mach's circuit, the capacitor A is charged by an influence machine until the breakdown voltage of spark gap I is exceeded. Upon the discharge closing the circuit at I, the charge on A is shared with B and a blast wave emanates from I. By varying the resistance between B and C, the time for C to charge to the breakdown voltage of spark gap II can be varied. Spark gap II is the light source for photographing the blast wave generated at I. Mach used a thin, water-filled tube as resistor, varying the immersion depth of the electrodes.

A schlieren photograph made with this arrangement is shown in Figure 14. The resolution in this 100-year-old picture is not as good as one using

Mach's delay - circuit

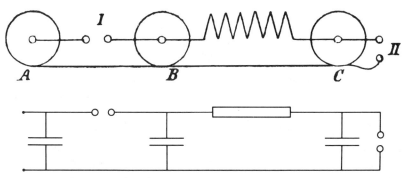

Figure 13   Mach's delay circuit (Mach & Grüss 1878). (*Below*) Same circuit in modern symbols.

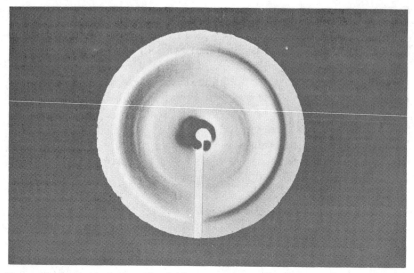

Figure 14 Schlieren photograph of blast wave emanating from spark gap at center (Castner 1892).

modern, high-speed electronic circuits and modern lenses and photographic film, but all features of interest are evident.

7. BALLISTIC EXPERIMENTS

In view of his having been the developer of the spark light source and delay circuits, it is not surprising that Ernst Mach performed the first basic experiments leading to the visualization of projectiles in flight. It is interesting how Mach himself describes his motivation in undertaking ballistic experiments. In a lecture in 1897 (Mach 1923), he told his audience:

> In the year 1881 I heard a lecture by the Belgian ballistician Melsens. He proposed that high speed projectiles push a considerable mass of compressed air ahead of them. In Melsen's opinion the compressed air may cause mechanical, explosion-like effects on the body it strikes. I wanted to investigate this idea experimentally and to make the processes perceptible, if they exist. My desire to do so was the more intense because all the means were available. I had previously used and tested them in other experiments.

It was his idea to photograph a high-speed projectile in flight in a darkened room by means of an electric spark of extremely short duration. The spark should be both light source and shutter. For detection of the compressed gas ahead of the projectile, the Toepler schlieren method was appropriate.

The first successful photographs of projectiles were made in 1886 by P. Salcher and S. Riegler, following Mach's suggestions. Salcher was a professor at the Naval Academy at Fiume (today, Rijeka, Yugoslavia). In the same year, Mach and Salcher described the event to the Vienna Academy. The leading page of the publication appears in Figure 15; this is a reproduction from Mach's personal bound volume of the academy proceedings. The two photographic prints—probably the oldest existing pictures of the bow wave ahead of a supersonic object—were apparently glued onto the page by Mach himself. This document can be considered as the beginning of the field of supersonic aerodynamics.

Intense discussion between Salcher and Mach took place by mail. Salcher reported the progress of the work weekly. In our library are 140 letters from Salcher, the first one written 14 February 1885. Unfortunately Mach's letters have never been located; from reading just one side of the correspondence, one can judge that Mach must have suggested tests and answered Salcher's questions. The first photographs that Salcher made of supersonic bullets displayed the characteristic bow wave. In his letter to Ernst Mach dated 23 May 1886, Salcher sketched what he saw in the photographs. This sketch is shown in Figure 16. To one familiar with the physics of shock waves, it is obvious how familiar Mach already was with these new supersonic phenomena. He interpreted the bow wave at once as the envelope of disturbances originating from the projectile, and he supposed it to be a shock front. Figure 17 displays the well-known construction illustrating the idea. The sine of the half-angle of the vertex of the cone is the ratio of the sound speed to the projectile speed, i.e. the reciprocal of the Mach number.

Two unsolved ballistics problems of that time could be answered by knowing of the existence of the bow shock ahead of a supersonic projectile.

Artillerists knew that two bangs could be heard downrange from a gun when high-speed projectiles were fired, but only one from low-speed projectiles. It was realized that in addition to the bang from the muzzle of the gun, an observer downrange would hear the arrival of the bow shock.

The second problem can be traced back to the Franco-Prussian war of 1870–1871. It was found that the new French Chassepôt high-speed bullets caused big crater-shaped wounds. The French were suspected of having used explosive projectiles and therefore of having violated the International Treaty of Petersburg prohibiting the use of explosive projectiles. As mentioned earlier, the Belgian Melsens had tried to refute the suspicion. Mach now gave the complete and correct explanation. The explosive type wounds were caused by the high-pressure air between the bullet's bow wave and the bullet itself.

Kaiserliche Akademie der Wissenschaften in Wien.

Sitzung der mathematisch - naturwissenschaftlichen Classe vom 10. Juni 1886.

(Sonderabdruck aus dem akademischen Anzeiger Nr. XV.)

Das w. M. Herr Regierungsrath Prof. E. Mach in Prag übersendet ~~folgende~~ vorläufige Mittheilung: „Über die Abbildung der von Projectilen mitgeführten Luftmasse durch Momentphotographie."

Auf Mach's Bitte haben die Herren Professoren Dr. P. Salcher und S. Riegler in Fiume einen von Mach und Wentzel mit negativem Erfolg ausgeführten Versuch (Vergl. Akadem. Anzeiger 1884, Nr. XV und Sitzungsberichte 1885, Bd. 92, II. Abth., S. 636) mit grösseren Projectilen und grösseren Geschwindigkeiten (Infanteriegewehr, 11 Mm. Geschoss, 440 M. Geschwindigkeit) wiederholt, und haben das Resultat mit voller Schärfe erzielt. Die Luftmasse erscheint als ein das Projectil einhüllendes Rotationshyperboloïd, dessen Achse in der Flugbahn liegt. An den Bildern zeigen sich noch manche Einzelheiten, deren sichere Interpretation sich auf weitere Versuche gründen muss.

Figure 15 Reproduction of Mach's personal copy of the publication announcing the first successful photograph of a supersonic projectile in flight. The photographic prints were probably affixed by Mach himself.

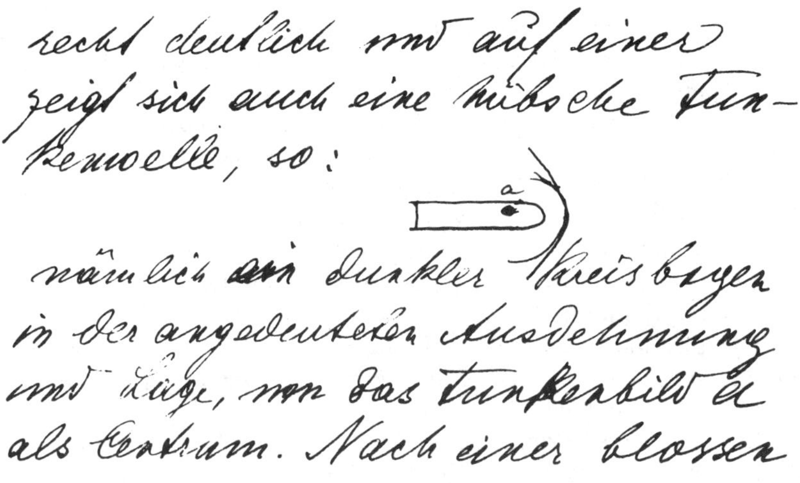

recht deutlich und auf einer
zeigt sich auch eine hübsche Fun-
kenwelle, so :

nämlich ein Dunkler Kreisbogen
in der angedeuteten Ausdehnung
und Lage, von das Funkenbild ā
als Centrum. Nach einer blossen

Figure 16   Sketch of bullet and its bow shock in Salcher's letter to Mach.

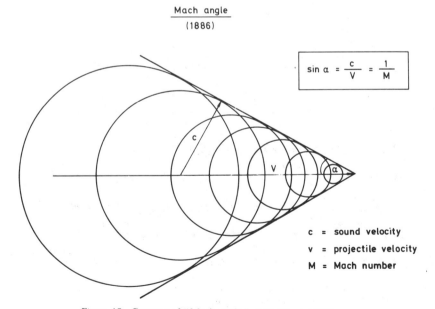

Mach angle
(1886)

$$\sin \alpha = \frac{c}{V} = \frac{1}{M}$$

c = sound velocity
v = projectile velocity
M = Mach number

Figure 17   Concept of "Mach angle," 1886 (Mach 1923).

Photographing of bullets in flight was continued by Salcher in Pola, Yugoslavia, and by Mach in Meppen, the ballistic range of the Krupp Steel Company. More details became visible, e.g. the Mach lines originating from the rough projectile surface, and the turbulent wake pattern behind the body. Mach gave the explanations for these (Mach & Salcher 1887, 1889a). Figure 18 shows the sketch of Mach lines in a letter from Salcher dated 21 May 1886. More details are in letters dated 26 June 1886, 31 March 1888, and 25 January 1889.

After Ernst and Ludwig Mach developed the interferometer to increase the amount of information beyond that obtained from Toepler schlieren photographs, they presented quantitative data on the change in density and pressure of the air behind the bow shock. One of the first photographs is shown in Figure 19.

The experimental arrangement giving such excellent photographs of supersonic projectiles in the 1880s may be of interest. An illustration in an 1889 publication is reproduced in Figure 20. A hand-cranked influence machine served as the high-voltage source. This machine had to be disconnected from the capacitor, a Leyden jar, after the capacitor received the required charge, and the trigger and spark-gap light source had to be connected to it. Moreover, the camera shutter had to be opened, the gun had to be fired, and finally the camera shutter had to be closed after the projectile arrived and triggered the spark light source.

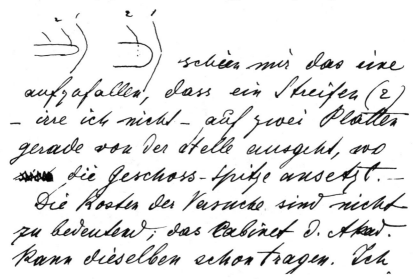

Figure 18 Sketch of "Mach lines" in Salcher's letter dated 21 May 1886. Salcher notes that "a stripe (2)—if I'm not mistaken—goes out exactly from the place where the nose of the bullet begins."

Figure 19   Interferogram of flow field around a supersonic projectile (Mach 1923).

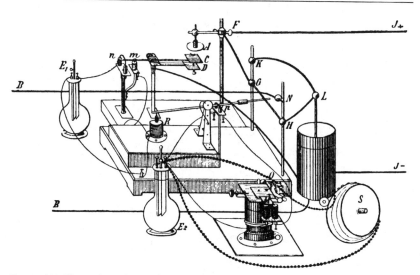

Figure 20   Illustration of experimental arrangement used by Mach in photographing projectiles in flight at Meppen (Mach 1889, Mach & Salcher 1889a).

Referring to Figure 21, in which Mach's circuit is presented in terms of today's symbols, the capacitor L is charged to a predetermined voltage by influence machine $J - J+$. The voltage is determined by the weight applied to the vane electrometer ACD; The vane C is attracted more strongly to A as the voltage of A rises, and when the attraction reaches the desired value, contact n closes, exciting relay R which switches the capacitor L from the influence machine to the light source gap and the trigger gap B. The relay R also closes contact p so that the camera shutter and igniter of the gun are triggered.

Parenthetically, I should like to remark that Mach was a very cautious experimenter. In connection with the procedures at the firing site where these photographs were made, he wanted to avoid any chance of unintentional firing, and introduced the following precaution: Instead of the relay actually firing the gun, it rang a bell. Upon hearing the bell, an artillerist fired the gun. Mach expressed pleasure at how well this precaution worked and was fascinated by the short reaction time of the soldier.

Returning to the circuit and the arrangement of the apparatus, the trigger for the spark light source consisted of allowing the projectile to short out two parallel wires in series with the capacitor and the spark gap. Exposure times of about 1 microsecond were obtained—a remarkable achievement.

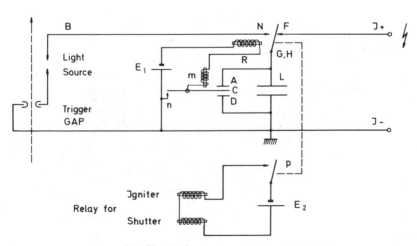

Simplified Circuit - Diagram

of

MACH'S Light Source

Figure 21   Circuit of arrangement shown in Figure 20.

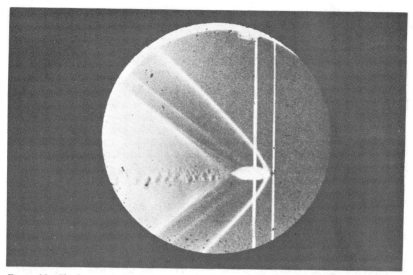

Figure 22 Shadowgram of projectile. Light source triggered by the projectile shorting the two wires seen in the photo (Castner 1892, Mach & Salcher 1887, Mach 1892).

It bothered Mach that important aspects were not visible because of the trigger wires appearing in the picture (see Figure 22). He therefore devised the delay mechanism shown in Figure 23. When the projectile's bow wave enters the open end of tube r, it progresses as a shock wave to the nozzle at the far end. When it reaches the end, a jet of air blows the flame of a burning candle through an orifice in a metal plate electrically connected to the tube and the spark gap. The ionized gas from the candle flame forms a conducting path to discharge the capacitor through the light source. By the choice of the length of the delay tube, the time delay between the projectile passing the mouth of the tube and the triggering of the light source can be regulated. It is chosen to catch the bullet in the visual field of the camera.

These devices are good examples of the experimental skill of Ernst Mach and illustrate that he was a physicist as well as a philosopher.

8. SUPERSONIC JET

Mach & Salcher (1889b) reported another gas-dynamic phenomenon. Tests in a torpedo plant in Fiume showed the formation of wave patterns in an air jet expanding into the atmosphere. Figure 24 is an excerpt of a letter Salcher wrote to Mach, dated 19 April 1888. Salcher described the succession of lines crossing the jet—Salcher used the word "Lyra" for them—and the changing pattern with increasing reservoir pressure. At

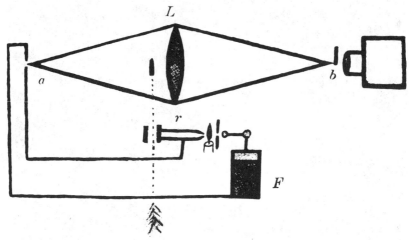

Figure 23 Optical arrangement and delay circuit (Mach 1923). Camera behind knife edge (*b*), lens (*L*), and source (*a*) comprise a schlieren arrangement. Delay tube (*r*), candle, orifice, and metal sphere comprise the trigger switch to discharge the capacitor (*F*) through the light source gap.

first Mach explained them as stationary acoustic waves, but later he recognized that some of the lines had to be shock waves. It is appropriate that the father of relativity theory recognized that instead of a projectile moving through air at rest, the body could be at rest and the air could move past with a speed greater than sound, producing the same shock-wave phenomena. Today this arrangement is called a "supersonic blow-down wind tunnel."

In one publication (Mach & Salcher 1889b), the authors state:

> At the time of the experiments involving photographing projectiles in flight, Salcher had the idea of investigating the reverse case of the air moving and the test body at rest in order to verify the results obtained.
>
> If the jet has a big enough cross-section, a phenomenon analogous to the bow shock becomes visible at the fixed body.

In spite of this early recognition of the possibility of the supersonic wind tunnel as an experimental facility, others denied that supersonic flow could exist in an exhausting jet. Finally Prandtl in 1904 convinced the doubters with his analytical treatment of the supersonic jet.

9.   CONCLUDING REMARKS

The dominant theme in Mach's experimental research in gas dynamics and ballistics was the shock wave. With his methodical approach, his knowledge of optical apparatus, and his all-round experimental ability,

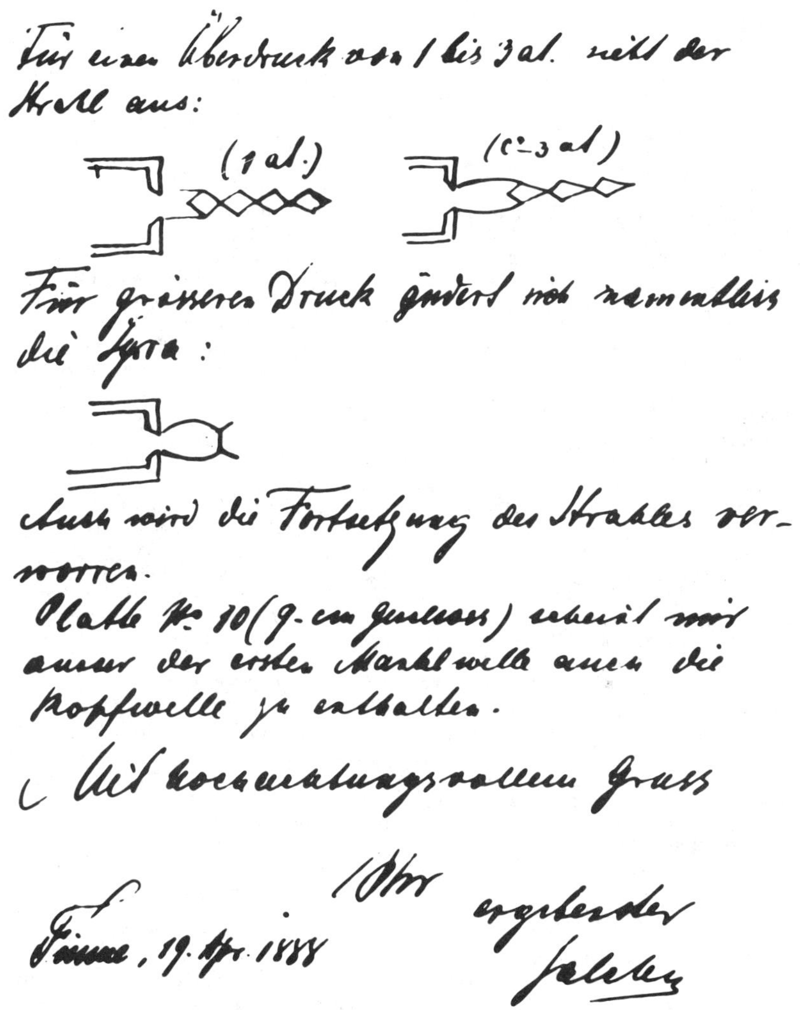
Für einen Überdruck von 1 bis 3 at. sieht der Strahl aus:

(1 at.) (c. 3 at.)

Für grösseren Druck ändert sich namentlich die Spitze:

Auch wird die Fortsetzung des Strahles verworren.

Platte No. 10 (9-cm Geschoss) scheint mir ausser der ersten Machschen Welle auch die Kopfwelle zu enthalten.

Mit hochachtungsvollem Gruss

Ihr ergebenster Salcher

Fiume, 19. Apr. 1888

Figure 24 Sketch of jet patterns in Salcher letter dated 19 April 1888.

he made accessible a new field in physics—the field of supersonic flow of gases. His experiments opened the door to supersonic flight and modern ballistics, determining the trend for several decades. His optical arrangements are still regarded as optimum for visualizing supersonic phenomena (Merzkirch 1970).

Thus it is not surprising, in answer to the question posed at the beginning of this article, that the name of Ernst Mach has been chosen for a research institute devoted to optical visualization methods for its experimental work in fluid dynamics and ballistics.

Literature Cited

Ackeret, J. 1929. *Schweiz. Bauztg.*, Vol. 94

Antolik, K. 1875. Das Gleiten elektrischer Funken. *Pogg. Ann. Phys. Chem.*, pp. 14–37

Ben Dor, G., Takayama, K., Kawauchi, T. 1980. The transition from regular to Mach reflexion and from Mach to regular reflexion in truly non-stationary flows. *J. Fluid Mech.* 100:147–60

Bershader, D. 1981. Apparatus. In *Methods of Experimental Physics, Vol. 18, Fluid Dynamics*, ed. R. J. Emrich, Pt. 9. New York: Academic

Bleakney, W., Weimer, D. K., Fletcher, C. H. 1949. *Rev. Sci. Instrum.* 20:807

Castner, J. 1892. Momentphotographie im Dienste der Ballistik, Momentphotographien von E. und L. Mach. *Prometheus* 91:615–18

Herrmann, D. B. 1966. Ernst Mach und seine Stellung zur Doppler-Theorie. *Symp. aus Anlass des 50. Todestages von Ernst Mach, Ernst-Mach-Inst.*, Freiburg, W. Germany, pp. 171–85

Keenan, P. C., Seeger, R. J. 1944. Analysis of intersections of shockwaves. *US Navy Bur. Ordnance Explos. Res. Rep. 15*

Kutterer, R. E. 1966. Die Beiträge von Ernst Mach zur Ballistik. *Symp. aus Anlass des 50. Todestages von Ernst Mach, Ernst-Mach-Inst.*, Freiburg, W. Germany, pp. 96–113

Mach, E. 1860. Über die Änderung des Tones und der Farbe durch Bewegung. *Sitzungsber. Akad. Wiss. Wien.* 41:543–60

Mach, E. 1861. Über die Controverse zwischen Doppler und Petzval bezüglich der Änderung des Tones und der Farbe durch Bewegung. *Z. Math. Phys.* 6:120–26

Mach, E. 1875. *Grundlinien der Lehre von den Bewegsempfindungen*. Leipzig: Verlag Wilhelm Engelmann

Mach, E. 1878a. Neue Versuche zur Prüfung der Doppler'schen Theory der Ton- und Farbeänderung durch Bewegung. *Sitzungsber. Akad. Wiss. Wien* 77:299–310

Mach, E. 1878b. Über den Verlauf der Funkenwellen in der Ebene und im Raume. *Sitzungsber. Akad. Wiss. Wien* 77:819–38

Mach, E. 1889. Über die Schallgeschwindigkeit beim scharfen Schuss nach von dem Krupp'schen Etablissement angestellten Versuchen. *Sitzungsber. Akad. Wiss. Wien* 98:1257–76

Mach, E. 1892. An account of scientific applications of photography. *J. Camera Club*, pp. 110–12

Mach. E. 1923. Über Erscheinungen an fliegenden Projektilen. *Populärwiss. Vorlesungen, 5. Auflage, Leipzig* 18:356–83

Mach, E., Fischer, A. 1873. Die Reflexion und Brechung des Schalles. *Sitzungsber. Akad. Wiss. Wien* 67:81–88

Mach, E., Grüss, G. 1878. Optische Untersuchung der Funkenwellen. *Sitzungsber. Akad. Wiss. Wien* 78:476–80

Mach, E., Salcher, P. 1887. Photographische Fixierung der durch Projektile in der Luft eingeleiteten Vorgänge. *Sitzungsber. Akad. Wiss. Wien* 95:764–80

Mach. E., Salcher, P. 1889a. Über die in Pola und Meppen angestellten ballistisch-photographischen Versuche. *Sitzungsber. Akad. Wiss. Wien* 98:41–50

Mach, E., Salcher, P. 1889b. Optische Untersuchung der Luftstrahlen. *Sitzungsber. Akad. Wiss. Wien* 98:1303–9

Mach, E., Simonides, J. 1879. Weitere Untersuchung der Funkenwellen. *Sitzungsber. Akad. Wiss. Wien* 80:476–86

Mach, E., Sommer, J. 1877. Über die Fortpflanzungsgeschwindigkeit von Explosionsschallwellen. *Sitzungsber. Akad. Wiss. Wien* 75:101–30

Mach, E., Tumlirz, O., Kögler, C. 1878. Über die Fortpflanzungsgeschwindigkeit der Funkenwellen. *Sitzungsber. Akad. Wiss. Wien* 77:7–32

Mach, E., von Weltrubsky, J. 1878. Über die Formen der Funkenwellen. *Sitzungsber. Akad. Wiss. Wien* 78:551–60

Mach, E., Wentzel, J. 1885. Ein Beitrag zur Mechanik der Explosionen. *Sitzungsber. Akad. Wiss. Wien* 92:625–38

Mach, E., Wosyka, J. 1875. Über einige mechanische Wirkungen des elektrischen Funkens. *Sitzungsber. Akad. Wiss. Wien* 72:44–52

Merzkirch, W. 1966. Die Beiträge Ernst Machs zur Entwicklung der Gasdynamik. *Symp. aus Anlass des 50. Todestages von Ernst Mach, Ernst-Mach-Inst.*, Freiburg, W. Germany, pp. 114–31

Merzkirch, W. 1970. Mach's contribution to the development of gasdynamics. In *Boston Studies in the Philosophy of Science 6*, pp. 42–59. Dordrecht: Reidel

Whitham, G. B. 1957. A new approach to problems of shock dynamics. I. Two-dimensional problems. *J. Fluid Mech.* 2:145–71

Ann. Rev. Fluid Mech. 1983. 15:29–45

FLUID MECHANICS OF GREEN PLANTS

Richard H. Rand

Department of Theoretical and Applied Mechanics, Cornell University, Ithaca, New York 14853

INTRODUCTION

Why is the study of the biomechanics of green plants important? First, it has been estimated that plant life comprises 99% of the Earth's biomass (Bidwell 1974). Second, green plants are virtually the only ultimate source of food for animals through photosynthesis (the process of conversion of solar energy to stored chemical energy).

A biofluid-mechanical overview of a typical green plant is shown in Figure 1. See Nobel (1974) for an extensive self-contained quantitative introduction and order-of-magnitude analysis; for a shorter quantitative introduction, see Merva (1975). Meidner & Sheriff (1976) have written a short introduction that uses engineering concepts with a minimum of mathematics, and Canny (1977) has written a brief nonmathematical introduction for fluid mechanicians.

The leaves are the site of photosynthesis. This process requires sunlight, CO_2, and water, and produces glucose (a simple sugar) and oxygen. Sugars manufactured in the leaves are translocated to other parts of the plant via the vascular phloem tissue. Water and minerals absorbed in the roots are brought up to the leaves via the vascular xylem tissue. The upward xylem flow (called the transpiration stream) is driven by evaporation at the leaves, while the largely downward phloem flow is thought to be driven by concentration differences created locally by active transport (e.g. the Munch hypothesis; see Bidwell 1974).

Studies of each of these parts of the plant have involved special fluid mechanics problems based on the particular physiological function and geometry. This article introduces the reader to the concepts and problems

29

0066-4189/83/0115-0029$02.00

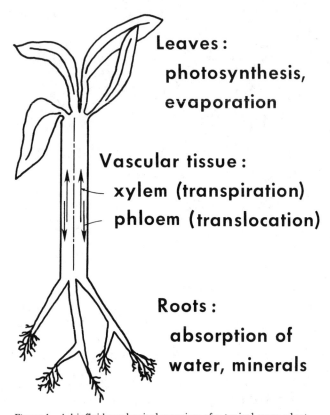

Leaves:
photosynthesis,
evaporation

Vascular tissue:
xylem (transpiration)
phloem (translocation)

Roots:
absorption of
water, minerals

Figure 1   A biofluid-mechanical overview of a typical green plant.

that are unique to the fluid mechanics of plants and reviews the mathematical literature on this subject.

FLOW IN THE VASCULAR TISSUE

Let us compare the vascular system of plants with the more familiar human vascular system. In contrast to the human circulatory system, the vascular system of plants is open (Figure 1) and includes extensive branching at both the leaves and roots. Unlike the blood vessels of human physiology, the conduits of plants are formed of individual plant cells placed adjacent to one another. During cell differentiation the common walls of two adjacent cells develop holes (called pits or pores; see Esau 1965), which permit fluid to pass between them. The xylem contains tracheids and vessel elements (Figure 2) that die after reaching maturity, while the phloem contains sieve elements that remain metabolically active.

Blood vessels are often modeled as elastic tubes since their deformation may be significant due to the pulsatile nature of the flow. In plants, however, the flow is quasi-steady and the vascular cells (like all plant cells) have stiff cell walls, making a rigid-tube model appropriate. Reynolds numbers for flow in the human aorta and in the xylem of a plant are respectively about 2000 and 0.02 (see Table 1). This means that slow viscous (creeping) flow (Happel & Brenner 1965), in which the inertia terms are neglected in the Navier-Stokes equations, is a reasonable model for flow in the plant vascular system.

The plant physiologist needs to know the pressure drops involved in flow through the vascular tissue. Such questions arise, for example, in the evaluation of various conjectured mechanisms for driving the phloem flow.

The fluid mechanics of phloem flow has been considered by Rand & Cooke (1978) and Rand et al. (1980). As shown in Figure 2, this involves flow through a series of cylindrical sieve tubes separated by perforated sieve plates. Due to the mathematical complexities of slow viscous flow, only the relatively unrealistic axisymmetric case of a single pore has been considered (Figure 3). The results of the analysis were compared with

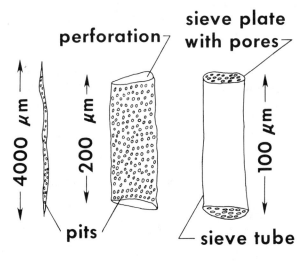

Figure 2   Fluid-conducting cells in the vascular tissue of plants (after Esau 1965). Tracheids and vessel elements are found in the xylem, while sieve elements are found in the phloem. Here and in the rest of this paper, the dimensions given are typical but do not represent statistical averages.

Table 1 Comparison of Reynolds numbers for flow in a xylem vessel and the human aorta

	Xylem vessel	Human aorta
Velocity U (cm/sec)	0.1	40
Radius R (cm)	0.002	1.5
Kinematic viscosity v (cm²/sec)	0.01	0.03
Reynolds number UR/v	0.02	2000

Poiseuille's law (which provides the standard approach currently used by plant physiologists). Poiseuille's law, when applied to the sieve tube and the pore in series, was found to underestimate the exact pressure drop by about a factor of two.

Let us now consider the flow in the xylem. Flow between two neighboring xylem tracheid cells occurs through pits (Figure 2). A typical bordered pit (Figure 4) consists of a circular border that arches over the pit cavity and contains a closing membrane. The closing membrane is composed of a thick central region, which is relatively impermeable to the flow of fluid, and a thin perforated peripheral region through which flow is possible. In nature the bordered pit is found in both open and closed states. In the open state, flow is possible from one tracheid to another, while in the closed state virtually no flow occurs through the pit.

This problem was studied by Chapman et al. (1977) by assuming an ideal fluid and using conformal mapping. The thin peripheral region of the closing membrane was modeled as linear springs, and equilibrium for a given flow rate was obtained by balancing the net hydrodynamic force on the central region of the closing membrane with the elastic restoring force of the peripheral region. Figure 5 shows the results of this analysis. It was found that for a given flow rate through the pit there are two equilibrium displacements, one stable and the other unstable. As the flow rate is increased to a value larger than the maximum permissible (see Figure 5), the pit snaps shut. Thus the pit functions as a valve to limit the flow in the xylem pathway.

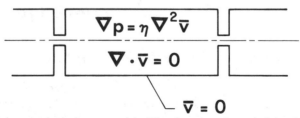

Figure 3 Axisymmetric single-pore model of flow in a sieve element of the phloem tissue (Rand & Cooke 1978). The field equations correspond to the steady creeping motion of an incompressible fluid.

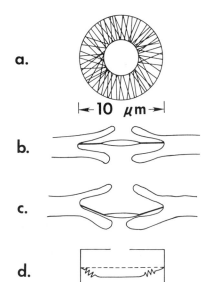

a.

⊢10 μm⊣

b.

c.

d.

Figure 4 Schematic diagram of a bordered pit found in xylem tracheid cells. (*a*) Top view. The closing membrane is composed of a thick central region and a thin peripheral region. (*b*) Side view. A circular border arches over the pit cavity and contains the closing membrane. Pit is open. (*c*) Pit is closed. (*d*) Two-dimensional hydrodynamical model (Chapman et al. 1977). The dashed and solid lines represent initial and displaced positions respectively.

A problem related to flow in the vascular system concerns observed daily changes in stem diameter accompanying changes in the rate of transpiration. The phenomenon is explained in terms of a decrease in the water content of cells near the xylem tissue resulting from an increase in the rate of transpiration. In order to understand this problem, we must consider the concept of water potential.

Water in plants moves as a result of gradients in chemical concentration (cf. Fick's law), hydrostatic pressure, and gravitational potential. Plant physiologists have found it convenient to deal with these diverse effects by using a single quantity, the water potential ψ (Nobel 1974):

$$\psi = p - RTc + \rho gz, \tag{1}$$

where

p = hydrostatic pressure (bar),

R = gas constant = 83.141 cm^3-bar/mole K,

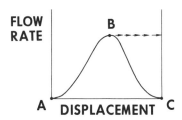

FLOW RATE

A   DISPLACEMENT   C

Figure 5 Results of the analysis of the model in Figure 4*d* (Chapman et al. 1977). Points A and C correspond to zero and maximum displacement respectively. As the flow rate is increased, the displacement of the membrane is increased until point B, after which the pit snaps shut (arrows). The equilibrium states on curve BC are unstable.

T = temperature (K),

c = concentration of all solutes in assumed dilute solution (mole/cm³),

ρ = density of water (g/cm³),

g = acceleration of gravity = 980 cm/sec²,

z = height (cm).

Here ψ is in bars, a convenient unit commonly used in plant studies for measuring pressure. One bar equals 10^6 dyne/cm² and is approximately equal to one atmosphere.

An individual plant cell consists of a cell wall surrounding a cell membrane (the plasmalemma), inside of which lies the cell protoplasm (Figure 6). In order for the cell to be in equilibrium with its surrounding medium, the water potential inside the cell must equal the water potential outside the cell. However, since the plasmalemma is able to maintain a concentration difference between the interior and the exterior of the cell, the hydrostatic pressure inside the cell can be larger than that outside the cell [from Equation (1)]. This situation (of which there is no parallel in the case of animal cells) is resolved by the elastic extension of the plant cell wall, creating a "turgor" pressure inside the cell.

Let us return now to the problem of the daily changes in stem diameter due to dehydration during transpiration. In order for the transpiration stream to flow, there must be a negative gradient in water potential from the roots to the leaves. This gradient reduces the value of the water potential at all points in the xylem (compared with values corresponding

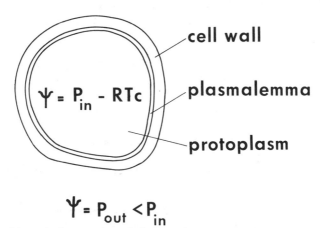

Figure 6 Schematic diagram of a typical plant cell. For equilibrium, the water potentials ψ inside and outside the cell must be equal.

to zero transpiration). This in turn causes a decrease in water potential inside a typical cell near the xylem tissue throughout the stem, and accordingly reduces the cell's turgor pressure and the associated elastic extension of the cell wall. As a result, the size of the cell and the diameter of the stem are decreased.

Molz & Klepper (1972) studied this problem by assuming radial diffusion of water potential, a concept first discussed by Philip (1958a,b,c). They obtained good agreement with experimental observations and were able to explain an observed hysteresis loop in the stem diameter-leaf water potential relationship. Their work was extended by Parlange et al. (1975), who considered a variable diffusion coefficient and a corresponding nonlinear diffusion equation.

A related and important concept is the distinction between the symplasm and the apoplasm. The symplasm consists of all the protoplasm (inside the plasmalemma) of all the living cells of the plant, together with the plasmodesmata (thin strands of cytoplasm that go from the interior of a given cell, through the cell wall, and into the interior of a neighboring cell). In terms of point set topology, the symplasm is thought to be a connected set. The apoplasm consists of those regions of the plant that contain water and are not in the symplasm. In particular the apoplasm includes the xylem (which consists of dead cells), as well as the fluid in the cell walls of all the cells of the plant. Flow in the symplasm has been estimated to involve a resistance about 50 times as large as that in the apoplasm (Meidner & Sheriff 1976, p. 51).

The flow of water in the parallel symplasm and apoplasm pathways has been described by a pair of coupled diffusion equations (Molz 1976; see also Molz & Ikenberry 1974 and Molz & Hornberger 1973). The coupling represents the flow between the symplasm and the apoplasm and depends upon various resistances in the model. Molz (1976) has applied these equations to a boundary-value problem representing the immersion of a sheet of tissue initially in equilibrium into a bath of pure water.

Aifantis (1977) has decomposed the flow in the apoplasm into two components representing flow in the xylem vessels and flow in the cell walls. His treatment, based on the modern theory of continuum mechanics, neglects viscous effects and results in two coupled diffusion equations. Unger & Aifantis (1979) have applied this theory to a boundary-value problem representing flow in a cylindrical stem.

Flow in the plasmodesmata of the symplasm has been studied by Blake (1978). An individual plasmodesma has an internal diameter of about 0.05 μm and a length of about 1 μm. This work represents the smallest scale yet considered in the biofluid mechanics of plants.

FLOW IN THE LEAF

The structure of the leaf can be explained in terms of its function. The thinness of the leaf enables CO_2 to diffuse from the ambient atmosphere into the leaf interior, where it is utilized in photosynthesis in the chloroplasts of the mesophyll cells (Figure 7). The familiar branching pattern of the vascular tissues in the leaf serves to irrigate the mesophyll cells in order to replace water that has been lost through evaporation. Water loss is generally thought to be undesirable, especially in times of drought (although the cooling effect of evaporation may be of significant value). The outer layer of leaf cells (the epidermis; see Figure 7) is covered with a layer of waxy material called cutin that tends to prevent water loss.

CO_2 and water vapor respectively enter and leave the leaf through small holes in the epidermis called stomata (Meidner & Mansfield 1968). An individual stomate is composed of two specialized guard cells (Figure 8) which, through their elastic deformation under hydrostatic loading, can affect the width of the stomatal pore. Thus stomata can act like valves to limit water loss when CO_2 is not needed for photosynthesis. For example, stomata are generally closed at night when the absence of sunlight prevents photosynthesis.

Cooke et al. (1976) have considered the elastostatics of a stomatal guard cell by using a linear anisotropic thin-shell model and a finite-element analysis. An increase in hydrostatic pressure in the guard cell tended to open the stomatal pore, while an increase in neighboring subsidiary cell pressure tended to close the pore. It was shown that the elliptic shape of the stomate (Figure 8a,b) is critical for opening and that other features (such as wall thickening and radial stiffening) could help the opening process, but were not essential. In particular, it was shown analytically that a circular torus model would close rather than open

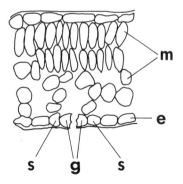

Figure 7 Schematic diagram of a transverse section of a leaf (after Nobel 1974). A representative value for leaf thickness is 300 μm. (m) mesophyll cell, (e) epidermal cell, (g) guard cell, (s) subsidiary cell.

upon inflation. This work was extended by Cooke et al. (1977) to include nonlinear effects.

The flow of water along the transpiration stream in the leaf proceeds through the branching xylem system to the xylem termini, and then continues through the apoplastic mesophyll cell walls to those mesophyll cells near the stomatal pore where evaporation occurs. Stroshine et al. (1979) have studied flow in the leaf xylem. This involves consideration of branching xylem vessels of various sizes as well as a diffusive flow between the xylem and the leaf symplasm (the interior of the mesophyll cells). It was concluded that the large vascular bundles offer relatively little resistance to flow compared with intermediate and small bundles.

The site of evaporation is the menisci in the mesophyll cell walls. These liquid-air interfaces are bounded by the strands of cellulose that constitute the cell wall. A representative interfibrillar space has a "diameter" of about 0.01 μm (Nobel 1974, p. 51). The pressure difference across a spherical meniscus is given by

$$\Delta p = 2\sigma/r, \tag{2}$$

where

σ = surface tension coefficient
= 73 dyne/cm for an air-water interface at 20°C,

r = radius of curvature of the meniscus (cm).

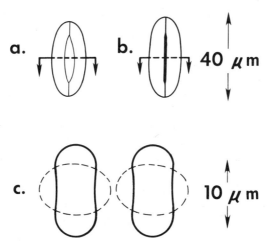

Figure 8 Schematic diagram of an elliptical stomate. (a) View looking onto the leaf surface. Pore open. (b) Pore closed. (c) Transverse view of finite-element model of two guard cells (Cooke et al. 1976). Note the change of scale. Dashed and solid curves represent initial and deformed configurations respectively.

Here Δp is about 300 bars, and since one bar is equal to a gravitational head of about ten meters, this effect accounts for the ascent of water to the tops of the highest trees. Of course this requires that the continuous fluid column reaching from the roots to the leaves be under considerable tension. Although the theoretical tensile strength of a perfect column of water greatly exceeds 300 bars (Hammel & Scholander 1976, p. 18; Nobel 1974, p. 52), the presence of small air bubbles and other imperfections reduces the observed tensile strength in laboratory experiments. Nevertheless, the plant is evidently able to grow a vascular system relatively free from air bubble defects.

The dynamics of a spherical evaporating meniscus has been studied by Rand (1978a). The analysis involved a nonlinear differential-integral equation and predicted damped oscillatory motions for a certain range of parameter values.

Gaseous Diffusion

The evaporating water proceeds by gaseous diffusion through the stomatal pore (if it is open), through a still air layer adjacent to the leaf, and into the ambient atmosphere. In a similar fashion, CO_2 diffuses into the leaf interior where it is absorbed into the wet mesophyll cell walls. Gaseous diffusion in the leaf has received a great deal of attention since it was first studied from a mathematical point of view by Brown & Escombe (1900). By modeling the leaf surface as a plane septum with a circular hole and the pore as a circular cylinder, they were able to explain the experimentally observed relatively large rates of transpiration from leaves (comparable to evaporative fluxes from an equal-sized body of water). A recent review of leaf diffusion models (Cooke & Rand 1980) contains many references in addition to those that follow.

Bange (1953) used an approximate analysis in order to consider a realistic geometry for the leaf interior as well as a still air layer outside the leaf. He found that as the wind speed increased, i.e. as the thickness of the still air boundary layer decreased, the stomata played an increasingly important role in controlling gaseous fluxes. Although a wider pore always results in a larger flux, this effect was shown to be negligible for relatively thick boundary layers.

Cooke (1967) considered diffusion through an elliptical pore. Using a relationship involving complete elliptic integrals, he showed that a slightly open stomate can permit relatively large diffusion rates. For example, an ellipse with a major to minor axis ratio of 20 has a discharge rate that is 39% of that of a circle of diameter equal to the major axis! Cooke (1969) considered the interaction effects between neighboring stomatal pores. Using separation of variables, he showed that the flux depends on both the spacing between stomata on the leaf surface and the boundary-layer thickness. Substantial increases in flux due to interaction effects could occur for closely spaced stomata with a relatively thin boundary layer.

Holcomb & Cooke (1977) extended this work by using the analogy between diffusion and the flow of electric current in an aqueous electrolyte solution. They built an electrolytic tank (copper sulphate in a copper and plexiglass container) and used it to study the effects of pore eccentricity, stomatal spacing, boundary-layer thickness, and pore depth.

Parlange & Waggoner (1970) used conformal mapping to study diffusion through a two-dimensional slit. They compared their results with the formula of Brown & Escombe (1900) and found the latter to be most accurate for thin, deep slits.

Current treatments of gaseous diffusion in the leaf (see, for example, Nobel 1974) utilize a one-dimensional model which, by analogy with Ohm's law, involves a series of resistances, each associated with a portion of the pathway. Parkhurst (1977) compared a three-dimensional field-equation approach with the commonly used one-dimensional resistance model and found that the latter involved an error of 44%.

Webster (1981) has applied the concept of the effectiveness factor to leaf diffusion in order to gauge the extent to which assimilation is diffusion limited. This factor is defined as the ratio of the actual assimilation rate to the assimilation rate that would occur in the absence of any CO_2 concentration gradients. An effectiveness factor of unity indicates that assimilation is kinetically limited, while a value considerably smaller than unity indicates that losses due to diffusion are significant.

Nearly all studies of leaf diffusion have assumed steady-state diffusion. Gross (1981), however, included time-dependent terms in order to estimate the time scale of the gaseous diffusion process. He found equilibrium to be essentially attained in less than one second.

The gaseous diffusion of water vapor and CO_2 differ in one important respect: although the CO_2 diffuses into the deep interior of the leaf to be absorbed by the mesophyll cells, several independent experimental investigations have shown that water vapor evaporates only from those cell walls near the stomatal pore (Tyree & Yianoulis 1980). This phenomenon has been explained by considering the physical chemistry of equilibrium between the liquid and gaseous phases at the cell wall (Rand 1977a,b). Since the cell-wall liquid is a dilute solution in which CO_2 is the solute and water the solvent, CO_2 satisfies Henry's law while water vapor satisfies Raoult's law. When stated as boundary conditions for the diffusion problem, these principles give substantially different predictions for CO_2 and water vapor, in qualitative agreement with the experimental observations. In related work, the diffusion of CO_2 in sun versus shade (i.e. thick versus thin) leaves was studied by Rand (1978b).

After diffusing as a gas to the mesophyll cell walls, CO_2 continues to diffuse as a solute to the chloroplasts in the cell interior. Sinclair et al. (1977) and Sinclair & Rand (1979) have modeled this process by assuming spherical cell geometry and Michaelis-Menten reaction kinetics (Thornley 1976). The resulting nonlinear ordinary differential equation for CO_2 concentration as a function of radial position was solved approximately by perturbation methods. Expressions for the rate of CO_2 assimilation by a single cell were obtained in terms of cell size and biochemical parameters.

This spherical cell model was incorporated into a more comprehensive model for CO_2 assimilation by Rand & Cooke (1980). The model took account of the gradual absorption of CO_2 into the mesophyll cell walls as CO_2 diffuses inward (i.e. diffusion with a distributed sink), as well as the effects of variation in cell-packing density. An approximate formula for CO_2 flux into the leaf in terms of basic geometrical and biochemical parameters was obtained by perturbations.

Stomatal Oscillations

A problem related to the gaseous fluxes in the leaf concerns the dynamic behavior of the stomatal apparatus. Experimental observations have revealed that the width of the stomatal pore often oscillates, typically with a period ranging from 10 to 50 min. Delwiche & Cooke (1977) modeled this phenomenon by balancing water fluxes between the guard cell, the subsidiary cell, and the rest of the plant. The gaseous flux through the stomatal pore acts like a feedback element and is responsible for the oscillation, which may be described as follows: Water evaporating from the wet mesophyll and subsidiary cell walls diffuses through the stomatal pore to the leaf exterior. This water is replaced both by a flux from the roots via the xylem, and by a flux from the guard cells to the subsidiary cells. The resulting decrease in hydrostatic pressure in the guard cells causes the stomatal pore width to decrease (Cooke et al. 1976). A smaller pore width slows the rate of evaporation, increasing the water potential in the mesophyll and causing water to accumulate there. In response to this accumulation, water diffuses back to the guard cells, increasing their hydrostatic pressure and increasing the pore width. The model takes the form of an autonomous system of two first-order ordinary differential equations for p_g and p_s (the pressures in the guard and subsidiary cells). The resulting flow in the p_g-p_s plane exhibits a limit cycle (Figure 9).

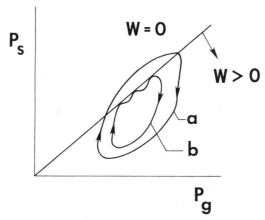

Figure 9 Limit cycles representing stomatal oscillations in the p_g-p_s plane. Here p_g and p_s represent hydrostatic pressures in guard and subsidiary cells respectively, and w represents pore width. The region above the straight line corresponds to a closed pore ($w = 0$). Arrow shows direction of increasing pore width. (*a*) Stomatal oscillation due to hydraulic feedback loop only (Delwiche & Cooke 1977). (*b*) Stomatal oscillation due to both hydraulic and CO_2 feedback effects (Upadhyaya et al. 1980a,b).

This work was extended by Rand et al. (1981) by embedding the original system of Delwiche & Cooke (1977) into a one-parameter family of systems. It was found that as the parameter (which represents the concentration of the osmotically active solutes in the guard cell) is varied, the dynamical properties exhibited by the system change (Figure 10). The system was shown to contain a Hopf bifurcation (Marsden & McCracken 1976) that involved the genesis of an unstable limit cycle. The oscillatory behavior was seen as a kind of dynamical bridge between the open and closed pore equilibrium states.

Upadhyaya et al. (1980a,b) extended the Delwiche & Cooke (1977) model by including CO_2 feedback effects. This involved modeling the guard cell biochemistry in order to include a CO_2 sensor in the system.

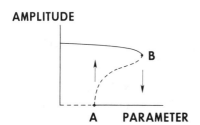

Figure 10 Changes in the amplitude of the stomatal oscillation of Figure 9*a* due to changes in a system parameter (Rand et al. 1981). Zero amplitude corresponds to equilibrium behavior. The dashed and solid lines correspond to unstable, and stable motions respectively. At point A an unstable equilibrium point becomes stable and throws off an unstable limit cycle (a Hopf bifurcation). At point B a stable and an unstable limit cycle coalesce. Arrows represent jump phenomena.

The model displayed a limit cycle oscillation, which involved a 2 min CO_2-based oscillation superimposed on the 20-min water-based oscillation previously discussed (Figure 9), in agreement with the experimental observations of other investigators.

In order to explore the effects of coupling between neighboring stomata, Rand et al. (1982) studied a model of the dynamics of a system of N coupled stomatal oscillators. This work was based on the possibility that even on the same leaf, some stomata may be open while others are closed. Although in principle one may envision *waves* of stomatal-opening moving across the leaf surface, analysis of the model predicted that for a wide range of parameter values a uniform leaf would exhibit a stable, spatially uniform, synchronized behavior.

Why do stomata oscillate? That is, in terms of Darwinian evolution, of what advantage to the plant are stomatal oscillations? Upadhyaya et al. (1981) investigated this question by comparing gaseous fluxes through a stomatal pore in an open equilibrium state with fluxes through an oscillating pore. For typical values of the system parameters, they found that stomatal oscillations tend to conserve water under relatively dry atmospheric conditions. However, this savings in moisture content occurs at the expense of a reduction in the CO_2 assimilation rate.

FLOW IN THE ROOT

Although we are concerned only with the role of the root as an organ for absorbing water and minerals from the soil, note that the root also serves to store carbohydrates and to anchor the plant in the soil.

Figure 11 shows a schematic diagram of a transverse section of a root. Water is absorbed from the soil through the many root hairs (the presence of which greatly increases the absorbing surface area of the root) and flows radially inward across a region of storage tissue called the cortex, toward the xylem in the centrally located vascular tissue. Between the cortex and the vascular tissue, however, lies the endodermis, a single layer of cells that are separated from one another by an impermeable barrier called the casparian strip. Water must pass through the symplasm of the endodermal cells in order to enter the vascular tissue. Thus the endodermis and casparian strip locally divide the apoplasm into two disconnected regions. Although the role of the endodermis is uncertain, it may function as a filter, selectively absorbing minerals, and it may be the site of observed changes in the plant's resistance to water flux, permitting absorption to occur more readily when the soil is less moist. Once the absorbed water reaches the xylem it flows axially. See Newman (1976) for a summary of flow in the root.

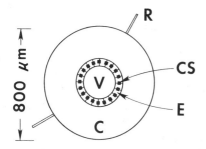

Figure 11   Schematic diagram of a transverse section of a root about one cm from the root tip (after Nobel 1974). (R) root hair, (C) cortex, (E) endodermis, (CS) casparian strip, (V) vascular tissue.

Unlike the leaf, the root has received relatively little attention from fluid mechanicians. The usual approach has been to use a lumped system resistance-capacitance electric circuit analog (see, for example, Seaton & Landsberg 1978). Although such models yield reasonable estimates for overall plant water fluxes, they do not take account of the geometry of the root. Of greater fluid mechanical interest are the following models, which involve a field-theory approach.

Molz (1975) considered radial diffusive flow in a cylindrical root surrounded by a cylindrical region of soil. Continuity of water potential and of water flux were assumed at the soil-root interface. The study indicated that water potential gradients in the soil are small compared with those in the root, except under very dry soil conditions.

Landsberg & Fowkes (1978) considered both radial absorption and axial diffusion of water along the length of a root. Their model predicted the value of the plant water potential at the base of the plant necessary to sustain a given flow rate through a root system with given characteristics. An expression was obtained for the optimal root length such that the overall root resistance to water is minimized. It is interesting to note that the mathematical statement of this problem is identical to that used to describe the assimilation of CO_2 in the intercellular air pathway of a leaf (Rand 1977a,b).

CONCLUDING REMARKS

As in other branches of biomechanics, research work on plants involves greater emphasis on modeling than does work in more traditional areas of mechanics. The researcher is presented with the biological description of the phenomenon to be studied and must invent an appropriate boundary-value problem to represent it.

Moreover the work is by its nature interdisciplinary. The interaction between mechanics and fields such as agricultural engineering or plant physiology is essential, both to generate the relevant problems and to evaluate the significance of the solutions.

Finally, we note that a glance at the contents of a current biomechanics journal or conference reveals that almost all the research work concerns animal systems, and most of that the human body. In view of the importance of plants to our planet and to our society, we might ask why the biomechanics of plants has received so little attention. Perhaps it is a case of chauvinism by species; work by our species has proceeded most rapidly on biological systems most similar to ourselves, and plants are very different. In any case it is clear that the body of knowledge of biomechanics, a relatively new field, is still developing new branches and we expect to see more attention given to the study of plants in the future.

Literature Cited

Aifantis, E. C. 1977. Mathematical modelling for water flow in plants. In *Proc. 1st Int. Conf. Math. Modeling, Rolla, Mo.*, ed. X. J. R. Avula, 2:1083–90

Bange, G. G. J. 1953. On the quantitative explanation of stomatal transpiration. *Acta Bot. Neerl.* 2:255–97

Bidwell, R. G. S. 1974. *Plant Physiology.* New York: MacMillan

Blake, J. R. 1978. On the hydrodynamics of plasmodesmata. *J. Theor. Biol.* 74:33–47

Brown, H. T., Escombe, F. 1900. Static diffusion of gases and liquids in relation to the assimilation of carbon and translocation in plants. *Philos. Trans. R. Soc. London Ser. B* 193:223–91

Canny, M. J. 1977. Flow and transport in plants. *Ann. Rev. Fluid Mech.* 9:275–96

Chapman, D. C., Rand, R. H., Cooke, J. R. 1977. A hydrodynamical model of bordered pits in conifer tracheids. *J. Theor. Biol.* 67:11–24

Cooke, J. R. 1967. Some theoretical considerations in stomatal diffusion: a field theory approach. *Acta Biotheor.* 17:95–124

Cooke, J. R. 1969. The influence of stomatal spacing upon diffusion rate. *ASAE Pap. No. 69-525.* Am. Soc. Agric. Eng., St. Joseph, Mich.

Cooke, J. R., Debaerdemaeker, J. G., Rand, R. H., Mang, H. A. 1976. A finite element shell analysis of guard cell deformations. *Trans. ASAE* 19:1107–21

Cooke, J. R., Rand, R. H. 1980. Diffusion resistance models. In *Predicting Photosynthesis for Ecosystem Models*, ed. J. D. Hesketh, J. W. Jones, 1:93–121. Boca Raton, Fla: CRC Press

Cooke, J. R., Rand, R. H., Mang, H. A., Debaerdemaeker, J. G. 1977. A nonlinear finite element analysis of stomatal guard cells. *ASAE Pap. No. 77-5511.* Am. Soc. Agric. Eng., St. Joseph, Mich.

Delwiche, M. J., Cooke, J. R. 1977. An analytical model of the hydraulic aspects of stomatal dynamics. *J. Theor. Biol.* 69:113–41

Esau, K. 1965. *Plant Anatomy.* New York: Wiley

Gross, L. J. 1981. On the dynamics of internal leaf carbon dioxide uptake. *J. Math. Biol.* 11:181–91

Hammel, H. T., Scholander, P. F. 1976. *Osmosis and Tensile Solvent.* New York: Springer

Happel, J., Brenner, H. 1965. *Low Reynolds Number Hydrodynamics.* Englewood Cliffs, NJ: Prentice Hall

Holcomb, D. P., Cooke, J. R. 1977. An electrolytic tank analog determination of stomatal diffusion resistance. *ASAE Pap. No. 77-5510.* Am. Soc. Agric. Eng., St. Joseph, Mich.

Landsberg, J. J., Fowkes, N. D. 1978. Water movement through plant roots. *Ann. Bot.* 42:493–508

Marsden, J. E., McCracken, M. 1976. *The Hopf Bifurcation and its Applications.* New York: Springer

Meidner, H., Mansfield, T. A. 1968. *Physiology of Stomata.* New York: McGraw-Hill

Meidner, H., Sheriff, D. W. 1976. *Water and Plants.* New York: Wiley

Merva, G. E. 1975. *Physioengineering Principles.* Westport, Conn: AVI Publ.

Molz, F. J. 1975. Potential distributions in the soil-root system. *Agron. J.* 67:726–29

Molz, F. J. 1976. Water transport through plant tissue: the apoplasm and symplasm pathways. *J. Theor. Biol.* 59:277–92

Molz, F. J., Hornberger, G. M. 1973. Water transport through plant tissues in the presence of a diffusable solute. *Soil Sci. Soc. Am. Proc.* 37:833–37

Molz, F. J., Ikenberry, E. 1974. Water
transport through plant cells and cell
walls: theoretical development. *Soil Sci.
Soc. Am. Proc.* 38:699–704

Molz, F. J., Klepper, B. 1972. Radial propagation of water potential in stems. *Agron.
J.* 64:469–73

Newman, E. I. 1976. Water movement
through root systems. *Philos. Trans. R.
Soc. London Ser. B* 273:463–78

Nobel, P. S. 1974. *Introduction to Biophysi-
cal Plant Physiology.* San Francisco: Free-
man

Parkhurst, D. F. 1977. A three-dimensional
model for CO_2 uptake by continuously
distributed mesophyll in leaves. *J. Theor.
Biol.* 67:471–88

Parlange, J. Y., Turner, N. C., Waggoner, P.
E. 1975. Water uptake, diameter change,
and nonlinear diffusion in tree stems.
Plant Physiol. 55:247–50

Parlange, J. Y., Waggoner, P. E. 1970.
Stomatal dimensions and resistance to
diffusion. *Plant Physiol.* 46:337–42

Philip, J. R. 1958a. The osmotic cell, solute
diffusibility, and the plant water economy. *Plant Physiol.* 33:264–71

Philip, J. R. 1958b. Propagation of turgor
and other properties through cell aggrega-
tions. *Plant Physiol.* 33:271–74

Philip, J. R. 1958c. Osmosis and diffusion
in tissue: half-times and internal gradients. *Plant Physiol.* 33:275–78

Rand, R. H. 1977a. Gaseous diffusion in
the leaf interior. *Trans. ASAE* 20:701–4

Rand, R. H. 1977b. Gaseous diffusion in
the leaf interior. In *1977 Biomech. Symp.*,
ed. R. Skalak, A. B. Schultz, 23:51–53

Rand, R. H. 1978a. The dynamics of an
evaporating meniscus. *Acta Mech.*
29:135–46

Rand, R. H. 1978b. A theoretical analysis
of CO_2 absorption in sun versus shade
leaves. *J. Biomech. Eng. Trans. ASME*
100:20–24

Rand, R. H., Cooke, J. R. 1978. Fluid dynamics of phloem flow: an axisymmetric
model. *Trans. ASAE* 21:898–900, 906

Rand, R. H., Cooke, J. R. 1980. A comprehensive model for CO_2 assimilation in
leaves. *Trans. ASAE* 23:601–7

Rand, R. H., Storti, D. W., Upadhyaya, S.
K., Cooke, J. R. 1982. Dynamics of cou-
pled stomatal oscillators. *J. Math. Biol.*
In press

Rand, R. H., Upadhyaya, S. K., Cooke,
J. R. 1980. Fluid dynamics of phloem
flow. II. An approximate formula. *Trans.
ASAE* 23:581–84

Rand, R. H., Upadhyaya, S. K., Cooke,
J. R., Storti, D. W. 1981. Hopf bifurca-
tion in a stomatal oscillator. *J. Math.
Biol.* 12:1–11

Seaton, K. A., Landsberg, J. J. 1978. Resistance to water movement through wheat
root systems. *Aust. J. Agric. Res.*
29:913–24

Sinclair, T. R., Goudriaan, J., Dewit, C. T.
1977. Mesophyll resistance and CO_2 compensation concentration in leaf photosynthesis models. *Photosynthetica*
11:56–65

Sinclair, T. R., Rand, R. H. 1979. Mathematical analysis of cell CO_2 exchange
under high CO_2 concentrations. *Photosynthetica* 13:239–44

Stroshine, R. L., Cooke, J. R., Rand, R. H.,
Cutler, J. M., Chabot, J. F. 1979.
Mathematical analysis of pressure cham-
ber efflux curves. *ASAE Pap. No.
79-4585.* Am. Soc. Agric. Eng., St. Joseph,
Mich.

Thornley, J. H. M. 1976. *Mathematical
Models in Plant Physiology.* New York:
Academic

Tyree, M. T., Yianoulis, P. 1980. The site of
water evaporation from sub-stomatal cavities, liquid path resistances and hydroac-
tive stomatal closure. *Ann. Bot.* 46:175–93

Unger, D. J., Aifantis, E. C. 1979. Flow in
stems of plants. In *Proc. 3rd Eng. Mech.
Div. Specialty Conf. ASCE,* Austin, Tex.,
pp. 501–4

Upadhyaya, S. K., Rand, R. H., Cooke, J.
R. 1980a. Stomatal dynamics. In *1980
Advances in Bioengineering,* ed. V. C.
Mow, pp. 185–88. New York: ASME

Upadhyaya, S. K., Rand, R. H., Cooke, J.
R. 1980b. A mathematical model of the
effects of CO_2 on stomatal dynamics.
ASAE Pap. No. 80-5517. Am. Soc. Agric.
Eng., St. Joseph, Mich.

Upadhyaya, S. K., Rand, R. H., Cooke, J.
R. 1981. The role of stomatal oscillations
on plant productivity and water use efficiency. *ASAE Pap. No. 81-4017.* Am.
Soc. Agric. Eng., St. Joseph, Mich.

Webster, I. A. 1981. The use of the effectiveness factor concept in CO_2 diffusion
in the leaf interior. *Ann. Bot.* 48:757–60

Ann. Rev. Fluid Mech. 1983. 15:47–76

SNOW AVALANCHE MOTION AND RELATED PHENOMENA

E. J. Hopfinger

Institut de Mécanique, Université de Grenoble, Grenoble, France

INTRODUCTION

Avalanches are natural phenomena that have a mysterious aura because they can strike unexpectedly and violently. The catastrophe at Val d'Isère, where in 1970 thirty-nine people were killed while having breakfast in a seemingly safe building, is a typical example. Similar or worse disasters have occurred during this century in other countries. The development of ski resorts brings more and more people into contact with the mountain environment and its inherent dangers, creating a need for new roads and better road protection. Increasing energy demands require the construction of new dams and electric power transmission lines. The costs of road protection and, in particular, the yearly costs arising from repairs of damage inflicted upon different structures and from the destruction of forests can be considerable. By drawing attention to these problems, I hope to palliate certain beliefs that avalanche research is of an esoteric nature.

Avalanche research has been pursued in areas of practical importance, such as forecasting, avalanche zoning, artificial release, snowpack stability, and avalanche dynamics (see, for instance, the review by La Chapelle 1977). Among these, the latter has an important fluid-mechanics component; the present review emphasizes therefore the motion and the impact of avalanches on obstacles. Avalanches occur essentially in two forms. One is referred to as a *dense-snow avalanche* (in the avalanche literature also called a *flowing avalanche*, a translation of "Fliesslawine"), which is a large-bulk-density, granular-snow gravity flow. The other is referred to as an *airborne powder avalanche*, which is driven by the extra weight of small snow particles (< 1 mm) suspended in the air. We are thus dealing

47

with complex granular-material flow and turbidity currents or clouds on inclines of complex geometry. Many of the problems raised by these flows have only incomplete solutions.

The first attempt to formulate a general theory of avalanche motion was made by Voellmy (1955), and this theory is still largely used by Western scientists and engineers (see, for instance, Leaf & Martinelli 1977). Voellmy assumes that the dynamics of avalanches can be approximated by the use of the usual assumptions of open-channel hydraulics with the addition of kinetic friction and a variation of density. Salm (1966) attempted to introduce the particulate nature of the flow by making the friction coefficients particle-size dependent. In essence, he used different friction coefficients for dense-snow avalanches and powder avalanches, but these values still needed to be determined from observations. Carrying the analogy with hydraulics even further, Kulikovskii & Eglit (1973), Gregoryan & Ostroumov (1977), and Brugnot & Pochat (1981) developed depth-averaged equations for the unsteady flow of dense-snow avalanches that allowed for variations in geometry and snow incorporation along the avalanche track. These equations have been solved numerically, and the results show that they form a good starting point for dense-snow avalanches. Further improvement is most likely to come from research on the flow of granular material, such as studied by Savage (1979). Airborne powder avalanches, on the other hand, can be treated as turbidity currents or clouds moving down inclines, as has been suggested by Losev (1965) and by Tochon-Danguy & Hopfinger (1975). Recent studies of inclined gravity currents (Britter & Linden 1980) and of buoyant clouds on inclined boundaries using solutes and also particulate matter for driving the flow (Beghin et al. 1981) are therefore relevant contributions. The answers to many questions, however, remain speculative. For example, how is snow incorporated into the current along the avalanche track? And how does sedimentation occur, and what are the effects of channeling? Geologists have studied some of these aspects in the context of submarine sediment gravity flows (see, for instance, Middleton & Hampton 1976) but in these circumstances as well, results are fragmentary and mainly qualitative.

A large body of information on avalanches and snow mechanics can be found in the proceedings of the four major conferences on this subject: the International Symposium on Scientific Aspects of Snow Avalanches, Davos, 1965; the Snow Mechanics Symposium, Grindelwald, 1974; the Symposium on Applied Glaciology, Cambridge, 1976, and the Symposium on Snow in Motion, Colorado, 1979. Useful practical information can be found in Perla & Martinelli (1976). Previous review articles on avalanches have been written by Mellor (1968, 1978), La Chapelle

(1977), and Perla (1980). Of these, Mellor (1978) discusses avalanche motion in a detailed manner. Some of what follows can therefore also be found in Mellor (1978), but the present review takes a different approach and contains sufficient new material to set it apart from these existing publications.

CLASSIFICATION CRITERIA

Avalanches occur under widely different conditions and can take various forms. The classification of avalanche phenomena is therefore useful in that it introduces a common vocabulary and satisfies the practical needs of communication.

Morphological Classification Features

The principal morphological features by which avalanches are classified (see, for instance, De Quervain 1973) are the form of motion, the free-water content in the snow cover, the location of the lower boundary of the flow, the track geometry, and the type of rupture of the snow cover. The dynamics of avalanches depends on all these features and it is therefore useful to discuss them in more detail.

Most avalanche-classification schemes recognize the existence of two distinct limiting forms of motion: (a) motion close to the ground and following ground contours of a snow mass of large bulk density (0.1–0.4 g cm^{-3}), referred to as a *flowing* or *dense-snow avalanche*; (b) motion of a snow cloud, consisting of a snow-particle suspension of low bulk density ($\leq 10^{-2}$ g cm^{-3}), called an *airborne powder avalanche* or more simply a *powder avalanche*. The terminology *mixed avalanche* is sometimes used to designate an avalanche in an intermediate state. Powder avalanches require dry snow (no free water content), whereas dense-avalanche flow can occur under wet or dry snow conditions. Since the material properties of wet and dry snow are widely different (wet snow is generally cohesive and snowball formation is possible) the distinction between *wet snow* and *dry snow avalanches* is useful. Some wet snow avalanches extend down to the ground, whereas others, and generally dry snow gravity flow, have the lower boundary within the snow cover. This introduces the distinction between *ground* and *surface avalanches*. When the avalanche descends a gully or ravine it is called a *channelled avalanche*, in opposition to an *open-slope avalanche*. The type of rupture of the snow cover depends on the state of intergranular cohesion. In loose snow a point fracture occurs (a *loose-snow avalanche*), whereas sufficient intergranular cohesion favors line fracture and the resulting motion is called a *slab avalanche* (soft slab or hard slab). Further important features of avalanches are the extent of rupture of the snow cover and the initial mass of snow released.

Magnitude of Released Snow Masses

Point failure is frequently observed during snow storms in loose snow of low density and on slopes steeper than about 45°. Because of their localized nature, small masses of snow are involved and these avalanches are therefore a steep-slope phenomenon. (However, in wet, cohesionless, late-spring snow, point initiation can occur on less steep slopes.) Line fracture, on the other hand, can release large snow masses and the resulting slab avalanche can take hazardous proportions. The initial thickness of slab avalanches is of order 1 m and the crown fracture line has a length of order 10^2 m and occasionally 10^3 m (Perla 1978); the bulk density of the snow cover released can vary between 0.08–0.4 g cm^{-3}. A typical slab-avalanche bed showing the crown surface is illustrated in Figure 1. The flanks may be sawtooth shaped, as can be seen in the figure.

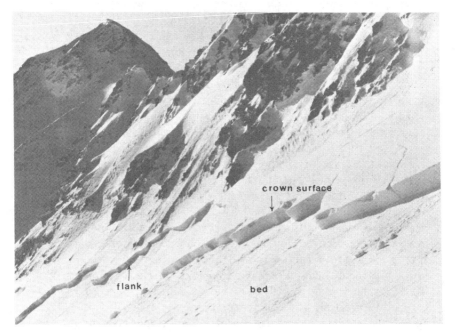

Figure 1 Example of a "crown fracture" of a slab avalanche. (Photo courtesy of the Swiss Federal Institute for Snow and Avalanche Research.)

Genetic Factors

There have also been attempts to classify avalanches according to genesis but these have not met with much acceptance. However, genetic factors of avalanche formation (in particular, precipitation, wind velocity, temperature variations in air and snow, and stratification of old snow cover) are of crucial importance to avalanche forecasting. Because rupture criteria for the snow cover are not understood, forecasting relies heavily on statistical tools (see, for instance, Obled & Good 1980).

RELEASE MECHANISMS

Armstrong (1977) and others have monitored metamorphic changes in snow structure, precipitation, and avalanche occurrence. Their data indicate that "temperature-gradient" metamorphism, giving rise to weak substratum layers consisting of coarse crystals with very weak intergranular bonds (called *depth hoar*), is the dominant cause of slab-avalanche formation.[1] This confirms the earlier finding of Perla (1971), who observed a large number of slab avalanches showing, in the majority of cases, the presence of depth-hoar crystals at the bed surface. The actual triggering is usually caused by load increase during storms. Other possible causes of avalanche initiation are ski loads, falling cornices, earthquakes, and of course artificial triggering. Grain growth (depth-hoar formation) is caused by water-vapor transport from the warm (0°C) ground to colder upper layers (Yen 1969) where it is deposited and freezes on the crystal faces. Thermal convection driven by the unstable temperature gradient (about $10-20°$ C m^{-1}) can enhance metamorphosis, as was pointed out by Palm (in Hopfinger et al. 1979). The thickness of the depth-hoar layers is of order 1 cm inside the snow cover and of order 10 cm adjacent to the ground (Perla 1978).

Rheological Properties of Snow

Although substratum weakening seems to be an important factor in reducing snowpack stability (Perla & La Chapelle 1970), the mechanisms of fracture initiation and fracture propagation are speculative. Snow is a complex material and its constitutive relations are not known and are not unique. There have been suggestions, however, that cohesive dry snow (before bond breakage has occurred) could be described by nonlinear or

---

[1] Dr. B. Salm pointed out to me that buried "surface hoar" is at least as frequent a cause for avalanche release as is depth hoar.

linear viscoelastic models (Brown et al. 1973, Desrues et al. 1980). A large body of information on rheological properties of snow can be found in the recent review by Salm (1982). A viscoelastic model has also been used by Lang (1977) to describe small-amplitude elastic waves in snowcover that is in a creeping-flow state. He suggests that these waves are related to the sawtooth pattern often observed at the flanks of a slab avalanche.

DENSE-SNOW OR FLOWING AVALANCHES

A dense dry-snow avalanche is a cohesionless granular material flow, whereas wet-snow avalanches are more characteristic of debris flow. Figure 2 shows a view of a dense dry-snow avalanche.

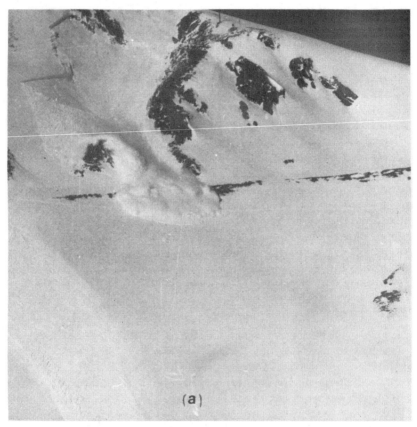

Figure 2 Sequential photos of a dry-snow avalanche at Col du Lautaret (France). (*a*) Avalanche issues from a gully, showing the large lateral spread; (*b*) avalanche has grown in size and is now accompanied by a trailing cloud. (Photos courtesy of CEMAGREF, Grenoble.)

Properties of Dense - Snow Avalanches

The bulk density of dry-snow avalanches is in the range 0.05–0.3 g cm^{-3} and wet-snow avalanches have a bulk density of 0.3–0.4 g cm^{-3}. Closely packed snow has a bulk density of approximately 0.6 g cm^{-3} (Wakahama & Sato 1977) and pores cease to interconnect completely when $\rho \approx 0.8$ g cm^{-3} (Mellor 1977). Bulk density is defined by $\rho = \nu\rho_p$, where ν is the volume fraction of solid material and ρ_p its density; for snow, ρ_p can be taken as the density of ice ($\rho_p = 0.917$ g cm^{-3}). The packed-bed, volume fraction of solids is then $\nu_\infty \simeq 0.7$. The volume fraction corresponding to the bulk densities of a dense-snow avalanche is therefore less than 0.35, which is well below the volume fraction ν_0 at which the internal friction angle of cohesionless granular material goes to zero. According to Bagnold's (1966) results on cohesionless material, ν_0 should be about

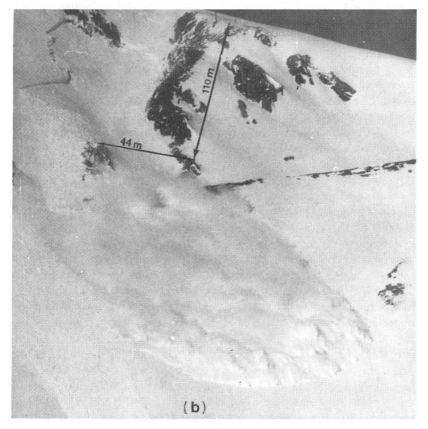

(b)

Figure 2   Continued.

0.5–0.6. Dense dry-snow avalanches are therefore, in general, in a "fluid" state, except on low slopes when they come to a stop. Initially, avalanches of this type caused by line fracture tend to slide like a rigid body, but intergranular bonds are rapidly broken and the mean shear gives rise to an instability and hence particle agitation, which causes further fragmentation. An exception to this might occur in the initial stage of hard-slab avalanches. Such avalanches could resist fragmentation and could therefore contain blocks comparable in size to the avalanche depth. For ground avalanches, the analogy with granular-material flow also suffers from the existence of large snow clods and from the fact that the material is cohesive and allows snowball formation.

A developed, dense-snow avalanche is roughly wedge-shaped, with the largest height (a few meters) near the front and the height trailing off toward the rear (Figure 2b). A wet-snow avalanche tends to dive into the snow cover because the frontal pressure leads to snow compaction and hence an increase in density. If the snow is wet throughout its depth, for instance, the avalanche can work its way down to the ground and literally plough through the snow cover (see Voellmy 1955). The velocity of such ground avalanches is in the range 10–30 m s^{-1}. On the other hand, the front of a dense dry-snow avalanche (as well as a damp-snow avalanche) overrides the snow cover in the track (Schaerer 1975) because it has a flow density lower than the snow-cover density. Such avalanches encounter less resistance and hence move with about twice the velocity of ground avalanches (30–60 m s^{-1}). The decrease in bulk density may be attributed partly to a normal stress effect, which tends to expand the flow, and also to air entrainment at the front and along the upper edge. At the same time, however, snow is entrained from below. A mechanism of entrainment similar to the "spilling breaker" model of Longuet-Higgins & Turner (1974) can be envisioned as a way of explaining the development of dense dry-snow avalanches. Because of simultaneous snow and air entrainment, the avalanche grows in size with distance, with the bulk density decreasing slowly. Snow entrainment is ultimately limited by the snow-cover depth. It should be noted that the direct momentum effect of air entrainment in dense dry-snow avalanches, important in powder avalanches, is negligible. An indication of the magnitude of air entrainment was given by Schaerer (1975). He compared the depth of the flowing avalanche with the depth after the avalanche has come to rest, with the resulting values given in Table 1. A general feature of dense dry-snow avalanches and to some extent of damp snow is the appearance of a cloud, as seen in Figure 2b. This cloud can turn into a powder avalanche if the track is long enough for the cloud to take on large proportions.

Table 1 Properties of dense-snow avalanches (from Schaerer 1975)

Snow type	Grain size (diameter in cm)	Depth ratio $(h/h_r)^a$	Estimated flow bulk density (g cm^{-3})
Dry	≤ 10	4	0.06–0.1
Damp	1–20	3	0.1–0.15
Wet	≥ 10	1.5	0.3–0.4

[a] h is the flow height, h_r the snow depth after the avalanche has come to rest.

Avalanche Velocity

The most commonly used formula for calculating velocity is based on a balance between downslope gravitational force, $\rho g S R C$, and frictional force per unit length, $f \rho U^2 C$, where C is the wetted perimeter and f is the friction coefficient. The result is Chézy's formula, first used in the context of avalanches by Voellmy (1955):

$$U = (\xi R S)^{1/2}, \tag{1}$$

where R is the hydraulic radius, S the slope ($S = \sin\theta$), and the friction coefficient is related to the coefficient in (1) by $\xi = g/f$. For ground avalanches, observations suggest that $\xi \simeq 400$–600 m s^{-2} and for dry-snow avalanches, $\xi \simeq 10^3$–1.8×10^3 m s^{-2} (Schaerer 1975, Martinelli et al. 1980). Since the hydraulic radius is equal to the flow height h on open slopes and is proportional to h in channeled flows, Equation (1) indicates that $U \propto h^{1/2}$. If it is assumed that h is proportional to the released snow-cover depth h_0, which is a reasonable assumption, the front velocity should scale as $h_0^{1/2}$.

Support in favor of $U \propto h_0^{1/2}$ is given by the remarkable compilation of avalanche velocities in terms of snow cover depth by A. Roch, published in Voellmy (1955). These data are reproduced[2] in logarithmic form (to show more effectively the 1/2 slope) in Figure 3. On steep slopes, where kinetic friction can be neglected[3], Chézy's formula seems to express the correct overall physical laws. For lower slopes the more general expression obtained by Voellmy (1955) (given here with the flow height replaced by the hydraulic radius) is

$$U = [\xi R(S - \mu\cos\theta)]^{1/2} \tag{2}$$

[2] I am grateful to A. Roch for making the original data available to me.
[3] It is frequent in engineering calculations to take into account kinetic friction on all slopes (see, for instance, Buser & Frutiger 1980). Considering the low solid fraction in dry-snow avalanches, the need for including μ is not clear.

where µ, the kinetic friction coefficient, can take values between 0.1 and 0.5 depending on flow conditions (mainly density). Slow-moving avalanches were observed by Schaerer to come to a stop on slope angles $\theta \approx 27°$, which gives $\mu = 0.5$ as an upper limit. It is interesting to note that for cohesionless spherical-particle granular material flow the internal friction angle is about 25° (Savage 1979).

Carrying the analogy with granular material flow even further leads to doubts as to whether Chézy's formula is appropriate for dense-snow avalanches. The results obtained by Savage show, for instance, an increase in velocity by a factor of two when, for essentially the same flow depth, the angle of inclination was increased from 32° to 39°. In fact, the flow depends strongly on fractional solid content near the boundary, which tends to decrease with increasing slope angle. A lower solid fraction gives rise to greater velocity and increased normal stress effect (Bagnold 1966) and this in turn reduces the friction. However, for sufficiently low solid fraction ($v < v_0$, as is the case in most avalanche flows on steep slopes), it is likely that a more definite relation between velocity and flow height would exist. This conjecture is supported by the data on avalanche velocities shown in Figure 3.

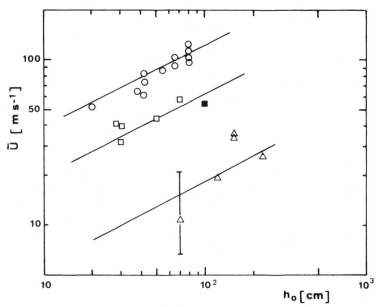

Figure 3 Average avalanche velocity as a function of snow depth for the three avalanche types; ○, powder avalanches; □ and ■, dry snow avalanches (■ corresponds to the avalanche shown in Figure 2); △, ground avalanches. Note that in the case of powder and surface avalanches h_0 refers to the depth of newly deposited snow, and in the case of ground avalanches to natural snow-cover depth. (Data made available by A. Roch.)

Numerical Models

Because of the close analogy between dense-avalanche flow and open-channel hydraulics, equations similar to those used for calculating flood waves have been developed to calculate avalanche movement. The use of numerical techniques to solve these equations permits one to take into account the effect of channeling and slope changes. The governing equations for one-dimensional flow used by Kulikovskii & Eglit (1973), Gregoryan & Ostroumov (1977), and Brugnot & Pochat (1981) are of the form

$$\partial P/\partial t + \partial Q/\partial x = 0, \tag{3}$$

$$\frac{\partial Q}{\partial t} + \frac{\partial(Q^2/P)}{\partial x} + gP\frac{\partial h}{\partial x}\cos\theta = gP\left(\sin\theta - \mu\cos\theta - \frac{fQ^2}{P^2R}\right), \tag{4}$$

where $P = \rho A$ is the concentration (A is the cross-sectional area), $Q = PU$ is the flux at position x and time t, and h is the flow depth. In deriving these equations it is assumed that the pressure is hydrostatic and that the free surface is horizontal in the cross-stream direction.

The most difficult problem arises in connection with a model for the jump conditions at the front. The simplest approach is to use the classical hydraulic-jump front conditions so that in a reference frame fixed in a front that is moving with velocity U_f, continuity and momentum give

$$U_f(P - P_s) = Q, \tag{5}$$

$$Q(U_f - U) = \frac{g\cos\theta}{n+1}(Ph - P_sh_s). \tag{6}$$

In this derivation it is assumed that the shock thickness is small so that pressure forces dominate gravitational force. Furthermore, the stress required to disrupt the snow cover ahead of the front, taken into account by Kulikovskii & Eglit (1973), is neglected here because of a lack of knowledge concerning its importance. In Equation (6), $n = 1$ for unconfined flow, and for semicircular or triangular gullies, $n = 2$ (Brugnot & Pochat 1981). The subscript "s" refers to the incorporated snow cover ahead of the leading edge; for the incorporated snow depth h_s, we have $h_s \leqslant h_0$. Equations (5) and (6) are used as downstream (ahead of the front) boundary conditions. The initial conditions are given by the total mass of snow initially set into motion and the upstream (far behind the front) boundary conditions are given by zero mass flux.

The use of these numerical models for engineering calculations relies on a good knowledge of friction coefficients, frontal conditions, and on the way the bulk density changes with velocity (or other parameters). Stereogrammetric photography of moving avalanches (facilitated by artificial release), as proposed by Bryukhanov (1967), Van Wijk (1967),

Samoilov (1976), and recently by CEMAGREF[4] (1980), permits a reasonable calibration of the model. Recent developments in the use of microwave radar systems for measuring avalanche velocity and height are promising in this context (Gubler 1981). Brugnot & Pochat (1981) showed that the observed variations in front velocity caused by changes in geometry are well predicted by the numerical model. However, sudden changes in terrain topography, as often happens when an avalanche issues from a ravine onto an open slope (see Figure 2a), are not adequately treated.

Fundamental studies of granular-material flow would help improve avalanche models. In particular, results on stress-strain relations as a function of solid fraction, on solid fraction (bulk density) dependence on velocity, and on cloud formation would be of immediate interest.

Run-Out Conditions

It is important in practice to be able to calculate where an avalanche will come to a stop. In the numerical models it is assumed that an avalanche ceases to flow when kinetic friction exceeds the sum of inertia, gravity, and pressure forces. The increase in bulk density with decreasing velocity is an important parameter in the problem, and it is at this point that ad hoc assumptions needing substantiation have been introduced. Buser & Frutiger (1980) calculated run-out distances using Voellmy's equation for a large number of fast-moving avalanches. A value of $\mu = 0.16$ gave satisfactory results. Salm (1966) examined the effect of an obstacle on run-out distance and showed that the additional energy dissipation involved could bring an avalanche to a quicker stop.

AIRBORNE POWDER AVALANCHES

An airborne powder avalanche is essentially a turbidity current flowing down an incline. A sketch of a powder avalanche showing possible snow and air entrainment mechanisms is presented in Figure 4, and a developed powder avalanche photographed at Rogers Pass (Canada) is shown in Figure 5. The great flow height of powder avalanches of the order of 100 m can clearly be seen in this photo. Powder avalanches occasionally move long distances over horizontal ground, and there have been informal reports of a possible uphill flow on a facing slope. It has long been recognized (Voellmy 1955) that their front velocity, which is of the order

---

[4] CEMAGREF: Centre National du Machinisme Agricole, du Génie Rural, des Eaux et des Forêts.

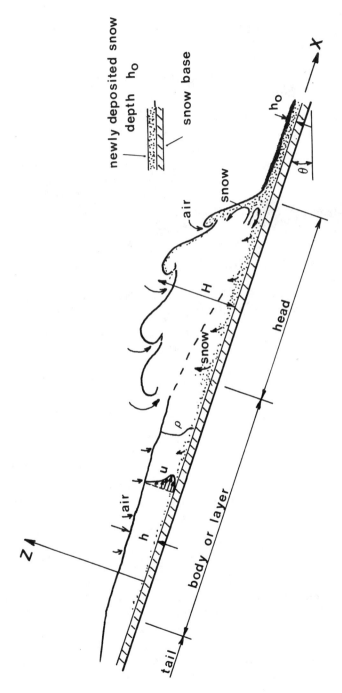

Figure 4 Sketch of a powder avalanche, indicating the snow and air entrainment mechanisms.

Figure 5 Head of a powder avalanche at Rogers Pass (Canada), showing the great flow depth of these flows. (Photo courtesy of P. Schaerer.)

of 100 m s^{-1} (see Figure 3), bears no simple correlation with the slope angle.

Cloud Formation Mechanisms

Before discussing powder-avalanche dynamics we would like to know how a dilute snow suspension is formed out of a dense gravity flow. Middleton & Hampton (1976), Pykhov & Longinov (1972), and Pykhov (1973) discuss some of the aspects of transformation from dense sediment flow into suspension flow in the context of submarine currents. Some of these ideas can be used to explain dust cloud formation in avalanches, but there are certain mechanisms particular to avalanches that are associated with the fact that the density of the ambient fluid is negligible compared with the bulk density of the flowing snow mass.

According to Voellmy (1955), a slab avalanche has initially a constant flow height until it reaches a velocity of approximately 10 m s^{-1}, at which speed intergranular bonds break down and fine particles are gradually suspended by the turbulent motions in the flow itself. A velocity of about 10 m s^{-1} has also been suggested by Mellor (1978) on the basis of observed snow particle diffusion into the air when a wind blows over a snowfield. A rough criterion for the occurrence of particle diffusion is that the friction velocity at the air-snow boundary $u_* =$

$(\tau/\rho_a)^{1/2}$, where τ is the shear stress in the air boundary layer, be about equal to the free-fall velocity of the snow particles, w_p. Since the friction velocity is about 5% of the mean velocity, particles of diameter 0.1–0.2 mm can be suspended in the air when the avalanche velocity is ≥ 10 m s^{-1}. The diameter of particles suspended in snow dust clouds is frequently in this size range. The rough criterion of 10 m s^{-1} for snow particles of 0.1–0.2 mm to remain suspended also satisfies the condition $w_p \lesssim U\sin\theta/10$ (Johnson 1963).

The mechanisms of cloud formation ascribed to turbulent flow (Voellmy) or to particle transport by the overlying air boundary layer (Mellor) would explain the existence of a trailing cloud on top of a dense dry-snow avalanche (Figure 2). This cloud can eventually overtake the dense snow body if the track is long and the cloud gains in volume. The passage over obstacles could also enhance cloud formation. However, in dry snow with weak intergranular bonds, powder avalanches sometimes form in a more explosive way shortly after release and it is conjectured here that a combined kinematic-dynamic shock, the foundations of which are discussed by Lighthill & Whitham (1955), is one possible mechanism to explain sudden cloud formation.

When the snow cover ruptures and starts to slide, no-slip conditions on the lower boundary would generate a strong mean shear and hence a normal stress; this would rapidly rupture intergranular bonds. The bulk density of the snow cover is such that the solid fraction is well below the value for the flow to take a fluidlike state (see the section on Dense Avalanches). Thus, a relation between U and flow height or between the flux Uh and the concentration h (in Lighthill & Whitham's terminology) is likely to exist. When it is further assumed that Chézy's formula (Equation 3) holds, the kinematic wave speed is $(3/2)U$, and long dynamic waves have a celerity $(gh)^{1/2}$ and therefore move downstream with velocity $U + (gh)^{1/2}$. Thus when the Froude number $U/(gh)^{1/2} > 2$, kinematic waves move at a higher speed than dynamic waves and the formation of a moving shock (a bore) of increasing strength is possible. This corresponds to Dressler's (1949) criterion, $\xi\sin\theta/g > 4$, for instability of a steadily flowing inclined layer and consequent roll-wave formation to occur.

A velocity of 10 m s^{-1} to which an avalanche accelerates in a very short time (Voellmy 1955) corresponds already to $F > 2$, so that F can quickly reach a value of 5 or more in the starting zone. The wave length λ of roll waves is estimated from Brock's (1967) linear theory as

$$\frac{\sin\theta\lambda}{F^2h} \approx 0.2\text{–}0.5,$$

giving for $F = 5$ and $\sin\theta = 0.5$, $\lambda/h \approx 10\text{–}25$.

These are reasonable numbers and would point to the possible existence of roll waves that are regions of strong turbulence with roller eddy motion, capable of strong air intake similar to that in a hydraulic jump. The main uncertainty lies in the amplification rate of the wave disturbances; rapid amplification would be required for shocks to occur shortly after release.

I have been able to locate as many as three rollers (or features that seem to be rollers), roughly equally spaced and spanning the whole width of a snow slab in the starting zone, in the film *Avalanche Control* distributed by the US Forest Service. This is a comforting observation, but it is not of course proof of the existence of roll-wave instability. Inhomogeneities in cohesion, as well as in the bulk density of the snow cover (of order 10% according to Perla 1978), that are frequently caused by wind transport of snow could also cause the snow slab to bulge in a wavy fashion. Furthermore, the downslope snow cover, which is more stable, can act as a barrier (called "Stauchwall") to the moving snow slab and cause roller formation and (if conditions are favorable) cloud development.

Gravity Current Structure of Powder Avalanches

Gravity currents on slopes and their relation to avalanches are only briefly mentioned in the review of gravity currents by Simpson (1982). Simpson's review emphasizes the widespread interest of gravity currents and crosses the boundary between disciplines. Here, I discuss inclined gravity currents in more detail and point out the direct relation with powder avalanches.

De Quervain's (1966) conjectures about the velocity and density profiles in powder avalanches have a striking resemblance to the profiles measured in an inclined steady plume by Ellison & Turner (1959). Tochon-Danguy & Hopfinger (1975) extended the inclined-plume analysis to powder avalanches, which in some cases can take the form of an inclined starting plume (i.e. consisting of a head and a body—see Figure 4). The body velocity can then be determined from a steady-plume analysis, because the head has no effect on the body flow but the body flow determines the front velocity (Turner 1973, p. 192). By integrating the momentum equation and using continuity and entrainment assumptions, Tochon-Danguy & Hopfinger obtained an expression for the steady-state mean body velocity (assuming that there is no snow incorporation along the track) that is valid also for large density difference. The expression obtained is here given in the form

$$U^3 = \frac{S_2 g_0' Q_0 \sin\theta}{E + (\rho_b/\rho_a)C_D},$$

(7)

which differs from the equation given in Tochon-Danguy & Hopfinger (1975) in that the term involving the Richardson number was neglected because its value is small compared with E. Furthermore, the equality expressing excess mass conservation, $g'hU = g_0'Q_0$, was used (where $g' = g(\rho - \rho_a)/\rho_a$).[5] In Equation (7), $g_0' = g(\rho_0 - \rho_a)/\rho_a$, Q_0 is the initial volume flux h_0U_0 per unit width, S_2 is a shape factor, the value of which is $0.6 \leq S_2 \leq 0.9$, C_D is the shear stress coefficient on the lower boundary, which depends weakly on the Reynolds number and has a typical value of $(1.3–3)\times10^{-3}$, and E is the experimentally determined rate of air entrainment, given by dh/dx. The subscript "0" refers to initial conditions of the snowcover, "a" to ambient condition (air), and "b" to near-bed conditions in the avalanche.

The velocity given in (7) is characteristic of the body or layer velocity, whereas observations give the rate of advance of the leading edge. Since the front velocity, U_f, is determined by the flow in the body, one would expect the relation $U_f \sim (g_0'Q_0)^{1/3}$ to hold also for the velocity of the leading edge. In a Boussinesq fluid (small density difference), this relation has been established experimentally by Wood (1965), Tochon-Danguy & Hopfinger (1975), and Britter & Linden (1980). Britter & Linden investigated the whole slope-angle range ($0° \leq \theta \leq 90°$) and were able to show that when $\theta \gtrsim 5°$ the front velocity can be approximated by

$$U_f = (1.5 \pm 0.2)(g_0'Q_0)^{1/3} \tag{8}$$

On zero slope the ratio $U_f^3/(g_0'Q_0) \simeq 2$ (Simpson & Britter 1979). The interesting result is that the velocity is only weakly dependent on slope angle and when $U_f/(g_0'Q_0)^{1/3}$ is plotted against slope angle θ, the scatter in the experimental data is practically of the same order as the variation with slope angle (Figure 6). However, a weak maximum in velocity is observed at $\theta \approx 20–40°$. When the density difference is not small, the shear-stress effect on the lower boundary becomes relatively more important [because C_D is weighted by ρ_b/ρ_a (see Equation 7)] and this would somewhat reduce velocities at low slope angles compared with what is shown in Figure 6.

The spatial growth rate of a gravity current is a function of the normal value of the Richardson number, $Ri_n = g'h\cos\theta/U^2$, to which the flow adjusts (Turner 1973, p. 181) and experiments show that when the density difference is small the growth rate is a linear function of slope angle. The experiments by Ellison & Turner (1959) give an empirical expression for E that can be cast in the form $E \simeq 9.5 \times 10^{-4}(\theta + 5)$ that can be used to calculate velocity from Equation (7). The head grows at a

---

[5] Note that in powder avalanches, ρ is the effective density, given by $\rho = \nu\rho_p + (1 - \nu)\rho_a$; because ν is small, the density of the carrier fluid (air) cannot be neglected.

faster rate and Britter & Linden obtained $dH/dx \simeq 4 \times 10^{-3}\theta$. ($x$ is measured from the virtual origin to the position where H is maximum.)

If Equation (7) or (8) is applied to calculate powder avalanche velocity, it must be assumed that $U_0 \approx U$ so that $g_0'Q_0 = g_0'h_0U$. Then with $g_0' = 500$–800 m s^{-2}, Equation (7) gives a mean body velocity $U \approx (60$–$95)(h_0)^{1/2}$ m s^{-1} when h_0 is substituted in meters. The velocity values obtained are comparable with observed avalanche velocities given in Figure 3. However, front velocities calculated from Equation (8) are somewhat low. The most likely explanation is that because of snow entrainment along the track, the avalanche front velocity is a larger fraction of the body velocity, and correlatively the head growth is at a lower rate than that observed in laboratory currents. The growth-rate values obtained for two-dimensional flow also no longer hold when the flow enters a gully or ravine, or issues from one onto open slope. For instance, Lüthi (1981) has shown experimentally that a gravity current issuing from a confined zone spreads laterally at an angle of about 60°.

Velocity and density profiles in an inclined gravity current have been measured by Hopfinger & Tochon-Danguy (1977). These measurements show a close resemblance to the profiles conjectured by De Quervain (1966). In the body, maximum velocities near the lower wall are 2–2.5 times the front velocity, and just behind the leading edge the maximum velocity is about 1.5 U_f. The density just behind the front is large (considerably larger than the average density) and has a maximum at about 0.1 H. These results are important for the interpretation of impact forces of avalanches.

Cloud Movement on Slopes — Low Density Clouds

An inclined gravity current requires that the buoyancy supply be maintained. This limiting state would not always seem to be a reasonable assumption for avalanches. It is therefore of interest to examine the other extreme case of a finite volume release, in which it is assumed that the whole flow has an identifiable volume.

Beghin et al. (1981) extended the theory of thermals to buoyant cloud convection on inclines. This gives a good description of the flow in the range $5° \leq \theta \leq 90°$. Experimentally they showed that the shape of the cloud can be approximated by a half-ellipse and that the cloud accelerates in a characteristic time $t_c \sim (4V_0^{1/2}/g_0'E\sin\theta)^{1/2}$, followed by a deceleration of the flow. The front velocity in the decelerating state, normalized by the square root of the released buoyancy and of the distance from the virtual origin to the leading edge $[(g_0'V_0)/x_f]^{1/2}$, where $V_0 = h_0l_0$ is the released volume per unit width, is plotted as a function of slope angle in Figure 6. It is seen that the slope-angle dependence is more pronounced than in the gravity-current counterpart. A velocity maximum is again observed at low slope angles (about 10–40°). Since the inherent experimental scatter in front velocity is large (due to the very distorted structure of the front and to variability in structure), the slope-angle dependence can be neglected for practical purposes and a mean value taken for the front velocity, which in the slope angle range $5° \leq \theta \leq 50°$ is of the form

$$U_f = (2.6 \pm 0.2)(g_0'V_0)^{1/2}x_f^{-1/2}. \tag{9}$$

At larger slope the velocity is lower, and at 90° the coefficient is about 1.5. At zero slope the flow is a finite-volume-release gravity current. The front velocity in this case is still given by (9), but the coefficient has a

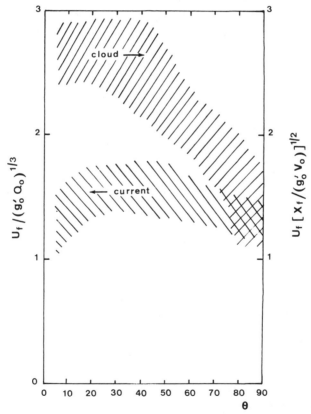

Figure 6 Nondimensional front velocity of gravity currents and clouds on inclines as a function of slope angle θ in degrees (data taken from Britter & Linden 1980 and Beghin et al. 1981).

value of 1.23 and x_f is measured from the physical origin (Beghin et al. 1981).

When typical values for g_0', V_0, and x_f are substituted into Equation (9), the velocity of an avalanche cloud is obtained. As expected, when an avalanche has the form of a cloud, velocities are considerably lower than when it has the form of a gravity current.

The spatial growth rate of a cloud, dH/dx, is larger by a factor of 4/3 than the growth rate of the head of an inclined gravity current. An important point is that no difference in growth rate between saline clouds and sand-suspension clouds is noticed.

Large Density - Difference Effects

In the laboratory experiments described above, the density difference between the flow and the ambient fluid was small compared with the mean density. This is generally not true in powder avalanches, which casts doubt on the quantitative usefulness of these laboratory results in the context of avalanches.

Experiments carried out by Hopfinger & Beghin (1980) with clouds of suspensions of natural barium sulphate having density $(\rho_0 - \rho_a)/\rho_a \simeq 1$ showed little difference when compared with clouds of density $(\rho_0 - \rho_a)/\rho_a \ll 1$; the maximum velocity normalized by $(g_0'V_0^{1/2})^{1/2}$ was found to be slightly lower and no apparent change in flow structure occurred, although the growth rate was somewhat higher. Kuenen (1951) experimented with clay- and sand-suspension turbidity currents of density $(\rho_0 - \rho_a)/\rho_a \lesssim 1$ and arrived at similar conclusions. He also noticed that for sand suspensions, velocities tended to be lower because of sedimentation. Ricou & Spalding (1961) showed that the inertial effect on entrainment in steady plumes is weak, and Hopfinger & Beghin (1980) were able to obtain a theoretical expression for the growth rate of high density clouds in the form

$$\frac{dH}{dx} = E'(\rho/\rho_a)^{1/2}. \tag{10}$$

For the maximum mass-center velocity, they obtained

$$U_m \simeq 0.6 \left(g_0'V_0^{1/2} \frac{k^{1/2}\sin\theta}{E'(1+k_v)} \right)^{1/2} (\rho_a/\rho_0)^{1/4}, \tag{11}$$

where E' is the spatial growth of a cloud in the Boussinesq limit, $k = H/L$, and k_v is the virtual-mass coefficient, equal to $2H/L\ (=2k)$ for an elliptic cylinder of major axis L and minor axis $2H$ (Batchelor 1967, p. 431). Values of k and E' are given in Beghin et al. (1981). Equation (11) indicates that the nondimensional velocity of a large density cloud is lower by a factor of $(\rho_a/\rho_0)^{1/4}$ compared with its value in the Boussinesq limit $(\rho_a/\rho_0 \neq 1)$. The choice of the value of ρ_0 is, however, not evident when this equation is applied to a snow avalanche cloud. Escudier & Maxworthy (1973) neglected the inertial effect on entrainment in their similarity analysis of large density-difference thermals, but included the virtual mass effect. Since the air entrainment rate is little modified by inertial effects, a high-density current or cloud will lose relatively less momentum to the entrained fluid than a low density cloud and will therefore be less affected by air entrainment. It should consequently move with larger nondimensional velocity. This momentum effect, however, is included in Equations (7) and (11) by defining g_0' to equal $g(\rho_0 - \rho_a)/\rho_a$, rather than $g(\rho_0 - \rho_a)/\rho_0$ as would be indicated by an extrapolation of the expression of g_0' from a Boussinesq fluid.

Snow Incorporation Mechanisms and Autosuspension Criteria

An important question concerns the rate and mechanisms of snow incorporation along the avalanche track. Snow entrainment increases the inertia of the flow and the excess mass, and hence affects the velocity. In a gravity-current-type avalanche this leads to an increased buoyancy flux and causes the avalanche to accelerate slowly down the slope, as is the case in a katabatic wind on a uniformly cooled slope (Ball 1956). Note however that the inertia associated with mass flux has a retarding effect.

In a cloud structure, predominant snow entrainment is likely to occur mainly toward the rear of the cloud where wall pressures can be low and appreciable suction is possible on steeper slopes. This is indicated in Figure 7, which shows the results of wall-pressure measurements made beneath saline clouds moving down slopes of 10° and 30° slope angle (Hopfinger et al. 1978).

Appreciable snow entrainment toward the rear would probably feed the cloud in the same way that the body feeds the head of a gravity current. This would make the cloud move faster and decrease its spatial growth. It is possible that because of snow entrainment, a body forms at the back and increases in length as the flow proceeds downslope. The conjecture is that a snow-entraining cloud develops into a gravity current, taking the form sketched in Figure 4, but eventually transforms back into a cloud structure after the cessation of snow entrainment.

It has been mentioned by Shoda (private communication) that snow from the snowcover ahead of the nose of a powder avalanche is occasionally lifted up by the avalanche and flows over the head, where it is incorporated. This entrainment mechanism has also been demonstrated in a gravity current that is made to move downslope over a thin dense layer (Hopfinger & Tochon-Danguy 1977). In this case the head appears to be more regular, with clear Kelvin-Helmholtz billows at the outer edge of the head.

The problem of autosuspension and sedimentation in turbidity currents has received a great deal of attention in the context of submarine sediment transport, where it causes the erosion of underwater canyons. Critical and comprehensive reviews of submarine turbidity currents and subaqueous sediment transport have been given, for instance, by Longinov (1971) and by Middleton & Hampton (1976). A numerical model has been developed by Komar (1977), and experimental studies of turbidity currents in connection with submarine canyons were initiated by Kuenen (1938, 1956). Johnson (1963) used energy arguments similar to Bagnold's (1962) to arrive at a rough criterion for the neglect of particle settling velocity in turbidity currents on slopes. This criterion requires that the mean settling velocity $\langle w_p \rangle$ of all the particles present be $\leq U \sin \theta / 10$, indicating that on small slopes a suspension is more difficult to sustain and would no longer be possible on horizontal boundaries. The criterion does not hold, however, when slope angles are very small ($\leq 3°$), and turbidity currents of suspended particles with very small settling velocity are possible on horizontal or weakly inclined boundaries (Michon et al. 1955, Middleton 1966, Tesakar 1969).

Sedimentation Conditions and Avalanche Collapse

Middleton (1966) showed that the head structure of a turbidity current is very similar to that of a saline gravity current, even when sedimentation occurs. He suggests that deposition of sediment occurs mainly in the body, where the turbulence intensity is smaller. The work of Britter & Simpson (1978) and Simpson & Britter (1979) on the dynamics of a gravity-current head on a horizontal boundary shows that mixing is more intense in the head than in the following layer, which feeds mass into the head. The implications of these observations for powder avalanches are that snow deposition on small slopes first occurs in the body, causing it to slow down (because the driving force is reduced). This in turn accelerates sedimentation and further slowdown by a runaway process. The head, which finds itself amputated of its buoyancy supply, then behaves like a cloud that moves onto a horizontal or weakly inclined boundary; namely, it will spread out and take the shape of a finite-volume-release gravity current. Front velocities associated with this spreading are much lower than the initial front velocity, which means that the suspension cannot be maintained for long. The rapidity of the whole sedimentation process depends on the size distribution of the suspended particles. But even assuming the size distribution is given,

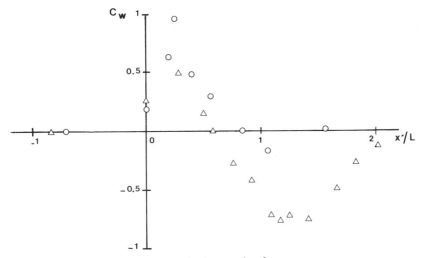

Figure 7 Wall pressure coefficient $C_w = (p - p_s)/\frac{1}{2}\rho U_f^2$ as a function of distance from the leading edge, x'/L, measured in a cloud; $\bigcirc, \theta = 10°$; $\triangle, \theta = 30°$. L is the cloud length, p_s the hydrostatic pressure.

the present state of knowledge does not permit a quantitative answer concerning the sedimentation rate. Progress on this aspect would be important not only for the prediction of avalanche collapse, but also for petroleum exploration off the continental shelf. (One wants to know how far out into the sea organic material can be transported by turbidity currents.)

The formation of a hydraulic jump at the foot of a slope when a sudden slope change occurs, as is observed in katabatic winds for instance (Ball 1957), would also enhance sedimentation just downstream of the jump. Even though hydraulic jumps in avalanches have not been reported, there is reason to believe that they occur when the avalanche has a body and encounters a sudden slope change.

IMPACT FORCES

From structural damages, Voellmy (1955) evaluated corresponding maximum dynamic pressures of 100–200 kN m^{-2} for dense-snow avalanches and about 50 kN m^{-2} for powder avalanches. Roch (1962) measured with a simple impact device pressures of 10^2–10^3 kN m^{-2} in mixed and powder avalanches. More recently, sophisticated instruments have been used and fantastically high impact pressures have been reported. La Chapelle (1977) mentions measurements made in the Japanese Alps by Shimizu and his colleagues (see Shimizu et al. 1980) which indicate values of $(0.47-2)\times10^3$ kN m^{-2}, depending on snow fall, and a maximum impulse of 3.9×10^3 kN s m^{-2}. The instruments were located at Shiai-dani on the reinforced concrete understory of a three-story construction barracks where in 1938, quoting La Chapelle, "this barracks was struck by an avalanche which carried away the superstructure on a 600 m aerial trajectory across an intervening ridge, across the Kurobe gorge, and smashed it against a rock cliff on Mt. Okukane, carrying to their deaths all 73 occupants."

Impact pressure measurements by Kotlyakov et al. (1977) in the Khibins, USSR, indicate average values in the range of 450–650 kN m^{-2} with a maximum value of 10^3 kN m^{-2}. This avalanche was of the dense-dry-snow type.

Calculation of Dynamic Pressures

Attempts to explain such high values on theoretical grounds have given rise to much speculation, based occasionally upon compressible-fluid-flow and shock-formation phenomena (see, for instance, Bryukhanov et al. 1967). In general, however, the dynamic pressure can be computed from

$$p = r\rho u^2, \tag{12}$$

where the coefficient r can take values between 0.5–2 depending on the type of avalanche. For a dense-snow avalanche, which does not tend to flow around the barrier, the collision is analogous to that of a deformable body with a rigid surface. The force per unit area is then just given by the rate of change of total linear momentum which is $mu(u/\ell)$, where m is the mass per unit area ($m = \rho\ell$) and ℓ is the length of the deformable body. This leads to $r = 1$. A larger coefficient is obtained when snow compaction is taken into account. Mellor (1968) used a one-dimensional shock analogy that considers the compaction of snow on impact; the resulting density jump then propagates upstream as a plastic wave (Wakahama & Sato 1977) with velocity c. Using mass and momentum conservation across the wave front in the usual form, we have

$$\rho_1(c - u_1) = \rho_2(c - u_2), \tag{13}$$

$$\Delta p = \rho_1(c - u_1)(u_2 - u_1). \tag{14}$$

Taking $u_2 = 0$, $\rho_1 = \rho$, and $u_1 = U$ (the avalanche velocity before impact), the expression obtained for the impact pressure is

$$p = \rho U^2(1 - c/U), \tag{15}$$

or equivalently,

$$p = \rho U^2\left[1 - \rho/(\rho_2 - \rho)\right]. \tag{16}$$

Equation (16) is referred to in the Russian literature as Khristianovich's formula (Khristianovich 1965). Both the bulk density ρ_2 and the plastic-wave speed have been measured by Wakahama & Sato (1977). They show that both quantities are increasing functions of ρ. The values measured by Wakahama & Sato are $0.5 \leq \rho_2 \leq 0.8$ g cm^{-3} for $0.18 \leq \rho \leq 0.3$ g cm^{-3}, indicating that for large ρ the compacted-snow density can exceed the packed-bed density (0.6 g cm^{-3}). In avalanches, the measurements of Kotlyakov et al. (1977) showed that $\rho_2 \approx 2\rho$, which gives $r \approx 2$ in Equation (12). The average impact pressure of an avalanche moving with 50 m s^{-1} front velocity and of bulk density 0.2 g cm^{-3} would then be 10^3 kN m^{-2}. Simple theoretical considerations therefore can explain the very high values measured by Shimizu and co-workers and also by Kotlyakov et al. in the Khibins.

An avalanche can probably carry large snow clods of high density, which can move away from the lower boundary because of a lift effect (Magnus effect) caused by rotation imposed by the mean shear. This might explain observed peak pressures, which are often a factor of two to three larger than the average pressure (Kotlyakov et al. 1977, Schaerer 1973).

In contrast to dense-snow avalanches, powder avalanches behave like Newtonian fluids. (They flow around the obstacle and in this case the dynamic pressure is given by $p = \frac{1}{2}\rho u^2$, which implies that $r = 1/2$.) Note that here ρ includes the density of the air ($\rho = \nu\rho_p + (1 - \nu)\rho_a$). When u is taken to be the velocity of the leading edge and ρ the average flow density, typical pressures would be 10–20 kN m^{-2}. Laboratory experiments by Hopfinger & Tochon-Danguy (1977) showed that the velocity inside the avalanche can be (1.5–2.5) times the front velocity, and close to the ground the effective density is considerably larger than the average density, by a factor of 2–4. Measured pressures in the avalanche can therefore be about a factor of 10 larger than the pressure calculated with front velocity and average density. Maximum pressure is expected to occur at about 0.1 H, which corresponds to maximum values in velocity and density (Hopfinger & Tochon-Danguy 1977).

It has also been conjectured by Hopfinger & Tochon-Danguy that the unsteady load could be of the same order as the drag. Inertia forces on structures have been evaluated for the case of tornadic winds by Wen (1975), and the arrival of a powder-avalanche front could have a similar effect.

The foregoing discussions indicate that dense-dry-snow avalanches are more hazardous than powder avalanches. However, the considerably greater flow height of powder avalanches compared with dense-snow flow and the fact that powder avalanches overrun obstacles, make the construction of passive defense structures for powder avalanches a major problem. Any low obstacle or barrier will have the effect of accelerating the flow just above it, as is well known from incompressible flow over obstacles. When such a defense structure is built in front of electric transmission-line masts or cable-car pillars, greater damage is likely to be caused. The mast can, in this case, literally be sheared off at the height of the barrier.

Avalanche Wind

It has been observed that structural and tree damage is also caused by the avalanche wind preceding the avalanche front. Tree damage has also been reported in situations where the avalanche has actually come to a stop before reaching the trees. Mountaineers, particularly in the Himalayas, have had their tents blown away by wind accompanying the passage of a nearby avalanche. Side and frontal winds (and possibly suction effects caused by avalanches) can be explained by the flow induced by a moving body. Tochon-Danguy & Hopfinger (1975) measured the flow induced by the head of a gravity current and gave a velocity of 0.2 U_f at a distance of 0.6 H downstream (ahead) of the front. Shen & Roper (1970) used potential-flow theory to calculate the drag and lift on a bridge caused by the wind of an avalanche flowing through the ravine below. It is more difficult, though, to explain damage ahead of an avalanche that has come to rest. The avalanche would have to come to rest in an abrupt manner, which is possible in dense avalanches but hardly in powder avalanches. Mellor (1968) evaluated the distance at which a mass of air, moving by its inertia, would have slowed down to half its initial velocity due to shear at the lower boundary and frontal resistance. The value obtained was about 1 km. If turbulent mixing were taken into account, which is the principal mechanism in retarding the flow, a considerably smaller distance would be obtained. However, when part of an avalanche has come to a stop due to impact on a low barrier, a vortex can be formed (see "air wave" below) capable of traveling long distances.

The avalanche wind is often referred to as "air blast" (see, for instance, Perla & Martinelli 1976) and this has sometimes been interpreted as a shock front. Compressibility effects in the air remain negligible, however, because even the fastest moving avalanche has a velocity of only about 1/3 the sound speed. (Inside the avalanche flow, the sound speed can be appreciably lower than in air because of the greater inertia of solid particles, but it is unlikely that compressibility effects will affect the dynamics.)

The Russian literature refers frequently to "air waves" (Grigoryan 1975, Urumbayev 1974, Kotlyakov et al. 1977), which result from an avalanche being stopped or deflected by a barrier. Kotlyakov et al. observed an air-wave formation associated with a dry avalanche hitting a dam. This wave front moved at greater speed than the initial avalanche. The observation can be explained by an acceleration of the flow above the dam and the generation of a vortex, as mentioned above in the context of impact pressures in powder-snow avalanches. Generally, a fast moving dry-snow avalanche is accompanied by a cloud riding on top of it; when the avalanche hits the barrier the dense portion is stopped whereas the cloud behaves like a fluid flowing around an obstacle. The acceleration effect is illustrated by an inclined gravity current impinging on a three-dimensional obstacle, shown by Hopfinger & Tochon-Danguy (1977).

GENERAL CONCLUSION

Snow avalanches are complex, turbulent gravity flows with characteristic properties ranging from grain and debris flow to fully turbidite flow. Field observations are intrinsically difficult and give only limited information; this is why progress in understanding the mechanisms has been slow. The fluid mechanics of related problems, in particular of granular-material flow and of inclined gravity currents, can give theoretical insight and help to develop realistic models of avalanche flow. A close relation between many aspects of snow avalanches and of submarine slides or turbidity currents has been noted, and a closer contact between the two scientific disciplines would be fruitful.

ACKNOWLEDGMENTS

I wish to thank T. Maxworthy, R. W. Griffiths, and K. Hutter for helpful comments and G. Brugnot for discussions concerning the practical aspects of this work. I am particularly grateful to B. Salm for giving a critical appraisal of the paper. His detailed remarks were most valuable.

Literature Cited

Armstrong, R. L. 1977. Continuous monitoring of metamorphic changes of internal snow structure as a tool in avalanche studies. *J. Glaciol.* 19:325–34

Bagnold, R. A. 1962. Auto-suspension of transported sediment: turbidity currents. *Proc. R. Soc. London Ser. A* 265:315–19

Bagnold, R. A. 1966. The shearing and dilatation of dry sand and the "singing" mechanism. *Proc. R. Soc. London Ser. A* 295:219–32

Ball, F. K. 1956. The theory of strong katabatic winds. *Austral. J. Phys.* 9: 373–86

Ball, F. K. 1957. The katabatic winds of Adélie Land and King George V Land. *Tellus* 9:201–8

Batchelor, G. K. 1967. *An Introduction to Fluid Mechanics.* Cambridge: Cambridge Univ. Press

Beghin, P., Hopfinger, E. J., Britter, R. E. 1981. Gravitational convection from instantaneous sources on inclined boundaries. *J. Fluid Mech.* 107:407–22

Britter, R. E., Linden, P. F. 1980. The motion of the front of a gravity current travelling down an incline. *J. Fluid Mech.* 99:531–43

Britter, R. E., Simpson, J. E. 1978. Experiments on the dynamics of a gravity current head. *J. Fluid Mech.* 88:223–40

Brock, R. R. 1967. Development of roll waves in open channels. *Tech. Rep. KH-R-16*, Keck Lab. Hydraul., Calif. Inst. Technol., Pasadena, Calif. 226 pp.

Brown, R. L., Lang, T. E., St. Lawrence, W. F., Bradley, C. C. 1973. A failure criterion for snow. *J. Geophys. Res.* 78: 4950–58

Brugnot, G., Pochat, R. 1981. Numerical simulation study of avalanches. *J. Glaciol.* 27:77–88

Bryukhanov, A. V. 1967. A study of the mechanisms of snow avalanche movement under various geographic conditions, using a special stereophotogrammatic technique. In *Snow Avalanches in the Khibins*, pp. 269–334 (In Russian)

Bryukhanov, A. V., Grigoryan, S. S., Miagkov, S. M., Plam, M. Ya., Shurova, I. Ya., Eglit, M. E., Yakimov, Yu. L. 1967. On some approaches to the dynamics of snow avalanches, In *Physics of Snow and Ice: International Conference on Low Temperature Science, 1966*, ed. H. Ōura, Vol. 1, Pt. 2, pp. 1223–41. Sapporo: Inst. Low Temp. Sci., Hokkaido Univ.

Buser, O., Frutiger, H. 1980. Observed maximum run-out distance of snow avalanches and the determination of the friction coefficients μ and ξ. *J. Glaciol.* 26:121–30

CEMAGREF. 1980. La stéréophotogrammétrie à cadence rapide d'avalanche. Cahier 39, No. 8

De Quervain, M. R. 1966. Problems of avalanche research. *Symp. Int. sur les Aspects Scientifiques des Avalanches de Neige, 5-10 Avril, 1965, Davos, Suisse. AIHS Publ.* 69:15–22

De Quervain, M. 1973. Eine internationale Lawinenklassifikation. *Z. Gletscherk. Glazialgeol.* 9(1–2):189–206

Desrues, J., Darve, F., Flavigny, E., Navarre, J. P., Taillefer, A. 1980. An incremental formulation of constitutive equations for deposited snow. *J. Glaciol.* 25:289–307

Dressler, R. F. 1949. Mathematical solution of the problem of roll waves in inclined channels. *Comm. Pure Appl. Math.* 2:149–94

Ellison, T. H., Turner, J. S. 1959. Turbulent entrainment in stratified flow. *J. Fluid Mech.* 6:423–48

Escudier, M. P., Maxworthy, T. 1973. On the motion of turbulent thermals. *J. Fluid Mech.* 61:541–52

Grigoryan, S. S. 1975. Mechanics of snow avalanches. *Symposium Mécanique de la Neige, Grindelwald, Avril 1974, IAHS-AISH Publ.* 114:355–68

Grigoryan, S. S., Ostroumov, A. V. 1977. Mathematical simulation of the process of motion of a snow avalanche (summary only). *J. Glaciol.* 19:664–65

Gubler, H. U. 1981. Messungen an Iliesslawinen, *EISLF Rep. No. 600*, Davos, Switz.

Hopfinger, E. J., Atten, P., Busse, F. 1979. Instability and convection in fluid layers: a report on Euromech 106. *J. Fluid Mech.* 92:217–40

Hopfinger, E. J., Beghin, P. 1980. Buoyant clouds appreciably heavier than the ambient fluid on sloping boundaries. *Proc. Int. IAHR Symp. Stratified Flows, 2nd, Trondheim*, pp. 495–504

Hopfinger, E. J., Layat, S., Roche, J.-P. 1978. Mesures des pressions dynamiques. *C. R. Deux. Renc. Int. Neige et Aval., ANENA, Grenoble*, pp. 183–87

Hopfinger, E. J., Tochon-Danguy, J. C. 1977. A model study of powder snow avalanches. *J. Glaciol.* 19:343–56

Johnson, M. A. 1963. Turbidity currents. *Sci. Prog.* 198:257–73

Komar, P. D. 1977. Computer simulation of turbidity current flow and the study of deep-sea channels and fan sedimentation. In *The Sea*, 6:603–21. Chichester, Engl: Wiley

Kotlyakov, V. M., Rzhevskii, B. N., Samoilov, V. A. 1977. The dynamics of avalanches in the Khibins. *J. Glaciol.* 19:431–39

Khristianovich, S. 1965. Rukovodstvo po snegolavinnym rabotam (A guide to snow-avalanche works) In *Sredneasiat. Nauchno-Issled. Gidrometeorol. Inst. 1965.* Leningrad: Gidrometeoizdat

Kuenen, Ph. H. 1938. Density currents in connection with the problem of submarine canyons. *Geol. Mag.* 1938:241–49

Kuenen, Ph. H. 1951. Properties of turbidity currents of high density. *Soc. Econ. Paleontol. Mineral. Spec. Publ.* 2:14–33

Kuenen, Ph. H. 1956. The difference between sliding and turbidity flow. *Deep-Sea Res.* 3:134–39

Kulikovskii, A. G., Eglit, M. E. 1973. Two-dimensional problem of the motion along a slope with smoothly changing properties. *PMM J. Appl. Math. Mech.* 37:792–803. Transl. from *Prikl. Mat. Mekh.* 37:837–48

La Chapelle, E. R. 1977. Snow avalanches: a review of current research and applications. *J. Glaciol.* 19:313–24

Lang, T. E. 1977. Wave pattern of flowing snow slabs. *J. Glaciol.* 19:365–73

Leaf, C. F., Martinelli, M. 1977. Avalanche dynamics. *USDA Forest Serv. Res. Pap. RM 183.* 51 pp.

Lighthill, M. J., Whitham, G. B. 1955. On kinematic waves. I. Flood movement in long rivers. *Proc. R. Soc. London Ser. A* 229:281–316

Longinov, V. V. 1971. The problem of turbidity currents in the lithodynamics of the ocean. *Oceanology* 11:303–10

Longuet-Higgins, M. S., Turner, J. S. 1974. An "entraining plume" model of a spilling breaker. *J. Fluid Mech.* 63:1–20

Losev, K. S. 1965. Some problems of the motion of avalanches and other analogous phenomena. In *Trudy Pervogo Vsesoyuznogo Soveshchaniya po Lavinam.* Leningrad: Gidrometeoizdat, pp. 112–16. Transl., 1969, in *Can. Natl. Res. Counc. Tech. Transl. 1383,* pp. 50–55

Lüthi, S. 1981. Experiments on non-channelized turbidity currents and their deposits. *Mar. Geol.* 40:M59–68

Martinelli, M., Lang, T. E., Mears, A. I. 1980. Calculations of avalanche friction coefficients from field data. *J. Glaciol.* 26:109–19

Mellor, M. 1968. Avalanches. *Cold Regions Sci. Eng. Monogr. III-A3d.* Hanover, N. H.:US Army Cold Reg. Res. Eng. Lab. 215 pp.

Mellor, M. 1977. Engineering properties of snow. *J. Glaciol.* 19:15–65

Mellor, M. 1978. Dynamics of snow avalanches. In *Rockslides and Avalanches. I. Natural Phenomena,* ed. B. Voight, pp. 753–92. Amsterdam: Elsevier

Michon, X., Goddet, J., Bonnefille, R. 1955. *Etude théorique et expérimentale des courants de densité, Tomes I, II.* Lab. Natl. Hydraul., Chatou, Fr.

Middleton, G. V. 1966. Experiments on density and turbidity currents. A. Motion of the head. *Can. J. Earth Sci.* 3:523–46

Middleton, G. V., Hampton, M. A. 1976. Subaqueous sediment transport and deposition by sediment gravity flows. In *Marine Sediment Transport and Environmental Management,* ed. D. J. Stanley, D. J. P. Swift, p. 197–218. New York: Wiley

Obled, C., Good, W. 1980. Recent developments of avalanche forecasting by discriminant analysis techniques: a methodological review and some applications to the Parsenn Area (Davos, Switzerland). *J. Glaciol.* 25:315–46

Perla, R. I. 1971. Characteristics of slab avalanches. *Proc. Symp. Snow Ice Relat. Wildlife Recreation,* ed. A. O. Haugen, pp. 163–83. Ames: Iowa Coop. Wildlife Res. Unit

Perla, R. I. 1978. Failure of snow slopes. In *Rockslides and Avalanches. I. Natural Phenomena,* ed. B. Voight, pp. 731–52. Amsterdam: Elsevier

Perla, R. I. 1980. Avalanche release, motion, and impact. In *Dynamics of Snow and Ice Masses,* ed. S. C. Colbeck, pp. 397–456. New York: Academic

Perla, R. I., La Chapelle, E. R. 1970. A theory of snow slab failure. *J. Geophys. Res.* 75:7619–27

Perla, R. I., Martinelli, M. 1976. *Avalanche Handbook. USDA Forest Serv. Agric. Handb.* 489. 238 pp.

Pykhov, N. V. 1973. The movements of sediments over an inclined sea floor after disturbance of their stability. *Oceanology* 13:893–96

Pykhov, N. V., Longinov, V. V. 1972. On the methods of computing the parameters of turbidity currents. *Oceanology* 12:761–71

Ricou, F. P., Spalding, D. B. 1961. Measurements of entrainment by axisymmetrical turbulent jets. *J. Fluid Mech.* 11:21–32

Roch, A. 1962. Mesure de la force des avalanches. Winterber. 1960/61. *Eidgenöss. Inst. Schnee- und Lawinenforsch.* 25:124–36

Salm, B. 1966. Contribution to avalanche dynamics. *Symp. Int. sur les Aspects Scientifiques des Avalanches de Neige, 5–10 Avril 1965, Davos, Suisse. AIHS Publ.* 69:199–214

Salm, B. 1982. Mechanical properties of snow. *Rev. Geophys. Space Phys.* 20:1–19

Samoilov, V. A. 1976. Stereophotogrammetric surveys of avalanches in the Khibins. *Mater. Glyatsiolog. Issled. Khron. Obsuzhdeniya.* 28:128–37

Savage, S. B. 1979. Gravity flow of cohesionless granular materials in chutes and channels. *J. Fluid Mech.* 92:53–96

Schaerer, P. A. 1973. Observations of avalanche impact pressures. *USDA Forest Serv. Gen. Tech. Rep. RM-3*, pp. 51–54

Schaerer, P. A. 1975. Friction coefficients and speed of flowing avalanches. *Symposium Mécanique de la Neige. Actes du Colloque de Grindelwald, Avril 1974. IAHS-AISH Publ.* 114:425–32

Shen, H. W., Roper, A. T. 1970. Dynamics of snow avalanche (with estimation for force on a bridge). *Bull. Int. Assoc. Sci. Hydrol.* 15:7–26

Shimizu, H., Huzioka, T., Akitaya, E., Narita, H., Nakagawa, M., Kawada, K. 1980. A study on high speed avalanches in the Kurobe canyon. *J. Glaciol.* 26:141–51

Simpson, J. E. 1982. Gravity currents in the laboratory, atmosphere and ocean. *Ann. Rev. Fluid Mech.* 14:213–34

Simpson, J. E., Britter, R. E. 1979. The dynamics of the head of a gravity current head. *J. Fluid Mech.* 88:223–40

Tesakar, E. 1969. *Uniform turbidity current experiments.* Civil Eng. Thesis., Techn. Univ. Norway, Trondheim. 170 pp.

Tochon-Danguy, J. C., Hopfinger, E. J. 1975. Simulation of the dynamics of powder avalanches. *Symposium Mécanique de la Neige. Actes du Colloque de Grindelwald, Avril 1974. IAHS-AISH Publ.* 114:369–80

Turner, J. S. 1973. *Buoyancy Effects in Fluids.* Cambridge Univ. Press

Urumbayev, N. A. 1974. Results of an investigation of the nature and effect of the air waves of an avalanche. In *Glyatsiologiya Sredney Azii. Laviny (Glaciology of Central Asia. Avalanches),* ed. Yu. D. Moskalev, Sredneaziat. Region. Nauchno-Issled. Gidrometeorol. Inst. Tr. 15(96):31–38

Van Wijk, N. C. 1967. Photogrammetry applied to avalanche study. *J. Glaciol.* 6:917–33

Voellmy, A. 1955. Über die Zerstörungskraft von Lawinen. *Schweiz. Bauz.* 73: no 12, 159–62; no. 15, 212–17; no. 17, 246–49; no. 19, 280–85. Transl., 1964, in *USDA Forest Serv. Alta Avalanche Study Cent. Transl. No. 2*

Wakahama, G., Sato, A. 1977. Propagation of a plastic wave in snow. *J. Glaciol.* 19:175–83

Wen, Y. K. 1975. Dynamic tornadic wind loads on tall buildings. *J. Struct. Div., Proc. ASCE* 101:169–85

Wood, I. R. 1965. Studies in unsteady self-preserving turbulent flows. *Univ. New South Wales, Austral. Water Res. Lab. Rep. No. 81*

Yen, Y. 1969. Recent studies on snow properties. In *Advances in Hydroscience,* Vol. 5, pp. 173–214. New York: Academic. 214 pp.

Ann. Rev. Fluid Mech. 1983. 15:77–96

ON THE THEORY OF THE HORIZONTAL-AXIS WIND TURBINE

Otto De Vries

National Aerospace Laboratory NLR, Amsterdam, The Netherlands

INTRODUCTION

The extraction of energy from the wind is an old idea, one used by sailing ships and windmills for many centuries. The development of ancient windmills was based on empiricism and engineering skill. The development of the fluid mechanics, or more specifically the aerodynamics, of windmills (*wind turbines* in modern usage) is more recent. The study of the aerodynamics of wind turbines was begun after World War I by Betz (1926) and Glauert (1935a) and got a new impulse after the "energy crisis" of 1973–74. Nowadays, it is a worldwide field of research, stimulated and guided by national research programs in the US, Sweden, Denmark, The Netherlands, Great Britain, and many other countries.

Scope of the Present Review

A complete review of the aerodynamic aspects of wind-energy conversion should encompass the following:

1. The characteristics of the natural wind, such as annual wind-velocity distributions, wind shear, turbulence, gusts, effects of local terrain conditions, and siting.
2. The theory of wind-driven turbines, operating in a homogeneous and nonturbulent flow.
3. The influence of the natural wind (turbulence, wind shear) and turbine-induced irregularities (yawing misalignment, blade-tower interaction) on turbine performance and blade loading.

77

0066-4189/83/0115-0077$02.00

4. The influence of wake interaction and decay in the terrestrial boundary layer in connection with large arrays of wind turbines.

The present review is limited to item (2.) for the case of horizontal-axis turbines. These turbines are singled out because the analysis is now in a transition phase from relatively simple to increasingly sophisticated. For a more extensive survey of the aerodynamic aspects, see De Vries (1979).

We begin with a description of the basic aerodynamic conversion process. It is then shown that the application of the laws of conservation of mass, momentum, and energy leads to a rough indication of the maximum possible energy output of a wind turbine. This is followed by a discussion of calculation methods for wind turbines, from the simple blade-element theory to the more sophisticated vortex and panel methods. The review concludes with some remarks on the experimental confirmation of the theory. This is important, since the ability to predict the characteristics of wind turbines for design purposes is the main motive for developing more sophisticated and complicated calculation methods.

Types of Wind Turbines

Many different types of wind turbines have been invented. A distinction can be made between turbines driven mainly by drag forces and turbines driven mainly by lift forces. A distinction can also be made between turbines with axes of rotation parallel to the wind direction (horizontal) and with axes perpendicular to the wind direction (vertical). The efficiency of wind turbines driven primarily by drag forces is low (see, for example, De Vries 1979, Chap. 4, pp. 5–6) when compared with the lift-force-driven type. Therefore, all modern wind turbines are driven by lift forces. The most common types are the horizontal-axis wind turbine (HAWT) and the vertical-axis wind turbine (VAWT; see Figure 1). The latter has the advantage that its operation is independent of the wind direction, whereas a HAWT has to be yawed into the wind direction.

Aerodynamic Characteristics of Wind Turbines

The aerodynamic operation of a wind turbine can be characterized by the following overall quantities: the rotor torque Q, the rotor drag D, the angular velocity Ω, and the power output $P = \Omega Q$. By dimensional analysis, these quantities can be made dimensionless as follows:

$$\lambda = \Omega R / U_0 \qquad \text{(tip-speed ratio)}, \qquad (1)$$

$$C_Q = Q / (\tfrac{1}{2} \rho U_0^2 R S_{\text{ref}}) \qquad \text{(torque coefficient)}, \qquad (2)$$

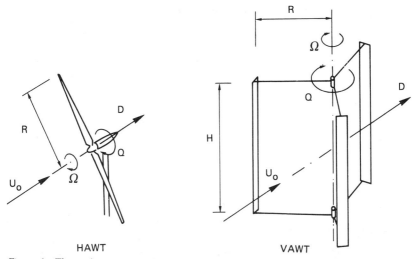

Figure 1   The main types of lift-driven wind turbines, namely the horizontal-axis wind turbine (HAWT) and the vertical-axis wind turbine (VAWT). Also indicated are the wind velocity U_0, the angular velocity Ω, the aerodynamic torque Q, and the rotor drag D.

$$C_P = P/(\tfrac{1}{2}\rho U_0^3 S_{\text{ref}}) \qquad \text{(power coefficient)}, \qquad (3)$$

$$C_D = D/(\tfrac{1}{2}\rho U_0^2 S_{\text{ref}}) \qquad \text{(rotor-drag coefficient)}, \qquad (4)$$

with R the maximum radius of the rotor, U_0 the undisturbed wind velocity, ρ the density of the air, and S_{ref} the cross section swept by the rotor blades ($= \pi R^2$ or $2HR$). It should be noted, however, that there is not yet a generally accepted standardization of symbols in the field of wind-energy conversion.

AERODYNAMIC CONVERSION PROCESS

The basic process of converting kinetic energy from the wind into mechanical energy of the rotor can be described in two ways, namely from the point of view of the rotor (the driving force) and from the point of view of the wind stream (loss of flow energy). Both points of view are, of course, connected by the laws of conservation of mass, momentum, and energy. The most direct way to set up a calculation method for a wind turbine is to consider the driving force.

The Driving Force

The rotor is driven by a component of the lift force. The lift force and the profile drag at each section of the rotor blade depend on the velocity of the relative flow (U_{rel}), the chord length (c), the angle of attack (α), and the dimensionless coefficients C_ℓ and C_d, which are functions of α (Figure 2). The section lift and drag (force per unit span of airfoil) can be expressed by

$$\text{section lift} = C_\ell c \tfrac{1}{2} \rho U_{rel}^2, \tag{5}$$

$$\text{section drag} = C_d c \tfrac{1}{2} \rho U_{rel}^2. \tag{6}$$

The section lift can also be related to the circulation Γ around the airfoil, namely

$$\text{section lift} = \rho \Gamma U_{rel}, \tag{7}$$

which implies

$$\Gamma = \tfrac{1}{2} C_\ell c U_{rel}. \tag{8}$$

The theory of wing sections is a well-established part of aerodynamics. (For the theory and a compilation of airfoil data, see Abbott & von Doenhoff 1959; for a modern approach, see Eppler & Somers 1978.) The lift force is perpendicular to U_{rel} and the profile drag is parallel to it. The lift force can only have a positive driving component (i.e. a component in the direction of motion; see Figure 2) when three conditions are fulfilled, namely (a) $\theta \neq 0$, (b) $C_\ell \neq 0$, and (c) θ and C_ℓ have the same sign. The profile drag always reduces the driving force.

When a HAWT operates with Ω constant and U_0 both constant and parallel to the axis of rotation, the flow relative to the rotor blade is steady. When the setting angle between the blade element and the plane of rotation (i.e. $\theta - \alpha$; see Figure 2) has the correct value, the driving force on the blade element is positive and constant during the complete revolution of the rotor. This situation is different from the case of a VAWT, where the lift force varies periodically from positive to negative

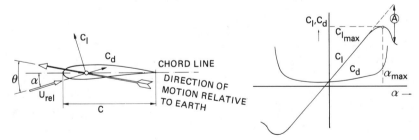

Figure 2 Airfoil characteristics in steady flow (schematic). C_ℓ is the section lift coefficient, C_d the section drag coefficient, c the section chord length, U_{rel} the flow velocity relative to airfoil, α the angle of attack, and θ the angle between U_{rel} and the direction of motion (relative flow angle). A: lift loss due to stall.

during a revolution of the rotor. The driving component of the lift force is, however, always positive or zero.

Loss of Flow Energy

The energy content of an incompressible flow is expressed by the so-called "Bernoulli constant,"

$$H = p + \tfrac{1}{2}\rho U^2, \tag{9}$$

where p is the static pressure and U the local velocity. In the case of flow of an incompressible and inviscid fluid, only forces perpendicular to the local velocity are possible, and the equation of motion for a fluid particle is

$$\frac{DH}{Dt} = \frac{\partial H}{\partial t} + \mathbf{U} \cdot \nabla H = \frac{\partial p}{\partial t}. \tag{10}$$

Equation (10) shows that the energy content of a fluid particle can only be changed by an unsteady static pressure variation working on the fluid particle along its path through the plane of the rotor. In the case of a HAWT, the flow is steady in a coordinate system fixed to the rotor, and $DH/Dt = 0$. This apparent paradox can be solved by adding the centrifugal and Coriolis forces to the equation of motion in the rotating coordinate system. Transformation of this modified equation to an earth-fixed coordinate system leads to the well-known turbine equation

$$\Delta H = \rho \mathbf{U} \cdot (\mathbf{\Omega} \times \mathbf{r}) = \rho U_{\tan\infty} \Omega r_\infty, \tag{11}$$

where ΔH is the total-head loss across the rotor, $\mathbf{r}$ is the distance to the axis of rotation, $U_{\tan\infty}$ is the tangential velocity component behind the rotor at radius r_∞.

Equation (11) is valid for a single rotor. In the case of two counter-rotating rotors, the equation must be applied to each rotor separately, in the ideal case leading to a value of $U_{\tan\infty} = 0$ behind the second rotor. Equation (11) is not valid for a VAWT, because in that case no coordinate system can be found in which the flow is steady.

APPLICATION OF CONSERVATION LAWS

When a wind turbine is surrounded by a "control" surface, the laws of conservation of mass, momentum, and energy, applied to the volume enclosed by the control surface, yield relations between the time-averaged aerodynamic forces on the turbine inside the volume and the velocities and static pressures on the control surface. The flow inside the volume does not appear in the equations and, therefore, most of the results are valid for both a HAWT and a VAWT. On the other hand, specific results can be obtained only by making assumptions about the flow inside the volume.

As is shown below, relatively simple assumptions lead to important conclusions about the maximum turbine performance.

Single-Streamtube Analysis

Betz (1926) considers only average axial velocities in a streamtube through the rotor disk S_{ref} (no radial and tangential velocities; see Figure 3) and applies the energy and linear momentum equation to that streamtube. He assumes that only the rotor absorbs energy and he neglects the viscous and turbulent stresses at the control surfaces. This leads to the following equations for C_P and C_D:

$$C_P = 4a(1-a)^2 \quad \text{and} \quad C_D = 4a(1-a), \tag{12}$$

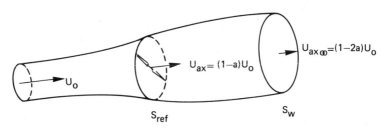

SINGLE STREAMTUBE

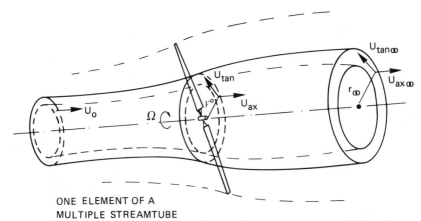

ONE ELEMENT OF A
MULTIPLE STREAMTUBE

Figure 3 Control surfaces for the application of the conservation laws.

where $a = (U_0 - U_{ax})/U_0$ is the axial induction factor and U_{ax} is the axial velocity through the rotor. Equation (12) shows that the maximum power output is obtained when

$$a_{opt} = 1/3 \quad \text{and} \quad (C_P)_{opt} = 16/27, \tag{13}$$

which is the well-known Betz limit.

An important by-product of Betz's analysis is a relation between the average axial velocity at the rotor (U_{ax}) and the velocity downstream in the wake $(U_{ax\infty})$:

$$U_{ax} = \tfrac{1}{2}(U_0 + U_{ax\infty}), \quad \text{or} \quad U_{ax\infty} = (1 - 2a)U_0. \tag{14}$$

The conservation of mass then leads to a relation between the downstream cross section of the wake (S_w) and the rotor disk area (S_{ref}),

$$S_w/S_{ref} = (1 - a)/(1 - 2a), \tag{15}$$

which reveals a physically unrealistic behavior at $a \geq \tfrac{1}{2}$.

Multiple - Streamtube and Angular - Momentum Analysis

It is often assumed that Equations (12) through (15), derived for a streamtube containing the entire rotor (single streamtube), also apply to an element of this streamtube (multiple-streamtube theory; see Figure 3). This notion becomes especially useful when tangential velocity components are included. In the case of a HAWT, the equation for the angular momentum relates the torque δQ on the blade elements in a streamtube element to the tangential velocity $U_{tan\infty}$. The connection with Equation (11) becomes clear by noticing that $\delta P = \Omega \delta Q$ and $\delta P = \Delta H U_{ax} 2\pi r \, dr = \Delta H U_{ax\infty} 2\pi r_\infty \, dr_\infty$ (mass conservation). This presupposes an axisymmetric flow. Glauert (1935a) takes a zero tangential velocity in front of the rotor and introduces the tangential induction factor a' for the tangential velocities at the rotor and downstream (Figure 3):

$$U_{tan} = a'\Omega r \quad \text{and} \quad U_{tan\infty} = 2a'\Omega r. \tag{16}$$

This leads, with some additional assumptions, to the relation

$$C_P = (2/\lambda)^2 \int_0^\lambda (1 - a)a'X^3 \, dX, \tag{17}$$

where $X = \Omega r/U_0 = \lambda r/R$ is the local tip-speed ratio. By assuming that the different streamtube elements behave independently of each other, it is possible to optimize the integrand for each X separately. For a fast running turbine ($\lambda \gg 1$), this leads to the result that $a_{opt} \to 1/3$, whereas for a slow running turbine ($\lambda_{opt} \cong 1$) or for elements close to the hub ($r/R \ll 1$), a_{opt} decreases. Substituting the result in Equation (17) yields $(C_P)_{opt}$ as a function of λ_{opt}, which is presented in Figure 4. The power

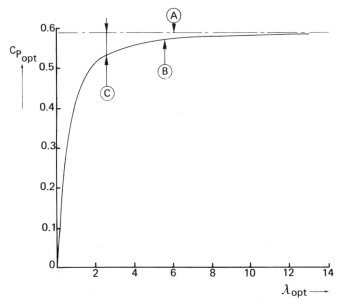

Figure 4 The optimum power coefficient for a HAWT according to Betz (1926) and Glauert (1935a). A: Betz optimum; B: Glauert optimum; C: loss due to wake rotation.

loss with respect to the optimum of Betz, $(C_P)_{\text{opt}} = 16/27$, can be interpreted as the kinetic energy of the rotation left in the wake. It appears from radial equilibrium that the wake rotation must be connected with a static pressure deficit in the wake, which is neglected by Glauert in the linear-momentum equation. This seems justified in the case of an aircraft propeller in cruise condition, with its small perturbation velocities, but is questionable in the case of a low-λ HAWT (De Vries 1979, Appendix C).

It is noted in passing that in the case of a VAWT, δQ is perpendicular to U_0 and the axisymmetric flow model cannot be applied. This excludes the possibility of using the angular-momentum equation to estimate in a relatively simple way the energy loss caused by the vorticity shed into the wake.

CALCULATION METHODS FOR A HORIZONTAL-AXIS WIND TURBINE

In the preceding discussion, overall estimates of the optimum power output of an isolated wind turbine were made, but it was not possible to determine the optimum shape of the rotor from these general considerations. In addition, it is important to obtain accurate predictions of the off-design conditions, i.e. $\lambda \neq \lambda_{opt}$, especially in connection with, for example, turbine control. Both kinds of calculations require more sophisticated methods. These methods have been developed, following in the footsteps of aircraft-propeller theory.

Before describing these calculation methods, we survey the different flow states of a rotor in a wind stream.

Survey of Flow States

The different flow states of a rotor in a wind stream, shown in Figure 5, have been taken from a presentation by Stoddard (1977). Only the "windmill brake state" ($0 \leq a < \frac{1}{2}$) and the "turbulent-wake state" ($\frac{1}{2} \leq a < 1$) are of interest for a wind turbine, because in those cases $C_P \geq 0$.

The simple momentum theory fails when $a \geq \frac{1}{2}$, because the trailing vortex structure is unstable at the large wake expansion when $a \to \frac{1}{2}$. Helicopter rotor tests show that the turbulent-wake state can be physically realized, but a good mathematical model is still lacking (see, for

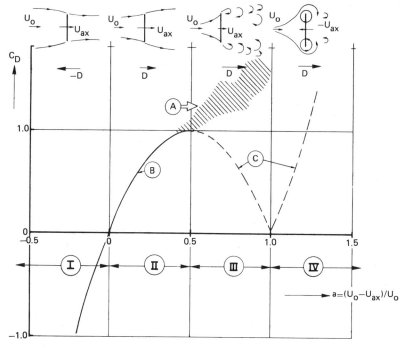

Figure 5 Survey of the different flow states of a horizontal-axis wind turbine, characterized by the average axial velocity U_{ax} through the rotor disk and the rotor drag D. A: helicopter experiments; B: Equation (12); C: momentum theory fails; I: propeller state; II: windmill brake state; III: turbulent wake state; IV: vortex ring state.

example, Zimmer 1972). There is some evidence that the effect of the turbulent-wake state starts already at $a = 0.4$, which is rather close to the optimum condition ($a = 1/3$).

At low induced velocities, the wind turbine is often affected by flow separation ($\alpha > \alpha_{max}$; see Figure 2), which presents a calculational problem. This contrasts sharply with a propeller at cruise condition, which operates at small α when the induced velocities are small. The optimum condition for a wind turbine is thus closely surrounded by the turbulent-wake state and conditions with blade stall, as illustrated in Figure 7 and discussed below.

Optimum Rotor Shape

The application of the conservation laws and some additional assumptions led Glauert to a formulation of the flow conditions in the plane of a rotor at optimum power output [see discussion of Equation (17) and Figure 4]. The simplest way to relate the flow in the plane of the rotor to the geometry of and the forces on the rotor is to use the existing blade-element theory for aircraft propellers, of which an excellent review is given by Glauert (1935b). The analysis does not determine the shape of the blade uniquely, but only the product of the local chord and the local lift coefficient for a chosen λ_{opt} and number of blades B. There is a good reason to choose a constant C_ℓ along the blade span (viz. at $(C_\ell/C_d)_{max}$). This choice leads to a strong variation of the chord length along the span (taper), as is illustrated in Figure 6. Since $C_\ell = $ constant means $\alpha = $ constant, this also implies that the blade-setting angle $i = \theta - \alpha$ varies strongly along the blade span (twist).

The analysis of Glauert (1935a) does not take the profile drag C_d and the effect of a finite number of blades (tip correction factor) into account. This limitation has been recently removed by a number of investigators, such as Wilson & Lissaman (1974), Wilson & Walker (1976), and Griffith (1977). For some different approaches toward the determination of the optimum shape, the reader is referred to Weber (1975), Stewart (1976), and Giordano (1979). The latter bases his approach on a different optimization for an aircraft propeller (Giordano 1974, Hirsch 1948). Whether improved calculation methods lead to different optimum blade shapes is an open question. Nevertheless, the aerodynamic optimum shape certainly is highly tapered and twisted, which is impracticable from the manufacturer's point of view. More practicable shapes require a calculation method for arbitrary blade shapes and a full range of tip-speed ratios. Such methods are discussed next.

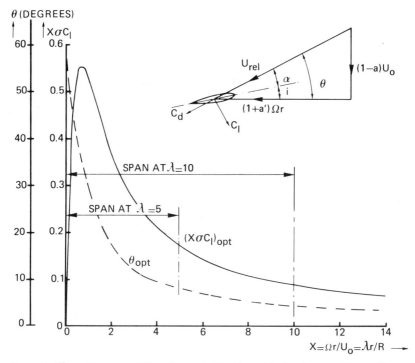

Figure 6   The optimum conditions for a wind turbine with $C_d = 0$, according to Glauert. $X\sigma C_\ell = (1/2\pi)B\lambda(c/R)C_\ell$; $\sigma = Bc/(2\pi r)$ is the local solidity ratio, B the number of rotor blades, i the blade setting angle, α the angle of attack, and θ the relative flow angle.

Blade - Element Theory

The blade-element theory assumes that the different spanwise elements are independent of each other and that the forces on the blade elements can be determined from the local flow conditions and the corresponding two-dimensional wing-section data (Figure 2). The local flow conditions are estimated by either momentum or vortex considerations, or both. Though originally developed for aircraft propellers (Glauert 1935b), the blade-element theory has been applied to wind turbines by Glauert (1935a), Wilson & Lissaman (1974), Holme (1981), and others.

A typical result of such an application is given in Figure 7. In this figure, the C_p vs. λ curve indicated by "C" is calculated with "ideal" airfoil characteristics, i.e. $C_d = 0$ and $dC_\ell/d\alpha = $ constant. The optimum is less than the Betz limit, because of (*a*) wake rotation, (*b*) tip losses (i.e. a finite number of blades instead of axisymmetric flow), and (*c*) hub losses (i.e. the blade terminates at r_{hub} instead of at $r = 0$).

Blade stall affects C_P by both the decrease of C_ℓ and the increase of C_d. This is clearly demonstrated by the area "D" and "E" at low λ. At large λ, C_d is small (α is small), but nevertheless the C_P loss is large, because this loss is almost proportional to $C_d\lambda^3$. Figure 7 also shows an empirical correction for the turbulent wake state "F," as discussed by Miller et al. (1976) and De Vries & Den Blanken (1981, Appendix C). The example of Figure 7 clearly illustrates that the optimum is closely surrounded by blade stall and a turbulent-wake state.

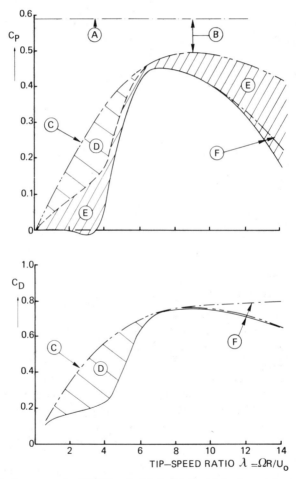

Figure 7 Performance of a HAWT, calculated with the blade-element theory ($r_{hub}/R = 0.15$, $B = 2$, blade with taper and twist). A: Betz limit; B: decrease of optimum; C: "ideal" profile characteristics; D: loss due to lift decrease (stall); E: loss due to profile drag; F: increase due to turbulent wake state (empirical correction).

With respect to the C_D vs. λ curve, the profile drag has hardly any effect and the lift loss due to blade stall is responsible for the marked decrease of C_D at low λ. The turbulent wake state might have a large effect at high λ and negative blade-setting angles.

For some critical remarks on the underlying assumptions of the blade-element theory, the reader is referred to De Vries (1979, Chap. 4, pp. 19–20).

Vortex Theory and Panel Methods

Instead of estimating the induced velocities from momentum equations and a number of additional assumptions, a more exact model can be formulated, analogous to the theory for a wing with a finite span. In this respect we can mention the "lifting-line" theory of Prandtl (1918), the "lifting-surface" theory of Weissinger (1947), and the "panel" method of Hess (1972) and others (Figure 8). The flow around the wing is assumed to be irrotational everywhere (except in a thin layer of trailing vorticity), and the velocity can therefore be calculated from a perturbation potential $\phi(x, y, z)$:

$$\mathbf{U} = \mathbf{U}_0 + \nabla\phi(x, y, z). \tag{18}$$

We assume in Equation (18) the following boundary conditions: (a) zero normal velocity on the wing surface and the trailing vortex sheet, (b) zero

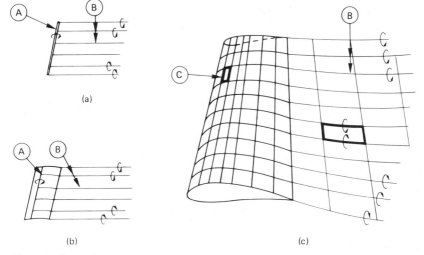

(a)

(b)   (c)

Figure 8 Increasing complexity in lifting-wing theory. (a) lifting-line theory (wing chord → 0), (b) lifting-surface theory (wing thickness → 0), (c) panel method. A: bound vortex; B: trailing vortex sheet; C: wing contour panel with singularity distribution.

pressure difference across the vortex sheet, and (c) smooth flow at the trailing edge of the wing (Kutta condition). It is often admissible to fulfill the boundary conditions for the trailing vortex sheet in a flat plane parallel to U_0 instead of at its actual position.

The flow around a steadily rotating HAWT in a steady wind stream U_0, parallel to the axis of rotation, can be considered as irrotational to first order. The velocities in an earth-fixed coordinate system can be calculated from an unsteady perturbation potential, namely

$$\mathbf{U} = \mathbf{U}_0 + \nabla \phi_u(x_u, r_u, \eta_u, t),\tag{19}$$

where $\mathbf{U}_0$ is the wind velocity, x_u, r_u, η_u are cylindrical coordinates, and t is time. In a rotating coordinate system fixed to the rotor, the flow is steady and rotational, but the perturbation velocities are still irrotational because this cannot be changed by a mere coordinate transformation. Thus

$$\mathbf{U}_{rel} = \mathbf{U}_0 + \mathbf{\Omega} \times \mathbf{r} + \nabla \phi(x, r, \eta),\tag{20}$$

where $\phi = \phi_u$ and $\eta = \eta_u - \Omega t$. Equation (20) demonstrates that although the flow in rotating coordinates is steady and rotational, the perturbation velocities can be calculated from a potential, in a way analogous to the airplane wing in rectilinear flight. However, as is illustrated in Figure 9, there is one important difference between wings and rotor blades, namely the shape of the trailing vortex sheet. In the case of negligible perturbation velocities, the shape is determined by $\mathbf{U}_0$ and $\mathbf{\Omega} \times \mathbf{r}$, but at large perturbations ($a = 1/3$), the shape of the trailing vortex sheet is determined by the induced velocities and cannot be given a priori. Therefore, the shape has to be determined by an iteration procedure ("relaxed-wake analysis") or has to be assumed beforehand ("fixed-wake analysis").

The potential-flow problem is solved by replacing the rotor blades and vortex sheets by singularity distributions (vortices, sources, doublets),

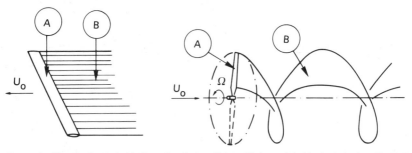

Figure 9 Vortex sheets behind a wing in rectilinear flight and behind a rotating HAWT (one blade shown). A: wing or blade with bound vorticity; B: trailing vortex sheet.

determining the strength of these distributions by fulfilling the boundary conditions, and then calculating the velocity field.

The lifting-line theory (Figure 8a) is the simplest approximation, because here the rotor blade is reduced to a single vortex line [the bound vortex $\Gamma(r)$; see Equations 7 and 8], and a sheet of continuous vorticity ($\gamma = -\partial\Gamma/\partial r$), or with the vorticity lumped into a number of discrete vortex lines. From this vortex system, we can calculate U_{rel} and α at the bound vortex. These values then determine the circulation by using Equation (8) and the aerodynamic characteristics of the airfoil.

The lifting-surface theory (Figure 8b) allows the possibility of including some effects of the finite chord length (see, however, van Holten 1975 for a remark on its applicability in a rotating system).

The panel methods (Figure 8c) are the most sophisticated, because they discretize the rotor blade and vortex sheet by a large number of (chordwise and spanwise) surface elements (panels) with a singularity distribution on the surface or inside the blade volume. The method approximates the pure potential-flow problem, and the effects of viscosity on C_{ℓ} and C_d have to be included by semi-empirical methods and possibly boundary-layer calculations. This is in contrast with the blade-element and lifting-line theories, in which two-dimensional airfoil data, measured in a wind tunnel, are used.

Panel methods for rotor systems have been developed for helicopters (Foley 1976, Rao & Schatzle 1977) and also for ship propellers (van Gent 1975), but the development of panel methods for wind turbines has only recently begun (see, for example, Suciu et al. 1977 and Preuss et al. 1980).

The main problem in the calculation of a HAWT is the effect of the expanding wake on the induced velocities. The author knows of no systematic studies on this effect. In this respect the studies on helicopter wakes are interesting, although they are not directly applicable to wind turbines, as follows from the discussion of (among others) Landgrebe & Cheney (1972).

EXPERIMENTAL CONFIRMATION OF THE THEORY

The blade-element theory for a HAWT has several theoretical shortcomings, which could be overcome by more advanced calculation methods. The computer codes, however, become more complex and also more time-consuming. Therefore, one must consider carefully whether or not the development of more sophisticated theories is worthwhile with respect to the total R & D costs of large wind-energy conversion systems.

The necessity for the more sophisticated theories has to be judged from a comparison between the calculated and measured performance of HAWTs. Surprisingly, many of the recently built HAWTs have been designed on the basis of aerodynamic theories developed between the years 1920–40, and hardly any experimental data are available.

Wind-Tunnel Tests Versus Field Tests

Reliable aerodynamic measurements on wind turbines in field tests are both difficult and time-consuming due to the stochastic character of the wind. Wind-tunnel tests on scaled models seem preferable because of the controlled test conditions and the rapid compilation of the desired data. Field tests are, of course, still indispensable for testing the control system, etc., under real atmospheric wind conditions.

A complete simulation of the full-scale conditions in a wind tunnel (apart from turbulence and wind shear) is obtained when the following dimensionless quantities are the same for the full-scale turbine and for the wind-tunnel model: (a) the tip-speed ratio (λ), (b) the Reynolds number (Re_c), and (c) the tip Mach number (Ma_t). As is discussed below, these conditions cannot be realized except in a compressed-air wind tunnel, but this introduces additional difficulties.

For a reliable comparison with calculated results, it is not necessary to obtain the full-scale Re_c, but values above a certain minimum have to be reached, e. g. $Re_c > 3 \times 10^5$. The Reynolds number is defined by and can be estimated from

$$Re_c = U_{rel}c/v \cong Ma_t R(a/v)(c/R)(r/R), \tag{21}$$

where $Ma_t = U_{rel}/a \cong \Omega R/a$ is the tip Mach number, a is the velocity of sound in air, and v is the kinematic viscosity of air. Equation (21) shows that for a given rotor shape $[c/R = f(r/R)]$ and for given properties of air (a, v), Re_c depends on Ma_t and the size of the rotor R.

For actual HAWTs, Ma_t is limited to 0.25 or 0.30 because of blade strength (centrifugal loads) and noise production. For a wind-tunnel test, Ma_t cannot be increased too much; otherwise, compressibility effects occur that are not present on the actual HAWT. Therefore, the size of the rotor is the only parameter left to secure a sufficiently high Re_c. The small size of the test sections of most low-speed wind tunnels, aggravated by wake-blockage effects in closed test sections (Alexander 1978, De Vries & Den Blanken 1981, Appendix B), reduces the maximum size of the models and in that way the attainable Re_c. How far open-jet or slotted-wall test sections are free of wake-blockage effects is still a point of discussion. Tests in large low-speed wind tunnels seem to be the safest solution.

Small-Scale Wind-Tunnel Test Results

An example of a comparison between results of a small-scale rotor test in a wind tunnel (viz. a rotor with a diameter of 0.75 m in a closed test section of 3 m × 2 m) and results of calculations with a blade-element theory are given in Figure 10 (De Vries & Den Blanken 1981). The

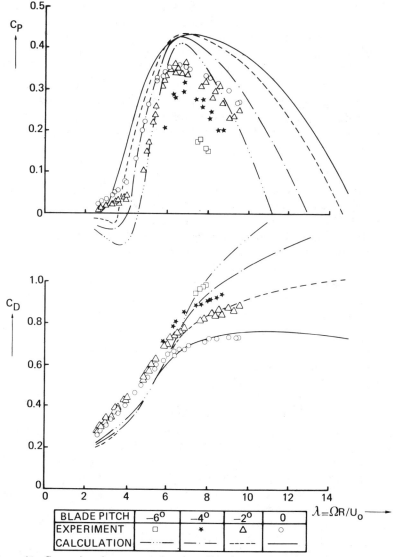

Figure 10 Comparison between results of small-scale rotor tests in a wind tunnel and calculations with the blade-element theory (De Vries & Den Blanken 1981).

Reynolds number used during these tests was $Re_c \cong 3.5 \times 10^5$ at $Ma_t \cong$ 0.55. This illustrates the small margin available for attaining sufficiently high values of Re_c in a normal low-speed wind tunnel.

Moreover, the wake-blockage corrections were large. Zero blade pitch angle was chosen at the calculated optimum condition, whereas a negative blade pitch meant an increased angle of attack of the blade elements. Also the spanwise distribution of ΔH, U_{ax}, and U_{tan} was measured close behind the rotor disk. This appeared to be a useful tool for the interpretation of discrepancies between test results and calculations.

The large discrepancy between the measured and calculated $(C_P)_{opt}$ is due to the strong underestimation of the profile drag. This is not yet fully explained, but it might be connected with the small size of the model. The good agreement between measured and calculated C_D suggests that the lift is well predicted. A comparison with Figure 7 indicates that the discrepancy of C_P as well as C_D at low λ stems from an underestimation of the lift beyond the stall.

The main conclusions of the investigation were the following:

1. The blade-element theory predicts the overall characteristics well, notwithstanding differences in the spanwise distribution of ΔH, U_{ax}, and U_{tan}. It is expected that the anomaly in the profile drag for attached flow disappears in large-scale tests at higher Re_c.
2. The two-dimensional profile data are no longer applicable at stalled rotor conditions. The C_ℓ values remained unexpectedly high beyond the stall.
3. The tip-correction methods developed for propellers seem inadequate in the case of wind turbines.
4. The experiments did not extend far enough into the turbulent wake state to verify the validity of the empirical relation used in the calculation method.

Full-Scale Tests

It is a remarkable fact that experimental data on the aerodynamic performance of HAWTs are so very scarce (Rohrbach 1976). There are some field-test results (see, for example, Smedman-Högström 1978 and Gustavsson & Törnkvist 1978) but they show appreciable scatter, and the inaccuracies in C_P vs. λ are masked by presenting the data in the form of P vs. U_0. Moreover, C_D is not measured, which means an important diagnostic element is missing in the evaluation of the calculation method. Though sufficient as acceptance tests for the HAWT as an electricity-generating device, the field-test data are insufficient for use as a basis for deciding whether or not to develop more sophisticated calculation methods.

CONCLUDING REMARKS

The situation described above for the HAWT contrasts sharply with the situation for the VAWT with curved (troposkien) rotor blades. For that type of turbine, comprehensive test results already exist (South & Rangi 1975, Blackwell et al. 1976, Sheldahl & Blackwell 1977, Worstell 1979, Sheldahl 1981). For a stimulating discussion of whether or not the VAWT is aerodynamically inferior to the HAWT, the reader is referred to Maydew & Klimas (1981).

It is hoped, that in the near future more complete and reliable data on the aerodynamic performance of the HAWT will become available to stimulate and motivate the further development of its aerodynamic theory.

Literature Cited

Abbott, I. H., von Doenhoff, A. E. 1959. *Theory of Wing Sections; Including a Summary of Airfoil Data.* New York: Dover. 693 pp.

Alexander, A. J. 1978. Wind tunnel corrections for Savonius rotors. *Proc. 2nd Int. Symp. Wind Energy Syst.*, Vol. 1. Div. E, pp. 69–80. Cranfield, Bedford, Engl.: BHRA Fluid Eng. 448 pp.

Betz, A. 1926. *Wind Energy and its Exploitation by Windmills.* Göttingen: Vandenhoeck & Ruprecht. 64 pp. (In German)

Blackwell, B. F., Sheldahl, R. E., Feltz, L. V. 1976. Wind tunnel performance data for the Darrieus wind turbine with NACA 0012 blades. *SAND76-0130*, Sandia Lab., Albuquerque, N.Mex. 61 pp.

De Vries, O. 1979. Fluid dynamic aspects of wind energy conversion. *AGARDograph No. 243*, AGARD, Neuilly-sur-Seine, France. 150 pp.

De Vries, O., Den Blanken, M. H. G. 1981. Second series of wind-tunnel tests on a model of a two-bladed horizontal-axis wind turbine; influence of turbulence on the turbine performance and results of a flow survey behind the rotor. *NLR TR 81069 L*, Natl. Aerosp. Lab. NLR, Amsterdam, The Netherlands. 204 pp.

Eppler, R., Somers, D. M. 1978. Low speed airfoil design and analysis. *NASA CP-2045*, Pt. 1, pp. 73–99. NASA Langley Res. Cent., Hampton, Va. 468 pp.

Foley, W. M. 1976. From daVinci to the present—a review of airscrew theory for helicopters, propellers, windmills and engines. *AIAA Pap. No. 76-367*, AIAA, New York, N.Y. 12 pp.

Giordano, V. 1974. The circulation distribution for an optimum propeller according to Betz and to the solution of Goldstein. *l'Aerotechnica Missile e Spazio*, No. 2, pp. 112–26 (In Italian)

Giordano, V. 1979. *The aerodynamic design of windscrews.* Presented at AIDAA Meet., Milan, 22–26 Oct.

Glauert, H. 1935a. Windmills and fans. *Aerodynamic Theory*, ed. F. W. Durand, Vol. 4, Div. L, pp. 324–41. Berlin: Springer. 434 pp.

Glauert, H. 1935b. Airscrew theory. *Aerodynamic Theory*, ed. F. W. Durand, Vol. 4, Div. L, pp. 170–293. Berlin: Springer. 434 pp.

Griffith, R. T. 1977. The effect of aerofoil characteristics on windmill performance. *Aeronaut. J.* 81:322–26

Gustavsson, B., Törnkvist, G. 1978. Test results from the Swedish 60 kW experimental wind power unit. *Proc. 2nd Int. Symp. Wind Energy Syst.*, Vol. 1, Div. D, pp. 1–8. Cranfield, Bedford, Engl.: BHRA Fluid Eng. 448 pp.

Hess, J. L. 1972. Calculation of potential flow about arbitrary three-dimensional lifting bodies. *MDD Rep. No. J 5679-01*, McDonnell-Douglas Corp., St. Louis, Mo. 160 pp.

Hirsch, R. 1948. Definition and calculation of single and counter-rotating optimum airplane propellers. *Publ. Sci. Tech. Minist. Air No. 220*, pp. 83–95. Paris: Minist. de l'Air. 191 pp. (In French)

Holme, O. A. M. 1981. *Detailed Aerodynamic Analysis of Horizontal-Axis Wind Turbines*, VKI Lect. Ser. 9. Rhode-St-Genèse, Belgium: von Karman Inst. Fluid Dyn. 39 pp.

Landgrebe, A. J., Cheney, M. C. 1972. Rotor wakes—key to performance prediction. *AGARD CP-111*, Div. 1, pp. 1–19. AGARD, Neuilly-sur-Seine, France. 466 pp.

Maydew, R. C., Klimas, P. C. 1981. Aerodynamic performance of vertical and horizontal axis wind turbines. *J. Energy* 5:189–90

Miller, R. H., Martinez-Sanches, M., Dugundji, J., Larrabee, E. E., Chopra, I., Humes, I., Chung, S. Y., Gohard, J. C., Edwards, W. T. 1976. Wind turbine performance. *NSF/RA-760569*, Div. II, pp. 54–58. RANN/NSF, Washington DC. 431 pp.

Prandtl, L. 1918. Theory of the lifting wing, I. *Nachr. Ges. Wiss. Göttingen, Math. Phys. Kl., 1918*, pp. 451–77 (In German)

Preuss, R. D., Suciu, E. O., Morino, L. 1980. Unsteady potential aerodynamics of rotors with applications to horizontal-axis windmills. *AIAA J.* 18:385–93

Rao, B. M., Schatzle, P. R. 1977. Analysis of unsteady airloads of helicopter rotors in hover. *AIAA Pap. No. 77-159*, AIAA, New York, N.Y. 11 pp.

Rohrbach, C. 1976. Experimental and analytical research on the aerodynamics of wind turbines. *COO-2615-76-T-1**, Hamilton Standard, Windsor Locks, Conn. 111 pp.

Sheldahl, R. E. 1981. Comparison of field and wind-tunnel Darrieus wind turbine data. *SAND80-2469*, Sandia Lab., Albuquerque, N.Mex. 21 pp.

Sheldahl, R. E., Blackwell, B. F. 1977. Free-air performance tests of a 5-metre-diameter Darrieus turbine. *SAND77-1063*, Sandia Lab., Albuquerque, N.Mex. 37 pp.

Smedman-Högström, A. S. 1978. Measurement of wind speed around a wind power plant in Sweden. *Proc. 2nd Int. Symp. Wind Energy Syst.*, Vol. 1, Div. B, pp. 73–84. Cranfield, Bedford, Engl.: BHRA

Fluid Eng. 448 pp.

South, P., Rangi, R. S. 1975. An experimental investigation of a 12 ft. diameter high speed vertical axis wind turbine. *LTR-LA-166*, NRCC, Ottawa, Canada. 16 pp.

Stewart, H. J. 1976. Dual optimum aerodynamic design for a conventional windmill. *AIAA J.* 14:1524–27

Stoddard, F. S. 1977. Momentum theory and flow states for wind mills. *Wind Techn. J.* 1:3–9

Suciu, E., Preuss, R., Morino, L. 1977. Potential aerodynamic analysis of horizontal-axis windmills. *AIAA Pap. No. 77-132*, AIAA, New York, N.Y. 10 pp.

van Gent, W. 1975. Unsteady lifting-surface theory for ship screws; derivation and numerical treatment of integral equation. *J. Ship Res.* 19:243–53

van Holten, T. 1975. *The computation of aerodynamic loads on helicopter blades in forward flight, using the method of the acceleration potential*. PhD thesis. Delft Univ. Techn., Delft, The Netherlands. 131 pp.

Weber, W. 1975. The optimum shape of rotating wings for horizontal-axis wind energy conversion systems. *Z. Flugwiss.* 23:443–47 (In German)

Weissinger, J. 1947. The lift distribution on swept-back wings. *NACA TM 1120*, NACA, Washington DC. 48 pp.

Wilson, R. E., Lissaman, P. B. S. 1974. Applied aerodynamics of wind power machines. *NSF/RA/N-74113*, Oreg. St. Univ., Corvallis. 118 pp.

Wilson, R. E., Walker, S. N. 1976. Performance-optimized horizontal-axis wind turbines. *Proc. Int. Symp. Wind Energy Syst.*, Div. B, pp. 1–28. Cranfield, Bedford, Engl.: BHRA Fluid Eng. 498 pp.

Worstell, M. H. 1979. Aerodynamic performance of the 17-metre-diameter Darrieus wind turbine. *SAND78-1737*, Sandia Lab., Albuquerque, N.Mex. 65 pp.

Zimmer, H. 1972. The rotor in axial flow. *AGARD CP-111*, Div. 8, pp. 1–16. AGARD, Neuilly-sur-Seine, France. 466 pp.

Ann. Rev. Fluid Mech. 1983. 15:97–122
Copyright © 1983 by Annual Reviews Inc. All rights reserved

THE IMPACT OF COMPRESSIBLE LIQUIDS

M. B. Lesser

University of Luleå, S-95187 Luleå, Sweden

J. E. Field

Cavendish Laboratory, University of Cambridge, Cambridge, England

INTRODUCTION

An understanding of the dynamics of liquids undergoing collisions with either solids or other liquids is needed in a number of technological situations. Some examples are splashes (Worthington 1908), impact of structures on liquid surfaces (Wagner 1932, Skalak & Feit 1966), and impact of liquid drops on erodible surfaces. As with many phenomena, liquid impact has both negative and positive aspects. The same mechanism that can damage an aircraft can be used to erode and cut materials ranging from paper, plastics, and shoe leather to hard rock. Because of its importance the study of liquid impact has been far-ranging with many interdisciplinary contributions.

The present review article concentrates on determining the maximum pressure-loading and other possible damage mechanisms that will erode a solid target. This implies that the early stages of the impact process are studied, during which the shock waves signaling a collision are spreading out through both the liquid and solid. Compressible behavior of the liquid and the motion of the solid target have to be taken into account. The problem involves, in general, free surfaces and the constitutive behavior of both liquid and target. The approach has been analytical, numerical, and experimental, with perhaps the largest gains made experimentally. In this review we first discuss the various theoretical models

97

that provide a framework for interpretation of the numerical and experimental results. Though the major emphasis is on the early-time compressible behavior of liquids striking solids, we occasionally refer to peripheral, though related, topics.

Background

Several reviews of liquid erosion problems have appeared in recent years, though the emphasis has been on the materials-science aspects of the subject. The reader can obtain a general orientation from the volume on erosion edited by Preece (1979). The article by Adler (1979a) on the mechanics of liquid impact contains important material from the viewpoint of application to material damage. Brunton & Rochester's article (1979) in the same volume covers similar material and is highly recommended for a general overview of the liquid erosion problem. Both articles contain extensive bibliographies with a scope that encompasses the more limited viewpoint taken in this review. A comprehensive bibliography on rain erosion with annotated references on work performed up to 1972 has been compiled by Sims & Trevett (1972). Other general sources are the book by Springer (1976), the series of five international conferences on erosion by solid and liquid impact (Fyall & King 1965, 1967, 1970, 1974, Field 1979), the jet-cutting conferences sponsored by the British Hydrodynamics Research Association (1972, 1974, 1976, 1978, 1980, 1982), and the proceedings of the ASTM conference on erosion (Adler 1979b).

THEORETICAL MODELS

Problem Definition

The general problem of liquid-solid impact is sketched in Figure 1. The liquid mass is considered to have a well-defined free surface and the solid is assumed to be essentially linear-elastic, at least during the initial stages of the impact process. Depending on the relative speeds and the constitutive behavior of the materials involved, the general impact will lead to a break-up of the continuum model through such effects as cavitation in the liquid and crack propagation in the solid. To make progress in an analytical treatment, it must be assumed that during the initial contact process the fluid remains a continuum and that the solid closely resembles a homogeneous and isotropic elastic material. During the initial phase of contact the viscous aspects of the flow are confined to the regions within a small distance from the boundaries; hence if $\sqrt{v/\ell C}$ is small enough (where ℓ is a typical drop dimension, v is the kinematic viscosity, and C is the ambient compressional wave speed), it is valid to ignore the viscosity terms in the equations of motion. This reasoning leads to an inviscid model of the compressible liquid with shocks moving into the liquid interior following the collision. Under certain conditions it can be expected that the viscoelastic properties of the target and the viscosity of the liquid will cause the shock's viscous structure to have a length scale of the order of the contact curvature, and for such cases special consideration must apply. This latter situation can also develop when numerical simulations make use of effective viscosity parameters. If the impact speed is sufficiently low for a given liquid, distinct shocks and high-speed jetting would not be expected. Engel (1955) in her early experiments used gravity to accelerate the drops to speeds in the several-meters-per-second range, and it is questionable if these experiments truly simulate the phenomena of higher-speed impacts. However, to the authors' knowledge there has been no investigation of the conditions for which the impact history will follow the pattern expected from traditional compressible-flow theory.

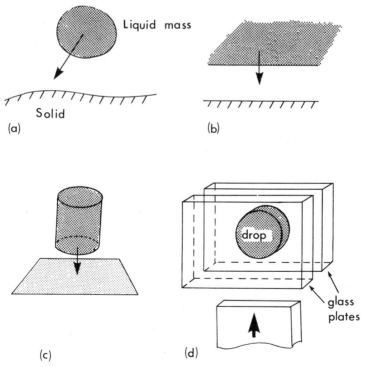

Figure 1 Liquid-impact configurations. (*a*) Liquid drop against solid. (*b*) Normal impact. (*c*) Cylindrical-jet impact. (*d*) Two-dimensional impact.

Figure 1 also shows typical configurations used in theoretical and experimental investigations. The cylinder has been used to simulate the impact of the head of a high-speed jet, while the "sandwich" configuration of a drop between two plates simulates a two-dimensional contact. As shown below, the latter approach combined with the use of a gel allows the study of highly idealized geometries.

Impact Stages

If a purely inviscid model is adopted, where the solid's motion is independent of strain-rate effects, the picture shown in Figure 2 can be expected to hold. Five situations that may develop during various stages of the collision are depicted. The first stage, a somewhat degenerate case, shows a pure normal impact and consists of a normal shock moving into both the liquid and solid. Glenn (1974) has carried out one of the most complete treatments of this problem in the context of liquid erosion as a preliminary to a numerical study. He made use of both a Tait equation-

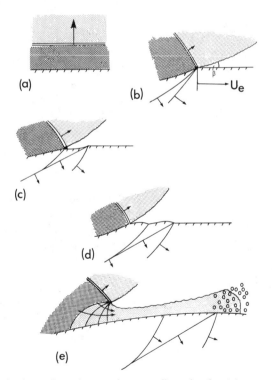

Figure 2 Shock configurations at the expanding edge for (*a*) normal impact, (*b*) tri-supersonic edge, (*c*) bi-supersonic edge, (*d*) simple supersonic edge, and (*e*) subsonic edge.

of-state for the liquid (the main interest is in water) and a polynomial fitted to equation-of-state data. Glenn also studied the one-dimensional model of the shock reflection and the resulting expansion produced on the rear free surface of the liquid. Similar considerations can be found in the underwater explosion literature reviewed by Holt (1977). The realization that high dynamic pressures can occur as a result of the compressive normal impact of a liquid was first noted by Cook (1928), who introduced the acoustic limit of the pressure, i.e. the "water-hammer" pressure, into the liquid-impact literature. In an earlier work Parsons & Cook (1919) conjectured that cavity collapse in a liquid could produce local shock loadings by means of the "water-hammer" mechanism, a possibility considered in greater depth by Benjamin & Ellis (1966), who also demonstrated its validity by means of high-speed photographic evidence. Cook (1928) also discussed the possible effects of the shape of the impacting surfaces; however, he did not seem aware of the role played by shock waves in this case. Early investigators did not realize that pressures well in excess of normal shock loadings are readily obtained during impact. The reasons for this are depicted in Figures $2b-e$, which show four possible situations at the expanding contact edge, i.e. the moving boundary separating wetted from unwetted target surface. The detailed geometry of the colliding surfaces determines the relevant process at any particular stage of impact.

For simplicity the reader can think of a spherical drop undergoing normal impact with a smooth plane surface. The "ideal" contact edge speed can be defined as the speed of the intersection between the oncoming drop and the surface if both are undisturbed by the impact. It depends on the impact speed and the angle between the free surface and the solid, and is given by

$$U_e = U_i \cot \beta,$$

where U_i is the impact speed and β the instantaneous edge angle.

For a simple elastic solid there will be two bulk waves, i.e. a dilational wave C_ℓ and a shear wave C_s. If C denotes the ambient-liquid sound speed, then $C_\ell > C_s > C$ for most materials of interest. In the tri-supersonic case ($U_e > C_\ell$; Figure $2b$) the liquid and solid surfaces ahead of the spreading contact remain undisturbed, while for the bi-supersonic ($U_e > C_s$; Figure $2c$) and supersonic case ($U_e > C$; Figure $2d$) the solid surfaces will be deformed ahead of the contact. We assume that the disturbance of the surface ahead of the expanding liquid does not drive the contact region into another regime. For a large enough angle the contact edge moves subsonically in relation to all wave speeds. When the shock detaches from the contact edge, the free surface near the wall then undergoes a sudden expansion, producing the jetting motion shown in Figure 2*e*.

These considerations show that the geometry of the contact, i.e. the initial shape of both the liquid and the solid, is crucial in determining both the qualitative and quantitative results of an impact. The mechanisms of Figure 2 that are seen as damaging to the solid are the high pressures developed behind the moving attached shock and liquid jetting over the disturbed surface of the solid. Contact of a liquid mass with a *depressed* surface can lead to a shock-focusing effect, which is a postulated cause (Field & Lesser 1982) of the central pit observed in many liquid-impact-eroded materials. Still another and somewhat less obvious damage mechanism is that of cavitation caused when the relief (expansion) waves from the free surface cross and produce large tensile stresses (negative pressures) in the liquid. The collapse of these cavities in the neighborhood of solid surfaces can cause the typical kind of erosion expected from such flows. This type of cavitation damage is in large part due to the liquid impact of the involuted cavity surface on the solid surface, i.e. again liquid impact but on a smaller spatial scale. Cavitation and its effects are discussed in reviews by Plesset & Prosperetti (1977) and by Mørch (1979). Also, in an early paper by Engel (1955), note was taken of the possible production of negative pressures by relief waves moving through the liquid and the resulting cavitation.

Incompressible Models

For completeness we briefly mention several works in which the liquid is considered incompressible. Cumberbatch (1960) considered a two-dimensional liquid wedge impacting a rigid plane. The motivation for this study was the slapping of free-surface waves against a dock. In his paper, Cumberbatch uses results found in the related problem of the entry of a solid wedge into a liquid. Another somewhat *ad hoc* analysis was performed by Savic & Boult (1955), who considered a potential flow solution in toroidal coordinates of a sphere of liquid. The radius of the sphere was adjusted to compensate for losses due to jetting, and the resulting approximate solution gave reasonable qualitative agreement with experimental work done by the same authors. Some connections between such incompressible models and the early-time compressible phase of the motion were examined by Lesser (1979) using a matched-asymptotic-expansion procedure.

Water Entry

In discussing the early phase of the motion, Skalak & Feit's (1966) paper on the water-entry problem has had a large influence on work in the liquid-impact area. In water-entry problems a body impacting a smooth plane surface is usually considered, while in liquid impact the concern is with a highly curved liquid mass striking a solid. The principal assumption of the Skalak & Feit work is that the impacting body can be replaced by a disturbance zone on the surface of the liquid that consists of the intersection of the body's cross section at any time after impact with the undisturbed free surface of the liquid. The problem is linearized in effect by treating the first term in an expansion in the body velocity over the liquid sound speed, i.e. in what can be called the impact Mach number. This region is considered to act like a piston that drives a linear compression wave into the liquid. The loading on the body is then calculated from the solutions of the wave equation for the pressure disturbance, with the normal derivative of the pressure supplying the needed boundary condition for the solution. For a two-dimensional case, e.g. wedge impact, the problem is further elucidated by making use of an analogy with supersonic airfoil theory. In this analogy the time axis is replaced by the main flow direction and the planform of the airfoil is given by the time history of the expanding contact edge. The supersonic phase of the motion corresponds to the case when the equivalent wing planform is outside the flow-direction Mach cones.

Two-Dimensional Calculations

Rochester (1979) made use of this analogy to study the pressure distribution resulting from the impact of a two-dimensional circular drop on a rigid target (Figure 1d). This configuration has recently been the focus of several comprehensive experimental investigations, which are discussed below. Simple integration over distributions of supersonic sources, in line with the above analogy, is sufficient to resolve the problem during the period of supersonic contact velocity. As the contact point slows down to the liquid sound speed, the edge pressure becomes singular due to the linearization which ignores convection of the wave pattern by moving liquid. Rochester does not carry the solution beyond this point. The main advantage of this calculation is that it provides an exact solution of the two-dimensional linearized, or acoustic impact, problem. The edge pressures predicted by the model are infinite for any finite impact speed when the edge contact velocity reaches the liquid sound speed. We shall see that this infinity is caused by ignoring the convection of the disturbance pattern at the edge, and that high edge pressures are to be expected.

Another problem in which the impact Mach number is considered small was treated by Lesser & Field (1974). The geometry considered was a two-dimensional wedge (Figure 3). A solution was found by using the fact that the problem has Ct as a length scale, so that the time variable can be eliminated by use of spatial coordinates divided by the distance traveled by the initial impact signal. The results of this calculation agree well with the numerical calculations for the cylinder obtained by Glenn (1974) for the period before the cylindrical geometry takes effect.

The case of a supersonic contact on a rigid plane was first treated by Heymann (1969). In this work he assumed that he was dealing with a curved drop and that he could approximate the local behavior around the contact point by assuming that both the drop edge and the attached shock at the edge were plane. It was later pointed out by Lesser (1981) that this is not an approximation at the edge but in fact the means by which boundary conditions must be developed at the edge in the solution of the full problem. Heymann also made a simplification in the problem by assuming that the shock-wave speed C^* was related to the normal flow speed by the relation $C^* = C + kV$, where k is a constant and V is the normal-velocity jump. This relation in effect determines an approximate equation-of-state for the liquid. This resulted in a relatively cumbersome set of algebraic relations that were solved iteratively by numerical means. In addition, Heymann assumed that when the derivative of the wall pressure became a maximum with respect to the shock angle the shock would separate from the contact point and "jetting" would commence. This allowed him to calculate, for a given impact speed, at what point a spherical drop would begin to develop a side-jetting motion. Experimental work, discussed below, appears to give a later time and hence a larger angle at the edge for jet development, even when allowance is made for observational difficulties. Field et al. (1979c,

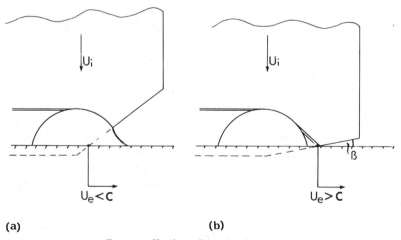

Figure 3   Shock configuration for wedges.

Field & Lesser 1982) have repeated Heymann's calculations for the rigid surface without making any special approximation in terms of the shock velocity or the condition for separation. In their work, the Tait equation-of-state was used, and a single algebraic relation for the wall pressure was developed in terms of the parameters $M_i = U_i/C$ and $M_e = U_e/C$, defined in terms of the ideal contact edge speed. They also introduced the idea of representing a collision in the (M_e, M_i)-parameter plane shown in Figure 4. In this representation the state at the contact edge for a curved impacting drop is given by a straight line such as "1," where M_e infinite corresponds to the initial normal impact. The angle β before shock separation is given by a line between the origin and the state point, such as OA. The pressure equation exhibits a double root, corresponding to a weak and a strong shock; however, the weak root is the significant one if the configuration is to develop from one of normal impact. At shock detachment the two roots coalesce and no real solution exists for the pressure in the remainder of the (M_e, M_i)-parameter plane. The boundary for this is shown for a rigid wall in Figure 4. The point at which the line M_i = constant crosses this boundary is the detachment point for the shock; thus at this point the pressure maximizes and one expects the commencement of jetting. The calculations of Field et al. appear to agree with Heymann in that the pressure at shock detachment is about three times the "water-hammer" pressure for $M_i \approx 0.1$ and that shock detachment takes place earlier than indicated by experimental evidence.

Lesser (1981) extended the model by treating the surface as yielding, i.e. by assuming the surface to possess a local mechanical admittance. This was achieved in two ways: first by assuming an elastic target during

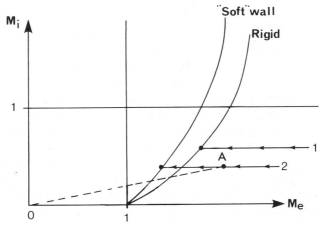

Figure 4 The (M_i, M_e) parameter plane; see text for details.

the period of triple-supersonic edge contact speed in which case the effective admittance can be calculated in terms of the material's elastic constants, and second by simply assuming an *ad hoc* effective impedance for the material. These calculations show a shift in the region of the parameter plane where shock detachment takes place. As it is unrealistic to expect the targets in the experiments to be perfectly rigid and as softer targets lead to large detachment angles, it would appear that the major cause for the discrepancy between theory and experiment is due to target deformation under the impact load. These results are built around the general equation for the attached shock-free surface configuration

$$\bar{p}^2 M_e^2 F(M_i \bar{p}) - (\bar{p} - M_i)^2 - M_e^2 - \bar{Y}^2 \bar{p}^2 (M_e^2 + M_i^2)$$
$$+ 2\bar{Y}\bar{p}(M_e^2 + M_i^2 - M_i \bar{p}) = 0,$$

where $F(\xi) = [1 - (1 + K\xi)^{-1/k}]/\xi, K = 7$ for water, $\bar{p} = p/\rho_0 U_i C$ with p the wall pressure at the expanding edge, and $\bar{Y}$ the normalized admittance of the solid, $\rho_0 CY$. The limit M_e infinite implies that $\beta = 0$, i.e. the normal-impact case, while $\bar{Y}$ vanishing is the rigid wall. Also $M_i \ll 1$ is the acoustic case, i.e. the impact velocity is small compared to the liquid sound speed.

The largest edge pressure develops just before shock detachment. Using an asymptotic analysis of the shock-detachment condition, Lesser shows in the rigid wall case that for small M_i

$$\bar{p} \sim (2M_i)^{-1/3} + 1.889 M_i^{1/3} + O(M_i)$$

and

$$\beta \sim \tan^{-1}\left(M_i / \sqrt{1 + 4.762 M_i^{2/3}}\right)$$

for the critical edge. For $M_i < 0.12$ these results are within 10% of the numerical solutions and also agree with Figure 7 of Heymann (1969).

The discrepancy between theory and experiment in predicting the edge angle at which jetting would commence has been conjectured by Heymann (1969) to be due to the nonrigid nature of the target. Engel (1972) countered this by noting that in experiments she obtained the same angle for sapphire and polymethylmethacrylate (PMMA), materials with quite different acoustic impedances. However, the effective impedance of the material at shock detachment is a complicated function of the solid's properties, having no simple relation to the ρC_ℓ value of the material. This can be inferred from Lesser's (1981) solution for the triple-supersonic impact. Further discussion of these points is given in the experimental section.

In the same paper, Lesser also develops a geometrical-acoustics model of the impact of arbitrarily shaped drops under the condition of supersonic edge contact. Only the rigid-wall case is developed. The basic idea used is that the envelopes of the wavelets developed at the supersonically expanding contact edge approximate the shock front. For the period before shock detachment the front is relatively uncurved, and if the pressure is developed as an expansion in distance from the front over its curvature, the initial term gives a good approximation to the wall pressure. The two-dimensional version of this theory is compared with Rochester's exact solution discussed above, and a three-dimensional version (for an axisymmetric drop) is compared with numerical calculations carried out by Rosenblatt et al. (1979a,b). The three-dimensional numerical calculation uses a finite-difference code with an effective artificial-viscosity method to take account of the developing shock waves. Thus there is an effective time delay in the calculation for the development of the shock system from the initial impact. The result is that the shock develops and escapes much later in the numerical solution than in Lesser's analytical solutions. The edge region where the shock detaches is calculated exactly in the analytical work and is a result of the conservation laws and the assumed equation-of-state. Thus it must be considered as a checkpoint for any numerical work. Unlike some previous work (Huang et al. 1971, 1973), Rosenblatt does reach the expected pressure levels; however, the time step is not small enough in the calculations to achieve the proper phasing of events, such as the shock detachment. Therefore the numerical work provides good agreement for the pressure levels, but it cannot be used to explain the disagreement between Heymann's original estimate of the shock detachment point and the experimental evidence. The numerical procedure used by Rosenblatt has been extended to a number of more complex cases, including crack development in the target (Rosenblatt et al. 1979b) and impact against a microcrack. In the interpretation of this work one must be aware of the limitations of the artificial-viscosity methods. The materials scientist interested in overall impact measures, such as maximum pressure, can probably use such results safely. However, a detailed understanding of the fluid-mechanical events requires more care.

Theoretical developments have reached the stage where the principal observed features of a relatively simple impact can be explained. These features are (a) the early impact pressure signature with its maximum at the expanding edge, and (b) the rapid side jetting. On the other hand, close quantitative agreement with experiment has not yet been achieved because of the difficulties in making precise measurements that fit the idealizations of the theory. The problem of calculating the side jetting in a rational manner also remains a challenge for the theorist. The geometrical optics approach provides the possibility of dealing with more realistic configurations and of taking into account more interesting target behaviors.

EXPERIMENTAL

Multiple-Impact Devices

A variety of techniques have been devised for studies of liquid-solid impact. Historically the first was the wheel-and-jet method, which has been used in various forms since the 1920s. In a typical configuration, specimens are attached to the edge of a disk so that they protrude radially. As the disk rotates, the specimens impact one or more cylindrical jets of liquid. The wheel and jets are usually in a chamber that is pumped to as low a pressure as possible to reduce aerodynamic distortion of the jets. Anyone who has used such an apparatus will know that specimen alignment is critical for obtaining reproducible data. The reasons for this have only recently emerged (Field & Lesser 1982). If the specimen is perfectly aligned with the face of the specimen parallel to the axis of the jet, initial contact is along a line OO′ (see Figures 5a and b). There is then compressible behavior with no jetting until the contact area has a width $2x$. For a brittle material, tensile failure occurs only outside this central region. This unfractured central strip corresponds to the circular region surrounded by circumferential cracks for spherical-drop impact. However, if the specimen is misaligned through an angle β (see Figure 5c), the contact is not instantaneous along OO′ but moves upwards with a velocity that is supersonic for sufficiently small β and subsonic for large β. In the first of these cases high edge pressures will be generated at the contact point but there will be no flow, while for large β

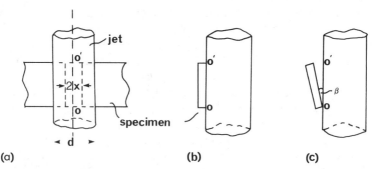

(a)   (b)   (c)

Figure 5   Impacts between specimen and jet in a wheel-and-jet apparatus. (*a*) Plan view. (*b*) Side view for a well-aligned specimen. (*c*) An angled specimen.

a high-velocity jet will move upward. The three cases of perfect alignment, small β, and large β are very different. A key point is that for the velocity used in this test the angle β that determines whether or not the contact point moves supersonically along OO′ is a few degrees at most. Hancox & Brunton (1966) give examples of damaged PMMA specimens eroded in this test, one for perfect alignment and one inclined with β = 9°. The first specimen shows cracks and flow damage outside an undamaged central strip. The angled specimen exhibits arclike cracks and erosion cutting across the central axis. Very different erosion rates were also recorded.

A development of the wheel-and-jet approach for higher-velocity testing are whirling-arm devices. Here the specimens are attached to a long arm rotated in an atmosphere of water drops. Velocities of up to Mach 3 in air have been achieved (Wahl 1970, Kunz 1974). A variety of linear test facilities exist: specimens have been launched ballistically from a gas gun through an atmosphere of drops (Mortensen 1974); specimens and complete components have been attached to rocket-driven sledges and propelled on rails through a simulated rain field (King 1965, Reinecke 1974, Schmitt et al. 1974, Sullivan & Hockridge 1974, Mortensen 1974); droplets have been accelerated to Mach 0.83 in a wind tunnel and impacted against targets; and specimens have been attached to an aircraft and flown at high speed through rain (Fyall et al. 1975).

Single-Impact Techniques

The attraction of producing controlled single impacts is that it allows detailed study of both liquid and solid. One approach is to fire a specimen mounted at the front of a projectile against a suspended or falling drop (Jenkins & Booker 1960, Fyall 1967, Fyall & Smith 1968, Adler & James 1979, Field et al. 1979a, Rickerby 1977, Blair 1981). The main difficulties with this technique are (a) decelerating the specimen after impact without adding further damage, and (b) distortion of the drop by gravity and the air blast ahead of the projectile. It is clear from the theory of liquid impact that if a drop is flattened it will have higher radii of curvature in the contact region and will behave effectively as a larger drop and thus be more damaging. Adler's tests are conducted in a helium atmosphere at an ambient pressure of 0.1 torr to minimize distortion. The low pressure necessitates using a drop made of a solution of water and ethylene glycol, with 80% water by volume, to reduce evaporation. The other workers have devised baffle systems to reduce the aerodynamic effects. Rickerby was able to suspend drops of up to 6-mm diameter using a water-glycerine solution with a few percent of glycerine. Adler (1979c) has analyzed various sources of published data on single-drop impact patterns on PMMA. For nominally similar-sized drops, the dimensions of the damage patterns were noticeably smaller for the low-pressure experiments or for those in which special care had been taken to reduce drop distortion.

An alternative approach devised by Brunton is to keep the specimen stationary and to project a jet of liquid against it (Bowden & Brunton 1958, 1961). A projectile is fired into a stainless steel chamber containing a small quantity of water sealed in by a neoprene disk. The projectile and neoprene move forward as a piston and extrude the liquid through a narrow orifice. The ratio of jet to projectile velocity is typically 3 to 5 times. This method has been used extensively for studies of the mechanisms of erosion (Bowden & Brunton 1961, Bowden & Field 1964, Field et al. 1974, 1979a,b, Gorham & Field 1976, Gorham et al. 1979). Although a basically simple and inexpensive technique, the design and machining of the nozzles are critical if coherent, smooth-fronted and reproducible jets are to be produced. Figure 6 is a single-shot photograph of a jet from a 0.8-mm nozzle. The umbrella of spray comes mainly from

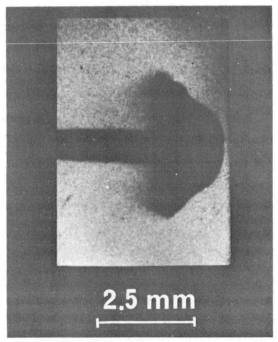

Figure 6 Single-shot photograph of a jet from a 0.8-mm nozzle. The central core is coherent liquid. The umbrella of spray is made up of micron-sized droplets, which do not contribute to the damage. Note the smooth, slightly curved, front profile (Field et al. 1979b).

the water in the parallel section of the orifice. It is made up of droplets of micron size that do not contribute to the damage. The jet-front velocity increases over a distance of ~ 10 mm. At distances of 2 to 3 times this, the onset of Taylor instability begins to disrupt the front surface. A stand-off distance of 10 mm is recommended in all experiments. The acceleration after leaving the orifice results from the liquid from the main part of the chamber taking a finite time to move through the slower-moving liquid from the orifice section. The essential point is that there is a central core of coherent liquid which at a stand-off distance of 10 mm has a smooth, slightly curved front profile. This has allowed the simulation of *drop* impact with these jets. The "equivalence" is reasonable because it is the front profile of the impacting liquid (drop or jet) that determines the important compressible stage of impact. Jets fired from a 0.4-mm orifice nozzle, for example, simulate impact with 2-mm diameter water drops for a wide range of velocities. Figure 7 shows a sequence of pictures of a jet from a 1.6-mm nozzle (equivalent to a ~ 10-mm diameter drop) impacting a PMMA specimen. Note the clearly defined pulse, of short duration, caused by the compressible stage of impact. Further

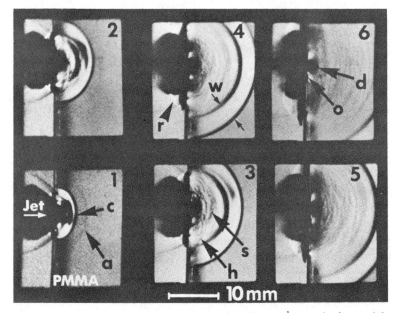

Figure 7 Shadowgraph sequence of the impact of a 750 m s^{-1} water jet from a 1.6-mm nozzle onto a PMMA block (1 μs per frame). (*a*) Stress front induced by detached air shock; (*r*) reflected air shock; (*c*) main compressive stress pulse of width w; (*h*) head wave; (*s*) shear front (poorly defined because of the nature of the optical system); (*d*) subsurface "shear" failure; (*o*) main ring crack (Field et al. 1979a).

discussion of the technique and of the jets used to simulate drop impact is given in Field et al. (1979a,b).

Hammitt et al. (1974), Kinslow (1974a,b) and Kinslow et al. (1974, 1976) have also used the jet method but appear to have taken less care over nozzle design and the optimization of stand-off distance. Their photographs show rather incoherent slugs of liquid, and Hammitt (1974) has reported a decrease in erosion with increasing impact velocity, a clear indication that jet break-up is occurring.

The Brunton single-jet apparatus was modified by Kenyon (1967) to produce multiple impacts. A spring-loaded hammer struck a metal diaphragm sealing the chamber containing the liquid. Repetition rates of 20–30 s^{-1} and jet velocities of up to 300 m s^{-1} were achieved. This method has been used by Timm & Hammitt (1969), Vickers & Johnson (1972, 1973), and Hammitt et al. (1974) for erosion studies. The jet quality and reproducibility is poor in these multiple-jet devices at present, but the concept is potentially of great practical interest and is worthy of detailed study.

A number of schemes have been proposed for creating high-speed jets of liquid for cleaning, cutting, and mining applications. For the most part, these methods utilize steady-state jets: however, transient or non-steady methods appear more favorable for attaining velocities that are supersonic with respect to the liquid. The transient method (Vortsekhovsky 1968, Cooley 1971, Edney 1976) typically involves firing a packet of liquid through a converging nozzle. In this process the front region of the liquid is rapidly compressed and accelerated. The jet velocity is at its maximum as it crosses the orifice plane and can be as high as 10 to 12 times the projectile velocity (Rhyming 1973) or, in absolute terms, up to ~ 5 km s^{-1}. This is higher than with the Brunton nozzle, but in the rain erosion simulation (discussed above) the coherence and front profile of the liquid are of prime importance, whereas in the mining application high velocity and coherence are the key factors. However, there are problems with projecting liquid jets at very high velocities. For example, if the jet velocity exceeds the speed of sound in the liquid, then decompression occurs after the liquid leaves the nozzle. This causes disruption of the jet. At a later stage, deceleration by air drag and Taylor instability cause further break-up. Field & Lesser (1977) have discussed these aspects of the mechanics of high-speed liquid jets.

An alternative approach for projecting a liquid mass at velocities up to ~ 5 km s^{-1} has been described by Sullivan & Hockridge (1974). A water-gelatine drop with 97% water is carried in a sabot fired by a helium gun. The sabot falls away, leaving the drop to impact the specimen.

Two-Dimensional Impacts

The idea of using disk-shaped bubbles or drops for cavitation or liquid-impact studies was first suggested by Brunton (1967). The technique was further developed and extended in subsequent research (Brunton & Camus 1970a,b, Camus 1971, Rochester & Brunton 1974a,b, Rochester 1979). In the case of drop impact, a small quantity of liquid was placed between transparent plates separated by a small distance: surface tension pulled the drop into a curved profile (Figure 1d). Impact was achieved by projecting a third plate between the two spaced plates. High-speed photography, with either a rotating-mirror camera or single-shot schlieren photography, was used to record the impact events on a microsecond scale. To obtain synchronization with the rotating-mirror camera, the impact plate was accelerated with an explosive detonator; this created velocities of up to ~ 100 m s^{-1}. The great advantage of this two-dimensional work was that it allowed processes occurring inside an impacted drop to be observed in detail without the refraction problems inherent with spherical drops.

An example of Camus' work (1971) for a 70 m s^{-1} impact is shown in Figure 8. The "sequence" is in fact a composite of three different experiments, since the single-shot schlieren approach was used to show the stressed regions more clearly. Figure 8a is from the very early stage of impact, where both solid and liquid are highly compressed in the contact region and there is no jetting. In Figure 8b a shock has detached and is moving up through the drop and jetting has started, though high-pressure regions (labeled p) still persist in the liquid. Finally, in Figure 8c the shock is about to reflect at the upper surface and jetting is more advanced. Note that in Figures 8b and 8c the interface between liquid and solid appears textured. Other sequences (Camus 1971) have confirmed that this effect is due to cavities that form during impact,

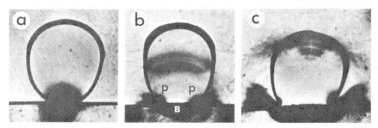

Figure 8 Impact with a two-dimensional drop at 70 m s^{-1}. Note the absence of flow in (a) and the shock structure, high-pressure lobes (labeled p), and jetting in (b) and (c). The texture appearance at the interface above B is due to the formation of cavitation bubbles. The drop has a diameter of ~ 2 mm (Camus 1971).

presumably when the release waves cross (see the theoretical section). These cavities may eventually collapse violently and cause damage. Cavities can also develop when the primary shock reflects at the upper liquid/air interface; however, these cavities are well away from the solid and do not contribute to the damage process.

A disadvantage of impact with a drop, from the viewpoint of both analysis and experimental interpretation, is that there is a constantly changing angle between the drop boundary and the solid (and β in Figure 2). This is overcome by using wedges of liquid having a constant, prechosen β. This idea was developed during the collaborative work of the present authors (Field et al. 1979c, Field & Lesser 1982). It involves casting thin sheets of a water-gelatine mixture and then cutting out the desired shape to be impacted.

Jetting

As discussed above, jetting does not take place immediately when a drop impacts a flat surface, since there is the initial regime where the contact expands supersonically. When jetting starts, its velocity can be several times the impact velocity. This is not surprising since the geometry is very similar to that which produces high-speed jets with shaped charges (Birkhoff et al. 1948) during high-velocity solid collisions (Walsh et al. 1953, Harlow & Pracht 1966) and during explosive welding (see, for example, Bahrani et al. 1967).

The velocity of the jets from under $\sim$ 2-mm drops was measured by Jenkins & Booker (1960). The ratio of jet to impact velocity was $\sim$ 2 at 1200 m s^{-1}, rising to $\sim$ 6 at 100 m s^{-1}. These results correlate well with those of Camus (1971) and Brunton & Camus (1970a,b), who recorded a ratio approaching 10 for their two-dimensional disk geometry at an impact velocity of 30 m s^{-1}. Qualitatively the trend of these results is reasonable, since with lower impact velocity the angle β for first flow would be smaller and the expected jet velocity higher.

It has been emphasized that accurate measurement of jet velocities is difficult. The two-dimensional geometries are easier to study, with the two-dimensional wedge method potentially the best, since it has a constant β. All measurements to date have been made on jets that have begun to break up into a spray of droplets (see Figures 7 and 8). Until much higher framing-rate photography can be applied to this problem, the above-quoted velocities must be treated with caution.

The high-velocity jetting is an important mechanism for material removal. In many situations the impact pressures distort the material surface by, for example, causing cracking, grain tilting, and slip or twinning, which leave steps on the surface. The flow then causes preferential erosion of these steps, which are typically a few hundred nanometers high (Bowden & Brunton 1961, Bowden & Field 1964, Field 1967). With many materials the jetting leaves ripple marks in an annular band around the crater (see, for example, Bowden & Brunton 1961, Adler 1979c). Their formation is similar to the wave structure, which occurs during explosive welding. Wilson & Brunton (1970) have modeled the interface geometry by driving two inclined liquid sheets together. They used high-speed photography to study the jet and wave formation.

Angle for Jetting

Bowden & Field (1964) suggested, based on linear acoustic reasoning, that jetting would start at an angle $\beta_r = \sin^{-1}M_i$ for impact with a rigid target. The shock detachment angle for a rigid target as computed by Lesser (1981) and Heymann (1969) gives an even smaller angle, e.g. for $M_i = 0.05$, $\beta_c = 2.25°$ as opposed to $\sin^{-1}M_i = 2.86°$. Experimental studies have invariably found larger angles than predicted by these theories. For example, Camus (1971) and Brunton & Camus (1970a,b) recorded angles between 10° and 20°, with an average of 11° in the velocity range 30–100 m s^{-1}. Hancox & Brunton (1966) deduced, from damage studies on PMMA with their wheel-and-jet apparatus, an angle of ~17° for impact at 60 m s^{-1}.

Various ideas have been advanced to explain this. An early suggestion by Hancox & Brunton (1966) that viscosity delays the onset of jetting is not convincing considering the velocities and pressures involved. Huang (1974) and Rochester & Brunton (1974b) have argued that flow is only observed when the jet velocity exceeds that of the expanding contact circle and that this will be greater at an angle β_c. Recently, one of us (Lesser 1981) has shown (see above) that the deformability of the target has a major effect on increasing β_c. As mentioned earlier, aerodynamic effects are also likely to affect experimental observations by distorting the contact surface of the drop to a greater effective radius. Finally, if the flow angle is deduced indirectly from damage studies the result will be an overestimate. The reasons for this are that when the jet first forms it may be too thin to have damage potential, and furthermore it needs to act on a perturbed surface. As Blowers (1969) has shown, the Rayleigh surface wave that gives the main out-of-surface displacements takes some time to reach a peak after the supersonic contact phase ceases.

Impact Pressures

The damage caused by liquid impact, and particularly the damage-threshold velocities required to initiate failure, are convincing evidence that compressible behavior occurs during impact. Bowden & Brunton

(1961) measured the magnitude and duration of the impact load caused by jet impact using a piezoelectric gauge. The peak load and its duration fitted well with the picture of compressible behavior, lasting until release waves from the sides of the jet reached the central axis. The possibility of pressures exceeding "water-hammer" pressures was suggested by Jolliffe (1966), who measured the size of dislocation-etch pit rosettes caused by drop impact and deduced pressures as high as $2\rho CU$. Further support for pressures of this magnitude was provided by Thiruvengadam (1967), who compared erosion threshold velocities from multiple-impact experiments with fatigue-stress endurance limits.

Qualitative evidence for high edge pressures has been provided by photographic studies made by Camus (1971), Gorham (1974), and Field et al. (1974). Figure 8*b*, for example, shows high-pressure regions in the liquid. Other sequences have shown high-pressure lobes associated with edge pressures in PMMA targets.

Experimental study of the detailed pressure field beneath an impacting drop is clearly complex, since the pressures are intense, highly localized, and transient. The first attempt was made by Johnson & Vickers (1973), who measured the normal- and shear-stress distribution under a 50-mm diameter water jet, impacted head-on by an aluminium-plate pressure cell at ~ 50 m s^{-1}. They recorded a pressure on the central axis of $\frac{2}{3}\rho CU$, rising to $1.5\rho CU$ at the edge of the jet. A more recent revaluation of these data (Salem et al. 1979) has suggested that the pressure results should be $0.2\ \rho CU$ and $0.43\ \rho CU$ respectively. In their own experiments using the Johnson & Vickers jet apparatus but with newly designed pressure gauges, Salem et al. failed to find high edge pressures.

These apparently contradictory results can be explained in terms of the theory of liquid impact, described above and in Field & Lesser (1982), which emphasizes the importance of the profile of the impacting surfaces. If a flat-ended cylindrical jet hits a surface end-on, it will create an initial uniform pressure that will then decay as the release waves from the boundaries move toward the central axis. However, in practice, jets are likely to have their front surfaces distorted, either because of their method of production or because of air drag (or both). If they become slightly curved, high edge pressures would be expected for the reasons given earlier. However, if the front of the jet distorts into a wedge (and there is clear evidence for this in Salem et al.'s photographic work), then whether or not high edge pressures result will depend on the impact velocity and wedge angle. In Salem et al.'s experiments the wedge angle β was $\sim 20°$, which for their velocity of impact (~ 35 m s^{-1}) will give a subsonic expanding contact, the possibility of jetting from the start, and no high pressures (as is indeed observed). In the earlier work of Johnson

& Vickers (1973), the authors describe their jets as "not flat, but slightly rounded," which is what would be inferred from their pressure results. The central pressure recorded was well below the expected "water-hammer" pressure. This suggests either that the response time of the gauges or their calibration was at fault, or alternatively that their jets were not coherent.

Rochester & Brunton (1974a,b, 1979) and Rochester (1979) have used two experimental techniques to record the pressure distribution for impact velocities in the range 10–140 m s^{-1}. In the first method, a thin, oblong-sectioned bullet was fired at a two-dimensional drop; in the second, a cylindrical bullet was fired into the side of a vertical liquid jet. Each bullet contained a small (~ 0.25-mm diameter) piezoelectric ceramic

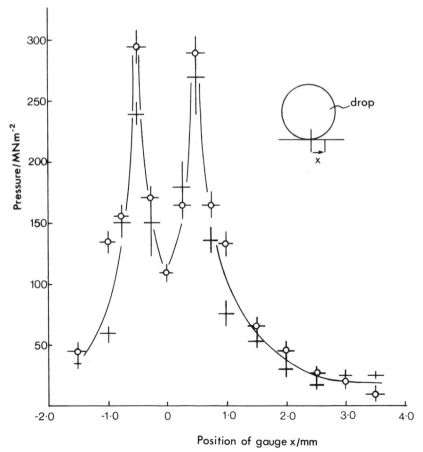

Figure 9 Peak pressure under a cylindrical water drop (after Rochester & Brunton 1979).

gauge embedded in the impact surface. Typical results are given in Figure 9. When allowance is made for compressibility of the gauge, the central and peak pressures are ρCU and $2.5\rho CU$ respectively.

The central pressure should ideally be the one-dimensional pressure for a 100 m s^{-1} impact, which after shock correction should be $\sim 1.13\rho CU$. Assuming Rochester & Brunton's gauge calibration was 13% too low throughout, the edge pressure would be $\sim 2.8\rho CU$. Considering the complexity of the experiments, this is remarkably close to the theoretical estimates.

Rochester & Brunton also recorded the shear on the surface and found it to be zero at the center, rising to a peak of 6MN m^{-2} at a distance of ~ 2 mm from the center. This is further out than the pressure peak and supports the earlier contention that the jet has its main damage potential sometime after the end of the supersonic contact regime.

Impact on a Wetted Surface

In many practical and test situations a layer of liquid will remain on the solid surface or be trapped in cracks and erosion pits. Brunton (1967) has shown that as the layer thickness increases, the normal pressure decreases. This is reasonable since for a thick layer the central pressure at the liquid-liquid interface would be half the "water-hammer" pressure. Brunton also found that the layer reduced the shear damage caused by jetting. It is interesting to note that Camus (1971) found that the air layer trapped between the colliding surfaces also reduces shear damage, since if similar velocity impacts are performed in vacuo or in helium (which escapes more readily) the shear damage is increased.

Liquid trapped in cracks or crevices can cause increased damage when the surfaces are impacted. Field (1967) and Adler & Hooker (1978) give examples. This is an area where further analysis and experiments would be beneficial.

CONCLUSIONS

In this review we have concentrated on the problem of liquid impact in the intermediate-speed range, that is in the range above the speed obtainable by falling, but below that of say hypersonic reentry. The major features of such an impact are high edge pressures, side jetting, and cavitation in the liquid. Theoretical work has led to a reasonable understanding of these phenomena by employing a model of edge shock detachment and subsequent pressure relief leading to both jetting and cavitation. By taking some of the properties of the target into account, an explanation has been found for the apparently late development of jetting observed by experimenters. The difficult free-surface problem, inherent in impacts with ideal edge-contact speeds below the sound speed in the liquid, has been treated analytically only in the regime of low impact Mach numbers. Some interesting conjectures (Glenn 1974) regarding the ratio of the side jet speed to the impact speed for the cylinder impact provide a challenge for both the experimenter and theoretician, as does the problem of calculating the side jet following shock detachment.

Major interest centers on predicting the maximum pressure loading during impact. For a situation such as a spherical drop striking a plane surface, the impact starts out as normal. This leads to the weak solution for the attached edge shock and the consequent highest pressure occurring just before shock detachment when the strong and weak solution coalesce. One can imagine the liquid spreading out on a curved surface in such a manner as to drive the edge-contact Mach number from a subsonic to a supersonic value. In such a situation it is not clear which root (strong or weak) may be relevant. Thus the possible occurrence of the strong branch of the solution poses another important problem.

The use of modern high-speed photographic methods, combined with ingenuity in choosing experimental configurations, has provided us with an excellent qualitative picture of the impact process. The employment of high-frequency-response crystal gauges has allowed some quantitative conclusions to be drawn; however, good agreement between a well-controlled experiment and theory remains to be achieved. We feel that the two-dimensional gel technique holds great promise for achieving this end.

Erosion data obtained from the various impact-test devices rarely correlate well. This article has emphasized the importance of the solid/liquid configuration at impact. Clearly, liquid distortion before collision, variations in specimen alignment, surface preparation, and the presence (or not) of a liquid layer on the specimen will all be factors in determining material loss and erosion.

Finally, it is hoped that the better understanding of liquid impact which has emerged in recent years, and which has formed the basis of this review, will benefit the technologies where liquid impact is important.

Acknowledgments

One of us (MBL) would like to thank Professor Jerome Sackman and the Civil Engineering Department of the University of California at Berkeley for providing a working environment for writing during the summer of 1981. We would both like to thank Drs. J. H. Brunton, J-J. Camus, D. A. Gorham, M. C. Rochester, and S. van der Zwaag for providing material for the figures.

Literature Cited

Adler, W. F. 1979a. The mechanisms of liquid impact. In *Erosion*, ed. C. M. Preece, pp. 127–84. New York: Academic

Adler, W. F., ed. 1979b. *Erosion: Prevention and Useful Applications, STP 664*, Am. Soc. Test. Mater., Philadelphia, Pa. 643 pp.

Adler, W. F. 1979c. Single drop impacts on PMMA. See Field 1979, Pap. 9. 10 pp.

Adler, W. F., Hooker, S. V. 1978. Rain erosion behavior of PMMA. *J. Mater. Sci.* 13:1015–25

Adler, W. F., James, T. W. 1979. Particle impact damage in ceramics. *ONR Tech. Rep. for Contract No. N00014-76-C-0744.* 144 pp.

Bahrani, A. S., Black, T. J., Crosland, B. 1967. Mechanics of wave formation in explosive welding. *Proc. R. Soc. London Ser. A* 296:123–36

Benjamin, T. B., Ellis, A. T. 1966. The collapse of cavitation bubbles and the pressures thereby produced against solid boundaries. *Philos. Trans. R. Soc. London Ser. A* 260:221–40

Birkhoff, G., MacDougall, D. P., Pugh, E. M., Taylor, G. I. 1948. Explosives with lined cavities. *J. Appl. Phys.* 19:563–82

Blair, P. W. 1981. *The liquid impact behavior of composites and some infra-red transparent materials.* PhD thesis. Cambridge Univ.

Blowers, R. M. 1969. On the response of an elastic solid to droplet impact. *J. Inst. Math. Its Appl.* 5:167–93

Bowden, F. P., Brunton, J. H. 1958. Damage to solids by liquid impact at supersonic speeds. *Nature* 181:873–75

Bowden, F. P., Brunton, J. H. 1961. The deformation of solids by liquid impact at supersonic speeds. *Proc. R. Soc. London Ser. A* 263:433–50

Bowden, F. P., Field, J. E. 1964. The brittle fracture of solids by liquid impact, by solid impact and by shock. *Proc. R. Soc. London Ser. A* 282:331–52

British Hydraulic Research Association. 1972, 1974, 1976, 1978, 1980, 1982. *Proc. Int. Symp. Jet Cutting Technol.*, BHRA, Cranfield, Engl.

Brunton, J. H. 1967. Erosion by liquid shock. See Fyall & King 1967, pp. 535–60

Brunton, J. H., Camus, J-J. 1970a. The flow of a liquid drop during impact. See Fyall & King 1970, pp. 327–46

Brunton, J. H., Camus, J-J. 1970b. Flow in cavitation and drop impact studies. *Proc.*

9th Int. Conf. High-Speed Photogr., ed. W. G. Hyzer, W. G. Chase, pp. 444–50. SMPTE, New York

Brunton, J. H., Rochester, M. C. 1979. Erosion of solid surfaces by the impact of liquid drops. In *Erosion*, ed. C. M. Preece, pp. 185–248. New York: Academic

Camus, J-J. 1971. *A study of high speed liquid flow in impact and its effect on solid surfaces.* PhD thesis. Cambridge Univ.

Cook, S. S. 1928. Erosion by water hammer. *Proc. R. Soc. London Ser. A* 119:481–88

Cooley, W. C. 1971. Correlation of data on erosion and breakage of rock by high pressure water jets. *12th Int. Symp. Rock Mech.*, ed. G. B. Clark, pp. 653–65. AIME, New York

Cumberbatch, E. 1960. The impact of a water wedge on a wall. *J. Fluid Mech.* 7:535–74

Edney, B. 1976. Experimental studies of pulsed water jets. *Proc. Int. Symp. Jet Cutting Technology, Chicago, Ill.*, Pap. B2, BHRA, Cranfield, Engl.

Engel, O. G. 1955. Water drop collisions with solid surfaces. *J. Res. Nat. Bur. Stand.* 54:281–98

Engel, O. G. 1972. Damage produced by high-speed liquid drop impacts. *J. Appl. Phys.* 44:692–704

Field, J. E. 1967. The importance of surface topography on erosion damage. See Fyall & King 1967, pp. 593–603

Field, J. E., ed. 1979. *5th Int. Conf. Erosion Solid Liq. Impact*, Cavendish Lab., Cambridge, Engl.

Field, J. E., Lesser, M. B. 1977. On the mechanics of high speed liquid jets. *Proc. R. Soc. London Ser. A* 357:143–62

Field, J. E., Lesser, M. B. 1982. Studies of two-dimensional liquid wedge impact and their relevance to the liquid-drop impact problem. *Proc. R. Soc. London Ser. A.* Submitted for publication

Field, J. E., Camus, J-J., Gorham, D. A., Rickerby, D. G. 1974. Impact damage produced by large water drops. See Fyall & King 1974, pp. 395–420

Field, J. E., Gorham, D. A., Rickerby, D. G. 1979a. High-speed liquid jet and drop impact on brittle targets. See Adler 1979b, pp. 298–318

Field, J. E., Gorham, D. A., Hagan, J. T., Matthewson, M. J., Swain, M. V., van der Zwaag, S. 1979b. Liquid jet impact and damage assessment for brittle solids. See Field 1979, Pap. 13. 11 pp.

Field, J. E., Lesser, M. B., Davies, P. N. H. 1979c. Theoretical and experimental studies of two-dimensional liquid impact. See Field 1979, Pap. 2. 8 pp.

Fyall, A. A. 1967. Single impact studies with liquids and solids. See Fyall & King 1967, pp. 563–86

Fyall, A. A., King, R. B., eds. 1965, 1967, 1970, 1974. *1st, 2nd, 3rd, 4th Int. Conf. Rain Erosion Assoc. Phenom.*, RAE Farnborough, Engl.

Fyall, A. A., Smith, P. A. 1968, Single impact studies of rain erosion mechanisms. *US Avionics Lab., Symp. Electromagn. Windows*, 1:55–77

Fyall, A. A., King, R. B., Higgs, G. R., Fawcett, J. H., Williams, H. C. J. 1975. The rain erosion characteristics of Concorde; flight trials in Singapore. *RAE Farnborough Tech. Rep. 75080*, pp. 1–84

Glenn, L. A. 1974. On the dynamics of hypervelocity liquid jet impact on a flat rigid surface. *Z. Angew. Math. Phys.* 25:383–98

Gorham, D. A. 1974. *High velocity liquid jets and their impact on composite materials*. PhD thesis. Cambridge Univ.

Gorham, D. A., Field, J. E. 1976. The failure of composite materials under high-velocity liquid impact. *J. Phys. D* 9:1529–41

Gorham, D. A., Matthewson, M. J., Field, J. E. 1979. Damage mechanisms in polymers and composites under high-velocity liquid impact. See Adler 1979b, pp. 320–42

Hammitt, F. G. 1974. Experimental and theoretical research on liquid drop impact. See Fyall & King 1974, pp. 319–45

Hammitt, F. G., Timm, E. E., Hwang, J. B., Huang, Y. C. 1974. Liquid impact behavior of various elastomeric materials. *STP 567*, Am. Soc. Test. Mater., Philadelphia, Pa., pp. 197–210

Hancox, N. L., Brunton, J. H. 1966. The erosion of solids by the repeated impact of liquid drops. *Philos. Trans. R. Soc. London Ser. A* 260:121–39

Harlow, F. H., Pracht, W. E. 1966. Formation and penetration of high-speed collision jets. *Phys. Fluids* 9:1951–59

Heymann, F. J. 1969. High-speed impact between a liquid drop and a solid surface. *J. Appl. Phys.* 40:5113–22

Holt, M. 1977. Underwater explosions. *Ann. Rev. Fluid Mech.* 9:187–214

Huang, Y. C. 1974. Three stage impact process in liquid impingement. See Fyall & King 1974, Preprint. 7 pp.

Huang, Y. C., Hammitt, F. G., Yang, W. J. 1971. Impact of spherical water drops on a flat rigid surface. *Univ. Mich. Dept. Eng. Rep. 033710–10.T*

Huang, Y. C., Hammitt, F. G., Yang, W. J. 1973. Hydrodynamic phenomena during high-speed collision between liquid drop and rigid plane. *J. Fluids Eng. Trans. ASME* 95:276–94

Jenkins, D. C., Booker, J. D. 1960. In *Aerodynamic Capture Particles*, ed. E. G. Richardson, pp. 97–103. Oxford: Pergamon

Johnson, W., Vickers, G. W. 1973. Transient stress distribution caused by water-jet impact. *J. Mech. Eng. Sci.* 15:302–10

Jolliffe, K. H. 1966. The applications of dislocation etching techniques to the study of liquid impact. *Phil. Trans. Roy. Soc. London Ser. A* 260:101–8

Kenyon, H. F. 1967. Erosion by water jet impacts. *AEI Ltd. Rep. TPIR 5587, Parts I, II*. 18 pp. 28 pp.

King, R. B. 1965. Rain erosion testing at supersonic speeds using rocket-propelled vehicles. See Fyall & King 1965, pp. 49–57

Kinslow, R. 1974a. Rain impact damage of supersonic radomes. *Tenn. Technol. Univ. Rep. No. TTU-ES-74-3*. 468 pp.

Kinslow, R. 1974b. Supersonic liquid impact of solids. See Fyall & King 1974, pp. 271–93

Kinslow, R., Smith, D. G., Sahai, V. 1974. High velocity liquid impact damage. *US Army Missile Command. Tech. Rep. TTU-ES-74-1*. 104 pp.

Kinslow, R., Scardina, J. T., Huang, J. 1976. Radome erosion by raindrop impact. *US Army Missile Command Tech. Rep. TTU-ES-76-1*. 82 pp.

Kunz, W. 1974. The design and construction of a rain-erosion test stand for velocities up to 1000 m s^{-1}. See Fyall & King 1974, pp. 187–98

Lesser, M. B. 1979. The fluid mechanics of liquid drop impact with rigid surfaces. See Field 1979, Pap. 1. 7 pp.

Lesser, M. B. 1981. Analytic solutions of liquid-drop impact problems. *Proc. R. Soc. London Ser. A* 377:289–308

Lesser, M. B., Field, J. E. 1974. The fluid mechanics of compressible liquid impact. See Fyall & King 1974, pp. 235–69

Mørch, K. A. 1979. Dynamics of cavitation bubbles and cavitation liquids. In *Erosion*, ed. C. M. Preece, pp. 309–55. New York: Academic

Mortensen, R. B. 1974. Advanced erosion test facility. See Fyall & King 1974, pp. 155–85

Parsons, C. A., Cook, S. S. 1919. Investigations into the causes of corrosion or erosion of propellers. *Trans. Inst. Nav. Archit.* 61:223–47

Plesset, M. S., Prosperetti, A. 1977. Bubble dynamics and cavitation. *Ann. Rev. Fluid Mech.* 9:145–85

Preece, C. M., ed. 1979. *Treatise on Materials Science and Technology. Vol. 16. Erosion.* New York: Academic. 450 pp.

Reinecke, W. G. 1974. Rain erosion at high speeds. See Fyall & King 1974, pp. 209–33

Rhyming, I. L. 1973. Analysis of unsteady imcompressible jet nozzle flow. *Z. Angew. Math. Phys.* 24:149–64

Rickerby, D. G. 1977. *High velocity liquid impact and fracture phenomena.* PhD thesis. Cambridge Univ.

Rochester, M. C. 1979. *The impact of a liquid drop and the effect of liquid properties on erosion.* PhD thesis. Cambridge Univ.

Rochester, M. C., Brunton, J. H. 1974a. The influence of the physical properties of the liquid on the erosion of solids. *STP 567,* Am. Soc. Test. Mater., Philadelphia, Pa., pp. 128–48

Rochester, M. C., Brunton, J. H. 1974b. Surface pressure distribution during drop impingement. See Fyall & King 1974, pp. 371–93

Rochester, M. C., Brunton, J. H. 1979. Pressure distribution during drop impact. See Field 1979, Pap. 6. 7 pp.

Rosenblatt, M., Ito, Y. M., Eggum, G. E. 1979a. Analysis of brittle targets from a subsonic water drop impact. See Adler 1979b, pp. 227–54

Rosenblatt, M., Ito, Y. M., De Angelo, L. 1979b. Numerical simulations of ceramic target response to water drop impacts including effects of surface flaws. See Field 1979, Pap. 4. 10 pp.

Salem, S. A. L., Al-Hassani, S. T. S., Johnson, W. 1979. Measurements of surface pressure distribution during jet impingement by a pressure pin transducer. See Field 1979, Pap. 7. 11 pp.

Savic, P., Boult, G. T. 1955. The fluid flow associated with the impact of liquid drops with solid surfaces. *Can. Nat. Res. Counc. Rep. MT-26,* pp. 43–83

Schmitt, G. F., Reinecke, W. G., Waldman, G. D. 1974. Influence of velocity, impingement angle, heating and aerodynamic shock layers on erosion at velocities of 1700 m s^{-1}. *STP 567,* Am. Soc. Test. Mater., Philadelphia, Pa., pp. 219–38

Sims, M. G., Trevett, P. W. 1972. *Bibliography of Rain Erosion.* RAE Library Bibliogr. No. 327. MOD, London. 115 pp.

Skalak, R., Feit, D. 1966. Impact on the surface of a compressible fluid. *J. Eng. Ind. Trans. ASME* 88B:325–31

Springer, G. S. 1976. *Erosion by Liquid Impact.* New York: Wiley. 264 pp.

Sullivan, R. J., Hockridge, R. R. 1974. Simulation techniques for hypersonic erosion phenomena. See Fyall & King 1974, pp. 109–54

Thiruvengadam, A. 1967. The concept of erosion strength. *STP 408,* Am. Soc. Test. Mater., Philadelphia, Pa., pp. 22–41

Timm, E. E., Hammitt, F. G. 1969. A repeated water jet device for studying erosion by water jet impacts. *Univ. Mich. Rep. 02643-1-PR*

Vickers, G. W., Johnson, W. 1972. Erosion damage on α-brass and Perspex due to repeated water jet impact. *Int. J. Mech. Sci.* 14:765–71

Vickers, G. W., Johnson, W. 1973. Some results in the erosion of prestressed materials due to water-jet impact. *J. Mech. Eng. Sci.* 15:295–301

Vortsekhovsky, B. V. 1968. Jet nozzle for providing high pressure heads. *Br. Pat. 1,109,286*

Wagner, H. 1932. Über Stoss-und Gleitvorgänge an der Oberfläche von Flussigkeiten. *Z. Angew. Math. Mech.* 12: 193–215

Wahl, N. E. 1970. Design and operation of Mach 3 rotating-arm erosion test apparatus. See Fyall & King 1970, pp. 13–38

Walsh, J. M., Shreffler, R. G., Willig, F. J. 1953. Limiting conditions for jet formation in high velocity conditions. *J. Appl. Phys.* 24:349–59

Wilson, M. P. W., Brunton, J. H. 1970. Wave formation between impacting liquids in explosive welding and erosion. *Nature* 226:538–41

Worthington, A. M. 1908. *A Study of Splashes.* New York: Longmans. Republished 1963 (New York: MacMillan)

Ann. Rev. Fluid Mech. 1983. 15:123–47

AUTOROTATION[1]

Hans J. Lugt

David W. Taylor Naval Ship Research and Development Center,
Bethesda, Maryland 20084

1. INTRODUCTION

The subject of autorotation has been a stepchild in fluid dynamics. It is
barely mentioned in textbooks and considered more a curiosity than a
topic for serious study. Some people have even labeled autorotation as
"toy aerodynamics," a notion that is certainly not justified. Once,
autorotation was expected to be of importance in the understanding of
flying, particularly in studying airfoils, but this idea died quickly. Later,
autorotation gained interest in various unrelated areas such as flight
dynamics of aircraft, aeroballistics, biology, meteorology, and sports. The
diversity of these application areas is part of the reason for the lack of a
uniform treatment, for the difficulty in collecting literature, and for the
variances in the definition of autorotation. Another reason for the lack of
a comprehensive treatment is the difficulty, both experimental and
theoretical, of obtaining qualitative data and in explaining autorotation
satisfactorily. Imagine trying to measure the surface pressure on a
rotating piece of cardboard falling in air! The theoretical difficulties are
indicated by the fact that autorotation is essentially a nonlinear phenomenon associated with vortex shedding. Analogies exist to vortex-induced
vibrations of bodies, a subject area that has attracted more attention in
the past than autorotation. All of these problems are addressed in this
review.

---

[1] The US Government has the right to retain a nonexclusive royalty-free license in and to
any copyright covering this paper.

2. DEFINITION

The concept of autorotation is not uniquely defined in fluid dynamics. Sometimes, aerodynamicists consider any continuous rotation of a body in a parallel flow without external sources of energy as autorotation. Under this definition, windmills, waterwheels, anemometers, and certain tree fruits and seeds are "autorotating" devices. These bodies are geometrically shaped in such a way that, whenever they are kept fixed in a fluid flow, a torque is created that initiates rotation as soon as the bodies are released. Examples of this kind of "pseudo-autorotation" are given in Figure 1.

On the other hand, a body can exhibit autorotation proper [in the classical sense of Riabouchinsky (1935), who introduced the term "autorotation" in 1906] only if one or more stable positions exist at which the fluid flow exerts no torque on the resting body. In this case, a sufficiently strong initial impulse is required before the fluid flow can sustain a continuous spinning of the body.

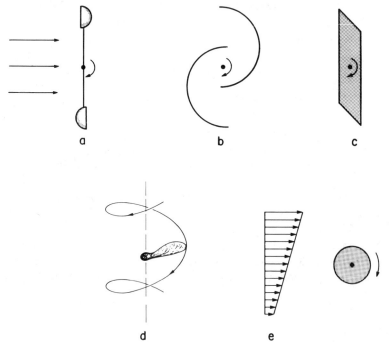

Figure 1   Pseudo-autorotation. (*a*) Cup-anemometer; (*b*) Savonius rotor (cross section); (*c*) Riabouchinsky plate (cross section); (*d*) maple seed; (*e*) cylinder rotating due to shear flow.

Although the axis of rotation may assume any orientation with respect to the flow, it is useful theoretically and didactically to distinguish two special cases of axis orientation for autorotation in the restricted, classical sense. In the first special case, the axis of rotation is parallel to the flow, and the body must be symmetrical with regard to the parallel stream so that no torque is present when the body is at rest. This state is generally stable. An initial impulse is then necessary to obtain autorotation. Typical examples are the Lanchester propeller and the spinning airfoil (Figure 2).

In the second special case, the axis is perpendicular to the parallel flow and the body need be symmetrical with respect to the parallel flow in only one stable position (Lugt 1980). In this position the flow exerts no torque on the resting body, and again an initial impulse is necessary to obtain autorotation. However, other positions may exist from which the body will start rotating when released with no initial impulse, provided the body can pick up and store sufficient angular momentum to overcome the adverse torque around the stable position. The fundamental difference between the two cases of autorotation may also be expressed in this way: While the rate of stable autorotation is essentially constant for bodies autorotating parallel to the flow (provided the wake of the body is fairly constant), the rate of autorotation in the other case is basically periodic.

Examples of bodies autorotating perpendicular to the flow are the falling rectangular piece of cardboard and the rotating dumbbell (Figure 3). Actually, the axis of the cardboard (or plate) may be free to move or may be fixed in a parallel stream. Other symmetrically shaped bodies that can autorotate are also shown in Figure 3. A word of caution is in order here with respect to symmetry properties. Whereas bodies displaying pseudo-autorotation have only axial symmetry (if any at all), the bodies in Figure 3 have both axial and plane symmetry. For example, the Savonius rotor in Figure 1 has a twofold axial symmetry but the triangle in Figure 3 has threefold axial and plane symmetry.

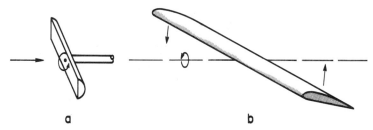

a   b

Figure 2   Bodies autorotating parallel to the flow. (*a*) Lanchester propeller; (*b*) autorotating airfoil.

For irregularly shaped bodies, the difference between pseudo-autorotation and autorotation proper is not always clear and can depend on the position of the axis of rotation and the flow situation. For instance, the cup anemometer in Figure 1 has a point between the geometric center and the cup facing the stream at which the body is in a stable equilibrium for a certain Reynolds-number range. Also, the autorotation of airplanes and reentry bodies can border on pseudo-autorotation. The situation is even further obscured when the axis of rotation can assume any orientation with respect to the flow (Figure 4). Hence it is sometimes unavoidable to trespass into the gray area between pseudo-autorotation and classical autorotation. In this review, the word autorotation is used in the classical sense.

The application areas of autorotation are manifold. The principles of the Lanchester propeller and the autorotating wing are basic for the

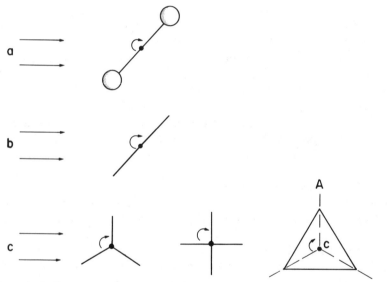

Figure 3 Bodies autorotating perpendicular to the flow. (*a*) Rotating dumbbell; (*b*) rotating plate; (*c*) examples of Magnus rotors. *A-C* line of lateral symmetry

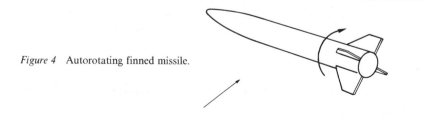

Figure 4 Autorotating finned missile.

understanding of spinning aircraft (Jones 1963). The free-falling auto-rotating plate is exemplified in nature by certain tree fruits and seeds (McCutchen 1977) and its fall characteristics are used in aeroballistics to predict dispersion patterns for the distribution of leaflets and bomblets (Burgess 1967) or to study fragmentation of the flying body. Autorotat-ing missiles with cruciform fins pose a control problem, and the same is true for autorotating reentry bodies and disposable nose sections of aircraft (Cohen et al. 1974, A. M. O. Smith 1953). Autorotating plates have also been proposed for deceleration and high-lift devices (Neumark 1963, Iversen 1969, Miller 1973). In meteorology the growth of hailstones of oblate spheroidal shape is affected by autorotation (Kry & List 1974b).

3.   AUTOROTATION PARALLEL TO THE FLOW

Autorotation of bodies about an axis parallel to the flow is in general easier to explain and more amenable to a quantitative description than autorotation perpendicular to the flow. The reason is that the fluid motion is steady in the stable state for a body-fixed reference frame. In almost all theoretical considerations, the two-dimensional strip method is used. Then, the local angle of attack changes due to the rotational velocity $V = R\Omega$ according to

$$\Delta\alpha = \tan^{-1}p, \qquad p = \frac{V}{U} \tag{1}$$

where U is the constant speed of the parallel flow, R the radius or half-span of the wing-type body, Ω the angular velocity, and p the roll parameter. The total angle of attack α of a flow against a blade element is then $\alpha_0 + \Delta\alpha$, with α_0 the angle of attack for $V = 0$.

The aerodynamic forces support rotation if $V/U = p > L/D$, where L and D are the lift and drag, respectively (Figure 5). To meet this criterion, the total angle of attack must be in the region of stall so that the slope $dL/d\alpha$ becomes negative and its absolute value according to Glauert (1919) so large that the condition

$$\frac{dL}{d\alpha} + D < 0 \tag{2}$$

is satisfied. The local condition (2) for a blade element need not be met everywhere along the span as long as the integration over the whole span results in a torque in the direction of rotation.

The Lanchester propeller and the rotating wing are now explained in detail on the basis of this steady-state strip theory.

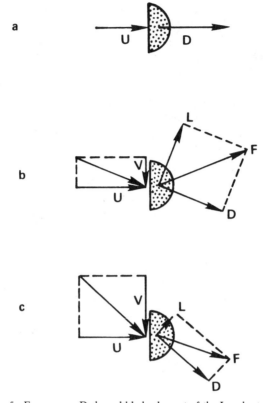

Figure 5 Forces on a D-shaped blade element of the Lanchester propeller.

3.1   Lanchester Propeller

The forces acting on the Lanchester propeller (Figure 2*a*) are depicted in Figure 5 for a D-shaped blade element. Without rotation, only a drag force *D* is present (*a*). A slowly rotating blade causes a lift *L*; the resultant *F* opposes rotation (*b*). Thus, any slow initial rotation ceases in time. If, however, the initial rotation is large enough, the resultant force can support rotation (*c*). The propeller starts autorotating and increases its angular velocity until a steady equilibrium with the damping force is reached.

Applying the steady-state strip theory to the Lanchester propeller, one must set $\alpha_0 = 90°$. The slope $dL/d\alpha$ for the total angle of attack $\alpha_0 \pm \Delta\alpha$ is negative in general (Den Hartog 1932). However, for small p, condition (2) is usually not met (partly due to bearing friction) until a critical initial value of p is reached.

The relation between initial impulse and final rate of autorotation may be illustrated more clearly by a method developed by Riabouchinsky (1935): A motor drives the propeller with constant angular velocity Ω, and the torque T acting on the propeller is measured as a function of p. The resulting curve is shown in Figure 6. When p is small or large, an outside positive torque is necessary to drive the propeller. Between these values, a p-range exists in which the outside torque is negative and thus requires a braking effect on the rotation of the propeller. Riabouchinsky called this range "autorotative." If the outside torque is removed, the propeller increases its angular velocity until $T = 0$, which is the state of autorotation (point A in Figure 6). The second state $T = 0$, which is the initial value of p needed to induce autorotation, is unstable. The conditions for stable autorotation are thus

$$T = 0, \qquad \frac{dT}{dp} > 0. \tag{3}$$

Although the functioning of the Lanchester propeller has been demonstrated many times, there appears to be only one experimental investigation that provided quantitative data, that by Parkinson (1964). He found that in the Reynolds-number range $2800 < \mathrm{Re} < 1.6 \times 10^5$, with R the tip radius, v the kinematic viscosity, and $\mathrm{Re} = RU/v$, the value p for stable autorotation is almost independent of Re and is between approximately 1.96 and 2.16 for a D-shaped blade. The initial value of p for the onset of autorotation is about 0.5. However, friction of the shaft must be taken into account.

Parkinson computed p by means of the steady-state strip theory and used experimental data for L and D. He obtained $p = 1.79$ for stable

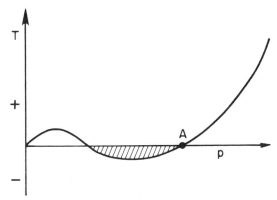

Figure 6 "Riabouchinsky curve" for the explanation of autorotation. A is the point of stable autorotation.

autorotation and $p = 0.45$ for the initial value. Hence the stable rate of rotation is underestimated by the theory by 9% to 17%. Clearly, the neglect of three-dimensional effects is the main reason for this discrepancy. Parkinson (1964) pointed out that centrifugal effects would shift the midsection of the propeller farther out with a torque favorable to autorotation and thus increase p.

3.2   Autorotating Wing

The explanation for a wing autorotating about a fixed axis parallel to the flow (Figure 2b) is basically the same as that for the Lanchester propeller. The angle of attack of a wing element on the downward-moving half of the wing is $\alpha_0 + \Delta\alpha$, whereas the angle of attack of the corresponding wing element on the upward-moving half is $\alpha_0 - \Delta\alpha$. Stall at $\alpha_0 + \Delta\alpha$, i.e. a negative $dL/d\alpha$, may cause a torque in the direction of rotation. But the problem here is more involved; since α_0 is not restricted to one value (90° for the Lanchester propeller), it is now a parameter, ranging from ca. 0° to values far into the stall region. Discussion of the influence of various values of α_0 on autorotation gives more insight into the nonlinear behavior of this phenomenon.

Extending Riabouchinsky's curve in Figure 6 to a family of curves with α_0 as a parameter, one obtains Figure 7. Glauert's (1919) theoretical data for a RAF 6 wing have been depicted as a typical example. Notice that autorotation can occur for $p \ll 1$ and that p is much smaller for the autorotating wing than for the Lanchester propeller. In Figure 8, the roll

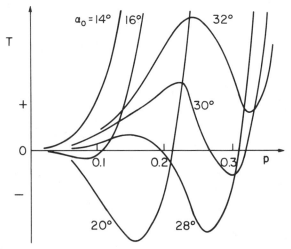

Figure 7   The torque of a rotating wing as a function of p and α_0 (adapted from Glauert 1919).

parameter p for autorotation is plotted versus the angle of attack α_0. The lowest value of α_0 for nonzero p is at 15°, close to the stall angle. Below this value the wing is stable for $p = 0$; beyond, it is unstable for $p = 0$. The latter result implies that a nonrotating wing can start autorotating at the slightest rotational disturbance. From approximately 26.5° to 31.5°, two equilibrium states for $p \neq 0$ occur as in the case of the Lanchester propeller. The smaller value of p is unstable, the larger stable.

The curves in Figures 7 and 8 differ, of course, for other wing profiles and flow situations. Negative $dL/d\alpha$, for instance, can occur for delta wings when the vortex lift drops because of vortex breakdown over the wing area (Küchemann 1978).

The computation of the characteristics of autorotating wings today is still based on the steady-state strip approach with the aid of empirical data for a wing element. According to Glauert (1919), the accuracy is about 15%, that is, on the order of Parkinson's (1964) findings for the Lanchester propeller.

Figures 7 and 8 and Glauert's condition (2) indicate that the problem of autorotation is nonlinear. This is not surprising since stall is the cause of autorotation. Figure 8, by the way, is a familiar curve in catastrophe theory (Zeeman 1977), and condition (2) can be derived from a power-series solution of the nonlinear-oscillator equation (Parkinson 1974).

The yawing motion of the fuselage of an airplane can lead to autorotation. This phenomenon is based on the same principle as that of the roll of a wing becoming self-sustained. This observation is important for the understanding of freely spinning airplanes, which is discussed in Section 5.

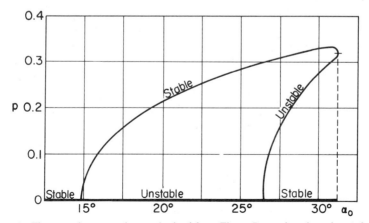

Figure 8 The rate of autorotation p, obtained from Figure 7, as a function of α_0 (adapted from Glauert 1919).

4. AUTOROTATION PERPENDICULAR TO THE FLOW

The fundamental difference between autorotation parallel to and autorotation normal to the flow is that the rate of stable autorotation is constant in the first case and periodic in the second. This difference affects the definition of autorotation: The torque T and angular velocity Ω (or p) in Equation (3) must be replaced by the corresponding quantities averaged over one cycle, where a body with n-fold axial symmetry has a cycle of period $2\pi/n$. Thus, the average torque $\bar{T}$ for use in the definition of stable autorotation is

$$\bar{T} = \frac{n}{2\pi} \int_0^{2\pi/n} T \, d\alpha, \tag{4}$$

where

$$T = I \frac{d\Omega}{dt} \tag{5}$$

with I the moment of inertia and t the time.

A consequence of the difference in characteristics between autorotation parallel to and normal to the flow is that in the latter case, when the axis is kept fixed, steady-flow models with empirical ("static") data cannot be used unless they are introduced in a "quasi-steady" sense. "Quasi-steady" means that the forces on the body vary so slowly with rotation that they can be computed at a particular instant as if the body were not rotating. Even then it is easy to see that a quasi-steady approach always results in $\bar{T} = 0$, except when static hysteresis or a sudden drop in the force coefficient at increased speed occurs. Static hysteresis means a hysteresis effect occurring from one steady state to another in contrast to "dynamic hysteresis" of an accelerated body (Ericsson & Reding 1969). For an autorotating body with freely moving axis, the quasi-steady model may be useful under certain restrictions (see Section 4.2). In general, a strip theory cannot be applied to bodies autorotating normal to the flow since the whole wake region periodic in time must be considered.

The following discussion of two examples (Sections 4.1 and 4.2) illustrates the role and applicability of the concepts introduced here in more detail.

4.1   Rotating Dumbbell

Not much is known about the rotating dumbbell depicted in Figure 3a. The explanation for autorotation is based on the sudden drop of the drag coefficient of a sphere at the transition from laminar to turbulent flow near $\text{Re} = 2RU/v = 4 \times 10^5$, with R the radius of the sphere (A. M. O. Smith 1953). The effective velocity of the sphere advancing in the main stream is larger than that of the retreating sphere, yet the latter sphere experiences a larger drag since it is still in the laminar flow range (Eiffel-Prandtl paradox). This quasi-steady approach for explaining the autorotation of dumbbells may be classified as one within the framework of strip theory: the two spheres are the only elements.

Two parallel, long circular cylinders attached to each other by a thin rod can be used instead of two spheres, since the drag curve for the cylinder exhibits the same sudden drop at about $\text{Re} = 4 \times 10^5$.

Except for an experiment with rotating dumbbells cited by A. M. O. Smith (1953), no empirical data are available, nor are technical applications known. The rotating dumbbell will probably remain a curious aerodynamic toy that easily demonstrates the Eiffel-Prandtl paradox.

4.2 Flat Plate Rotating Perpendicular to the Flow

A simple experiment demonstrates the capability of a flat plate to autorotate: Cut out a rectangular strip of paper, say 1 cm × 10 cm, and drop it by releasing the strip from a horizontal position, tilted slightly chordwise. The paper will fall along an oblique, straight (slightly undulating) trajectory with rapid rotation normal to the flight path (Figure 9). A deviation from the initial horizontal position (or small mass asymmetry), however, will result in autorotation along a helical flight path. Still, it is not difficult to keep the paper horizontal enough to provide directional stability.

A flat plate that is free to rotate about a fixed axis in the airstream of a wind tunnel can also autorotate under certain conditions.

Extensive experiments under both free-fall and wind-tunnel conditions were made by Dupleich (1941), Cheng (1966), Bustamante & Stone

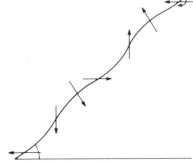

Figure 9 The path of a falling, autorotating plate.

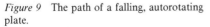

(1969), E. H. Smith (1971), and Glaser & Northup (1971). Most of the experimental data obtained by these researchers have been evaluated and summarized by Iversen (1979).

Free-fall and wind-tunnel data reveal that a certain moment of inertia I is required for a plate to autorotate. The roll parameter p increases with larger values of I but becomes almost independent of I for $I/\rho bc^4 > 1$, where ρ, b, and c are the density of the fluid, span, and chord length, respectively (Iversen 1979). In this region of independence, the difference between free-flight and wind-tunnel autorotation becomes indistinguishable.

With inertia effects no longer considered, the dependence of p on aspect ratio, plate thickness, and Reynolds number, as well as lift and drag behavior, remains to be discussed.

Iversen (1979) offers the following formula based on experimental data for the influence of the aspect ratio $A = b/c$ on p:

$$p = \left\{ \frac{A}{2+\sqrt{4+A^2}} \left[2 - \left(\frac{A}{A+0.595} \right)^{0.76} \right] \right\}^{2/3}, \tag{6}$$

where the relative thickness of the plate (defined by thickness/c) is smaller than 0.01. For $A > 5$, the influence of A on p can be ignored, and this restriction can be relaxed with the aid of endplates.

The relative thickness has a negligible effect on p for values less than 0.01. For thick plates and fat noncircular cylinders, the roll parameter p decreases. Autorotation of circular cylinders has never been observed.

The roll parameter is nearly independent of the Reynolds number for $\mathrm{Re} = cU/\nu > 1500$ and is about 0.8–1.0 for thin plates and large aspect ratios. Numerical and measured data as low as $\mathrm{Re} = 100$ indicate a drop in p for such small Re (E. H. Smith 1971, Lugt 1980). When $\mathrm{Re} \to 0$, autorotation ceases.

Almost all experiments have been made for subsonic flows. Bustamante & Stone (1969) found evidence for autorotation in supersonic flow, but little is known about this flow region.

The lift-to-drag ratio, which equals the cotangent of the glide angle, increases with increasing p. Iversen (1979) correlated the empirical data of Dupleich for L/D with p and other flow parameters in a graph that displays measured values for L/D in the interval (0.4, 1.6). A rule of thumb was presented earlier by Iversen (1969) in the form $L/D = c_L/c_D = 1.4p$, which is based on $p^2 = A/8$, $c_D = 2$, $c_L = A^{0.5}$ for $A < 8$ where $c_D = D/(\rho/2)U^2bc$, and c_L is defined correspondingly.

Instantaneous flow quantities, such as force and moment coefficients at a certain instant, and the surface pressure have been measured by

Cheng (1966) and E. H. Smith (1971). Cheng published for the first time the distribution of the surface pressure along the chord of a rotating plate and data for the aerodynamic torque. The main novel features in E. H. Smith's experiments are recorded data on the instantaneous lift and drag coefficients.

The experimental difficulties in obtaining reliable data should be noted. In addition to problems of reducing bearing friction, which every experimentalist in this field has encountered, measurement of the instantaneous aerodynamic loading on the rotating plate caused difficulties. Cheng (1966) used 13 pressure taps at midspan, a sliding pressure seal, and a pressure transducer. Essentially the same arrangement but with a different sliding-seal mechanism was used by Miller (1973, 1979). E. H. Smith (1971) mounted strain gauges on one of the bearing supports to measure lift and drag. Data on the instantaneous distribution of the surface pressure and torque for a freely falling plate do not exist. Their acquisition would be a challenge for any experimentalist!

Theoretical explanations of the autorotating plate started with Maxwell (1890) in 1853 for the case of the freely falling plate. He recognized that the center of mass and the center of aerodynamic forces do not coincide, giving rise to a torque. He then assumed that this torque can be divided into a quasi-steady part and a contribution due to rotation. The latter, Maxwell believed, would always have a damping effect. From $\alpha = 0°$ to $90°$ the torque on the plate supports rotation ("supporting period"), whereas from $\alpha = 90°$ to $180°$ the torque acts on the plate in opposition to rotation ("retarding period"). Maxwell recognized that the torque in the supporting period is larger than in the retarding period because the translational velocity of the plate for $\alpha = 0°$ is larger (the drag is smaller) than that for $\alpha = 90°$. The higher translational speed around $\alpha = 0°$ causes a larger torque in the supporting period. This driving net torque would be balanced, according to Maxwell, by the damping effect of the rotation.

Maxwell's argument is only partially correct, and this only when the amplitude of the undulatory path of the plate's axis (Figure 9) is sufficiently large. However, with increasing moment of inertia the amplitude decreases, and the difference between free-fall and fixed-axis autorotation vanishes. The quasi-steady model ceases to be valid, and the rotational effect must be considered to support autorotation. This concept is explained in the paragraphs that follow (Lugt 1980).

In potential-flow theory, the torque coefficient $c_M = T/(\rho/2)U^2(c/2)^2$ acting on a plate with a fixed axis depends only on α (and not explicitly on Ω), whether it rotates or not:

$$c_M = -\pi \sin 2\alpha, \tag{7}$$

where α is either a constant or time-dependent, i.e. $\alpha = \Omega t$. The average torque is zero, and the plate always autorotates. This solution is not useful for explaining autorotation, but it reveals the importance of the existence of the supporting and retarding periods caused by the asymmetric locations of the stagnation points.

Viscous effects must be taken into account. Numerical computations with the Navier-Stokes equations reveal the following relationship between instantaneous torque and flow patterns: In Figure 10 the torque coefficient c_M of a thin elliptic cylinder is displayed as a function of α for various values of p at Re = 200 (p constant in time). Drastic changes for different p occur in the retarding period near $\alpha = 135°$. If one averages c_M and plots it versus p, one obtains Riabouchinsky's curve of Figure 6 (Lugt 1980). Clearly, the deviation of curves in Figure 10 from the sinusoidal behavior in the retarding period is related to the "autorotative" region in Figure 6.

A comparison of computed and photographed flow patterns reveals that this "autorotative" region coincides approximately with the synchronization of vortex-shedding frequency and rate of rotation, a manifestation of the nonlinear "lock-in" effect. The situation is sketched in Figure 11b for the decisive α-range of about 135°. For values of p outside the autorotative region, the frequency of vortex shedding is either smaller

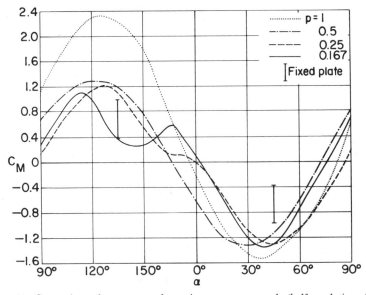

Figure 10 Comparison of c_M versus α for various p over one cycle (half-revolution of the plate) at Re = 200 (from Lugt 1980).

(Figure 11a, superharmonic modes) or larger (Figure 11c, subharmonic modes) than the rate of rotation. The superharmonic modes differ by integral values from each other, a fact that has a simple physical explanation: a rapidly rotating plate traps vortices before it releases them after several revolutions. The trapping of vortices over a small range of p constitutes a lock-in at higher harmonics, a phenomenon observed in vortex-induced vibration (Ericsson 1980; see also Section 6). In the subharmonic range, the vortex-shedding frequency approaches that of a nonrotating plate when $p \to 0$ and becomes independent of the slow rate of rotation.

For autorotation, the flow patterns are essentially independent of the Reynolds number and are even similar for the laminar and turbulent states. The inertial forces due to the periodic change of the flow field do not give sufficient time for the diffusion of vorticity or the small-scale turbulent eddies to have a significant influence on the overall process.

The reason for autorotation becomes evident when the surface pressure in the retarding period is compared with that in the supporting period for various p. Near $\alpha = 135°$, the vortex at the retreating edge (which is very pronounced in wind-tunnel movies made by the Aerodynamics Department at the University of Notre Dame) has just separated from the body for $p \approx 0.25$ (Figure 11b), and the flow around the edge looks similar to the flow past a plate parallel to the stream. The surface pressure and vorticity are almost symmetric on the two rear halves of the plate (Figure 12), which means that the surface pressure is quite constant on both sides of the plate. In contrast, for higher as well as lower p, the front surface pressure near the retreating edge is very low (in qualitative agreement with potential flow) and thus counteracts autorotation. Computations confirm that the average torque in the retarding period has a minimum in the autorotative range (area under the c-curve for $p = 0.25$ in the α-range $[90°, 0°]$ in Figure 10). This strange behavior does not occur in the supporting period, in which the average torque is almost constant. Details of the delicate balance between the torque contributions of the retarding and supporting periods are given in Lugt (1978, 1980).

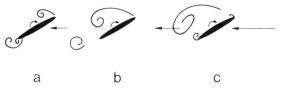

a   b   c

Figure 11   Sketch of vortex shedding about a power-driven rotating plate. (a) p is greater than that for autorotation; (b) lock-in range of p (autorotative); (c) p is slower than that for autorotation.

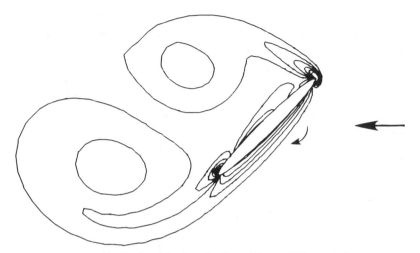

Figure 12   Equivorticity lines for Re = 200, $p = 0.25$, $\alpha = 135°$.

For two-dimensional freely falling plates and for autorotating plates of finite span (three-dimensional flow), computations are not known to exist. A decrease in autorotation due to three-dimensional effects, observed experimentally and expressed in Equation (6), is caused by the weakening of the shed vortices. Considering the vortices idealized by vortex tubes in a potential flow, one realizes that a body with finite span is surrounded by a flow in which the vortex tubes are either closed or must end on the body. They are bent at the ends of the plate, thus giving rise to a secondary flow. This flow interferes with the primary (two-dimensional) flow by weakening it in two ways: by decreasing the strength of the primary flow, and by speeding up the process of instability, since curved vortex tubes are highly unstable in general.

4.3   Other Autorotating Devices

The flat plate and the thin elliptic cylinder are by no means the only cylindrical bodies capable of autorotation. Other shapes like cruciform plate arrangements, triangles, and squares can autorotate (Figure 3c). Riabouchinsky (1935) experimented with triple and cruciform plates, and he found that the rate of rotation decreases with increasing number of plates. The number of plates at which autorotation ceases is not yet known.

These cylindrical bodies belong to the general class of Magnus rotors that are defined as cylindrical bodies of fairly arbitrary cross section, designed for fast spinning. The name is derived from the Magnus effect of producing lift through rotation. Magnus rotors also include bodies whose self-sustained spinning has been labeled in Section 2 as "pseudo-autorotation" (Figure 1). The numerous publications include papers by Flatau (1964), Yelmgren (1966), Stilley (1967), Gebman (1967), Brunk (1967), Iversen (1969), and Miller (1979).

Obviously, p is larger for asymmetric rotor devices of the types depicted in Figure 1 than for symmetric devices. The Savonius rotor appears to have the highest value of p among the Magnus rotors investigated (Yelmgren 1966, Iversen 1969), i.e. $p = 1.37$ for $A = 1.45$ and Reynolds numbers of the order 10^5. (A theoretical upper limit is $p = 2$, based on potential flow around a rotating circular cylinder and on the criterion of the occurrence of closed streamlines.) The Darrieus rotor is mentioned in this context as a device that is not a Magnus rotor. Its blades whirl around the center, similar to the cylindrical dumbbell but with very high values of p up to 7. (Strickland et al. 1979). It is conjectured that the Darrieus rotor also works for blades with fore-aft symmetry. The motion would then constitute autorotation proper.

Bodies need not be cylindrical to be able to autorotate. Riabouchinsky (1935) mentioned airships that can autorotate perpendicular to the flow. A. M. O. Smith (1953) was originally motivated in his research by the autorotation of nose sections of aircraft that can be released in an emergency. Reentry bodies and spacecraft can autorotate when diving into the atmosphere, a situation that must be avoided by all means. (Astronauts Glenn and Carpenter could prevent their Mercury capsule from autorotation only by opening the parachute.) Disk-shaped bodies can also autorotate, as demonstrated by a coin falling in water (Willmarth et al. 1964) or a discus thrown without a spin in the azimuthal direction (Soong 1976). Hailstones are also able to autorotate (Kry & List 1974a,b, List et al. 1973). Calculations for a rotating discus and for hailstones were based on the quasi-steady approach with the aid of empirical static data.

The problems and major issues today involving bodies autorotating perpendicular to the flow are summarized as follows:

1. Accurate performance prediction requires better knowledge and correlation of wind-tunnel and free-fall data for p, L/D, and directional stability as functions of the various parameters involved.
2. Local and instantaneous aerodynamic force characteristics are needed.
3. For freely falling autorotating bodies, the limits of applicability of the quasi-steady approach must be studied.
4. For powered Magnus rotors, sub- and superharmonic modes need to be determined.

5. AUTOROTATION AT ARBITRARY ANGLE TO THE FLOW

The difficulties encountered in Sections 3 and 4 in explaining and computing autorotation parallel to and normal to the flow are magnified considerably for bodies that autorotate at an arbitrary angle to the flow. Here, the differences in flow characteristics between rotation about a fixed and a freely moving axis (free flight) become decisive compared with the previously discussed cases. In free flight the body performs motions with six degrees of freedom, and gyroscopic effects are therefore present.

It has been stated in Sections 3 and 4 that the torque on a body rotating about a fixed axis depends on p as expressed by the Riabouchinsky curve (Figure 6). A qualitative polynomial representation of this relation is

$$c_M = c_1 p + c_3 p^3 + c_5 p^5, \tag{8}$$

where the coefficients c_i are constants. In certain regions a cubic relation suffices (Figure 7).

The physical meaning and the magnitude of the coefficients c_i can be obtained either by a strip theory with static-force data or by solutions of the Navier-Stokes equations for the special cases discussed in Sections 3 and 4.

Such an interpretation of the "global" aerodynamic coefficients c_i with detailed knowledge of the local flow field is not yet possible for the three-dimensional motion of bodies in free flight or rotating about fixed axes. These problems are simply too complex. Instead, only measured data of global force and moment components (like the moments of roll, yaw, and pitch) can be correlated to the aerodynamic coefficients. Before discussing this further, attention is drawn to the coupling among rolling, yawing, and pitching motions through unequal moments of inertia in free flight. For instance, the roll moment c_M is balanced in steady roll according to the Euler equations by the "inertia coupling"

$$c_M = (C - B)qr, \tag{9}$$

where q and r are the rates of pitch and yaw, and B and C are the dimensionless moments of inertia about the transverse y-coordinate and the vertical z-coordinate, respectively.

A common practice in aerodynamics is to consider only linear damping for c_M. Sometimes aerodynamic coupling analogous to the inertia coupling is included. Apparently, fifth-order polynomials of the form of Equation (8) are not necessary to demonstrate autorotation. However, it is clear that linear damping and quadratic coupling only must be applied with caution.

Another question arises on whether experimental aerodynamic static data suffice for correlation with the aerodynamic coefficients, or whether rotary data must be used, as shown for certain instances by Anglin (1978) and Langham (1978).

The following discussion is limited to two cases of great engineering interest: (*a*) the spinning and rolling of aircraft, and (*b*) the autorotation of finned missiles. It must be stressed, however, that these subject areas can be discussed only briefly within the scope of this review because of the vast amount of existing literature.

5.1   Autorotation of Spinning and Rolling Aircraft

When an airplane is stalled asymmetrically, either unintentionally or on purpose, it may descend nose-down in a rolling and yawing motion on a spiral path at an angle of attack between stall and 90° (Figure 13). This phenomenon is called "spin" (Jones 1963).

Airplanes have been troubled by spin from the time they came of age (about 1909). At the end of World War I the phenomenon of spin (and recovery from it) was understood in principle (Lindemann et al. 1918, Relf & Lavender 1918, Glauert 1919), although total avoidance of spin

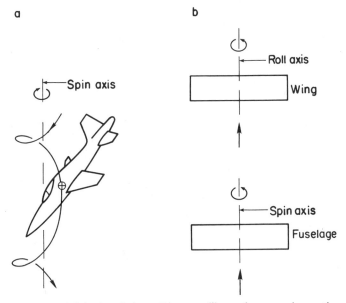

Figure 13   (*a*) Spinning airplane; (*b*) pure rolling and pure yawing motions.

was not achieved in the succeeding decades. The problem persists today for light commercial airplanes because of their relatively greater engine power (Bowman 1971, Anderson 1979), and it is even aggravated in modern fighter aircraft (Anglin 1978) because of their greater weight, swept wings, and jet engines (see below). After all, maneuverability and stability of aircraft are incompatible in a way.

Aerodynamicists distinguish three phases of spin: (a) spin entry, (b) equilibrium spin, in either a steady or oscillatory mode, and (c) spin recovery. Only the second phase with all control surfaces in neutral positions is autorotation proper. "Flat spin" and "steep spin" are distinguished by the angle θ of the longitudinal axis of the airplane with the vertical. In a flat spin the angle θ is larger than 80°; in a steep spin it is $45° < \theta < 75°$. According to the relation $r_0 S^2 = g \cot \theta$, where S, r_0, and g are the rate of spin around the spin axis (not the rate of roll), the radius of the helical path, and the gravitational constant, respectively, a small value of r_0 in flat spin results in a large value of S (which for θ close to 90° is essentially a yawing motion with S on the order of half a revolution per second). A larger value of r_0 in steep spin decreases S.

Spin may be considered roughly as a combination of two autorotation effects (Figure 13): roll of the wing, as described in Section 3.2, and yawing of the fuselage, where the fuselage may be regarded as a wing or a Lanchester propeller, especially for flat spin (Clarkson 1958, Clarkson et al. 1978).

The influence of the fuselage on spin becomes crucial in modern fighter planes with their long fuselage nose and heavy fuselage mass loadings. According to Neihouse et al. (1960), the moments of inertia about the y- and z-axes are an order of magnitude larger than those of World War II airplanes. In addition, swept wings have a negative influence on longitudinal and rolling stability. Steady spin at high angles θ can change at lower angles to oscillatory spin (Skow & Titiriga 1978) with typical "limit cycle" behavior (Mehra & Carroll 1979). Near stall the phenomenon of "wing rock" can occur, caused by aerodynamic roll moment hysteresis (Schmidt 1979). Also, spin entry is facilitated by asymmetric vortex shedding from the nose of the aircraft (Spangler & Mendenhall 1977) or by vortex breakdown over delta wings.

The design of modern aircraft has led to the discovery of another kind of autorotation that can occur during fast rolling maneuvers. Again, inertia forces of high-speed aircraft cannot be neglected, and they tend to swing the fuselage out of the flight path (Phillips 1948). This sideslipping causes coupling of longitudinal and lateral gyroscopic and aerodynamic characteristics, similar to spin. However, in contrast to spin, rolling motion involves much higher values of p and the flight path is essentially horizontal. Pinsker (1957) has shown that this rolling with sideslip can become self-sustained without asymmetrical moments enforced by the aileron; he called this situation "autorotation in roll." Nonlinear inertia cross-coupling and linear aerodynamic damping are sufficient to demonstrate the existence of autorotation in roll for a certain parameter range. However, a recent paper by Sachs & Fohrer (1980) shows that additional aerodynamic coupling, through introduction of rolling moments due to angle of attack and sideslip and of quadratic terms of pitching moments due to angle of sideslip, has a strong influence on autorotation. This is an example of the importance of the correct polynomial representation of the aerodynamic coefficients for the realistic description of autorotation.

Cohen & Nimri (1976) investigated rolling delta wings by including leading-edge vortices in their model.

5.2 Roll Speed-Up of Finned Missiles

Three regimes of steady rotation are distinguished for the rolling motion of finned missiles in free flight, depending on the angle of attack α, which is here defined as the angle between flow and horizontal missile axis in Figure 4 (Nicolaides 1957). For convenience, only the cruciform-fin configuration is considered.

For small angles of attack greater than $\alpha = 0°$, steady roll can be maintained only by canted fins. This region is called "roll slow-down," because p decreases with larger α until a situation is reached at which the fins are locked with the coning motion of the missile. This situation, which occurs for canted as well as uncanted fins, is therefore called "roll lock-in" or "lunar motion" and extends up to about $\alpha = 36°$. Beyond this value up to 90°, p drastically increases and reaches a value of 0.25 at 90° for uncanted fins with Re of the order of 10^5. This region of steady roll is called "roll speed-up" and is the only region of autorotation proper.

In contrast to "autorotation in roll" and equilibrium spin of aircraft for which yawing and pitching are essential to explain autorotation, finned missiles can autorotate without yawing and pitching. This fact and the extreme complexity of the flow field around a missile in free flight resulted in studies that are mostly devoted to a missile autorotating about a fixed axis in a wind tunnel.

Experiments have shown (Greene 1960) that roll speed-up about a fixed axis occurs whether the cruciform fins are attached to the body or disconnected, and even without a body. It was conjectured by Nicolaides (1957) that roll speed-up was similar to wing autorotation, although not exactly the same, and Greene (1960) suspected a "special kind of autorotation." Looking at the simplest case of $\alpha = 90°$, Lugt (1961) interpreted roll speed-up as the same phenomenon as plate autorotation normal to the flow (Figure 3b and c, Section 4.2).

As in the case of spin and roll of aircraft, analysis of roll speed-up can be done at present only by correlating experimental data with aerodynamic coefficients. Whereas for spin and roll of aircraft, inertia coupling is essential for steady autorotation and provides solutions with linear damping only (whether this is realistic or not), roll speed-up about a fixed axis requires the fifth-order polynomial of Equation (8) in general, or at least a third-order one for certain parameter regions. This follows from the interpretation of reducing roll speed-up to plate autorotation. It is interesting that Daniels (1970) and Cohen et al. (1974) needed third-order polynomials to correlate experimental data with the aerodynamic coefficients. Other work on nonlinear damping was reported by Uselton & Jenke (1977).

For finned missiles below $\alpha = 40°$, vortices shed from the body may interfere with the fins. In fact, Nicolaides (1957) used this argument to explain roll lock-in. A more recent paper on autorotation with body-vortex interference was published by Fiechter (1972). Papers on rotating missiles in free flight have included those by Murphy (1971) and Clare (1971).

6.   ANALOGY TO VORTEX-INDUCED VIBRATION

Den Hartog (1932) was probably first to call attention to the analogy between the Lanchester propeller and a special kind of vibration labeled as "galloping" and observed on cables. This vibration is characterized by the property that the reduced frequency (which is the ratio of oscillation frequency times characteristic length to parallel-flow velocity) is much smaller than unity, and hence a quasi-steady strip theory for computing the side force can be applied. In fact, this strip theory for blade elements is the same for the Lanchester propeller and galloping, including Glauert's condition in Equation (2). Den Hartog even suggested the Lanchester propeller as a simple test apparatus for studying various cable cross sections for their galloping behavior. The only difference between the Lanchester propeller and galloping is that the motion of the former is steady, whereas the latter is slowly oscillatory.

Analogies and similarities also exist between vortex-induced vibration for unrestricted values of the reduced frequency and rotation of plates normal to the flow:

1. Lock-in or synchronization near the natural (linear) resonance frequency.
2. Energy transfer from the fluid flow to the body.

3. Existence of sub- and superharmonic modes for forced vibration and rotation.

These properties of vibrating and rotating bodies in a parallel flow are typical of nonlinear coupled oscillators, which can be described, for instance, by the van der Pol equation for the side force of a vibrating body coupled with an equation for the body motion. A huge amount of literature on various types of these ordinary differential equations exists (Sarpkaya 1979), almost all of it dealing with translational vibration normal to the flow. A common description of rotational oscillation and rotation could be achieved by extending the restoring moment $c\alpha$ to $c\sin\alpha$ and thus obtaining pendulum-type equations. The author knows of no such studies.

The nonlinear-oscillator model is, of course, a mechanical analog for a qualitative description of a fluid-dynamical phenomenon. But it is not a model in the sense of an approximation to the aerodynamic field equations (which are partial differential equations). In that respect, the nonlinear-oscillator model will always be deficient, and the proper approach will be the solution of the field equations (Sarpkaya 1979). This statement is also true for the problems connected with autorotation.

ACKNOWLEDGMENTS

The author would like to thank Mr. H. Cheng from DTNSRDC for fruitful discussions and Mrs. A. Phillips for her contribution to the readability of this article. The preparation of the manuscript was supported by the Independent Research Program at DTNSRDC.

Literature Cited

Anderson, S. B. 1979. Historical overview of stall spin characteristics of general aviation aircraft. *J. Aircr.* 16:455–61

Anglin, E. L. 1978. Aerodynamic characteristics of fighter configurations during spin entries and developed spins. *J. Aircr.* 15:769–76

Bowman, J. S. 1971. Summary of spin technology as related to light general-aviation airplanes. *NASA-TN-D-6575.* 34 pp.

Brunk, J. E. 1967. Aerodynamics and flight mechanics of high performance autorotating glide-type bomblets. In *Proc. Conf. Dyn. Aerodyn. Bomblets*, Vol. 1. *Tech. Rep. AFATL-TR-67-195*, Eglin AFB, Fla.

Burgess, F. F., ed, 1967. *Proc. Conf. Dyn. Aerodyn. Bomblets*, Vol. 1. *Tech. Rep. AFATL-TR-67-195*, Eglin AFB, Fla.

Bustamante, A. G., Stone, G. W. 1969. The autorotation characteristics of various shapes for subsonic and hypersonic flows. *AIAA Pap. No. 69-132*

Cheng, H. 1966. *An experimental investigation of the autorotation of a flat plate.* MASc thesis. Univ. Br. Columbia. 103 pp.

Clare, T. A. 1971. Resonance instability for finned configurations having nonlinear aerodynamic properties. *J. Spacecr.* 8:278–83

Clarkson, M. H. 1958. Autorotation of fuselages. *Aeronaut. Eng. Rev.*, Feb., pp. 33–36

Clarkson, M. H., Malcolm, G. N., Chapman, G. T. 1978. A subsonic, high-angle-of-attack flow investigation at several Reynolds numbers. *AIAA J.* 16:53–60

Cohen, C. J., Clare, T. A., Stevens, F. L. 1974. Analysis of the nonlinear rolling motion of finned missiles. *AIAA J.* 12:303–9

Cohen, M. J., Nimri, D. 1976. Aerodynamics of slender rolling wings at incidence in separated flow. *AIAA J.* 14:886–93

Daniels, P. 1970. A study of the nonlinear rolling motion of a four-finned missile. *J. Spacecr. Rockets* 7:510–12

Den Hartog, J. P. 1932. Transmission line vibration due to sleet. *Trans. AIEE* 51:1074–86

Dupleich, P. 1941. Rotation in free fall of rectangular wings of elongated shape. *NACA Tech. Memo. 1201*

Ericsson, L. E. 1980. Kármán vortex shedding and the effect of body motion. *AIAA J.* 18:935–44

Ericsson, L. E., Reding, J. P. 1969. Unsteady airfoil stall. *NASA CR-66787*, Lockheed, Sunnyvale, Calif.

Fiechter, M. 1972. Kegelpendelung, Autorotation und Wirbelsysteme schlanker Flugkörper. *Z. Flugwiss.* 20:281–91

Flatau, A. 1964. An investigation of the rotational and aerodynamic characteristics of high aspect ratio rotors. *CRDL TM 1-4*, Edgewood Arsenals, Md.

Gebman, J. R. 1967. A description of the rotational characteristics of high aspect ratio modified rectangular rotors. In *Proc. Conf. Dyn. Aerodyn. Bomblets*, Vol. 1. *Tech. Rep. AFATL-TR-67-195*, Eglin AFB, Fla.

Glaser, J. C., Northup, L. L. 1971. Aerodynamic study of autorotating flat plates. *Rep. ISU-ERI-Ames 71037*. Eng. Res. Inst., Iowa State Univ., Ames

Glauert, H. 1919. The rotation of an aerofoil about a fixed axis. *Advisory Committee for Aeronautics, Rep. and Memo. No. 595*, March, pp. 443–47

Greene, J. E. 1960. An investigation of the rolling motion of cruciform-fin configurations. *NAVORD Rep. 6262*

Iversen, J. D. 1969. The Magnus rotor as an aerodynamic decelerator. *Proc. Aerodyn. Deceleration Syst. Conf.*, 2:385–95. Air Force Flight Test Cent., Edwards AFB, Calif.

Iversen, J. D. 1979. Autorotating flat-plate wings: the effect of the moment of inertia, geometry and Reynolds number. *J. Fluid Mech.* 92:327–48

Jones, B. M. 1963. The spin. In *Aerodynamic Theory*, ed. W. F. Durand, Vol. 5. New York: Dover. 347 pp.

Kry, P. R., List, R. 1974a. Aerodynamic torques on rotating oblate spheroids. *Phys. Fluids* 17:1087–92

Kry, P. R., List, R. 1974b. Angular motions of freely falling spheroidal hailstone models. *Phys. Fluids* 17:1093–1102

Küchemann, D. 1978. *The Aerodynamic Design of Aircraft*. Elmsford, NY: Pergamon

Langham, T. F. 1978. Correlation of experimental and theoretical steady-state spinning motion for a current fighter airplane using rotation-balance aerodynamic data. *AIAA Pap. 78-1373*, pp. 325–36

Lindemann, F. A., Glauert, H., Harris, R. G. 1918. The experimental and mathematical investigation of spinning. *Advisory Committee for Aeronautics, Rep. and Memo. No. 411*, March, pp. 716–29

List, R., Rentsch, U. W., Byram, A. C., Lozowski, E. P. 1973. On the aerodynamics of spheroidal hailstone models. *J. Atmos. Sci.* 30:653–61

Lugt, H. J. 1961. Self-sustained spinning of a cruciform fin system. *Proc. 5th US Navy Symp. Aeroballistics*. Vol. 1., Pap. No. 35, Nav. Ordnance Lab., White Oak, Md. 10 pp.

Lugt, H. J. 1978. Autorotation of plates. *David Taylor Nav. Ship Res. Dev. Cent. Rep. 78/058*. 104 pp.

Lugt, H. J. 1980. Autorotation of an elliptic cylinder about an axis perpendicular to the flow. *J. Fluid Mech.* 99:817–40

Maxwell, J. C. 1890. On a particular case of the descent of a heavy body in a resisting medium. *Scientific Papers*, p. 115. Cambridge Univ. Press

McCutchen, C. W. 1977. The spinning rotation of ash and tulip tree samaras. *Science* 197:691–92

Mehra, R. K., Carroll, J. V. 1979. Global stability and control analysis of aircraft at high angles-of-attack. *ADA-084938*. Sci. Syst. Inc., Cambridge, Mass.

Miller, M. C. 1973. A dynamic and aerodynamic analysis of an articulated autorotor decelerator system. *AIAA Pap. No. 73-463*. 15 pp.

Miller, M. C. 1979. Wind-tunnel measurements of the surface pressure distribution on a spinning Magnus rotor. *J. Aircr.* 16:815–22

Murphy, C. H. 1971. Response of an asymmetric missile to spin varying through resonance. *AIAA J.* 11:2197–2201

Neihouse, A. I., Klinar, W. J., Scher, S. H. 1960. Status of spin research for recent airplane designs. *NASA TR R-57*

Neumark, S. 1963. Rotating aerofoils and flaps. *J. R. Aeronaut. Soc.* 67:47–63

Nicolaides, J. D. 1957. On the rolling motion of missiles. *Bur. Ordnance Tech. Note No. 33*

Parkinson, G. V. 1964. On the performance of Lanchester's "Aerial Tourbillion." *J. R. Aeronaut. Soc.* 68:561–64

Parkinson, G. V. 1974. Mathematical models of flow-induced vibrations of bluff bodies. In *Flow-Induced Structural Vibrations*, ed. E. Naudascher, pp. 81–127. New York: Springer. 774 pp.

Phillips, W. H. 1948. Effect of steady rolling on longitudinal and directional stability. *NACA TN 1627*

Pinsker, W. J. G. 1957. Critical flight conditions and loads resulting from inertia cross-coupling and aerodynamic stability deficiencies. *NATO AGARD Rep. 107*

Relf, E. F., Lavender, T. 1918. The autorotation of stalled aerofoils and its relation to the spinning speed of aeroplanes. *Advisory Committee for Aeronautics, Rep. and Memo. No. 549*, Oct., pp. 448–52

Riabouchinsky, D. P. 1935. Thirty years of theoretical and experimental research in fluid mechanics. *J. R. Aeronaut. Soc.* 39:282–348, 377–444

Sachs, G., Fohrer, W. 1980. Einfluss der aerodynamischen Kopplung auf die Flugzeugdynamik bei schnellen Rollbewegungen. *Z. Flugwiss. Weltraumforsch.* 4:379–88

Sarpkaya, T. 1979. Vortex-induced oscillations. *J. Appl. Mech.* 46:241–58

Schmidt, L. V. 1979. Wing rock due to aerodynamic hysteresis. *J. Aircr.* 16:129–33

Skow, A. M., Titiriga, A. 1978. A survey of analytical and experimental techniques to predict aircraft dynamic characteristics at high angles of attack. *NATO AGARD Conf. Proc. No. 235*, Pap. 19. 37 pp.

Smith, A. M. O. 1953. On the motion of a tumbling body. *J. Aeronaut. Sci.* 20:73–84

Smith, E. H. 1971. Autorotating wings: an experimental investigation. *J. Fluid Mech.* 50:513–34

Soong, T. C. 1976. The dynamics of discus throw. *J. Appl Mech.* 98:531–36

Spangler, S. B., Mendenhall, M. R. 1977. Further studies of aerodynamic loads at spin entry. *Nielsen Eng. and Res. Rep. ONR-CR212-225-3*

Stilley, G. D. 1967. Unified stability criteria for autorotating glide bomblets. In *Proc. Conf. Dyn. Aerodyn. Bomblets*, Vol. 1. *Tech. Rep. AFATL-TR-67-195*, Eglin AFB, Fla.

Strickland, J. H., Webster, B. T., Nguyen, T. 1979. A vortex model of the Darrieus turbine: an analytical and experimental study. *J. Fluid Eng.* 101:500–5

Uselton, B. L., Jenke, L. M. 1977. Experimental missile pitch- and roll-damping characteristics at large angles of attack. *J. Spacecr.* 14:241–47

Willmarth, W. W., Hawk, N. E., Harvey, R. L. 1964. Steady and unsteady motions and wakes of freely falling disks. *Phys. Fluids* 7:197–208

Yelmgren, K. 1966. The autorotation of Magnus rotors. *Dept. Aerosp. Eng. Rep.*, Univ. Notre Dame, South Bend, Ind. 115 pp.

Zeeman, E. C. 1977. *Catastrophe Theory: Selected Papers 1972–1977*. Reading, Mass: Addison-Wesley. 675 pp.

Ann. Rev. Fluid Mech. 1983. 15:149-78

BREAKING WAVES ON BEACHES

D. H. Peregrine

School of Mathematics, University of Bristol, Bristol BS8 1TW, England

INTRODUCTION

Visitors to any coastline exposed to open water can see the dramatic transformation of surface waves that occurs as they advance onto a beach. The waves offshore have a relatively smooth water surface, whereas the waves arriving at the shoreline have rough white fronts of spray and bubbles. The transition between these two types of waves is the subject of this review; the term "wave breaking" is used here to describe the transition from a smooth wave to the quasi-steady state with a white-water front rather than to any particular instant within the transition.

The most prominent stage of wave breaking is the initial overturning motion of the wave crest that creates spray and white water, often by the forward projection of a jet of water. Much wave-breaking research has consisted of experiments to determine, for given offshore wave characteristics and beach slope, when and where waves first break and what type of initial breaking motion results. Galvin (1972) gives a review of such work.

The descriptive terms for breaker type use the initial motion to characterize them and are as follows (Galvin, 1968, 1972):

Spilling   White water appears at the wave crest and spills down the front face, sometimes preceded by the projection of a small jet.

Plunging   Most of the wave's front face overturns and a prominent jet falls near the base of the wave, causing a large splash.

149

Collapsing   The lower portion of the front face overturns and behaves like a truncated plunging breaker.

Surging   No significant disturbance of the smooth wave profile occurs except near the moving shoreline.

There often appears to be a smooth gradation between all these types of waves, which hints at the possibility of a one-parameter family of breaking "events," once allowance is made for the geometric scale of the wave. However, it is the beach that causes wave-breaking, and beach shapes vary widely, so that it is not difficult to find occasions when waves do not fit well into the above set of descriptions. Perhaps the most frequent such exception is the *shore break*, where the whole face of the wave from trough to crest becomes vertical with relatively little or no water in front of it. Very strong turbulent motions result; these waves seem to be among those that surfers call "sand-busters" or "dumpers."

Most experiments on breaking waves are influenced by the example of a single wave train specified in deep water by frequency ω, and amplitude a_0, and incident on a plane beach of slope α. Results of experiments of this type appear to depend on two parameters: the beach slope α and initial wave steepness, e.g. $a_0\omega^2/g$. Summary diagrams for wave height and water depth at breaking may be found in texts and manuals such as Wiegel (1964), Silvester (1974), Horikawa (1978), and Coastal Engineering Research Center (1977). Furthermore, some experimental results depend on a single parameter $a_0\omega^2/g\alpha^2$, discussed in the next section.

This review is oriented toward understanding the fluid dynamics of wave-breaking, rather than discussing the above type of experiment. The account of when and how waves break is entirely in terms of inviscid, initially irrotational flow, without concern for the air above the surface or surface tension. This idealization is proving sufficiently successful that other physical aspects of the problem are considered secondary and are discussed separately.

THE APPROACH TO WAVE-BREAKING

There are two major theoretical approaches to the problem of finding where waves break on a beach; both are only appropriate for beaches of gentle slope, and both originated well over 100 years ago.

Shallow-Water Steepening

Equations to describe finite-amplitude shallow-water waves are obtained by assuming that water-surface slopes are sufficiently gentle that water-particle accelerations are negligible compared with gravity. This assumption implies that the pressure at any point consists solely of the hydrostatic pressure due to the weight of water above that point. This theory was developed by Airy (1845) in response to Russell's (1834; see Miles 1980) observation of the existence of the solitary wave. Airy found that the equations can be put into a form showing that the front face of any wave of elevation propagating on water of uniform depth must steepen (see Lamb 1932, Sect. 187, or Lighthill 1978, Chap. 2). Airy, and many others since then (e.g. see Stoker 1957), took this steepening to imply that such shallow-water waves necessarily break; this despite the existence of the solitary wave, which is a wave of elevation that propagates unchanged on uniform water depth, i.e. without breaking.

Shallow-water steepening can be simply described in the case where a wave is propagating into uniform water conditions, e.g. still water of constant depth. Each portion of a wave with elevation ζ and horizontal water velocity u travels with the long-wave velocity corresponding to the total depth plus the water velocity, i.e. with velocity $[g(D + \zeta)]^{1/2} + u$, where D is the water depth and g the acceleration due to gravity. Thus the higher parts of a wave travel faster.

The result of wave steepening is that water accelerations increase to the point where they have a significant effect on the pressure, and this must be accounted for. The only analytically tractable case has been for waves of small amplitude; that is, a "near-linear" approximation is made and only the "first" nonlinear terms are included. The resulting equations are the Boussinesq equations (e.g. see Whitham 1974, Sect. 13.11). Among their solutions is the solitary wave in which the shallow-water steepening is exactly balanced by the effect of the water's acceleration, more commonly called a dispersive effect (see Miles 1980 for a survey of solitary waves).

The balance between shallow-water steepening and the effect of water acceleration is expressed in terms of the Ursell number (Ursell 1953)

$$\mathrm{Ur} = H/D\sigma^2, \tag{1}$$

where H is wave height and σ is a measure of maximum gradients, e.g. $D\partial/\partial x$. If, as is often convenient, σ is replaced by D/L, where L is the wavelength, the natural interpretation of large Ur corresponding only to shallow-water steepening is not correct, since long-wavelength waves frequently look like a train of solitary waves with long flat troughs between crests. However, small values of Ur reliably indicate that any shallow-water approximation is inappropriate (e.g. $HL^2/D^3 < 4\pi^2$).

When solutions of the Boussinesq equations are calculated for waves longer than solitary waves, shallow-water steepening occurs, but this is then countered by wave curvature and maximum elevation increasing so as to form a sequence of undulations without the occurrence of breaking. For example, a wave that is a smooth change in water depth between depth D and $D + \Delta D$ develops into a set of waves called an undular bore (see Peregrine 1966, or Fornberg & Whitham 1978, where equivalent solutions of the Korteweg-de Vries equation are given).

Despite the existence of nonbreaking solutions, a simple smooth "step-up" wave of height ΔD may break because of shallow-water steepening. If $\Delta D/D$ is greater than 0.7, then the wave continues to steepen and rapidly breaks, forming a turbulent bore (known as a hydraulic jump if there is sufficient current to hold it stationary). For the range $0.3 < \Delta D/D < 0.7$, undulations occur but the leading wave breaks (Binnie & Orkney 1955). Near the lower limit of this range, breaking may not occur for a relatively long time (Favre 1935).

From the above behavior of waves over a flat bed we can deduce when shallow-water steepening is the primary cause of breaking for waves on a beach. The waves must have propagated to a portion of the beach where (a) their length is much longer than the water depth, (b) their slopes are still gentle, and (c) their height is almost as great as the depth.

A periodic solution of the shallow-water equations on a plane beach is given by Carrier & Greenspan (1958). It can be matched away from the shore with the linear wave solution (Keller 1961), and for those cases where the beach has only gentle surface slopes it gives an accurate solution for the perfect reflection of incident periodic waves. Linear water-wave theory has solutions that correspond to perfect reflection for any beach slope, but the assumptions of linear theory do not hold at the shoreline unless the beach slope there is $O(1)$. The Carrier-Greenspan solution provides a local solution near the shoreline for gentle slopes.

A wave with near-total reflection satisfies the description of a surging wave. The Carrier-Greenspan solution gives a limit to such waves, since as the amplitude increases a vertical surface slope is predicted at the shoreline. Meyer & Taylor (1972) show that there is a solution corresponding to total reflection of waves if

$$a_0(\pi/2\alpha)^{1/2}\omega^2/g\alpha^2 \leqslant 1/2, \tag{2}$$

where the factor $(\pi/2\alpha)^{1/2}$ connects the amplitude offshore, a_0, with the amplitude of the Carrier-Greenspan solution. See Guza & Bowen (1976) for further details and experimental measurements agreeing with this result.

Parameters equivalent to $a\omega^2/g\alpha^2$, which appears in (2), have been found useful for correlating surf-zone properties as well as determining whether or not waves may break. For a discussion comparing several surf-zone properties, see Battjes (1974), where Iribarren & Nogales (1949) are credited with first using such a parameter to divide breaking and nonbreaking waves. As might be expected from the above, it is the properties of waves that are close to this nonbreaking condition that are well correlated by using the parameter in (2) (e.g. see Guza & Bowen 1976). Its relevance to the above discussion of equations for shallow-water waves is demonstrated by Munk & Wimbush (1969), who consider it as a ratio of $a\omega^2/\alpha$ (a measure of water acceleration up the beach slope) to αg (the component of gravity parallel to the beach). Battjes gives further physical interpretations.

Refraction and Waves of Limiting Steepness

For calculating the refraction of waves approaching a beach, the two most common assumptions are that (a) there is no reflection of the waves, and (b) the beach slope is so gentle that the waves are like plane periodic waves on water of constant depth. In the simplest case of steady waves normally incident on a beach, the amplitude variation is obtained from the constancy of wave-action flux, which is equivalent to constant energy flow in the absence of currents. The first example of this method is for linear long waves (Green 1838) and in that case it leads to Green's Law, that wave amplitude is proportional to $D^{-1/4}$.

The position of wave breaking is estimated by introducing a limit to wave steepness. The existence of a limiting wave steepness for traveling waves has been known since Stokes (1880) studied the flow near the crest of such a wave, and the limiting waves may now be calculated with considerable accuracy for any depth (Williams 1981).

In practice, coastal engineers use formulas such as

$$(Hk)_{\max} = 0.89 \tanh kD, \tag{3}$$

due to Miche (1944), where $H = 2a$ is the wave height, $k = 2\pi/L$ is the wave number, and L is the wave length. Van Dorn (1978) shows that Equation (3) is a reasonable fit to experimental data from beaches of gentle slope. Since small values of kD are most frequently encountered at breaking, both in experiments and in nature, a limiting ratio, D/H, is more often used. This ratio is called a "breaker index" and is given values in the range 1.1–1.3.

Limiting-steepness waves have provided a starting point for theoretical studies of wave breaking. Longuet-Higgins (1980a) includes an account of the part they have played in the study of breaking waves. However, development of accurate solutions for periodic waves of lesser steepness (Schwartz 1974, Longuet-Higgins 1975, Cokelet 1977b, Longuet-Higgins & Fox 1978, reviewed in Schwartz & Fenton 1982) has shown that wave phase velocity and most other integral properties of waves such as energy flow have their maximum, for given mean depth and wave number, for waves of less than maximum steepness. When accurate wave solutions are used in a refraction calculation, this results in (a) two possible steep-wave solutions for a restricted range of water depths, and (b) in solutions ceasing to exist for shallower water (e.g. Sakai & Battjes 1980, Stiassnie & Peregrine 1980, and Ryrie & Peregrine 1982). [There is a second type of double-valued solution in Ryrie & Peregrine (1982), which Peregrine (submitted for publication) discusses in more detail, showing the second solution is rarely relevant.] The first two of these papers show reasonable agreement for wave height and velocity with the detailed measurements of Hansen & Svendsen (1979), except within one wave length of breaking.

In experiments, waves on beaches do not retain the symmetry about their crests that a periodic wave train on uniform depth has. Hansen & Svendsen (1979) include measures of asymmetry and profile measurements that illustrate this point. The same authors (Svendsen & Hansen 1978) give a perturbation analysis of cnoidal waves on a sloping bottom. Their comparisons with experiment look satisfactory in the range $25 \leqslant$ Ur $\leqslant 300$, and they find a maximum asymmetry for Ur ~ 50. Here we are using Ur $= HL^2/D^3$.

Any solution depending on local plane-wave solutions, or perturbations to them, implies that wave properties can adjust to changes in depth. If the time scale of such an adjustment is long, then there must be a correspondingly slow variation of depth. An adjustment involving adjacent waves can occur more readily in deep water, where any disturbance is communicated through the half-space of fluid, than in shallow water, where disturbances propagate along a strip of fluid with velocities that cannot be much greater than $(gD)^{1/2}$. An indication of this difference is the way the ratio of group velocity to phase velocity approaches unity as $D/L \to 0$. In long waves, wave crests behave like independent entities. Thus periodic waves can be accurately represented by a train of solitary waves [e.g. see Stiassnie & Peregrine 1980, Witting (unpublished) 1981, and Williams 1981].

The "rate of adjustment" of a train of solitary waves to disturbances can be estimated from the interaction between a nearly equal pair of solitary waves for which there is an exact solution of the Korteweg-de Vries equation (Whitham 1974, Equation 17.21). (The Korteweg-de Vries equation is an appropriate approximation to the Boussinesq equations.) By using such an estimate, Stiassnie & Peregrine (1980) show that in this regime waves have time to interact with their neighbors only if

$$\alpha < 2(3)^{1/2}(H/D)^{3/2}\exp\left[-\tfrac{1}{2}(3\mathrm{Ur})^{1/2}\right]. \tag{4}$$

This is a very severe requirement on α, since Ur should be greater than $O(50)$ for the train-of-solitary-waves approximation and if Ur = 50, the exponential in expression (4) equals 0.002.

This suggests that for practical beach slopes a slowly varying periodic-wave solution is inappropriate once Ur > 50, and consideration of each wave crest as an independent entity may be better. An appropriate solution to examine is that for a solitary wave. It has often been remarked that waves on beaches resemble solitary waves (e.g. Munk 1949). Departures from a true solitary-wave profile as it propagates over varying depth have been analyzed for (a) wave reflection by Peregrine (1967) and (b) direct perturbations by Kaup & Newell (1978). Miles (1979) discusses the changes that occur, Grimshaw (1979) provides a theoretical framework for a more general study, and Ippen & Kulin (1955) and Street & Camfield (1966) report experimental results. Chan & Street (1970) give a numerical solution. To date no work allows for seaward flow between each crest.

All the perturbations of a solitary wave as it moves over differing depths are at the back of the wave. These can grow and give rise to further wave crests (Madsen & Mei 1969), and very long waves on gentle beaches show this phenomenon (Gallagher 1972). Perhaps it is more relevant that in the regime described by the Boussinesq equations the forward face of any wave of elevation tends to become like a solitary wave, unless it is interacting with other significant waves.

Freilich (1982) has made a comparison between calculations with the Boussinesq equations and observation of ocean waves arriving at a beach in the region before they break. A very satisfactory agreement was obtained between measured spectra at depths of 10 m and 3 m when the 10 m spectrum was used as input to a spectral representation of Boussinesq's equations to calculate that at 3 m.

Instabilities

When approximate methods, such as refraction methods or integration of the Boussinesq equations, are used for waves on beaches they fail or are unreliable as waves approach the steepness that limits periodic waves, or the maximum solitary-wave height. This is usually interpreted to imply that breaking soon follows, as is observed to be the case (except where the possibility of multiple-crest formation must also be considered). If waves at this point are close to symmetrical, the study of wave-train instabilities is relevant.

For symmetrical periodic traveling waves, Longuet-Higgins (1978b) has found that deep-water waves become unstable with a rapidly growing instability once their steepness, ak, exceeds 0.406. This should be compared with a maximum steepness of 0.443 and the steepness of maximum phase velocity, which is 0.436. The perturbation giving this instability involves a reduction in amplitude of alternate crests and an increase in amplitude of the remaining crests. Longuet-Higgins finds that the perturbation has no contribution from the first harmonic in its Fourier description and is stationary relative to the wave; hence he associates the instability with the maximum of the first harmonic of the basic periodic solution, which Schwartz (1974) shows to be at $ak = 0.412$.

However, Longuet-Higgins (1978a) shows that there is almost certainly an instability associated with the maximum of a wave's phase velocity. Further support for this instability comes from Cleaver's (1981) stability analysis of Longuet-Higgins & Fox's (1977) local solution for flow at and near the crest of steep waves. Cleaver finds an instability that becomes more stable if matching conditions from less-than-maximum-steepness waves are included.

Numerical computations (Longuet-Higgins & Cokelet 1978) clearly show that the first-mentioned instability rapidly leads to wave breaking. It is possible that the energy transfer between wave crests is simply sufficient to increase the local wave steepness to a point where the second instability, associated with maximum phase velocity, can grow to breaking. There is a corresponding slackening followed by a final increase in the growth rate shown in Figure 16 of Longuet-Higgins & Cokelet (1978), together with a localization of the perturbation, as shown in their Figure 14. Thus it appears that all waves close in form to the steepest possible wave suffer from one or more rapidly growing instabilities. It is appropriate to refer to them as "Longuet-Higgins instabilities."

There are other instabilities of wave trains. The modulational instability of Benjamin & Feir (1967) is investigated by Longuet-Higgins (1978b) and Cleaver (1981) and does eventually lead to wave breaking (Benjamin 1967, Longuet-Higgins & Cokelet 1978). However, (a) it is likely that the final breaking process can be ascribed to a Longuet-Higgins instability, (b) the growth rate of the instability is relatively small, and (c) it diminishes as the water depth is decreased, so it is unlikely to be of much importance on beaches.

An analysis by McLean et al. (1981) of three-dimensional perturbations extends Longuet-Higgins (1978a,b). The Benjamin-Feir instability is found to extend to waves of steepness $ak = 0.39$ before disappearing. The Longuet-Higgins instability that involves alternate wave crests is found to exist, with small growth rates, at all wave steepnesses and to have a maximum growth rate for oblique perturbations. Su et al. (1982a) report experiments in which steep two-dimensional deep-water waves develop a three-dimensional breaking pattern that is qualitatively similar to what one would expect for this instability. The work of McLean et al. (1981) does not extend to sufficiently steep waves to examine the phase-velocity-maximum instability. For values of $kD = O(1)$, the Benjamin-Feir instability diminishes in importance (Cleaver 1981), but experiments (Su et al. 1982b) and theory (McLean 1982) show that the alternate-crests instability becomes more important.

For shallow-water depths, the modulational and alternate-crests instabilities are likely to be unimportant because they involve transfer of energy between each crest and its neighbors. Natural breaking waves are frequently observed to be uniform along their crests for considerable distances, so it is also unlikely that three-dimensional instabilities are important in this context. This leaves the constant-phase-velocity instability as the one most likely to be relevant to wave breaking, though even that may often be of little or no significance, since waves on a beach are being subjected to a finite disturbance.

WAVE OVERTURNING

In a majority of wave-breaking events on beaches, an element of the water surface becomes vertical; a portion of the surface then overturns, projects forward, and forms a jet of water. Such overturning may be small or large compared with the wave and is well developed in plunging breakers. Observation shows it occurs both when waves are nearly symmetrical (perhaps liable to instability) and quite asymmetrical (usually due to shallow-water steepening).

Overturning looks very similar whatever its scale, suggesting that there may be some similarity solution that gives a local description of the overturning motion. The demonstration by Cleaver (1981) that the local flow at the crest of a steep progressing wave suffers from an instability supports this notion. It is possible that the full nonlinear development of the most unstable linear perturbation gives such a similarity solution. However, the disturbances that provoke breaking are not infinitesimal, nor are the waves initially symmetrical. Numerical solutions for overturning waves give only a little support to this idea.

Experimental measurements of wave profiles and of velocity distributions as waves begin to break are numerous. Recent publications are by Van Dorn (1978), Hansen & Svendsen (1979), Kjeldsen & Myrhaug (1980), Flick et al. (1981), and Hedges & Kirkgöz (1981). Reference to these papers and the surveys by Galvin (1972) and Cokelet (1977a) will reveal most earlier work. Field measurements are less comprehensive. Papers to refer to are by Iwagaki et al. (1974), Thornton et al. (1976), Suhayda & Pettigrew (1977), Weishar & Byrne (1978) and Hotta &

Mizuguchi (1980). However, since the flow is unsteady, measures of acceleration or pressure are desirable; these have become available from numerical solutions for wave overturning.

The first demonstration of a numerical solution for wave overturning is by Longuet-Higgins & Cokelet (1976) and has been followed by others—Longuet-Higgins & Cokelet (1978), Cokelet (1978), Peregrine et al. (1980), Vinje & Brevik (1980), McIver & Peregrine (1981), and Srokosz (1981). The comments here are based on these papers and further results from P. McIver and A. New of Bristol University. Except for the work of Vinje & Brevik (1981) and A. New, all results are for deep-water waves; however, the finite-depth computations appear to be similar. A variety of disturbances are used to provoke breaking, yet superficially the overturning motions resemble each other and natural waves unless there is an appreciable standing-wave component. Velocity and acceleration fields, however, show more variation. Figure 1 illustrates a wave breaking in finite water depth.

A detailed study of a few overturning-wave solutions (Peregrine et al. 1980) reveals three features of the overturning motion, all of which are apparent *before* the face of the wave has a vertical tangent (see Figure 2).

(i) Water-particle velocities exceed the wave velocity. This property has often been quoted as a criterion for wave breaking. In fact the "wave velocity" is not well determined, since its shape is unsteady and each point such as the highest point or point of maximum slope has a different, time-varying velocity. Computations indicate that it is realistic to expect velocities up to twice the phase velocity of a linear wave.

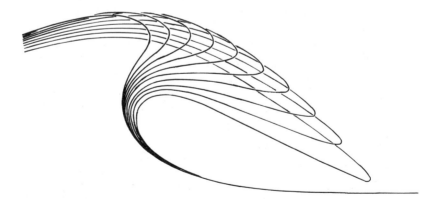

Figure 1 A sequence of computed wave profiles for a periodic wave after an "impulsive depth perturbation" from 0.20 to 0.07, where the wavelength is unity. (Figure supplied by A. New, Bristol University.) Profiles are plotted with a uniform velocity imposed to match the trough profiles.

(ii) In a thin region on the front of the wave, water accelerations exceed the acceleration of gravity. The existence of such a region was unexpected; with hindsight, it is clear that appreciable accelerations are necessary to accelerate the water near the surface that is projected forward in a jet. Computed solutions show accelerations greater than $5g$ in the subsequent development of the overturning, and it seems likely that such accelerations occur in natural waves.

(iii) An extensive, poorly defined region on and beneath the back slope of the wave has low water accelerations. This region ensures that the high pressure gradients necessary to produce the acceleration in region (ii) can exist. Hydrostatic pressure and wave asymmetry suffice to provide this "support" to region (ii). Such support can have a more precise physical form in other circumstances. For example, if a rigid vertical plate is moved horizontally "sweeping up" a layer of water, that water rises up the plate and is then projected forward. The plate will support whatever pressure is necessary to accelerate the water.

Study of the dynamics of overturning reveals nothing special about the instant when the wave first has a vertical tangent, or the emergence of the overhanging jet. The following comments provide a partial explanation. The dynamic boundary condition at the free surface is physically "weak": the pressure is constant, it is not driving the flow, all acceleration is due to the water's inertia and to gravity. The kinematic boundary condition is that the surface is a material surface moving with the water. In almost every velocity field, material surfaces become strongly distorted and folded. Thus, for motion with a free surface, one can expect the surface

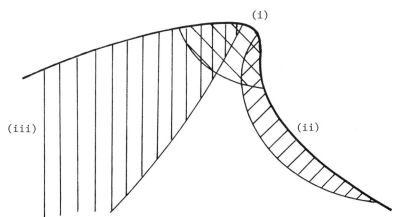

Figure 2 Sketch of three dynamically significant regions in a wave approaching breaking. (i) Particle velocity greater than phase velocity of steepest wave. (ii) Water accelerations greater than gravity. (iii) Water accelerations less than one-third gravity (after Peregrine et al. 1980).

to develop thin sheets in any region in which the water motion is so energetic that gravity is of little importance. The argument above indicates that overturning should not be interpreted as a singularity of the surface or the flow in the water. This argument is insufficient in itself to explain wave breaking. It is also relevant to the later spray-forming stages of the breaking process.

The velocity field in the water can be represented by flow singularities outside the region occupied by water. This proves to be surprisingly effective (McIver & Peregrine 1981, 1982), and only a few singularities are required. This approach was investigated when the author discovered that in certain instances most of the velocity field of the near-vertical front face of the wave could be described by a single line sink.

There are some analytic solutions that contribute to understanding aspects of an overturning wave. Longuet-Higgins (1980b) gives a solution having the shape of a rotating hyperbola with asymptotes that enclose a steadily reducing angle. McIver & Peregrine (1981) verify with numerical computation that it is a suitable model for flow near the tip of a projected jet, but in order to obtain sufficient resolution near the sharply curved tip, an atypical example with zero-gravity was examined. Longuet-Higgins (1981a) investigates the simplest combination of Stokes's (1880) solution for the crest of the steepest wave and a branch point of the velocity potential, which gives some realistic-looking profiles for the surface as it becomes vertical.

The high water accelerations on the face of the wave reveal the importance of that area. A. L. New (conference lecture, 1981) fitted ellipses to large portions of the face in some computed solutions and photographs of waves. He found that there was a good fit and the ratio of the axes of the ellipses was always close to $\sqrt{3}$. An analytic solution that has fluid round the exterior of an ellipse has been found (New, personal communication). New's results in turn stimulated Longuet-Higgins (1981b, 1982) to find further analytic solutions that satisfy the boundary conditions, and one of these, P_3, fits the underside of breaking waves remarkably well. It is a self-similar solution; in suitably oriented Cartesian coordinates the surface is given by

$$(x, y) = At^4\left(-3\mu^2 + \tfrac{1}{3}, -\mu^3 + 2\mu\right), \tag{5}$$

where A is a constant (see Figure 3). Comparisons between analytic solutions and computational results are sufficiently good (McIver & New, personal communication) to encourage further investigation.

If satisfactory analytic solutions are found, they should help to provide a measure of the "strength" of an overturning event. A quantitative measure is desirable since the established descriptive approach is entirely qualitative, and there are significant differences other than those of geometric scale relative to the rest of the wave. The emerging jet can have quite different velocities relative to the wave (e.g. see Srokosz 1981). For example, most of the face of a wave may become vertical but the jet can be very small with a low relative velocity, whereas in another case the jet can have sufficient relative velocity to plunge far ahead of the wave crest.

Another property that varies is the direction of the jet, e.g. if there is a strong reflected wave the jet can be projected almost vertically. In a field study, Weishar & Byrne (1978) found that the horizontal distance the plunging jet traveled and the time it took to fall were both greater than the simple free-fall trajectory that Galvin (1969) obtained by assuming horizontal projection at the wave-crest velocity. Their suggestion that this may in part be due to a vertical component of projection is consistent with computed solutions that show variations in the direction of projection of the jet.

The duration of the overturning flow is an aspect of the motion about which very little is known. For example, can it be defined? It is important to define some such quantity in order to be able to quantify energy and momentum loss from the wave motion.

There is likely to be some interdependence among these properties, especially for waves on beaches, where breaking most commonly occurs for the single reason that waves propagate into diminishing water depths.

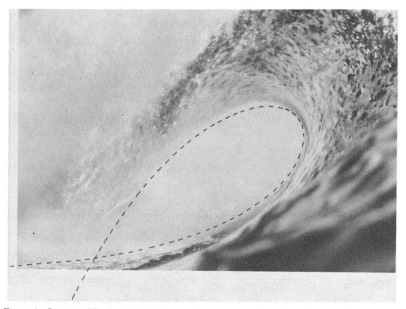

Figure 3 Longuet-Higgins (1981b) P_3 solution superposed on a breaking wave (photograph courtesy of *Surfing Magazine*).

Overturning does not necessarily occur in spilling breakers. There is reasonable doubt about the initiation of spilling breaking. In many cases it commences with a relatively small overturning. Miller (1976) illustrates this with a photograph in which overturning on the scale of 5 mm is occurring. However, in other cases white water seems to appear without surface slopes becoming near-vertical. A possible mechanism in these cases is Rayleigh-Taylor instability. This occurs if a sufficiently large portion of the surface has an acceleration component directed into the fluid greater than the component of gravity in that direction. The development of Rayleigh-Taylor instability usually leads to a highly convoluted surface that is also characteristic of a breaking wave.

There is, however, no clear evidence for the occurrence of Rayleigh-Taylor instability in water-wave motion. As mentioned above, white water appears to arise without overturning. Note that photography of wave profiles through the transparent side of a flume is an unreliable guide in this context because the wall boundary layer often "breaks" in a different manner from the main part of the waves, and bubbles in the boundary layer can obscure a small jet in mid flume (e.g. see comments in Ippen & Kulin 1955).

Steadily propagating wave solutions have no accelerations greater than $\frac{1}{2}g$, though the maximum standing wave has an acceleration of g at its crest. In computations, marginally destabilizing accelerations have been found in circumstances with an appreciable standing-wave component. The large scale of the numerical discretization relative to the size of these regions, and the presence of numerical instabilities, allow no definite conclusions to be drawn (McIver & Peregrine 1982).

Another possible "cause" of wave breaking is the steepening and breaking of short waves as they are overtaken by the crest of a longer wave. In a wind-driven sea, this occurs frequently and it is also seen on beaches. There are many factors involved in the interaction of short waves with long waves. Those interested may consult Garrett & Smith (1976) and Peregrine (1976, Sect. IIF). Dagan (1975) analyzes the case of linearized short waves on a steep free surface, which is equivalent to an instability analysis of spatial growth. For waves on beaches the simultaneous breaking of both long and short waves can occur, and it is unlikely that short waves are more than incidental to the breaking of larger waves.

THE EVOLUTION OF A PLUNGING BREAKER

Once the jet from an overturning wave hits the water, at the plunge point, water splashes up, sometimes to a height greater than the original wave. From the plunge point onward the breaking process appears to degenerate rapidly into a chaotic motion of air and water. However, close and careful inspection often reveals a surprising amount of order within the wave. A misleading impression of the flow is easily obtained, since in any one direction a single drop or bubble is sufficient to interrupt a line of sight.

The subsequent evolution of the wave depends on the position of the *plunge point*, i.e. the place and instant of time where the falling jet touches the undisturbed surface. If it is near the crest of the wave the resulting splash may be directed down the wave and it becomes a spilling breaker. At the other extreme, which is most likely on a steep beach, the jet may travel beyond the base of the wave; if it lands in water of negligible depth the jet is simply redirected up the beach and constitutes the major part of the run-up due to the wave. Such an event can lead to an excursion of the shoreline that greatly exceeds that due to a more normal wave impeded by backwash from its predecessor. The more usual intermediate case is now discussed.

The plunging jet closes over the air beneath it to form a tube around which there is considerable circulation. Air pressure usually prevents the rapid collapse of this tube, and even without air pressure the circulation around the tube implies that there is a minimum radius at which centrifugal acceleration balances an inward pressure gradient. The non-circular initial state and three-dimensional instabilities both contribute to this tube having a relatively short life. Sometimes the trapped air vents through the surface with a sudden spout of spray. Sawaragi & Iwata (1974) indicate that the tube or vortex descended toward the bed in their experiments, which included measurements of several relevant quantities.

The splash-up commences from the plunge point. At the present time only a few visual and photographic observations of the splash-up have been made and its mechanism and the origin of the water in the splash-up are not clear. The water must come partly from the jet and partly from previously undisturbed water; the division might be even or tend toward one extreme.

One view is that the jet "rebounds" (Figure 4a). At the other extreme, it can be considered to penetrate the surface and then, because of its forward motion and downward momentum, it acts like a solid surface and "pushes up" a jet of previously undisturbed fluid. This is illustrated by a sketch in Figure 4b and by a photograph in Figure 5. The photograph shows a wedge of water free from bubbles that looks as if it has been pushed upward. An intermediate possibility is sketched in Figure 4c.

Peregrine (1981) describes an initial attempt to analyze a simplified model of the splash-up. A thin uniform jet falls onto a thin layer of water resting on a rigid surface. A one-dimensional model is used but is inadequate because it does not consider the vorticity at the interface between the two fluid regions.

Figure 6 is a photograph of a large jet, caused by overturning, falling onto a thin fast-flowing backwash; a splash is emerging above a thin layer of air. Peregrine's (1981) analysis indicates that the relative velocity between the two bodies of water is too great for any simple splash solution to be relevant. The dynamics of this example are obscure.

The type of questions a model of the splash-up might answer could also be examined experimentally. For example, is it possible for the

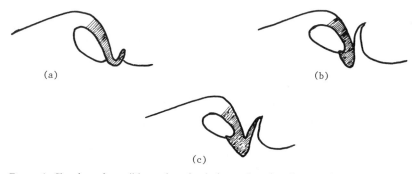

Figure 4 Sketches of possible modes of splash-up after the plunge point. Water in and from the falling jet is shaded.

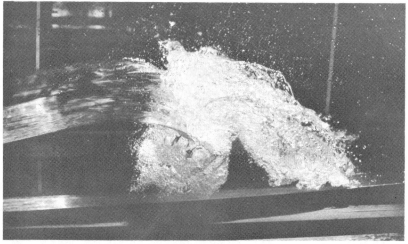

Figure 5 Wedge of clear water apparently pushed up by the plunging jet of a wave breaking from the left on a 1 : 35 beach. Photograph taken at ISVA, Technical University of Denmark. (The upper sloping black strip is a support for the flume.)

splash-up to fall back onto the initiating jet? Can part of the jet "return" under the tube of trapped air and augment (or disrupt) the circulation around it?

Although the range of possible motions is wide, on gently sloping beaches a plunging jet usually causes a splash to be projected forward over another tube of air to hit undisturbed water at a second plunge point, with the cycle starting again with another splash-up. Galvin (1969) calls this second plunge point the splash touchdown point and gives measurements of the distance between the first two plunge points. Miller (1976) draws attention to these cycles of plunge and splash-up that entrap tubes of air and give rise to strong vortex-like motions in each cycle.

Several splash-up cycles can occur before the turbulence, which is usually evident from the plunge point onward, destroys the organized nature of the motion and a bore results. The two photographs in Figure 7 illustrate three cycles and how the motion in the flume is sufficiently deterministic that it may be reproduced on different occasions. These cycles also occur in natural waves, as is shown in Figure 8, where the plunge point for each cycle is visible as a cleft in the surface.

The vortex-like motions from each cycle all have the same direction of rotation, and hence high rates of shear exist between them. This and the high shears arising at each plunge imply that the turbulent intensity is

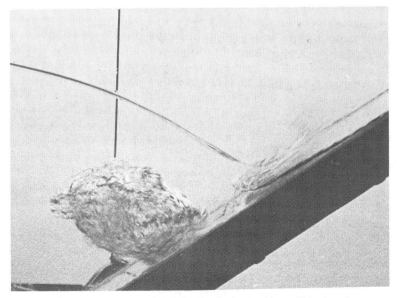

Figure 6   Splash of a large plunging jet into a thin rapid backwash.

very high. This is consistent with the rapid decrease in height and loss of energy from the wave motion in this region. Svendsen et al. (1978) call this the "outer region" of the surf zone and plot results from experiments on gently sloping beaches that indicate that many waves lose 50% of their breaking height while traveling a distance less than 10 times the water depth at breaking. Sawaragi & Iwata (1974) estimate the energy loss from the wave into the first vortex-like motion to be about 15–30% of the dissipation that occurs there.

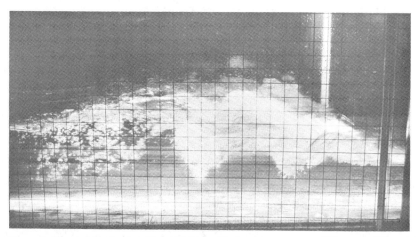

Figure 7 These two photographs are taken with the same wave conditions on different occasions and illustrate the order in the cycles of "plunge-splash-up" that occur after breaking. The exposure in the top photo is much longer than in the bottom. Bottom photograph courtesy of I. A. Svendsen. Both taken at ISVA, Technical University of Denmark.

Figure 8 A natural breaking wave illustrating the "clefts" that mark successive plunge points. Photograph taken at the Scripps Institution of Oceanography pier.

QUASI-STEADY BREAKING WAVES

Most breaking waves settle into a quasi-steady state after the plunge point and any ensuing plunge-splash-up cycles. Exceptions occur where the plunging jet or its succeeding splash penetrate shallow water and are deflected by the bed to become part of the run-up, or where irregularities of the beach cause breaking to be an intermittent process.

The quasi-steady state is one in which the wave form changes relatively slowly and has a strongly turbulent region on the face of the wave. If the turbulence is confined to a region near the crest of the wave, the wave is a "spilling breaker." On the other hand, if the whole face of the wave is turbulent it is a "bore" (or "turbulent bore" if considered in circumstances where undular bores may be occurring). There is, of course, a whole range of intermediate waves.

The mean flow in quasi-steady breaking waves includes a recirculating region, or "roller," since water can be seen "tumbling" down the front of the wave. However, the turbulent velocities in the roller exceed the mean velocities relative to the wave in the roller and so attention should be focused on the turbulence (Peregrine & Svendsen 1978). Right from the toe of the roller, turbulence can be seen to spread away from the surface. It is generated by the velocity difference between the undisturbed water and that tumbling down the wave's face. In this respect the mechanism for generation of turbulence is similar to the one occurring in the turbulent mixing layer that arises when parallel streams of differing velocities are allowed to meet. However, it is likely that active generation of turbulence is confined to those portions of the front of the wave with a significant slope of the mean surface.

A moderately detailed model of the mean flow in a steady bore has been devised (Madsen 1981, Madsen & Svendsen 1982) that gives a good account of many features, such as the roller and the bore's surface profile, for the steady case of a transition between two constant levels. Extension to the more general cases found on beaches would involve modeling the balance between (a) gravity causing turbulent water to fall forward, and (b) the wave's velocity relative to water in front, which tends to sweep the turbulent water over the wave's crest.

The classical bore model of a simple transition between uniform levels is helpful in studying the surf zone. However, the bores are often sufficiently weak, or the level behind them drops so quickly, that secondary undulations grow. On the other hand, Svendsen et al. (1978) find that their estimates of the rate of energy dissipation in bores on a beach are greater than the classical value. These properties, the dynamics of spilling breakers, and the unsteadiness of these waves are closely related and require more study.

Behind the region of active turbulence generation, the turbulence continues to spread, as is demonstrated by Banner & Phillips (1974) and Peregrine & Svendsen (1978) (see Figure 9). Detailed velocity measurements in this region by Battjes & Sakai (1981) and Duncan (1981) for a wave behind a two-dimensional hydrofoil and measurements by Stive (1980) of an ensemble of breaking waves in a laboratory beach confirm the views expressed by Banner & Phillips and Peregrine & Svendsen that the turbulence is similar to that in a two-dimensional turbulent wake. In Figure 9, the spread of turbulence is made visible by minute bubbles (rise velocity around 1 mm s^{-1}) that were originally floating on the water surface.

In considering the experiments with hydrofoils, it is tempting to extend the analogy with a wake and to use the momentum deficit associated with the wake as a parameter for breaker strength. However, if this is done a paradox appears once a turbulent bore is considered. The successful classical approach to finding the changes in height and velocity across a bore involves an assumption of conservation of momentum, and hence no momentum deficit.

This apparent paradox may be resolved by considering the dimensionless energy-momentum flux diagram introduced by Benjamin & Lighthill

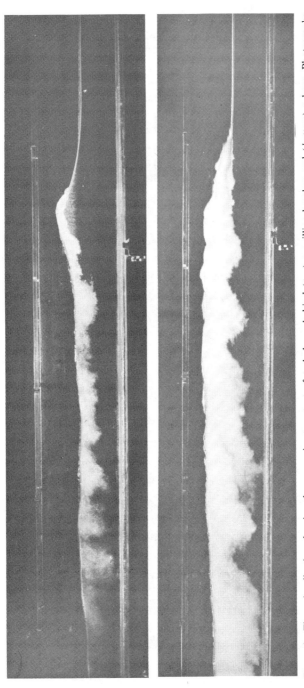

Figure 9 Flow visualization showing the approximate extent of turbulence behind (*top*) a spilling breaker and (*bottom*) a bore. Photographs courtesy of I. A. Svendsen, ISVA, Technical University of Denmark.

(1954) (see Figure 10). In the figure, the hydrofoil meets water with a subcritical velocity corresponding to a flow at A. The wave drag on the hydrofoil gives an equal and opposite force on the flow, reducing the momentum flux (or flow-force) and creating a flow at point B. If there is no flow separation, viscous forces and dissipation are negligible so that the flow's energy is unaltered. However, point B is outside the region of steady wave solutions, breaking occurs without further loss of momentum from the flow, and the resulting wave train corresponds to point C. The energy loss BC is the measure of the breaking. Any "momentum deficit" must be examined in this context; there is a change in mean depth and to analyze an experiment it is necessary to measure such changes as well as the wave train.

A bore meets water at a supercritical velocity, i.e. a flow relative to the bore corresponding to point D on the supercritical flow curve. The turbulence in the bore reduces the energy to point E on the subcritical flow curve, or else to an intermediate point F where a wave train forms behind the initial breaker. The effect of breaking in both these examples

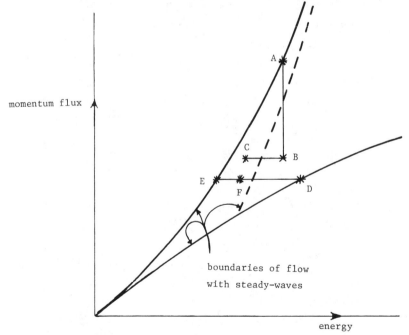

Figure 10 Energy-momentum flux diagram, after Benjamin & Lighthill (1954). The diagram is distorted for clarity. The two heavy lines represent steady uniform flows, the upper being subcritical, the lower supercritical. The broken line represents flows with stationary waves of maximum steepness.

is a loss of energy. Transfers of momentum between waves and mean flow are not considered since S is the total momentum flux.

A hydrofoil suffers no wave resistance in a two-dimensional supercritical flow since the flow force cannot be reduced; an exception occurs if it has sufficient drag to generate an upstream propagating bore that transforms the flow it meets into a subcritical one.

For breaking waves on beaches there is no reference frame in which the waves are steady, but the overall effect of wave breaking can be deduced in a similar manner. For example, in an idealized case on a horizontal bed consider a uniform wave train that suffers wave breaking (say, due to an instability) but then reforms into another uniform wave train. The conservation of mass and momentum then give, after averaging over a wave period,

$$U_1 D_1 + I_1 = U_2 D_2 + I_2 \tag{6}$$

and

$$(D_1 U_1 + I_1)(U_1 + I_1/D_1) + \tfrac{1}{2}gD_1^2 + S_{xx1} = (D_2 U_2 + I_2)(U_2 + I_2/D_2)$$

$$+ \tfrac{1}{2}gD_2^2 + S_{xx2}, \tag{7}$$

where U is the uniform current, D the depth, I the mass flow associated with the waves, S_{xx} the wave-momentum flux, density is taken equal to unity, and subscripts refer to conditions on each side of the breaking region. If the wave motion is defined in such a way that $I_1 = I_2 = 0$ [corresponding to Stokes's (1847) second definition of phase velocity], Equations (6) and (7) simplify and it can be readily shown that a loss of momentum from the wave motion leads to an increase of water depth if U is zero or if the flow is subcritical. Subcritical flow conditions are normal on a beach and the resulting depth increase, "wave set-up," is well known. Supercritical flows can occur, for example, on the shoreward side of shallow sand bars, and in such cases there is a decrease of depth due to wave breaking.

The above interpretation is only appropriate in shallow water. In deeper water the momentum loss from the waves is better interpreted as a surface shear stress acting on the mean flow. Duncan (1981) considers the detailed dynamics of the roller region to be best described as giving a surface shear stress, and he makes a first approximation for distinguishing between wave and mean motion. However, as McIntyre (1981) states, caution is necessary in ascribing momentum to the wave motion.

This consideration of mean flow changes and apparent surface stresses brings us to the topic of surf-zone dynamics, which is outside the scope of this review. However, the above discussion shows that a full under-standing of wave breaking should include properties such as the rate-of-momentum transfer from the wave motion to the mean flow.

OTHER PHYSICAL EFFECTS

Air

The direct dynamic effects of the air on breaking are probably slight unless there is a large, say $O(20 \text{ m s}^{-1})$, relative velocity between the air and the water, in which case there can be considerable spray formation before the plunge point. For steady waves, Dore (1978) shows that surface boundary-layer effects are sometimes significant. The indirect effects of air, as in bubbles (see below) and in cushioning the collapse of the tube beneath a plunging jet, are more important.

Surface Tension

The clearest effect of surface tension is on steep waves less than about 10 cm high. The development of a plunging jet and the entrainment of air are both inhibited. Such waves, right down to an amplitude of around 2 mm, can still develop strongly turbulent regions, so that it is sensible to retain the description "breaking" even though the water surface remains continuous. Banner & Phillips (1974) note how important this type of breaking wave can be for sustaining wind stress on the sea.

Drops and Bubbles

The combination of air and surface tension in drops and bubbles fashions much of what is visible of a breaking wave. Their dynamic effects on waves are mainly such as to increase the rate at which the motion becomes more disorganized and turbulent. The buoyancy of bubbles is irrelevant in the main breaking process, but has some in-fluence on the nature of the decaying turbulence left behind by breakers.

Perhaps the most important dynamic effect of bubbles is the way in which they, along with the air, cushion the impact of waves on any objects in the breaking zone of a beach. The velocity of sound in water can be reduced by an order of magnitude due to the presence of bubbles. It should be noted, however, that the typical size of bubbles is very different in salt and fresh water (e.g. see Scott 1975 and Monahan 1969, 1971). This is of particular importance if laboratory experiments are being used to model ocean conditions.

Vorticity

The main features of breaking waves generate ample vorticity. Here we mention the effects of preexisting vorticity, which has been generated by viscosity or turbulence in the water before the wave meets it.

The effect of a velocity shear near the surface, such as might be due to an on- or offshore wind, has been described by Banner & Phillips (1974) and Phillips & Banner (1974). An onshore surface shear tends to reduce the height of a wave at breaking; an offshore shear increases it. An extreme example is the surface-shear wave described by Peregrine (1974). It occurs in fast-flowing backwash and may reach a height several times the depth of water before collapsing.

If water is flowing toward the wave, the shear near the bed can have an overwhelming effect on the internal-flow pattern. The increase in pressure on the bed as a wave approaches can lead to flow separation from the bed. This has not been studied thoroughly, but it is clearly important for sediment transport. Matsunaga & Honji (1980) report experiments where flow separation at the base of a breaking wave is a major influence on the beach profile.

Measurements have been made of the flow pattern in hydraulic jumps, i.e. stationary bores. Resch & Leutheusser (1972) have shown how the inflow conditions affect its structure. A uniform inflow, which would correspond to conditions a bore would meet when propagating onto a weak current, leads to an internal-flow pattern as described above, with turbulence spreading downward into the incident flow. An inflow that has the fully developed profile of a steady turbulent flow gives rise to separation under the front of the wave and a substantially different flow field, though without any significant change in the surface profile. This latter case is relevant to waves that are almost brought to rest by the backwash, a common happening.

CONCLUDING REMARKS

In this review I have attempted to indicate how much, or little, is known about the dynamics of wave breaking. This means that breaking is treated as a specific event for each identifiable wave. Such an approach is desirable for understanding and accounting for natural waves. The variation from wave to wave is usually quite considerable. For example, some discussions of waves on beaches refer to a "break point." Figure 11 shows the distribution of the points at which breaking started from a 24-minute filmed sequence of waves on a North Devon beach. There is no one break point; there is instead a wide breaking zone. The beach from which this record comes was almost perfectly plane, with a slope of 1:60 at the time of filming. The majority of natural beaches have nonuniform profiles, often including bars.

Another source of variation is the response of the surf zone to the incident waves. Appreciable surface displacements and corresponding currents occur on gently sloping beaches; these are sometimes known as "surf beats" and are also ascribed to edge waves, but they may primarily be due to the envelope of amplitudes of the incident waves. The existence of these longer-scale motions means that successive waves enter different depth and current conditions.

To gain adequate understanding of wave breaking on beaches, it seems desirable to exploit the fact that most wave breaking occurs when waves are not far removed from the solitary wave in character, and to examine analytically, numerically, and experimentally the behavior of such waves on differing slopes against differing currents.

Only occasional reference has been made here to the large body of literature of measurements of wave properties on beaches. This is because properties such as the change of wave height with depth are better understood if considered in the context of surf-zone dynamics. For example, some experimenters measure the first few waves of a wave train in order to avoid the influence of reflected waves; if those waves are breaking waves, however, the chances are high that the "mean" level is steadily rising and the shoreward mass flow is significant, since those waves are establishing the set-up of mean level on the beach. In natural waves, such changes of level and current do not cease. Recent papers in this area are Guza & Thornton (1982) and Symonds et al. (1982).

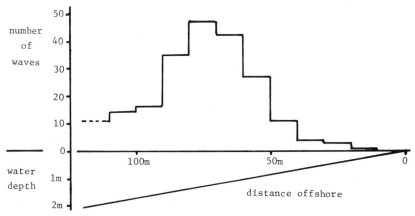

Figure 11   The spatial distribution of breaking waves in a time interval of 24 minutes on a natural beach of slope 1:60.

ACKNOWLEDGMENTS

The author acknowledges support from the La Jolla Foundation for Earth Sciences, which has permitted this review to be written close to a beach at the Institute of Geophysics and Planetary Physics, University of California, San Diego. Thanks are due for reprints, comments on a draft, and for discussions or other assistance to E. D. Cokelet, R. T. Guza, R. E. Flick, M. S. Longuet-Higgins, P. McIver, A. L. New, S. Pazan, I. A. Svendsen, and others.

Literature Cited

Airy, G. B. 1845. Tides and waves, Section 392. In *Encyc. Metropolitana*, 5:241–396

Banner, M. L., Phillips, O. M. 1974. On the incipient breaking of small scale waves. *J. Fluid Mech.* 65:647–56

Battjes, J. A. 1974. Surf similarity. *Proc. Conf. Coastal Eng., 14th*, pp. 466–79

Battjes, J. A., Sakai, T. 1981. Velocity field in a steady breaker. *J. Fluid Mech.* 111:421–37

Benjamin, T. B. 1967. Instability of periodic wave trains in non-linear dispersive systems. *Proc. R. Soc. London Ser. A* 299:59–75

Benjamin, T. B., Feir, J. E. 1967. The disintegration of wave trains on deep water. I. Theory. *J. Fluid Mech.* 27:417–30

Benjamin, T. B., Lighthill, M. J. 1954. On cnoidal waves and bores. *Proc. R. Soc. London Ser. A* 224:448–60

Binnie, A. M., Orkney, J. C. 1955. Experiments on the flow of water from a reservoir through an open channel. II. The formation of hydraulic jumps. *Proc. R. Soc. London Ser. A* 230:237–46

Carrier, G. F., Greenspan, H. P. 1958. Water waves of finite amplitude on a sloping beach. *J. Fluid Mech.* 4:97–109

Chan, R. K.-C., Street, R. L. 1970. Shoaling of finite-amplitude waves on plane beaches. *Proc. Conf. Coastal Eng., 12th*, pp. 345–61

Cleaver, R. P. 1981. *Some nonlinear properties of steep surface waves.* PhD thesis. Cambridge Univ.

Coastal Engineering Research Center, US Army. 1977. *Shore Protection Manual.* US Gov. Print. Off. 3 vols., 3rd ed.

Cokelet, E. D. 1977a. Breaking waves. *Nature* 267:769–74

Cokelet, E. D. 1977b. Steep gravity waves in water of arbitrary uniform depth. *Philos. Trans. R. Soc. London Ser. A* 286:183–230

Cokelet, E. D. 1978. Breaking waves—the plunging jet and interior flow field. In *Mechanics of Wave-Induced Forces on Cylinders*, ed. T. L. Shaw, pp. 287–301. London: Pitman. xiv+752 pp.

Dagan, G. 1975. Taylor instability of a non-uniform free-surface flow. *J. Fluid Mech.* 67:113–23

Dore, B. D. 1978. A double boundary-layer model of mass transport in progressive interfacial waves. *J. Eng. Math.* 12:289–301

Duncan, J. H. 1981. An experimental investigation of breaking waves produced by a towed hydrofoil. *Proc. R. Soc. London Ser. A* 377:331–48

Favre, H. 1935. *Ondes de translation.* Paris: Dunod. 215 pp. (4 plates)

Flick, R. E., Guza, R. T., Inman, R. L. 1981. Elevation and velocity measurements of laboratory shoaling waves. *J. Geophys. Res.* 86:4149–60

Fornberg, B., Whitham, G. B. 1978. A numerical and theoretical study of certain nonlinear wave phenomena. *Philos. Trans. R. Soc. London Ser. A* 289:373–404

Freilich, M. H. 1982. *Resonance effects on shoaling surface gravity waves.* PhD dissertation. Univ. Calif., San Diego

Gallagher, B. 1972. Some qualitative aspects of nonlinear wave radiation in a surf zone. *Geophys. Fluid Dyn.* 3:347–54

Galvin, C. J. 1968. Breaker type classification on three laboratory beaches. *J. Geophys. Res.* 73:3651–59

Galvin, C. J. 1969. Breaker travel and choice of design wave height. *J. Waterw. Harbors Div., Proc. ASCE* 95:175–200

Galvin, C. J. 1972. Wave breaking in shallow water. In *Waves on Beaches and Resulting Sediment Transport*, ed. R. E. Meyer, pp. 413–56. New York: Academic

Garrett, C., Smith, J. 1976. Interaction between long and short surface waves. *J. Phys. Oceanogr.* 6:925–30

Green, G. 1838. On the motion of waves in a variable canal of small depth and width. *Trans. Cambridge Philos. Soc.* 6:457–62

Grimshaw, R. 1979. Slowly varying solitary waves. I. Korteweg-de Vries equation. *Proc. R. Soc. London Ser. A* 368:359–75

Guza, R. T., Bowen, A. J. 1976. Resonant interactions for waves breaking on a beach. *Proc. Conf. Coastal Eng., 15th,* pp. 560–79

Guza, R. T., Thornton, E. B. 1982. Swash oscillations on a natural beach. *J. Geophys. Res.* 87:483–91

Hansen, J. B., Svendsen, I. A. 1979. Regular waves in shoaling water, experimental data. *Inst. of Hydrodyn. Hydraul. Eng., Tech. Univ. Denmark, Ser. Pap. 21.* (Erratum Oct. 1979). 20 + 212 pp.

Hedges, T. S., Kirkgöz, M. S. 1981. An experimental study of the transformation zone of plunging breakers. *Coastal Eng.* 4:319–33

Horikawa, K. 1978. *Coastal Engineering: an Introduction to Ocean Engineering.* New York: Wiley. xii + 402 pp.

Hotta, S., Mizuguchi, M. 1980. A field study of waves in the surf zone. *Coastal Eng. Jpn.* 23:59–79

Ippen, A. T., Kulin, G. 1955. Shoaling and breaking characteristics of the solitary wave. *Tech. Rep. No. 15,* MIT Hydrodyn. Lab., Cambridge, Mass. vii + 56 pp.

Iribarren, C. R., Nogales, C. 1949. Protection des ports, Section II. *Comm. 4, 17th Int. Nav. Congr.,* 31:80. Lisbon

Iwagaki, Y., Sakai, T., Tsukioka, K., Sawai, N. 1974. Relationship between vertical distribution of water particle velocity and type of breakers on beaches. *Coastal Eng. Jpn.* 17:51–58

Kaup, D. J., Newell, A. C. 1978. Solitons as particles, oscillators, and in slowly changing media: a singular perturbation theory. *Proc. R. Soc. London Ser. A* 361:413–46

Keller, J. B. 1961. Tsunamis—water waves produced by earthquakes. *Internat. Union Geodesy. Geophys. Monogr. 24,* pp. 154–66. Tsunami Hydrodyn. Conf., Honolulu

Kjeldsen, S. P., Myrhaug, D. 1980. Wave-wave interactions, current-wave interactions and resulting extreme waves and breaking waves. *Proc. Conf. Coastal Eng., 17th,* pp. 2277–303

Lamb, H. 1932. *Hydrodynamics.* Cambridge, Engl: Cambridge Univ. Press. xv + 738 pp. 6th ed.

Lighthill, J. 1978. *Waves in Fluids.* Cambridge, Engl: Cambridge Univ. Press. xv + 504 pp.

Longuet-Higgins, M. S. 1975. Integral properties of periodic gravity waves of finite amplitude. *Proc. R. Soc. London. Ser. A* 342:157–74

Longuet-Higgins, M. S. 1978a. The instabilities of gravity waves of finite amplitude in deep water. I. Super-harmonics. *Proc. R. Soc. London Ser. A* 360:471–88

Longuet-Higgins, M. S. 1978b. The instabilities of gravity waves of finite amplitude in deep water. II. Sub-harmonics. *Proc. R. Soc. London. Ser. A* 360:489–505

Longuet-Higgins, M. S. 1980a. The unsolved problem of breaking waves. *Proc. Conf. Coastal Eng., 17th,* pp. 1–28

Longuet-Higgins, M. S. 1980b. On the forming of sharp corners at a free surface. *Proc. R. Soc. London Ser. A* 371:453–78

Longuet-Higgins, M. S. 1981a. On the overturning of gravity waves. *Proc. R. Soc. London Ser. A* 376:377–400

Longuet-Higgins, M. S. 1981b. A parametric flow for breaking waves. *Int. Symp. Hydrodyn. Ocean Eng., Trondheim, Norway,* pp. 121–35

Longuet-Higgins, M. S. 1982. Parametric solutions for breaking waves. *J. Fluid Mech.* In press

Longuet-Higgins, M. S., Cokelet, E. D. 1976. The deformation of steep surface waves on water. I. A numerical method of computation. *Proc. R. Soc. London Ser. A* 350:1–26

Longuet-Higgins, M. S., Cokelet, E. D. 1978. The deformation of steep waves on water. II. Growth of normal-mode instabilities. *Proc. R. Soc. London Ser. A* 364:1–28

Longuet-Higgins, M. S., Fox, M. J. H. 1977. Theory of the almost highest wave: the inner solution. *J. Fluid Mech.* 80:721–41

Longuet-Higgins, M. S., Fox, M. J. H. 1978. Theory of the almost highest wave. Part 2. Matching and analytic extension. *J. Fluid Mech.* 85:769–86

Madsen, O. S., Mei, C. C. 1969. The transformation of a solitary wave over an uneven bottom. *J. Fluid Mech.* 39:781–91

Madsen, P. A. 1981. A model for a turbulent bore. *Inst. Hydrodyn. Hydraul. Eng., Tech. Univ. Denmark, Ser. Pap. 28.* 149 pp.

Madsen, P. A., Svendsen, I. A. 1982. Turbulent bores and hydraulic jumps. *J. Fluid Mech.* Submitted for publication

Matsunaga, N., Honji, H. 1980. The backwash vortex. *J. Fluid Mech.* 99:813–15

McIntyre, M. E. 1981. On the "wave-momentum" myth. *J. Fluid Mech.* 106:331–47

McIver, P., Peregrine, D. H. 1981. Comparison of numerical and analytical results for waves that are starting to break. *Int. Symp. Hydrodyn. Ocean Eng., Trondheim, Norway*, pp. 203–15

McIver, P., Peregrine, D. H. 1982. Motion of a free surface and its representation by singularities. *J. Fluid Mech.* Submitted for publication

McLean, J. W. 1982. Instabilities of finite-amplitude waves on water of finite depth. *J. Fluid Mech.* 114:331–41

McLean, J. W., Ma, Y. C., Martin, D. V., Saffman, P. G., Yuen, H. C. 1981. Three-dimensional instability of finite-amplitude water waves. *Phys. Rev. Lett.* 46:817–20

Meyer, R. E., Taylor, A. D. 1972. Run-up on beaches. In *Waves on Beaches and Resulting Sediment Transport*, ed. R. E. Meyer, pp. 357–411. New York: Academic

Miche, M. 1944. Le pouvoir réfléchissant des ouvrages maritimes exposés à l'action de la houle. *Ann. Ponts Chaussées* 121:285–318

Miles, J. W. 1979. On the Korteweg-de Vries equation for a gradually varying channel. *J. Fluid Mech.* 91:181–90

Miles, J. W. 1980. Solitary waves. *Ann. Rev. Fluid Mech.* 12:11–43

Miller, R. 1976. Role of vortices in surf zone prediction: sedimentation and wave forces. *Soc. Econ. Paleontol. Mineralog., Spec. Publ. No. 24*, pp. 92–114

Monahan, E. C. 1969. Fresh water whitecaps. *J. Atmos. Sci.* 26:1026–29

Monahan, E. C. 1971. Oceanic whitecaps. *J. Phys. Oceanogr.* 1:139–44

Munk, W. H. 1949. The solitary wave and its application to surf problems. *Ann. NY Acad. Sci.* 51:376–424

Munk, W., Wimbush, M. 1969. A rule of thumb for wave breaking over sloping beaches. *Oceanology* 6:56–59

Peregrine, D. H. 1966. Calculations of the development of an undular bore. *J. Fluid Mech.* 25:321–30

Peregrine, D. H. 1967. Long waves on a beach. *J. Fluid Mech.* 27:815–27

Peregrine, D. H. 1974. Surface shear waves. *J. Hydraul. Div., Proc. ASCE* 100:1215–27

Peregrine, D. H. 1976. Interaction of water waves and currents. *Adv. Appl. Mech.* 16:9–117

Peregrine, D. H. 1981. The fascination of fluid mechanics. *J. Fluid Mech.* 106:59–80

Peregrine, D. H., Svendsen, I. A. 1978. Spilling breakers, bores and hydraulic jumps. *Proc. Conf. Coastal Eng., 16th*, pp. 540–50

Peregrine, D. H., Cokelet, E. D., McIver, P. 1980. The fluid mechanics of waves approaching breaking. *Proc. Conf. Coastal Eng., 17th*, pp. 512–28

Phillips, O. M., Banner, M. L. 1974. Wave breaking in the presence of wind drift and swell. *J. Fluid Mech.* 66:625–40

Resch, F. J., Leutheusser, H. J. 1972. Reynolds stress measurements in hydraulic jumps. *J. Hydraul. Res.* 10:409–30

Ryrie, S., Peregrine, D. H. 1982. Refraction of finite-amplitude water waves obliquely incident to a uniform beach. *J. Fluid Mech.* 115:91–104

Sakai, T., Battjes, J. A. 1980. Wave shoaling calculated from Cokelet's theory. *Coastal Eng.* 4:65–84

Sawaragi, T., Iwata, K. 1974. Turbulence effect on wave deformation after breaking. *Coastal Eng. Jpn.* 17:39–49

Schwartz, L. W. 1974. Computer extension and analytic continuation of Stokes' expansion for gravity waves. *J. Fluid Mech.* 62:553–78

Schwartz, L. W., Fenton, J. D. 1982. Strongly nonlinear waves. *Ann. Rev. Fluid Mech.* 14:39–60

Scott, J. C. 1975. The role of salt in whitecap persistence. *Deep-Sea Res.* 22:653–57

Silvester, R. 1974. *Coastal Engineering.* Vols. 1, 2. Amsterdam: Elsevier. 457 pp., 338 pp.

Srokosz, M. A. 1981. Breaking effects in standing and reflected waves. *Int. Symp. Hydrodyn. Ocean Eng., Trondheim, Norway*, pp. 183–202

Stiassnie, M., Peregrine, D. H. 1980. Shoaling of finite-amplitude surface waves on water of slowly-varying depth. *J. Fluid Mech.* 97:783–805

Stive, M. J. F. 1980. Velocity and pressure field of spilling breakers. *Proc. Conf. Coastal Eng., 17th*, pp. 547–66

Stoker, J. J. 1957. *Water Waves.* New York: Interscience. 567 pp.

Stokes, G. G. 1847. On the theory of oscillatory waves. *Trans. Cambridge Philos. Soc.* 8:441–55 [*Math. Phys. Pap.* 1:197–229 (1880)]

Stokes, G. G. 1880. Considerations relative to the greatest height of oscillatory irrotational waves which can be propagated without change of form. *Math. Phys. Pap.* 1:225–28

Street, R. L., Camfield, F. E. 1966. Observations and experiments on solitary wave deformation. *Proc. Conf. Coastal Eng., 10th*, pp. 284–301

Su, M.-Y., Bergin, M., Marler, P., Myrick, R. 1982a. Experiments on nonlinear instabilities and evolution of steep gravity wave trains. Submitted for publication

Su, M.-Y., Bergin, M., Myrick, R., Roberts, J. 1982b. Experiments on shallow-water wave grouping and breaking. *Proc. 1st Int. Conf. Meteorol. Air-Sea Interaction Coastal Zone,* The Hague: K. Ned. Meteorol. Inst.

Suhayda, J. N., Pettigrew, N. R. 1977. Observations of wave height and wave celerity in the surf zone. *J. Geophys. Res.* 82:1419–24

Svendsen, I. A., Hansen, J. B. 1978. On the deformation of periodic long waves over a gently sloping bottom. *J. Fluid Mech.* 87:433–48

Svendsen, I. A., Madsen, P. A., Hansen, J. B. 1978. Wave characteristics in the surf zone. *Proc. Conf. Coastal Eng., 16th,* pp. 520–39

Symonds, G., Huntley, D. A., Bowen, A. J. 1982. Two-dimensional surf beat: long wave generation by a time-varying breakpoint. *J. Geophys. Res.* 87:492–98

Thornton, E. B., Galvin, J. J., Bub, F. L., Richardson, D. P. 1976. Kinematics of breaking waves. *Proc. Conf. Coastal Eng., 15th,* pp. 461–76

Ursell, F. 1953. The long-wave paradox in the theory of gravity waves. *Proc. Cambridge Philos. Soc.* 49:685–94

Van Dorn, W. G. 1978. Breaking invariants in shoaling waves. *J. Geophys. Res.* 83:2981–88

Vinje, T., Brevik, P. 1981. Numerical simulation of breaking waves. *Adv. Water Resour.* 4:77–82

Weishar, L. L., Byrne, R. J. 1978. Field study of breaking wave characteristics. *Proc. Conf. Coastal Eng., 16th,* pp. 487–506

Whitham, G. B. 1974. *Linear and Non-Linear Waves.* New York: Wiley-Interscience. xvii + 636 pp.

Wiegel, R. L. 1964. *Oceanographical Engineering.* Englewood Cliffs, NJ: Prentice-Hall. xi + 532 pp.

Williams, J. M. 1981. Limiting gravity waves in water of finite depth. *Philos. Trans. R. Soc. London Ser. A* 302:139–88

Ann. Rev. Fluid Mech. 1983. 15:179–99

INSTABILITIES, PATTERN FORMATION, AND TURBULENCE IN FLAMES

G. I. Sivashinsky[1]

Department of Mathematics and Lawrence Berkeley Laboratory, University of California, Berkeley, California 94720

1. INTRODUCTION

Considerable progress has recently been achieved in the understanding of the nature and character of spontaneous instabilities in premixed flames. The present survey is devoted to the latest theoretical results in this area, which have disclosed a deep affinity between flames and other nonequilibrium physical systems capable of generating regular and intrinsically chaotic structures.

1.1 *Premixed flames* It is well known that the rate of a chemical reaction (W) in a gaseous mixture is an increasing function of temperature; usually $W \sim \exp(-E/RT)$, where E is a constant, specific to the reaction and called its activation energy, and R is the universal gas constant. The larger E, the stronger the temperature-dependence of the reaction rate. Under normal conditions (atmospheric pressure, room temperature), the reaction rate in the majority of combustible mixtures is negligibly small. At sufficiently high temperatures, however, the reaction will take place at a substantial rate. When a combustible mixture is ignited at some point, e.g. by a spark, a rapid exothermic reaction is initiated; this causes the temperature to rise (via conduction) in the adjacent layer of the mixture, inducing a chemical reaction there, and so on. Thus the reaction, once begun, will spread through the mixture,

[1] On leave from the Department of Mathematical Sciences, Tel-Aviv University, Ramat-Aviv, P.O.B. 39040, Tel-Aviv 69978, Israel.

0066-4189/83/0115-0179$02.00

converting it into combustion products. This self-sustaining wave of an exothermic reaction is known as a *premixed flame*.

The thermal mechanism of flame propagation described above has been known for a long time (Mallard & Le Chatelier 1883), but it was not until the work of Zeldovich & Frank-Kamenetsky (1938) that a really sound theory was formulated for the problem of steady plane flame propagation. Typical profiles of temperature (T), concentration (C), and reaction rate (W) in a one-dimensional combustion wave are shown in Figure 1. T_u is the temperature of the unburned cold mixture, at which the reaction rate is negligibly small. C_u is the initial concentration of the reactant that is entirely consumed in the reaction (limiting reactant). T_b is the temperature of the burned gas, usually 5 to 10 times T_u. The thermal thickness ℓ_{th} of the flame is defined as D_{th}/U_u, where D_{th} is the thermal diffusivity of the mixture and U_u the propagation speed of the flame relative to the unburned gas.

As the reaction rate is strongly temperature-dependent $(E/RT_b \sim 20)$, the bulk of the chemical reaction occurs in a narrow temperature interval $\Delta T \sim RT_b^2/E$ around the maximum temperature T_b. This temperature region corresponds to a thin layer of width $\sim (RT_b/E)\ell_{th} = \ell_r$, outside which the chemical reaction may be neglected.

The propagation speed of the flame (U_u) is determined by balance between the quantity of heat liberated during the reaction and the heat required to preheat the fresh mixture:

$$U_u = \sqrt{\frac{2D_{th}}{T_b - T_u} \int_{T_u}^{T_b} W(C, T)\, dT}. \tag{1}$$

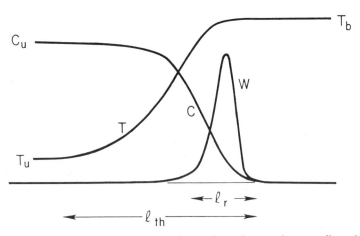

Figure 1   Profiles of temperature, concentration, and reaction rate in a one-dimensional combustion wave.

For one of the most rapidly burning mixtures $(2H_2 + O_2)$, one has $U_u \simeq 10$ m s^{-1} and $\ell_{th} \simeq 0.0005$ cm. For one of the most slowly burning mixtures (6% CH_4 + air), $U_u \simeq 5$ cm s^{-1} and $\ell_{th} \simeq 0.05$ cm.

The mathematical theory of steady plane flames is now—in principle—complete, and there are several monographs on the subject (e.g. Williams 1965, Frank-Kamenetsky 1969, Kanury 1975, Glassman 1977, Zeldovich et al. 1980b, Buckmaster & Ludford 1982). The situation is far less satisfactory in regard to nonsteady phenomena in curved flames. In particular, until recently there was a considerable gap between theoretical and experimental results concerning the stability of a premixed flame front. The marked progress achieved in this field over the past five years is largely due to the asymptotic methods that have penetrated combustion theory from other, more classical areas of fluid mechanics.

1.2 *Observed instabilities* For combustion to be actually maintained in the form of a steady plane reaction wave, the structure in question must be stable under small disturbances. Many flames are known to behave like one-dimensional reaction waves under normal laboratory conditions. However, some experiments have shown that there is a class of flames that prefer a characteristic two- or three-dimensional structure rather than a plane flame. It has long been known that the flame on a Bunsen burner may split up into triangular flamelets, which form—instead of the usual cone—a polyhedral pyramid that sometimes even rotates about its vertical axis (Smithells & Ingle 1882, Smith & Pickering 1928). It was later shown (Zeldovich 1944, Markstein 1949) that this manifestation of instability is not unique. In combustion in wide tubes, the flame frequently breaks up into separate cells, ~1 cm in size, in a state of constant subdivision and recombination (Figure 2). It was noticed that cellular structure tends to appear when the combustible mixture is deficient in the light reactant (e.g. rich hydrocarbon-air or lean hydrogen-air mixtures). Later it was pointed out that lean hydrocarbon-air mixtures are also not absolutely stable. Under the action of external large-scale disturbances, an originally smooth flame may exhibit sharp folds that are maintained under further deformation and extension of the flame (Figure 3). Moreover, recent experiments on the propagation of large-scale flames in unconfined vapor clouds have shown that cellular instability may also appear in lean hydrocarbon-air mixtures, with cell size ~10 cm (Lind & Whitson 1977, Ivashchenko & Rumiantsev 1978). Finally, we should mention one of the most recent results, which concerns a possibility of spinning propagation of luminous flames (Gololobov et al. 1981). Thus, even freely propagating flames represent an extremely rich physical system.

Figure 2   Rich propane-air cellular flame in state of chaotic self-motion. (Courtesy of P. Clavin, University of Provence, Marseille. Originally in Sabathier et al. 1981.)

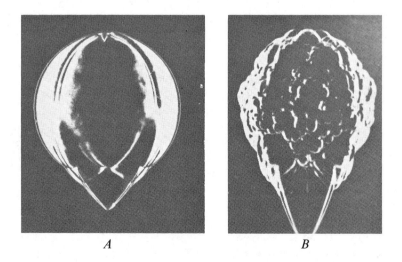

A   B

Figure 3   Free flame balls obtained in a low-speed laminar flow by spark ignition. (*A*) Cellularly stable lean butane-air flame. (*B*) Cellularly unstable lean hydrogen-air flame. (Courtesy of R. A. Strehlow, University of Illinois, Urbana, Illinois. Originally in Strehlow 1968).

1.3   *Fundamental assumptions of the theory*   A theoretical description of flame propagation requires simultaneous solution of an equation system including both the reaction-diffusion equations for each of the species and the equations of motion of the gaseous mixture. Since the propagation speed of the flame is significantly less than the speed of sound, effects due to the dynamic compressibility of the gas may be neglected. Hence, the density of the gas may be considered a function of temperature and concentrations only.

The approximation representing the combustion reaction as a simple scheme,

Fuel + Oxidant → Products,

is sufficient for a qualitative description of the majority of fluid-mechanical phenomena in premixed flames. It is convenient to consider the reactants as small additives to some "inert" gas; this justifies the use of the independent-diffusion approximation in calculating their diffusion rates. Moreover, this assumption enables one to consider the gas almost homogeneous and thereby to simplify the equation of state. The "inert" gas (=dilutant) in experiments is frequently nitrogen.

As the reaction rate is strongly temperature-dependent ($E/RT_b \gg 1$), the reaction zone may be considered infinitesimally narrow in comparison with the thermal thickness of the flame ($\ell_r \ll \ell_{th}$). Thus, if one is interested in dimensions of the order of ℓ_{th} or higher, it may be assumed that the reaction is concentrated on a certain surface—the flame front. The reaction rate (W) may therefore be replaced by a localized source, the intensity of which is determined by considering the processes that take place within the reaction zone. This problem may be solved in the spirit of Zeldovich & Frank-Kamenetsky (1938). For example, if the mixture is strongly nonstoichiometric (say, the concentration of the oxidant is much higher than that of the fuel), then the intensity of the localized reaction rate is proportional to $\exp(-E/2RT_f)$, where T_f is the temperature on the curved, nonsteady flame front, which may differ from T_b by a quantity of the order of RT_b^2/E.

2.   THERMODIFFUSIVE FLAME INSTABILITY

2.1   *One-reactant linear theory*   As early as the mid-1940s, Zeldovich (1944) proposed the following qualitative explanation for why cellular flames tend to form in mixtures that are deficient in the light reactant. Consider a curved flame front. It is clear that conduction of heat tends to decrease the curvature of the flame, i.e it is a stabilizing influence. Indeed, the parts of the chemical reaction zone that are convex toward the fresh mixture give out more heat than in a plane flame. The resultant cooling of the reaction zone slows down the forward-rushing parts of the flame. The concave parts, conversely, give out less heat than in the plane case and so the temperature increases, and hence also the reaction rate. The concave parts of the flame move forward at a higher speed than in a plane flame. Thus the surface of the curved front is smoothed out.

Diffusion, however, has the opposite effect. The parts of the reaction zone convex toward the fresh mixture receive more fuel than in a plane flame. The reaction rate in the convex parts increases and the front curvature becomes greater. Thus diffusion has a destabilizing effect.

Hence it is clear that if the molecular diffusivity D_{mol} of the limiting reactant is sufficiently greater than the thermal diffusivity D_{th} of the mixture, one can expect a plane flame to be unstable. In the opposite case, the flame front should be smooth.

The motion of a curved flame front is invariably accompanied by motion of the gas. However, it is evident from the foregoing qualitative arguments that hydrodynamic effects play a merely secondary role in the onset of thermodiffusive instability. For a theoretical analysis of the phenomenon, therefore, it makes sense to disregard them. Formally, this may be done by assuming that the density of the gas is constant. In that case, thermal disturbances in the flame cannot be transformed into hydrodynamic disturbances, and so the problem of hydrodynamics is completely divorced from that of combustion proper. In other words, when studying the motion of the flame one can consider the hydrodynamic field to be assigned in advance.

In the above model, linear analysis of the stability of a plane flame front to *long-wave* disturbances yields the following dispersion relation (Barenblatt et al. 1962):

$$\sigma = D_{th}\left[\tfrac{1}{2}\beta(1 - Le) - 1\right]k^2, \tag{2}$$

where $\beta = E(T_b - T_u)/RT_b^2$; $Le = D_{th}/D_{mol}$ is the Lewis number of the limiting reactant, which is assumed to be strongly deficient; σ is the rate-of-instability parameter; and $\mathbf{k}$ is the wave vector of the disturbance of the flame front, $F \sim \exp(\sigma t + i\mathbf{k}\cdot\mathbf{x})$. Thus in agreement with the previous qualitative analysis, the flame is stable only if the mobility of the limiting reactant is sufficiently low ($Le > Le_c = 1 - 2/\beta$). At $Le < Le_c$ the flame is unstable. In a typical flame, $\beta \simeq 15$, and so $Le_c \simeq 0.87$.

However, as was pointed out later (Sivashinsky 1977a), a flame, though possibly unstable to long-wave disturbances, is nevertheless always stable to short-wave disturbances. At $Le \simeq Le_c$ the dispersion relation incorporating the relaxation effect of short-wave disturbances is

$$\sigma = D_{th}\left[\tfrac{1}{2}\beta(1 - Le) - 1\right]k^2 - 4D_{th}\ell_{th}^2k^4. \tag{3}$$

Thus, when the flame is unstable ($Le < Le_c$), there is a wavelength λ_c corresponding to the highest amplification rate of small disturbances (maximum σ).

2.2 *Nonlinear theory*   What happens to the flame front after the development of progressive disturbances?

First, it is readily seen that the dispersion relation (3) may be expressed as a linear equation for the disturbance of the flame front:

$$F_t + D_{th}\left[\tfrac{1}{2}\beta(1 - Le) - 1\right]\nabla^2 F + 4D_{th}\ell_{th}^2 \nabla^4 F = 0. \tag{4}$$

This equation obviously yields exponential amplification of long-wave disturbances at $Le < Le_c$. In reality, this amplification will be checked by effects represented by certain nonlinear terms not present in Equation (4). The structure of these terms may be established via the following semiheuristic arguments.

Consider a curved flame front in the constant-density model. If the characteristic radius of curvature of the flame is significantly greater than its thermal thickness ℓ_{th}, then the propagation speed of the flame relative to the gas may be considered a constant equal to U_b. In a coordinate system at rest with respect to the undisturbed plane flame, the front $z = F(x, y, t)$ of such a curved flame is described by the eikonal equation:

$$F_t = U_b\left(1 - \sqrt{1 + (\nabla F)^2}\right). \tag{5}$$

Near the stability threshold $Le \simeq Le_c$, one expects that $(\nabla F)^2 \ll 1$. Hence

$$F_t + \tfrac{1}{2}U_b(\nabla F)^2 = 0. \tag{6}$$

Comparing this weakly nonlinear equation, which disregards effects due to distortion of the flame structure, with Equation (4), in which these effects are included, one reaches the reasonable conclusion that $\tfrac{1}{2}U_b(\nabla F)^2$ is precisely the nonlinear term missing from Equation (4). We thus obtain the following equation for the nonlinear evolution of the disturbed flame front:

$$F_t + \tfrac{1}{2}U_b(\nabla F)^2 + D_{th}\left[\tfrac{1}{2}\beta(1 - Le) - 1\right]\nabla^2 F + 4D_{th}\ell_{th}^2 \nabla^4 F = 0. \tag{7}$$

This equation is a rigorous asymptotic relation derived from the constant-density flame model, provided $Le - Le_c$ is a small parameter (Sivashinsky 1977b).

Thus the main nonlinear effect in the evolution of the disturbed flame front turns out to be of a purely kinematic nature. Qualitative arguments in favor of such a mechanism of nonlinear stabilization have been put forward by Manton et al. (1952), Markstein (1952), Petersen & Emmons (1961), Shchelkin (1965), and Zeldovich (1966), whose point of departure was the Huygens principle. Equation (7) may be put in a nondimensional, parameter-free form that is very convenient for numerical experimentation:

$$\Phi_\tau + \tfrac{1}{2}(\nabla\Phi)^2 + \nabla^2\Phi + 4\nabla^4\Phi = 0. \tag{8}$$

Numerical experiments on this equation have shown that when a plane flame is disturbed it ultimately evolves into a cellular flame, with characteristic cell size somewhat greater than λ_c. This structure is essentially nonsteady, the cells being in a state of continual *chaotic* self-motion (Michelson & Sivashinsky 1977; see also Figure 4). The chaotic behavior of cellular flames is well known from the classical experiments of Markstein (1949, 1964). This phenomenon was recently reconfirmed in experiments performed by Sabathier et al. (1981) under carefully controlled flow conditions, which prevented turbulence in the upstream flow (Figures 2 and 3).

Thus, despite its simplicity the one-reactant constant-density model proved sufficiently rich not only to provide an adequate description of the sensitivity of flame stability to the composition of the mixture and to predict the characteristic size of the cells, but also to describe the chaotic self-motion of the cells. Quite likely the model also describes polyhedral, rotating Bunsen flames (Buckmaster 1982b) and the apparently similar phenomenon of traveling waves that sometimes appear in place of chaotically recombining cells (Markstein 1964, Sabathier et al. 1981).

However, the range of validity of the constant-density theory is limited. According to the theory, the cell size should increase indefinitely as the Lewis number goes through its critical value Le_c. This is in clear contradiction to experimental observations, which indicate that the cell size at the stability threshold is finite (Markstein 1964, 1970). As we show below (Section 3.4), the effects due to the thermal expansion of the gas become significant near Le_c, and this completely alters the nature of the instability.

2.3 *Stability of nearly stoichiometric flames*   The case discussed above corresponds to a strongly nonstoichiometric mixture, where the depletion of the excess reactant can be neglected and its concentration considered constant. The model was extended to nearly stoichiometric flames by Sivashinsky (1980) and Joulin & Mitani (1981), who showed that the main functional relationships of the one-reactant theory remain intact provided that Le is interpreted as a suitably weighted average of the Lewis numbers of the fuel and the oxidant. This explains why cellular instability is observed in nearly stoichiometric mixtures even when there

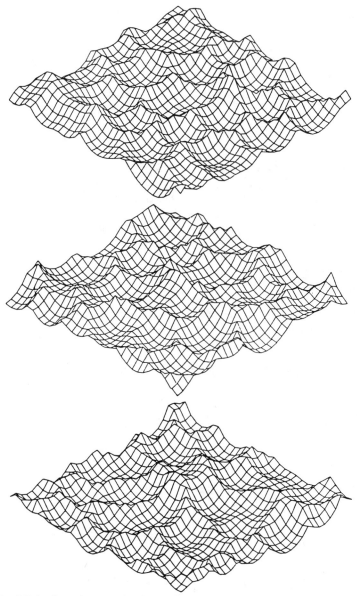

Figure 4 Cellular flame in a state of chaotic self-motion. Numerical solution of Equation (8) in $5\lambda_c \times 5\lambda_c$ square with periodic boundary conditions. Each surface represents the configuration of the flame front at three consecutive instants of time.

is a certain excess of the light component (Bregeon et al. 1978). A recently published paper of Mitani & Williams (1980) on nearly stoichiometric hydrogen-air cellular flames reveals that the predictions of the theory are also in qualitative agreement with the experimental data on cell size.

2.4   *Effects due to acceleration*   Since combustion is accompanied by thermal expansion of the gas, it is clear that buoyancy will exert a stabilizing effect on downward-propagating flames. As in the Boussinesq theory of natural convection, this effect may be incorporated through a *force* term, while remaining within the limits of the constant-density model (Matkowsky & Sivashinsky 1979). As a result, the dispersion relation (3) is augmented by the addition of a stabilizing term $g(1 - \varepsilon)/2U_b$, where $\varepsilon = \rho_b/\rho_u$ is the thermal expansion coefficient of the gas. The nonlinear equation (8) is modified to become

$$\Phi_\tau + \tfrac{1}{2}(\nabla\Phi)^2 + \nabla^2\Phi + 4\nabla^4\Phi + G\Phi = 0. \tag{9}$$

Here G is a nondimensional parameter proportional to the reciprocal of the Froud number. When $G > 1/16$, combustion is stable; this should be the case in sufficiently slow flames (Markstein & Somers 1953). Near the stability threshold, chaotic fluctuations disappear and the flame takes on a steady, almost regular, cellular structure with cell size λ_c (Figure 5).

2.5   *Effects due to heat loss*   It is known that in rich hydrocarbon-air mixtures cellular stability is likely to be observed near the flame propagation limit. Thus one might expect that inclusion of heat loss should expand the range of unstable Lewis numbers. Analysis of flame stability in the case of volumetric heat loss has fully corroborated this conjecture (Joulin & Clavin 1979, Sivashinsky & Matkowsky 1981). With heat loss included, the dispersion relation (3) becomes

$$\sigma = D_{th}\left[\frac{\tfrac{1}{2}\beta(1 - Le) - 1 - 2\ln\gamma}{1 + 2\ln\gamma}\right]k^2 - D_{th}\ell_{th}^2\left[\frac{4 + 2\ln\gamma}{\gamma^2(1 + 2\ln\gamma)}\right]k^4, \tag{10}$$

where γ is the ratio of the propagation speed of the nonadiabatic flame to that of the adiabatic flame (U_b). As one approaches the flame propagation limit ($\gamma \to 1/\sqrt{e}$), Le_c tends to unity, i.e. the instability region expands. Simultaneously one observes a sharp increase in the instability rate σ. Thus, for example, a downward-propagating adiabatically stable flame may become unstable when heat loss is taken into account.

2.6   *Effects due to stretching*   In practical situations the flame is frequently situated in a nonuniform flow field and is therefore subjected to large-scale flame stretch (Karlovitz et al. 1953). Recently, Law et al.

(1981), Ishizuka et al. (1982), and Ishizuka & Law (1982) have systematically investigated the extinction and stability limits of propane-air flames in stagnation-point flow, which imposes a well-characterized strain rate on the bulk flame. Their results show that while lean flames are absolutely stable, in rich mixtures flame-front instability of various configurations may appear, depending on the strain rate. When the flow rate is so slow that the flame is situated in the nearly one-dimensional flow field close to the burner surface, the instability is exhibited in the form of a cellular flame. If one increases the flow rate, and thereby the stretch, the flame recedes from the burner surface and moves further into the stagnation flow. It becomes star-shaped, with diametrically oriented ridges. Finally, with sufficiently strong blowing, all instabilities are suppressed and the flame becomes smooth. Thus stagnation-point flow stabilizes the flame.

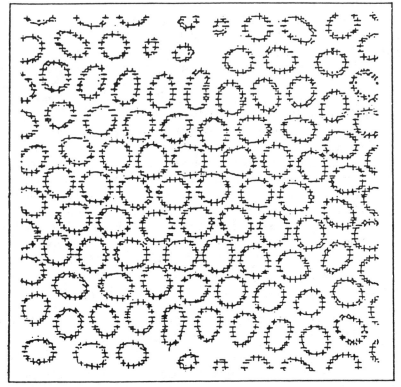

Figure 5 One of the level curves of a steady, nearly regular flame near the stability threshold ($G = 2/33$). Numerical solution of Equation (9) in $10\lambda_c \times 10\lambda_c$ square with periodic boundary conditions.

The mathematical problem of stability is quite unusual, in that the classical stability analysis appears somewhat misleading. For example, in the two-dimensional version of the problem the stagnation-point flow is $\mathbf{v} = (-qx, qy)$, where q is the flow rate parameter. The one-dimensional version of Equation (8) for the disturbance of a plane flame is modified to become (Sivashinsky et al. 1982)

$$\Phi_\tau + \tfrac{1}{2}\Phi_\eta^2 + \Phi_{\eta\eta} + 4\Phi_{\eta\eta\eta\eta} + \alpha(\eta\Phi)_\eta = 0, \tag{11}$$

where α is a nondimensional parameter proportional to q.

The solution of the linearized equation (11) may be expressed as a combination of functions of the type

$$\Phi = A(\tau)\exp(ik\eta e^{-\alpha\tau}), \tag{12}$$

where

$$A_\tau = (k^2 e^{-2\alpha\tau} - 4k^4 e^{-4\alpha\tau} - \alpha)A.$$

In the initial instant of time there are growing modes if $\alpha < 1/16$. However, as $\tau \to \infty$ the amplitude of the disturbance vanishes for any positive α. This is because with elapsing time any harmonic disturbance is stretched, while a disturbance of infinitely long wavelength ($k = 0$) is damped out as $\exp(-\alpha\tau)$. Thus the flame front would seem to be absolutely stable for any positive α. However, this does not agree with the cellular flame structure observed at small flow rates (i.e. small α).

The apparent contradiction is resolved by observing that the neglected nonlinear term represents mode interaction, which continually generates short-wave disturbances whose amplitudes may increase during certain time intervals. Hence, within the framework of the original nonlinear theory, one can expect a permanently excited state of the flame front if α is sufficiently small. Numerical experiments with the nonlinear equation (11) have confirmed that the flame front is cellularly unstable at $\alpha \leqslant 0.01$ and stable at $\alpha \geqslant 0.02$.

The nonlinear instability mechanism described above also occurs for an expanding spherical or cylindrical flame. In the latter case, for example, the evolution equation is (Sivashinsky 1979)

$$\Phi_\tau - \frac{1}{2\tau^2}\Phi_\vartheta^2 + \frac{1}{\tau^2}\Phi_{\vartheta\vartheta} + \frac{4}{\tau^4}\Phi_{\vartheta\vartheta\vartheta\vartheta} = 0, \tag{13}$$

where Φ is the disturbance of the expanding cylindrical front and ϑ is a suitably scaled polar angle. Introducing Cartesian coordinates on the flame surface ($\tau\vartheta = \eta$), one can bring Equation (13) to the following form:

$$\Phi_\tau - \tfrac{1}{2}\Phi_\eta^2 + \Phi_{\eta\eta} + 4\Phi_{\eta\eta\eta\eta} + \frac{1}{\tau}(\eta\Phi_\eta) = 0. \tag{14}$$

The last term in this expansion can be interpreted as a stretch generated by the expanding flame. Because of this stretch, linear analysis implies that the flame is absolutely stable; hence it is inadequate. The difficulty may be resolved by a device similar to that employed for the stagnation-point flame.

To conclude this section, we note that if Le is varied continuously with the flame stretch held fixed, the cell size remains finite at the stability threshold.

2.7 *Oscillatory and spinning flames* Analysis of the full dispersion relation for the constant-density flame model shows that if $\frac{1}{2}\beta(Le - 1) > 5$ the real part of the instability rate σ is positive, and so the flame may also be unstable at fairly large Lewis numbers (Sivashinsky 1977a). Since the imaginary part of σ does not vanish when the stability threshold is crossed, the new propagation mode induced by instability of a plane flame may consist of oscillations, traveling waves, or even spinning.

In real adiabatic gaseous flames, the instability region is not likely to be reached. However, as shown by Joulin & Clavin (1979), if there is volumetric heat loss the instability region is considerably expanded, and a real flame may well become oscillatorily unstable. The first announcement of spinning flame propagation was made recently by Gololobov et al. (1981) in connection with an investigation of the acetylene decomposition flames. Since such combustion is accompanied by high radiation heat loss (due to intensive sooting), the effect observed may well be the first corroboration of the theoretical possibility discussed above. Hitherto, oscillatory and spinning flames had been observed only in gasless combustion of condensed systems (Merzhanov et al. 1973). In these systems, the diffusivity of the fuel was zero ($Le = \infty$) and the instability region was easily reached. The pertinent theoretical analysis was presented in Matkowsky & Sivashinsky (1978) and Sivashinsky (1981).

Qualitatively new modes of oscillatory instability were discovered in investigation of flames stabilized on flat porous flame-holders. A steady theory of such burners was developed by Carrier et al. (1978), Ferguson & Keck (1979), and Clarke & McIntosh (1980). It was observed (Margolis 1980, 1981, Matkowsky & Olagunju 1981, Buckmaster 1982a) that conductive heat loss to the burner may be sufficient to bring on oscillatory instability even for relatively small Lewis numbers. As shown recently by Joulin (1981, 1982), flame oscillation is described by a delayed nonlinear differential equation of Hutchinson type:

$$\Phi_\tau + \tfrac{1}{2}B\{1 - \exp[-\Phi(\tau - 2)]\} = 0, \tag{15}$$

where Φ is the displacement of the flame front relative to its equilibrium position, and B is a number defined by the physicochemical parameters of the system.

If $B > \pi/2$, the equilibrium state ($\Phi = 0$) becomes unstable and the system begins to perform sawtooth oscillations. This oscillatory mode is in agreement with earlier numerical studies and certain experimental observations (Margolis 1980, 1981).

3. HYDRODYNAMIC FLAME INSTABILITY

3.1 *Linear theory*   In the discussion of thermodiffusive flame instability, we ignored the effect of thermal expansion of the gas and, by the same token, the interaction of the flame with the hydrodynamic disturbances that it generates. Darrieus (1938) and Landau (1944) made the first analysis of flame stability. They assumed that the flame is a density jump propagating at a *constant* speed in an incompressible, nonviscous, nonconducting fluid. This approach is quite legitimate if one is interested in disturbances of wavelength that considerably exceed the thermal thickness ℓ_{th} of the flame. Since the Darrieus-Landau model does not include any characteristic length, the instability rate σ must depend on the flame speed U_b and the disturbance wavevector $\mathbf{k}$ as follows:

$$\sigma = \Omega_0 U_b k, \qquad k = |\mathbf{k}|, \tag{16}$$

where Ω_0 is a nondimensional function of the parameter $\varepsilon = \rho_b/\rho_u$—the ratio of densities of the burned and unburned gas. For all $\varepsilon < 1$, $\Omega_0(\varepsilon) = \left(\sqrt{\varepsilon + \varepsilon^2 - \varepsilon^3} - \varepsilon\right)/(1 + \varepsilon)$ is positive, i.e. the flame is unstable to disturbances of all wavelengths. However, this is in conflict with experiment. Under normal laboratory conditions, one often observes smooth steady flames stabilized in quite wide tubes (diameter ~ 15 cm).

According to Equation (16), short-wave disturbances should increase at a higher rate than long-wave disturbances. But it is precisely for short-wave disturbances that the Darrieus-Landau model breaks down, since they induce distortion of the flame front structure and are liable to alter its propagation speed. For this reason, later work on hydrodynamic flame instability was aimed at correcting the Darrieus-Landau solution (16) in the region of short-wave disturbances. The first important progress in this direction was due to Markstein (1951). Markstein suggested that one characterize the effect of the distortions of flame structure by a certain phenomenological constant (having the dimension of length) that relates the flame propagation speed to the curvature of the front. The result is that a plane flame is stable to short-wave disturbances and unstable to long-wave disturbances. This would mean that a plane flame could only be observed when combustion takes place between walls that prevent the appearance of long-wave disturbances.

The Markstein theory yields stabilization of short-wave disturbances only when the phenomenological constant has a certain sign. However, it may also have the opposite sign due, say, to the possibility of thermodiffusive instability. In that case short-wave disturbances would provide an additional destabilizing factor.

The uncertainties inherent in the Markstein theory stimulated new research, in which flame instability was investigated by taking the flame structure into consideration (i.e. effects due to heat conduction, diffusion, viscosity, and chemical kinetics). However, the difficulties encountered in this process were so great that the investigators were forced to make various arbitrary assumptions concerning the structure of the disturbed flame front in order to avoid insurmountable mathematical problems. As a consequence, the results were in conflict not only quantitatively but even qualitatively. For example, until recently it was unclear whether viscosity is a stabilizing or destabilizing factor in flames (Markstein 1964).

Istratov & Librovich (1966) were the first to realize that the determination of corrections to the Darrieus-Landau solution is a singular perturbation problem. The long-wave disturbances of the flame front create hydrodynamic disturbances, extending on both sides of the front for a distance of the same order as the wavelength of the disturbance $(2\pi/k)$. Thus, the structure of the disturbed flame has at least two characteristic lengths, ℓ_{th} and $2\pi/k$ ($\gg \ell_{th}$). As there was then no systematic technique for the solution of such problems, a mathematically consistent implementation of this idea was achieved only recently by Frankel & Sivashinsky (1982) and Pelce & Clavin (1982). They obtained the following two-term expansion of the rate of instability parameter σ:

$$\sigma = \Omega_0 U_b k - \Omega_1 D_{th} k^2, \tag{17}$$

where

$$\Omega_1 = \frac{\varepsilon(1-\varepsilon)^2 - \varepsilon \ln \varepsilon (2\Omega_0 + 1 + \varepsilon)}{2(1-\varepsilon)[\varepsilon + (1+\varepsilon)\Omega_0]}$$
$$- \frac{\varepsilon(1+\Omega_0)(\varepsilon+\Omega_0)\beta(1-Le)}{2(1-\varepsilon)[\varepsilon+(1+\varepsilon)\Omega_0]} \int_0^{1/\varepsilon - 1} \frac{\ln(1+\xi)\,d\xi}{\xi}.$$

The correction to the Darrieus-Landau solution turned out to be independent of the Prandtl number. Thus, unlike the effects of heat conduction and diffusion, viscosity exerts a secondary effect on flame stability, which manifests itself apparently in higher-order terms of the expansion in powers of $k\ell_{th}$. These terms may, however, be significant when $\Omega_1 < 0$, when hydrodynamic flame instability combines with thermodiffusive instability, and also when Ω_1 is both positive and small.

It should be noted that when $\varepsilon = 0.2$, Ω_1 is negative if $\beta(1 - Le) > 2.66$. Thus, in comparison with the constant-density model, inclusion of thermal expansion of the gas somewhat narrows the thermodiffusive instability region.

When $\Omega_1 > 0$, Equation (17) implies the existence of a wavelength λ_c corresponding to the maximum amplification rate of small disturbances. If one assumes that the nonlinear evolution ultimately produces a structure with this characteristic length (Rayleigh principle), then one obtains $\lambda_c \simeq 100\ell_{th} \simeq 2$ cm for a typical slow flame ($\varepsilon = 0.2$, $\beta = 15$, $Le = 1.2$, $\ell_{th} = 0.2$ mm). Structures with this characteristic cell size should have been observed in the combustion of lean hydrocarbon-air mixtures in wide tubes (~ 10 cm). However, it is well known that this is not the case. The nature of the onset of hydrodynamic instability is clarified only after nonlinear effects have been incorporated.

3.2 *Nonlinear theory* Arguments analogous to those employed in the theory of thermodiffusive instability (see Section 2.2) yield the following evolution equation for the flame front:

$$F_t + \tfrac{1}{2}U_b(\nabla F)^2 = \Omega_1 D_{th}\nabla^2 F + \Omega_0 U_b I\{F\}, \tag{18}$$

where

$$I\{F\} = \frac{1}{4\pi^2}\int_{-\infty}^{\infty} |\mathbf{k}|\, e^{i\mathbf{k}(\mathbf{x}-\mathbf{x}')}F(\mathbf{x}',t)\, d\mathbf{k}\, d\mathbf{x}'.$$

When the thermal expansion of the gas is low (i.e. weak hydrodynamic instability), Equation (18) is a rigorous asymptotic relation, which can be derived from the full hydrodynamic equation system of the flame (Sivashinsky 1977b).[2]

Contrary to the predictions of the linear theory, numerical experiments on the one-dimensional version of Equation (18) in an interval of width $10\lambda_c$ have shown that during nonlinear evolution there appear on the flame front surface only one or two steady folds, strongly pointed toward the burned gas (Michelson & Sivashinsky 1977). If the folds lie at the ends of the interval, the flame will obviously present a smooth surface, convex toward the fresh mixture. Thus, hydrodynamic instability alone is sufficient to ensure that a flame in a wide tube will be curved (Uberoi 1959). A curved flame generates a gradient in the tangential component of the gas velocity along the front (i.e. stretch), and this is apparently what gives the configuration its remarkable stability (Zeldovich et al. 1980a; see also Section 2.6).

Figure 6 shows the results of the numerical solution to Equation (18) in a square $5\lambda_c \times 5\lambda_c$. Such folds are frequently formed when the flame

[2] In the limiting case of low thermal expansion, the gas flow is irrotational both ahead of the flame front and *behind* it. Thus, contrary to a periodically expressed opinion, the effect of vorticity generation on flame stability is not decisive.

crosses various kinds of obstacles, such as electrodes or a widely spaced wire grid (Markstein 1964, Palm-Leis & Strehlow 1969; see also Figure 3). In view of the stability of these folds (which are maintained even when the flame is stretched), it was suspected in the past that they are yet another variety of spontaneous flame instability (Markstein 1964, 1970).

The effects of diffusion and heat conduction are obviously most significant in the zone of the cusp, where the curvature of the front is extremely large. Outside this region the flame may be described perfectly well by the following truncated equation, which corresponds to the Darrieus-Landau model:

$$F_t + \tfrac{1}{2} U_b (\nabla F)^2 = \Omega_0 U_b I\{F\}. \qquad (19)$$

Since the Darrieus-Landau model does not involve any characteristic length, Equation (19) permits the existence of flames in which the distances between consecutive folds are arbitrarily large. A recent analysis of this equation by McConnaughey (1982) has shown that ∇F has a logarithmic singularity at the cusps of the folds. Thus, the flame front is infinitely sharp at the cusps and, consequently, the structure of such flames is essentially different from that of a Bunsen wedge.

3.3 *Thermal-expansion-induced cellular flames*   If the distance between the folds is increased indefinitely, the stabilizing effects of stretching and curvature are weakened and one expects new folds to appear. Thus, in order to investigate the fully developed hydrodynamic instability one must consider sufficiently large-scale flames. Experiments of this type were recently carried out by Lind & Whitson (1977) and Ivashchenko & Rumiantsev (1978) in connection with the investigation of accidental industrial explosions. Lind & Whitson experimented with lean hydro-

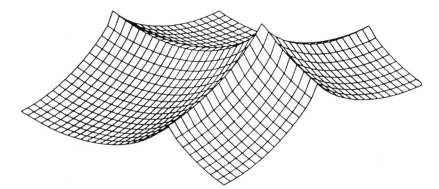

Figure 6   Thermal-expansion-induced steady folds. Numerical solution of Equation (18) in $5\lambda_c \times 5\lambda_c$ square with periodic boundary conditions.

carbon-air mixtures in 0.05-mm thick polyethylene film hemispheres of 5-m and 10-m radius. Ivashchenko & Rumiantsev carried out similar experiments in 0.05–0.08 mm thick rubber spherical shells of 2.5-m radius. It was noticed that the flame become rough as it expanded, taking on a "pebbled" appearance. This structure increased in size to about 0.4–1.0 m, with *finer* structure superimposed. For systems with markedly different burning velocities the measured space velocity was 1.6–1.8 times the expected value that was calculated from the normal burning velocity measured in the laboratory (Lind & Whitson 1977).

Ivashchenko & Rumiantsev also noted that when the sphere radius reached ~ 5 cm the flame became cellular, with cells about 1–2 cm in size. As the flame sphere grew the cell size increased, reaching ~ 6–10 cm for a sphere radius ~ 0.3–0.5 m. The maximum speed was 1.5–2 times the speed of the undisturbed spherical flame.

Stimulated by these experimental observations, Michelson & Sivashinsky (1982) undertook new numerical experiments on Equation (18) by considering a wider interval ($40\lambda_c$). They found that, alongside the deep folds on the flame, there indeed appeared a fine structure with a well-defined cell size (~ $5\lambda_c$). In the example cited previously, this implies a cell size of ~ 10 cm. Here the deep folds may evidently be associated with the large-scale structures observed by Lind & Whitson. The numerical experiments also indicate that the fine structure is non-steady. The cells continually and chaotically recombine, as occurs in thermodiffusive instability.

Of the earlier experimental observations of hydrodynamic instability in flames, we would like to mention the work of Simon & Wong (1953), who studied flames in a rich methane-air mixture filling a soap bubble of initial radius ~ 5 cm. When the flame radius was ~ 1.5 cm, the initially smooth flame front took on a cellular appearance, and simultaneously the flame was seen to accelerate. However, the relatively small volume of the mixture did not permit a sufficiently developed cellular instability, and the wrinkled flame did not reach the uniform propagation mode.

3.4   *Cellular flames near the thermo-diffusive instability threshold*   If Ω_1 is small (i.e. if Le is near Le_c), higher-order terms of the expansion in powers of $k\ell_{th}$ become important. This parameter range merits close attention, since it is here that the thermodiffusive model of a cellular flame breaks down (see Section 2.2). Moreover, almost any mixture may be brought to this range if the ratio of the reactant concentrations is suitably adjusted. However, computation of higher-order terms for the dispersion relation (17) is an extremely cumbersome task. Up to the present, this has been done only in the case of low thermal expansion, where one actually obtains a linear combination of the dispersion relation (16) and the relation (3) of the thermodiffusive theory (Sivashinsky

1977b). Numerical experiments on the corresponding fourth-order integrodifferential equation (Michelson & Sivashinsky 1977) have shown that if $\Omega_1 = 0$ (i.e. $Le = Le_c$), one obtains a chaotically recombining cellular structure with *finite* cell size of the order of λ_c. The cells, remaining finite, disappear for small positive Ω_1, i.e. when Le becomes greater than Le_c. Thus, near Le_c the dominating factor responsible for cellular instability may be of a purely hydrodynamic nature. We emphasize that the characteristic cell size in this case, as in the case of purely thermodiffusive instability, is determined by the fourth-order terms, regardless of whether Ω_1 is positive. Moreover, since $Le_c < 1$, the temperature variation along the front is quite similar here to that in purely thermodiffusive cells (Section 2). This is one of the essential differences between near-Le_c thermal-expansion-induced cells and the large-scale cells discussed in Section 3.3.

3.5 *Effects due to acceleration* Hydrodynamic instability of downward-propagating flames becomes weaker and may even be completely suppressed by buoyancy effects (see also Section 2.4). These effects were discussed in detail by Markstein (1964) in his phenomenological theory. The problem was recently reconsidered by Pelce & Clavin (1982), who consistently incorporated the effects of transport and chemical kinetics within the framework of a complete hydrodynamic flame model. In contradistinction to a freely propagating flame, a flame propagating in the presence of a stabilizing acceleration is stable to long-wave disturbances. Near the stability threshold, the unstable modes are concentrated near λ_c —the wavelength corresponding to the maximum amplification rate of σ. Consequently, hydrodynamic instability here should be manifested as cells of size $\sim \lambda_c$. This structure has indeed been observed in flames situated in a periodically varying acceleration field induced by vibrations of the gas flow (Markstein 1964, 1970). We note that cells of size ~ 1 cm appear here even in flames in lean hydrocarbon-air mixtures, which do not exhibit cellular instability under normal conditions.

A similar type of fixed-size structure may also be induced by oscillations of the flame-holder (Petersen & Emmons 1961) or by large-scale fluctuations of turbulent gas flow (Palm-Leis & Strehlow 1969).

ACKNOWLEDGMENT

These studies have been supported in part by the Director, Office of Energy Research, Office of Basic Energy Sciences, Engineering, Mathematical, and Geosciences Division of the US Department of Energy under contract DE-AC03-76SF00098. The hospitality of A. J. Chorin and A. K. Oppenheim is gratefully acknowledged

Literature Cited

Barenblatt, G. I., Zeldovich, Y. B., Istratov, A. G. 1962. On diffusional-thermal stability of a laminar flame. *Zh. Prikl. Mekh. Tekh. Fiz.* 4:21–26

Bregeon, B., Gordon, A. S., Williams, F. A. 1978. Near-limit downward propagation of hydrogen and methane flames in oxygen-nitrogen mixtures. *Combust. Flame* 33:33–45

Buckmaster, J. 1982a. Instability of the porous plug burner flame. *SIAM J. Appl. Math.* Submitted for publication

Buckmaster, J. 1982b. Polyhedral flames—an exercise in bimodal bifurcational analysis. *SIAM J. Appl. Math.* Submitted for publication

Buckmaster, J., Ludford, G. S. S. 1982. *Theory of Laminar Flames.* Cambridge Univ. Press.

Carrier, G. F., Fendell, F. E., Bush, W. B. 1978. Stoichiometry and flameholder effects on a one-dimensional flame. *Combust. Sci. Technol.* 18:33–46

Clarke, J. F., McIntosh, A. C. 1980. The influence of a flameholder on a plane flame, including its static stability. *Proc. R. Soc. London Ser. A* 372:367–92

Darrieus, G. 1938. *Propagation d'un front de flamme: assai de théorie des vitesses anomales de déflagration par développement spontané de la turbulence.* Presented at the Int. Congr. Appl. Mech., 6th, Paris, 1946

Ferguson, C. R., Keck, J. C. 1979. Stand-off distance on a flat flame burner. *Combust. Flame* 34:85–98

Frank-Kamenetsky, D. A. 1969. *Diffusion and Heat Transfer in Chemical Kinetics.* New York: Plenum

Frankel, M. L., Sivashinsky, G. I. 1982. The effect of viscosity on hydrodynamic stability of a plane flame front. *Combust. Sci. Technol.* In press

Glassman, I. 1977. *Combustion.* New York: Academic

Gololobov, I. M., Granovsky, E. A., Gostintsev, Y. A. 1981. On two combustion regimes at the limit of luminous flame propagation. *Fiz. Goreniya Vzryva* 17:28–33

Ishizuka, S., Law, C. K. 1982. An experimental investigation on extinction and stability of stretched premixed flames. *19th Symp. (Int.) Combust.* In press

Ishizuka, S., Miyasaka, K., Law, C. K. 1982. Effects of heat loss, preferential diffusion, and flame stretch on flame-front instability and extinction of propane/air mixtures. *Combust. Flame* 45:293–308

Istratov, A. G., Librovich, V. B. 1966. The influence of transport processes on the stability of a plane flame front. *Prikl. Mat. Mekh.* 30:451–66

Ivashchenko, P. F., Rumiantsev, V. S. 1978. Convective rise and propagation velocity of a large flame focus. *Fiz. Goreniya Vzryva* 40:83–87

Joulin, G. 1981. Conductive interactions of a wrinkled front with a cold flat burner. *Combust. Sci. Technol.* 27:83–86

Joulin, G. 1982. Flame oscillations induced by conductive losses to a flat burner. *Combust. Flame.* In press

Joulin, G., Clavin, P. 1979. Linear stability analysis of nonadiabatic flames. *Combust. Flame* 35:139–53

Joulin, G., Mitani, T. 1981. Linear stability analysis of two-reactant flame. *Combust. Flame* 40:235–46

Kanury, A. M. 1975. *Introduction of Combustion Phenomena.* New York: Gordon & Breach

Karlovitz, B., Denniston, D. W. Jr., Knappschaefer, D. H., Wells, F. E. 1953. Studies in turbulent flames. *4th Symp. (Int.) Combust.*, pp. 613–20

Landau, L. D. 1944. On the theory of slow combustion. *Acta Physicochim. URSS* 19:77–85

Law, C. K., Ishizuka, S., Mizomoto, M. 1981. Lean-limit extinction of propane/air mixtures in the stagnation-point flow. *18th Symp. (Int.) Combust.*, pp. 1791–98

Lind, C. D., Whitson, J. 1977. Explosion hazards associated with spills of large quantities of hazardous materials. Phase II. *Rep. No. CG-D-85-77. D.O.T. US Coast Guard Final Rep. ADA-047585*

Mallard, E., Le Chatelier, H. L. 1883. Combustion des melanges gazeux explosifs. *Ann. Mines, sér. 8.* 3:274–378

Manton, J., von Elbe, G., Lewis, B. 1952. Nonisotropic propagation of combustion waves in explosive gas mixtures and the development of cellular flames. *J. Chem. Phys.* 20:153–57

Margolis, S. B. 1980. Bifurcation phenomena in burner-stabilized premixed flames. *Combust. Sci. Technol.* 22:143–69

Margolis, S. B. 1981. Effects of selective diffusion on the stability of burner stabilized premixed flames. *18th Symp. (Int.) Combust.*, pp. 679–93

Markstein, G. H. 1949. Cell structure of propane flames burning in tubes. *J. Chem. Phys.* 17:428–29

Markstein, G. H. 1951. Experimental and theoretical studies of flame front stability. *J. Aeronaut. Sci.* 18:199–209

Markstein, G. H. 1952. Nonisotropic propagation of combustion waves. *J. Chem. Phys.* 20:1051–53

Markstein, G. H. 1964. *Nonsteady Flame Propagation*. Oxford: Pergamon

Markstein, G. H. 1970. Flames as amplifiers of fluid mechanical disturbances. *Proc. U.S. Natl. Congr. Appl. Mech., 6th,* pp. 11–33

Markstein, G. H., Somers, L. M. 1953. Cellular flame structure and vibrating flame movement in *n*-butane mixtures. *4th Symp. (Int.) Combust.,* pp. 527–35

Matkowsky, B. J., Olagunju, D. O. 1981. Pulsations in a burner-stabilized premixed plane flame. *SIAM J. Appl. Math.* 40:551–62

Matkowsky, B. J., Sivashinsky, G. I. 1978. Propagation of a pulsating reaction front in solid fuel combustion. *SIAM J. Appl. Math.* 35:465–78

Matkowsky, B. J., Sivashinsky, G. I. 1979. Acceleration effects on the stability of flame propagation. *SIAM J. Appl. Math.* 37:669–85

McConnaughey, H. V. 1982. *Topics in combustion.* PhD thesis. Cornell Univ., Ithaca, N.Y.

Merzhanov, A. G., Filonenko, A. K., Borovinskaya, I. P. 1973. New phenomena in combustion of condensed systems. *Proc. Acad. Sci. USSR, Phys. Chem. Sect.* 208:122–25

Michelson, D. M., Sivashinsky, G. I. 1977. Nonlinear analysis of hydrodynamic instability in laminar flames. Part II. Numerical experiments. *Acta Astronaut.* 4:1207–21

Michelson, D. M., Sivashinsky, G. I. 1982. Thermal expansion induced cellular flames. *Combust. Flame.* In press

Mitani, T., Williams, F. A. 1980. Studies of cellular flames in hydrogen-oxygen-nitrogen mixtures. *Combust. Flame* 39:169–90

Palm-Leis, A., Strehlow, R. A. 1969. On the propagation of turbulent flames. *Combust. Flame* 13:111–29

Pelce, P., Clavin, P. 1982. Influence of hydrodynamics and diffusion upon the stability limits of laminar premixed flames. *J. Fluid Mech.* In press

Petersen, R. E., Emmons, K. W. 1961. The stability of laminar flames. *Phys. Fluids* 4:456–64

Sabathier, F., Boyer, L., Clavin, P. 1981. Experimental study of a weak turbulent premixed flame. *Prog. Astronaut.*

Aeronaut. 76:246–58

Shchelkin, K. I. 1965. Instability of combustion and detonation in gases. *Usp. Fiz. Nauk* 87:273–302

Simon, D. M., Wong, E. L. 1953. Burning velocity measurement. *J. Chem. Phys.* 21:936

Sivashinsky, G. I. 1977a. Diffusional-thermal theory of cellular flames. *Combust. Sci. Technol.* 15:137–45

Sivashinsky, G. I. 1977b. Nonlinear analysis of hydrodynamic instability in laminar flames. Part I. Derivation of basic equations. *Acta Astronaut.* 4:1177–1206

Sivashinsky, G. I. 1979. On self-turbulization of a laminar flame. *Acta Astronaut.* 6:569–91

Sivashinsky, G. I. 1980. On flame propagation under conditions of stoichiometry. *SIAM J. Appl. Math.* 39:67–82

Sivashinsky, G. I. 1981. On spinning propagation of combustion waves. *SIAM J. Appl. Math.* 40:432–38

Sivashinsky, G. I., Matkowsky, B. J. 1981. On stability of nonadiabatic flames. *SIAM J. Appl. Math.* 40:255–60

Sivashinsky, G. I., Law, C. K., Joulin, G. 1982. On stability of premixed flames in stagnation-point flow. *Combust. Sci. Technol.* 28:155–59

Smith, F. A., Pickering, S. F. 1928. Bunsen flames with unusual structure. *Ind. Eng. Chem.* 20:1012–13

Smithells, S., Ingle, K. 1882. The structure and chemistry of flames. *J. Chem. Soc.* 61:204–17

Strehlow, R. A. 1968. *Fundamentals of Combustion.* Scranton, Pa: Int. Textbook Co., p. 254

Uberoi, M. S. 1959. Flow field of a flame in a channel. *Phys. Fluids* 2:72–78

Williams, F. A. 1965. *Combustion Theory.* Reading, Mass: Addison-Wesley

Zeldovich, Y. B. 1944. *Theory of Combustion and Detonation of Gases.* Acad. Sci. USSR (In Russian)

Zeldovich, Y. B. 1966. An effect that stabilizes the curved front of a laminar flame. *Zh. Prikl. Mekh. Tekh. Fiz.* 1:102–4

Zeldovich, Y. B., Frank-Kamenetsky, D. A. 1938. A theory of thermal propagation of flame. *Acta Physicochim. URSS* 9:341–50

Zeldovich, Y. B., Istratov, A. G., Kidin, N. I., Librovich, V. B. 1980a. Flame propagation in tubes: hydrodynamics and stability. *Combust. Sci. Technol.* 24:1–13

Zeldovich, Y. B., Barenblatt, G. I., Librovich, V. B., Makhviladze, G. M. 1980b. *Mathematical Theory of Combustion and Explosion.* Moscow: Nauka (In Russian)

Ann. Rev. Fluid Mech. 1983. 15:201-22

HOMOGENEOUS TURBULENCE

Jean Noël Gence

Ecole Centrale de Lyon, Université Claude Bernard Lyon I, 69130 Ecully, France

1. INTRODUCTION

When studying turbulent flows, it is customary to split all instantaneous random quantities into an averaged and a fluctuating part. Accordingly, the understanding of such flows implies the analysis of the different interactions between the mean and the fluctuating motion.

The mean flow field and the statistical moments associated with the fluctuating motion satisfy an infinite set of equations, among which the most useful are the following:

$$\frac{\partial \langle U_i \rangle}{\partial t} + \langle U_j \rangle \langle U_i \rangle_{,j} + \langle u_i u_j \rangle_{,j} = -\frac{1}{\rho} \langle p \rangle_{,i} + v \langle U_i \rangle_{,kk}$$

$$\frac{\partial}{\partial t} \langle u_i u_j \rangle + \langle U_k \rangle \langle u_i u_j \rangle_{,k} + \langle u_i u_j u_k \rangle_{,k}$$

$$= -\langle u_i u_k \rangle \langle U_k \rangle_{,j} - \langle u_j u_k \rangle \langle U_k \rangle_{,i} - \frac{1}{\rho} \left(\langle p u_i \rangle_{,j} + \langle p u_j \rangle_{,i} \right)$$

$$+ \left\langle \frac{P}{\rho} (u_{i,j} + u_{j,i}) \right\rangle + v \langle u_i u_j \rangle_{,kk} - 2v \langle u_{i,k} u_{j,k} \rangle$$

In order to simplify the above equations and to make it easier to understand the physics of the phenomena, different homogeneous turbulent flow models have been proposed. The simplest is the isotropic and homogeneous turbulence model introduced by G. I. Taylor, which does not involve a mean velocity gradient. But if real flows are to be described in a more realistic way, models with a uniform mean velocity gradient have to be considered; the fluctuating motion still exhibits the statistical homogeneity property even though it is no longer isotropic. Indeed, such models have proved extremely useful when developing prediction methods, even for flows commonly encountered in industrial applications.

0066-4189/83/0115-0201$02.00

A very detailed monograph on homogeneous turbulence was written by Batchelor (1952) and a generalization of the theory was proposed by Craya (1958), whose contributions were essentially devoted to flows with uniform mean velocity gradients. In what follows, we deliberately ignore all theoretical developments and confine our attention to the experimental aspects of homogeneous turbulent flows. We first discuss their general features and classify them according to their complexity:

1. Homogeneous and isotropic turbulence.
2. Homogeneous and anisotropic turbulence without a mean velocity gradient.
3. Homogeneous turbulence with a constant mean velocity gradient.

Throughout this paper, we use this classification as a guideline to discuss the different experiments that provide information thought to be essential to the understanding of the problem. At the same time, we also look at some of the research possibilities which appear as natural continuations.

2.   MAIN FEATURES OF HOMOGENEOUS TURBULENCE

2.1   Homogeneous Turbulence as a Model

In all real turbulent flows, the boundary conditions are such that the mean velocity gradient and all the moments associated with the fluctuating motion are space- and time-dependent. Both the experimental and theoretical study of turbulence are considerably simplified when all the moments of the fluctuating motion may be assumed to be invariant under any translation, i.e. when the concept of homogeneous turbulence is introduced. Although it is only a first approximation to the problem, this model retains most of the mechanisms that are active in real flows, except for the transfers connected with the spatial dependence of the moments.

The following is implied here:

1. Homogeneous turbulent flows must be boundless, which means that the mathematical problem associated with the study of this evolution is of the initial-value type.

2. All one-point concepts are space-independent and, in addition, all moments involving several points depend only on the vectors joining one of them to any of the others.

3. If a mean velocity gradient exists, it must be spatially uniform. Of course, under the assumption of incompressibility, the mean velocity field

is solenoidal. It also satisfies the following equation in the stationary case:

$$\langle U_j \rangle \langle U_i \rangle_{,j} = -\frac{1}{\rho} \langle P \rangle_{,i}. \tag{1}$$

As a consequence, the term on the left-hand side is solenoidal too. Craya (1958) has shown that the only mean flows compatible with the above conditions are a constant pure strain, a pure rotation, and plane mean flows like, for example, a constant shear. Under such conditions, the behavior of the one-point moments of the fluctuating field is governed by ordinary differential equations with respect to time, which are to be closed. Moreover, the spatial Fourier transform of the fluctuating field may be introduced, as well as the energy-spectrum concept, because of the absence of boundaries.

The homogeneous-flow assumption makes the analysis much simpler without jeopardizing its generality too seriously. It must be kept in mind, however, that since the Reynolds stress tensor is space-independent, there is no action by the fluctuating field on the mean flow, as indicated by Equation (1). Accordingly, the mean velocity gradient should be considered as a given external parameter, just like the viscosity of the fluid. Thus the only interactions that take place in homogeneous turbulence are summed up in Table 1. The characteristic scales associated with the fluctuating motion are intrinsic scales defined exclusively in terms of fluctuating quantities (see Corrsin 1963), such as the integral length scale and the velocity scale defined by

$$u' = \left(\langle q^2 \rangle / 3 \right)^{1/2}, \tag{2}$$

where $\langle q^2 \rangle$ is the trace of the Reynolds stress tensor.

Whatever the advantage of the foregoing model, satisfaction of the various homogeneity requirements imposes severe constraints on experi-

Table 1 Interactions occurring in homogeneous turbulence

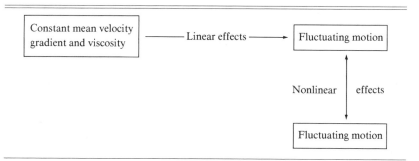

mental conditions, which are discussed in the next section, before any attempt is made at analyzing experimental data.

2.2   The Turbulent Box Concept

Usually, the homogeneous fluctuating motion that is generated upstream of a test duct is convected with a constant mean velocity $\langle U_1 \rangle$ parallel to its axis (denoted by x_1).

In order to satisfy the first condition of Section 2.1, the lateral and transverse length scales ought to be much smaller than the corresponding dimensions of the duct. Moreover, although the fluctuating motion cannot be strictly homogeneous along x_1, which is of course an evolution axis, it will be regarded as such if, as indicated by Corrsin (1963),

$$\frac{dL}{dx_1} \ll 1 \quad \text{and} \quad \frac{L}{Q}\frac{dQ}{dx_1} \ll 1, \tag{3}$$

where L is an integral length scale along x_1 and Q represents any statistical property associated with the fluctuating motion.

Instead of the unbounded space of the theory, and in accordance with the above two limitations, the experimentalist must consider a "turbulent box" whose dimensions are greater than the integral scales but smaller than the dimensions of the test duct, and which is convected along the x_1 axis with velocity $\langle U_1 \rangle$. In such a box, any statistical property of the fluctuating motion is considered as homogeneous, and its history, i.e. its time-dependence, is linked to the space-dependence in the duct by the relation

$$t \sim x_1/\langle U_1 \rangle,$$

which makes it possible to compare experiment and theory.

2.3   The Hierarchy of Homogeneous Flows

The different interactions to be found in homogeneous turbulence (see Table 1) are a priori strongly coupled. It is possible, however, to isolate some of the mechanisms involved if flows with and without mean velocity gradients are considered.

The latter class, which includes the so-called homogeneous isotropic flows, is the seat of mechanisms that are exclusively associated with the linear action of the viscosity and with the nonlinear self-interaction of the fluctuating motion. All other flows exhibit an additional interaction due to the linear influence of the constant mean velocity gradient on the fluctuating motion. They may be classified according to their complexity: pure plane strain, pure rotation, shear flows, and possibly more general mean plane flows that have not yet been experimentally investigated.

In what follows, a few typical experiments are discussed according to this hierarchy.

3.   HOMOGENEOUS TURBULENT FLOWS WITHOUT MEAN VELOCITY GRADIENT

In this section, we analyze successively isotropic and anisotropic turbulence, the main feature of the latter being to return toward isotropy. It is worth noting, however, that in both cases the turbulent kinetic energy decays with time according to the following equation:

$$\frac{\partial}{\partial t}\langle q^2/2 \rangle = -\langle \varepsilon \rangle, \tag{4}$$

where

$$\langle \varepsilon \rangle = \nu \langle u_{i,k} u_{i,k} \rangle \tag{5}$$

is obviously positive.

3.1   Homogeneous and Isotropic Turbulence

The statistical quantities describing the fluctuating motion are either isotropic tensorial functions or isotropic tensors; for example, the Reynolds stress tensor reduces to

$$\langle u_i u_j \rangle = \tfrac{1}{3}\langle q^2 \rangle \delta_{ij}.$$

Moreover, the behavior of the kinetic-energy spectrum $\mathcal{E}(K,t)$, where K is the modulus of the wave vector $\mathbf{K}$, is governed by a particularly simple equation

$$\frac{\partial}{\partial t}\mathcal{E}(K,t) + 2\nu K^2 \mathcal{E}(K,t) = \mathcal{S}(K,t), \tag{6}$$

where $\mathcal{S}(K,t)$ is associated with the nonlinear self-interaction of the fluctuating motion and represents a transfer between different wave numbers, since

$$\int_0^\infty \mathcal{S}(K,t)\, dK = 0. \tag{7}$$

Of course, (6) can only be solved if an assumption is made on $\mathcal{S}(K,t)$.

In most experiments, the turbulence is generated by a square-mesh biplane grid, with round or square rods, that is perpendicular to a flow of constant velocity $\langle U_1 \rangle$. The solidity of the grid, defined by the ratio of the area of the grid to the area of the section without the grid, is in general equal to 0.34. It is well known that at a distance x_1 of about 40 mesh lengths from the grid, the turbulence is homogeneous and quasi-isotropic in planes perpendicular to the x_1-axis. The homogeneity conditions (3) are well satisfied, since the integral length scale is of the order of the mesh length M and the rate of evolution of any statistical properties along x_1 is small enough. It is evident that one of the characteristic Reynolds numbers of such a flow is

$$\text{Re}_L = u'L/v,$$

where u' and L have already been defined in (2) and (3). Nevertheless, as noted by Corrsin (1963), it has been experimentally found that, for a fixed value of x_1/M, the ratios L/M and $u'/\langle U_1 \rangle$ are independent of the Reynolds number

$$\text{Re}_M = \langle U_1 \rangle M/v,$$

so that Re_L is proportional to Re_M which will therefore be the only characteristic Reynolds number used from now on. Experiment also proves that the component $\langle u_1^2 \rangle$ is greater than $\langle u_2^2 \rangle$ and $\langle u_3^2 \rangle$, which are generally equal, so that the isotropy condition on the Reynolds stress tensor is not fulfilled.

A major breakthrough was achieved by Comte-Bellot & Corrsin (1966, 1971), who improved the isotropy of the Reynolds stress tensor by using a small contraction ($c = 1.27$) located downstream from the grid. This contraction allowed them to reduce $\langle u_1^2 \rangle$ and to increase simultaneously the other two components so that the ratio $(\langle u_1^2 \rangle / \langle u_2^2 \rangle)^{1/2}$ was between 0.95 and 1.05. They also showed that the decay of $\langle u_1^2 \rangle$, as well as the decay of $\langle u_2^2 \rangle$, was given by a power law of the form

$$\langle u^2 \rangle / \langle U \rangle^2 = A \left(\frac{\langle U \rangle t}{M} - \frac{\langle U \rangle t_0}{M} \right)^{-n}, \tag{8}$$

where the exponent n lies between 1.15 and 1.33 in a Reynolds number range between 17,000 and 135,000. The fact that exponent n in (8) is different from 1 shows that the self-preserving property of the full turbulent spectrum is not attained during the decay. Comte-Bellot & Corrsin (1971) also reported data for the two-point space correlation functions of the u_1 velocity fluctuation in two mutually orthogonal directions, $f(r)$ and $g(r)$, and they observed that the theoretical condition imposed by isotropy,

$$g(r) = f(r) + \frac{r}{2} \frac{\partial f}{\partial r},$$

was well satisfied. This result enabled the authors to deduce with confidence the three-dimensional spectrum $\mathcal{E}(K, t\langle U \rangle / M)$ from the one-dimensional spectrum $E_{11}(K_1, t\langle U \rangle / M)$. Finally, they investigated the space-time-filtered correlation of the u_1 fluctuation

$$R_{11}[K_1; \langle U \rangle \Delta t, 0, 0, t_0; \Delta t]$$

from which, using isotropy relations, they deduced the filtered Eulerian time correlation at given K for a turbulent box

$$\mathcal{R}_{11}[K; t_0, \Delta t].$$

For each value of K, the area under the graph of $\mathcal{R}_{11}$ versus Δt is a measure of the temporal memory of the structures associated with the wave number K. The experiment of Comte-Bellot & Corrsin showed in particular that this lifetime is a decreasing function of K. Moreover, the authors established a spectrally local time-scaling factor $\tau^*(K)$ that is a "parallel" combination of the time characterizing (a) gross strain distortion by larger eddies, (b) wrinkling distortion by smaller eddies, (c) convection by larger eddies, and (d) gross rotation by larger eddies, i.e.

$$\tau^{*-1} = \left(K^2 \int_0^K \mathcal{E}(p)\,dp \right)^{1/2} + 2\left(\int_0^K p^2 \mathcal{E}(p)\,dp \right)^{1/2}$$
$$+ \left(K^4 \int_K^\infty p^{-2}\mathcal{E}(p)\,dp \right)^{1/2}.$$

The curves of $\mathcal{R}_{11}$ plotted against $\Delta t / \tau^*(K)$ then collapsed very well into a single curve.

The above-mentioned spectral results, which were obtained in a situation very close to isotropy, are among the best available data on this topic and may be used to test the closure assumptions expressing $\mathcal{S}(K, t)$ in terms of $\mathcal{E}(K, t)$. The transfer term $\mathcal{S}(K, t)$, just like any other term describing nonlinear effects, is unknown. Its determination from experiment is therefore of the utmost importance. It may be calculated from measurements of $\mathcal{E}(K, t)$ using Equation (6), as was first done by Uberoi (1963). Unfortunately, the reliability of his results is questionable, since the isotropy assumption is not satisfied in his experiment, due to the rather large value (1.2) of the ratio $(\langle u_1^2 \rangle / \langle u_2^2 \rangle)^{1/2}$. It became quite obvious that a more direct evaluation should be performed.

Such an evaluation was achieved by Van Atta & Chen (1969) in a grid turbulence whose mesh Reynolds number was equal to 25,000 and in which $(\langle u_1^2 \rangle / \langle u_2^2 \rangle)^{1/2}$ was about 1.12. The authors measured the two-point triple correlations of the velocity fluctuations

$$\langle u_j u_1 u_i' \rangle (r_1, 0, 0; t) = S_{j1i}(r_1, 0, 0; t),$$

and obtained $\mathcal{S}(K, t)$ from the isotropy relation

$$\mathcal{S}(K, t) = 4\mathcal{L}(K, t) - 2K\frac{\partial}{\partial K}\mathcal{L}(K, t),$$

where $- i\mathcal{L}(K_1; t)/K_1$ is the Fourier transform of $S_{i1i}(r_1, 0, 0; t)$. Errors, however, are to be expected for the small values of K, since it is well known that the turbulence is not isotropic in that range, as observed, for example, for the $E_{11}(K_1)$ and $E_{22}(K_1)$ spectra that do not satisfy the relation required by isotropy. The transfer term is linked to the two-point triple correlations, which evidently proves that the two-point probability density function (pdf) of the u_1 velocity fluctuation is not Gaussian. It then proves necessary to study in somewhat more detail the two-point moments of higher order than the second.

This information was obtained by Frenkiel & Klebanoff (1967a, b) and by Van Atta & Chen (1968) in similar experimental conditions. From time records of the fluctuation u_1 at one point and use of Taylor's hypothesis, these authors deduced the spatial correlation coefficient of order $m + n$ defined by

$$R^{m,n}(r_1) = \frac{\langle u_1^{(m)} u_1'^{(n)} \rangle (r_1, 0, 0; t)}{\langle u_1^2 \rangle^{(m+n)/2}}.$$

It was found that two-point correlations of even order (up to the eighth order) are nearly the same as those that would be measured if the two-point pdf was Gaussian, except for small values of r_1 for which a slight departure appears between the measured values of $R^{m,n}(r_1)$ and the corresponding value calculated from $R^{1,1}$ with the hypothesis of Gaussianity. The odd-order two-point correlations can be well predicted (up to seventh order) if the two-point pdf $P(U_1, U_1')$ is approximated by a Gram Charlier (i.e. non-Gaussian) expansion. In that set of measurements, attention has to be paid to the effect of high-pass filtering of the signals. This can be illustrated by the difficulty encountered by all authors in checking that two-point odd-order correlations vanish to zero at $r_1 = 0$, as they should for truly isotropic turbulence. This difficulty was not observed by Van Atta & Chen when they did not use a high-pass filter before the numerical treatment of the data, and they concluded that the discrepancy was due to the effects of filtering the low-frequency velocity fluctuations. This explanation is not very convincing because, as indicated by Bennett & Corrsin (1978), "the information lost due to high-pass filtering is at low frequency and may correspond to large values of r_1." It is then clear that other experiments are necessary to draw conclusions about the behavior of the odd-order two-point correlations in grid turbulence.

A more general result has been obtained by Van Atta & Yeh (1970), who investigated four-point moments of the u_1 fluctuation. The points were nevertheless aligned with the x_1 direction because of the use of Taylor's hypothesis. Again, except for small separations, the correlations can be well approximated by the Gaussian relation

$$\langle u_1 u_1' u_1'' u_1''' \rangle = \langle u_1 u_1' \rangle \langle u_1'' u_1''' \rangle + \langle u_1 u_1'' \rangle \langle u_1' u_1''' \rangle + \langle u_1 u_1''' \rangle \langle u_1' u_1'' \rangle.$$

This relation was proposed by Proudman & Reid (1954) as the "quasi-normal hypothesis," giving a functional relation between the transfer $\mathcal{S}(K, t)$ and the spectrum $\mathcal{E}(K, t)$. Ogura (1962) showed that negative parts in the spectrum $\mathcal{E}(K, t)$ are observed when using such an hypothesis, and Orszag (1970) proposed a correction that is known as the "eddy-damped quasi-normal approximation."

Besides the early contributions of Batchelor & Townsend (1948) and Batchelor & Stewart (1950), no really thorough analysis of small Reynolds number turbulent flows was achieved until Bennett & Corrsin's (1978) recent investigation. The authors studied very carefully the range of Reynolds numbers between 1300 and 1800. Because of the very low turbulence level, they corrected the measurements of all statistical quantities by taking the electronic noise into account, as well as the velocity fluctuations induced both by the free stream and the boundary layers. The u_1 fluctuation decayed according to the following empirical law:

$$\langle u_1^2 \rangle / \langle U_1 \rangle^2 = A[\langle U_1 \rangle t/M - \langle U_1 \rangle t_0/M]^{-2.51}.$$

This is in good agreement with theoretical predictions, when the interaction of the fluctuating motion is neglected compared to the effect of the viscosity. It is, however, important to note that the ratio $(\langle u_1^2 \rangle / \langle u_2^2 \rangle)^{1/2}$ grows continually with x_1 from 1.1 to about 1.2. Batchelor & Stewart (1950), who made the same observation, suggested that this anisotropy is linked to that of the large-scale structures that emerge in this "final period of decay." This result clearly shows that, no matter how useful it has proved, the generation of isotropic turbulence with a grid can only be achieved within limits and that "in spite of the obvious attendant measurements difficulties, it would be desirable to complement this traditional approach with experiments on isotropic box turbulence" (Corrsin 1963).

3.2   The Return to Isotropy of Homogeneous and Anisotropic Turbulence

When anisotropic turbulent flows are considered, the rate equation for the Reynolds stress tensor, which no longer reduces to Equation (4), may be written under the form suggested by Lumley & Newman (1977)

$$\frac{d}{dt}\langle u_i u_j \rangle = \left[\left\langle \frac{p}{\rho}(u_{i,j} + u_{j,i}) \right\rangle - 2\nu \langle u_{i,k} u_{j,k} \rangle + \frac{2}{3}\langle \varepsilon \rangle \delta_{ij} \right] - \frac{2}{3}\langle \varepsilon \rangle \delta_{ij},$$

(9)

where the traceless bracketed tensor of the right-hand side represents a *nonlinear* transfer between the Reynolds stress components, since the pressure fluctuation itself is given by

$$-\frac{\Delta p}{\rho} = u_{i,j}u_{j,i}. \tag{10}$$

In all the experiments that are discussed in the following, an initially homogeneous and anisotropic turbulence is created by the application of a constant strain to a grid-generated turbulence. Its evolution along a constant area duct located downstream of the straining duct is then studied.

Uberoi (1957) investigated the axisymmetrical case, where the principal components of the Reynolds stress tensor and the vorticity tensor were such that

$$\langle u_2^2 \rangle = \langle u_3^2 \rangle > \langle u_1^2 \rangle \quad \text{and} \quad \langle \xi_1^2 \rangle > \langle \xi_2^2 \rangle = \langle \xi_3^2 \rangle.$$

He noted that the principal components of the Reynolds stress tensor and the vorticity tensor have a tendency to become equal; from this observation, he inferred that in the absence of a mean velocity gradient, a nonisotropic turbulence returns toward isotropy. On the other hand, the characteristic time of the return toward isotropy of the components of the Reynolds stress tensor is greater than that of the components of the vorticity tensor, which is consistent with the fact that the big eddies have a larger time scale than the small ones. Finally, the characteristic time of return to isotropy of the Reynolds stress tensor is smaller than the time of decay defined by $\langle q^2 \rangle / 2 \langle \varepsilon \rangle$.

The above results were confirmed by Mills & Corrsin (1959), who also showed that the skewness factors associated with the fluctuations u_1 and $\partial u_1 / \partial x_1$ relaxed toward their isotropic value more rapidly than the Reynolds stress tensor. Comte-Bellot & Corrsin (1966) also observed that the tendency towards isotropy was slow in experiments where the grid was located downstream from the contraction. For example, the exponents n_1 and n_2, related respectively to the decay of $\langle u_1^2 \rangle$ and $\langle u_2^2 \rangle$, are very close ($n_1 \simeq 1.27$; $n_2 \simeq 1.24$).

The return to isotropy can also be investigated in the constant section duct that extends the distorting ducts discussed in the next section, as observed by Tucker & Reynolds (1968). Recently, Gence & Mathieu (1980) performed a similar experiment, but with experimental conditions that were quite different since the principal axes of the Reynolds stress tensor were made to rotate in the distorting duct upstream from the test section. They found that no rotation took place during the return to isotropy and concluded that this striking result could probably be generalized to any such process. It implies, in particular, that the underlined tensor in Equation (9) has the same principal axes as the Reynolds stress tensor.

In order to discuss and compare the previous experiments, it is essential to define a few basic parameters that characterize the return toward isotropy. This can be achieved by using Equation (9) and an equation for $\langle \varepsilon \rangle$ as proposed by Lumley & Newman (1977). The corresponding set of equations is closed by assuming that the unknown quantities are functions of $\langle \varepsilon \rangle$, $\langle u_i u_j \rangle$, and the viscosity ν. If the unknown functions are nondimensionalized, these terms can be expressed as functions of the nondimensional traceless symmetrical tensor b_{ij} defined by

$$b_{ij} = \langle u_i u_j \rangle / \langle q^2 \rangle - \delta_{ij}/3,$$

and of the Reynolds number Re_L, in which L is given by

$$L = \left(\langle q^2 \rangle / 3 \right)^{3/2} / \langle \varepsilon \rangle. \tag{11}$$

Experiment (e.g. Batchelor 1952, p. 106) shows that in quasi-isotropic grid turbulence, relation (11) is satisfied when L is the integral length scale of the fluctuation u_1 in the x_1 direction. It is then obvious that the independent scalar parameters characterizing the return to isotropy are Re_L and the independent invariants of the tensor b_{ij} defined by

$$\mathrm{II} = b_{ik} b_{ik} \quad \text{and} \quad \mathrm{III} = b_{ik} b_{k\ell} b_{\ell i} \tag{12}$$

In the above four experiments, the ratios

$$\frac{L}{\mathrm{II}} \cdot \frac{\partial \mathrm{II}}{\partial x_1} \quad \text{and} \quad \frac{L}{\langle q^2 \rangle} \cdot \frac{\partial \langle q^2 \rangle}{\partial x_1},$$

which clearly characterize the homogeneity of the flow along x_1, are of order 10^{-2} at the entrance of the test section and therefore satisfy requirement (3). Except in Mills & Corrsin's paper, no information is given regarding statistical homogeneity in planes perpendicular to x_1. The only available indication regarding the wall effects is provided by the value of the ratio L/D given in Table 2, where L is defined by (11) and D is the smallest dimension of the cross section.

Table 2 also gives the initial values of one of the main parameters associated with the energy decay and the return toward isotropy, i.e. the ratio of the two time scales τ_D and τ_R defined by

$$\tau_D = \langle q^2 \rangle \left(\frac{\partial \langle q^2 \rangle}{\partial t} \right)^{-1} \quad \text{and} \quad \tau_R = \mathrm{II} \left(\frac{\partial \mathrm{II}}{\partial t} \right)^{-1}.$$

It is to be noted that this ratio is approximately the same for the experiments of Uberoi, Mills & Corrsin, and Tucker & Reynolds, but is much smaller in the experiment of Gence & Mathieu, which essentially differs from the others by the sign of invariant III. This sign specifies the shape of the ellipsoid associated with the Reynolds stress tensor corresponding to two axisymmetric anisotropic situations, as indicated in Figure 1.

It would therefore be interesting to make systematic experiments with the same initial values of $\langle q^2 \rangle$, II, and Re_L, but opposite signs of III. Furthermore, there is a need for a complete investigation of all relevant statistical parameters measured in the same experiment.

4. CONSTANT MEAN VELOCITY GRADIENT FLOWS

The mathematical formalism describing constant mean velocity gradient flows was first developed by Burgers & Mitchner (1953) and Craya (1958). Denoting the constant mean velocity gradient by λ_{ij}, the rate

Table 2 Values of parameters

Experiment	Re_L	II	III	τ_D/τ_R	L/D
Uberoi (1957)	145	0.067	-7×10^{-3}	3.6	0.06
Mills & Corrsin (1959)	150	0.052	-5×10^{-3}	3.4	0.06
Tucker & Reynolds (1968)	320	0.08	-3.2×10^{-3}	2.82	0.24
Gence & Mathieu (1980)	450	0.08	$+4.2 \times 10^{-3}$	1.6	0.16

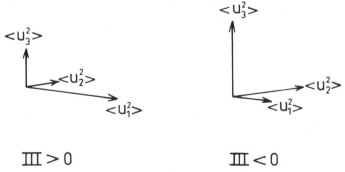

$$\mathrm{III} > 0 \qquad\qquad \mathrm{III} < 0$$

Figure 1   The physical meaning of the sign of invariant III.

equations for $\langle u_i u_j \rangle$ and for $\langle q^2 \rangle$, may be written

$$\frac{d}{dt}\langle u_i u_j \rangle = \underline{-\lambda_{ik}\langle u_k u_j \rangle - \lambda_{jk}\langle u_k u_i \rangle} + \left\langle \frac{P}{\rho}(u_{i,j} + u_{j,i}) \right\rangle$$

$$-2\nu\langle u_{i,k} u_{j,k} \rangle, \tag{13}$$

$$\frac{d}{dt}\langle q^2 \rangle = -2\lambda_{ij}\langle u_i u_j \rangle - 2\langle \varepsilon \rangle, \tag{14}$$

where the pressure fluctuation is given by

$$-\frac{\Delta P}{\rho} = 2\lambda_{\ell m} u_{m,\ell} + u_{\ell,m} u_{m,\ell} \tag{15}$$

This last equation shows that the pressure-deformation correlation term appearing in (13) is not only influenced by the nonlinear self-interaction of the fluctuating motion, but also by the linear action of the mean gradient.

The latter can be isolated from the former when the residence time of the turbulence in the test duct of length Λ is much smaller than the previously defined time τ_D. Using L given by (11), it is easy to show that this inequality reduces to (Batchelor 1952, p. 69)

$$u'/\langle U_1 \rangle < L/\Lambda. \tag{16}$$

Under these conditions, it is possible to neglect the nonlinear terms in (13), as has been done by Batchelor & Proudman (1954), who discussed the influence of a sudden distortion on an initially isotropic turbulence. In the same context, Deissler (1970) and Loiseau (1973) studied the action of a shear flow and Itsweire, Chabert & Gence (1979) investigated the action of a mean rotation on an initially nonisotropic turbulence. The general case of a uniform plane gradient has been investigated by Bertoglio et al. (1978a, b). Although they provide basic information for the modeling of the linear part of the pressure-deformation correlation in physical space, such calculations cannot describe real physical situations encountered in experiments where condition (16) is not generally satisfied and where all the different interactions specified in Table 1 take place simultaneously.

We now discuss pure strain, pure rotation, and constant shear.

4.1 Pure Constant Strain

Most of the experimental investigations dealing with pure constant strain are restricted to the plane case. The notation that is used in the following discussion is defined in Figure 2, which shows the directions of extension and contraction.

In all these physical situations, the action of the mean velocity gradient is characterized by the nondimensional product τ of its intensity D by the duration T of its application to the fluctuating motion.

The first reliable results were obtained by Townsend (1954), who experimentally studied the action of such a strain ($D = 9.6$ s^{-1}, $\tau = DT = 1.34$) on an initially quasi-isotropic grid turbulence. The upstream grid was located in a plane perpendicular to the $\langle u_1 \rangle$ direction. The distorting duct, which had a constant-area cross section, is such that its surface was the stream tube of a mean flow characterized by the given mean velocity gradient and by a constant mean velocity $\langle u_1 \rangle$ along the x_1 direction, as indicated in Figure 2. Townsend's main results qualitatively agree with the evolution predicted by the linear calculations of Batchelor & Proudman (1954) and can be summed up as follows:

1. The initially isotropic turbulence becomes nonisotropic under the influence of the mean strain, and the principal axes of the Reynolds stress tensor become identical to those of the strain. In particular, the component $\langle u_3^2 \rangle$ associated with the direction of contraction decays much less rapidly than in the isotropic case, while $\langle u_2^2 \rangle$ decays slightly faster. When the mean strain is applied during a long enough time, the ratio

$$K = \left(\langle u_3^2 \rangle - \langle u_2^2 \rangle \right) / \left(\langle u_3^2 \rangle + \langle u_2^2 \rangle \right)$$

is shown to grow and to reach an asymptotic value of about 0.42, as if the fluctuating field reached a nonisotropic equilibrium state.

2. The underlined term in (13), which in that case can be written

$$D\left(\langle u_3^2 \rangle - \langle u_2^2 \rangle \right),$$

is positive. Therefore the fluctuating motion receives energy from the mean motion.

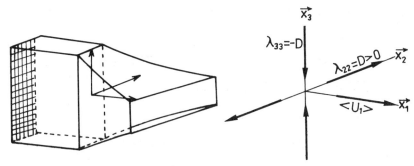

Figure 2 An example of distorting duct and the different axes of extension and contraction.

3. The eddies that are typical of the Taylor scale are also influenced by the mean gradient and do not remain in an isotropic state, as indicated by the ratio

$$K' = \left(\left\langle \left(\frac{\partial u_3}{\partial x_1} \right)^2 \right\rangle - \left\langle \left(\frac{\partial u_2}{\partial x_1} \right)^2 \right\rangle \right) \Big/ \left(\left\langle \left(\frac{\partial u_2}{\partial x_1} \right)^2 \right\rangle + \left\langle \left(\frac{\partial u_3}{\partial x_1} \right)^2 \right\rangle \right),$$

which would be zero in the isotropic case but here continuously grows with time up to a value of 0.4 for $\tau = 1.34$.

Two other experiments of the same kind were performed by Tucker & Reynolds (1968) ($D = 4.5$ s^{-1}; $\tau = 1.8$) and by Maréchal (1970) ($D = 19$ s^{-1}; $\tau = 2.6$). The first set of authors essentially confirmed Townsend's results. In addition, they showed that for higher values of τ, the ratio K could reach values as high as 0.6, which did not appear as an asymptotic limit. In the experiment of Maréchal, this ratio has a maximum value of 0.62 and slightly decays in the vicinity of the outlet section; this trend, however, might be due to the influence of the outlet of the duct. Moreover, Maréchal confirms the results of Townsend concerning the anisotropy of the fine structures and gives the evolution of the three integral scales $L_{11}^{(1)}$, $L_{11}^{(2)}$, and $L_{11}^{(3)}$ along the three coordinate directions, which can be deduced from the two-point u_1 correlations. In particular, $L_{11}^{(2)}$ grows while $L_{11}^{(3)}$ decays. The evolution of $L_{11}^{(1)}$ is more complicated, exhibiting a growth at the beginning of the distortion followed by a decay. This last behavior was suggested by the one-dimensional spectra of the u_1 fluctuation observed by Townsend.

In view of the behavior of $\langle u_2^2 \rangle$, $\langle u_3^2 \rangle$, $L_{11}^{(2)}$, and $L_{11}^{(3)}$, the plane distortion may be simulated, at least from a statistical point of view, by a crude model consisting of the elongation and the contraction of eddies aligned with the principal axes of the distortion (Figure 3).

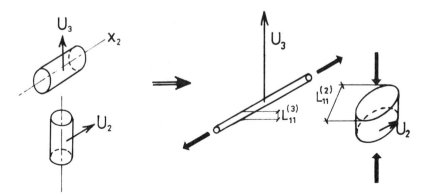

Figure 3   A crude model explaining the influence of the pure plane strain.

In these different experiments, the initial turbulence is not isotropic, but the principal axes of the Reynolds stress tensor are always aligned with those of the strain that acts upon it. Therefore, the initial statistical orientation of the fluctuating motion is not modified by the influence of the plane strain. It still remains to be seen what happens when a pure plane strain acts on an initially nonisotropic "oriented turbulence" in which one of the principal axes of the Reynolds stress tensor is aligned with x_1 and the other two make an initial angle φ_0 with those of the strain tensor. In order to answer this question, Gence & Mathieu (1979) designed a distorting duct in which a quasi-isotropic grid turbulence was subjected to two successive coplanar strains whose principal axes made an angle φ_0 (Figure 4). In such an experiment, the fluctuating motion, which is initially statistically oriented by the first strain, undergoes downstream the action of the second strain, whose principal axes are, of course, no longer aligned with those of the Reynolds stress tensor. It is observed that due to the influence of the second strain, the principal axes of the Reynolds stress tensor rotate around x_1 and exhibit a tendency to become asymptotically aligned with those of the strain tensor. The time scale of this process is of order D^{-1}. Besides, when φ_0 is greater than $\pi/4$, the underlined term in equation (13), which eventually recovers a positive value, starts taking on negative values, meaning that the kinetic energy of the fluctuating motion decays under the action of the mean gradient. Finally, during this first period the invariant II defined by (12) also decreases.

Besides the experiments with plane strains just discussed, it is worth mentioning Reynolds & Tucker's (1975) investigation of three-dimensional strains acting upon an initially quasi-isotropic turbulence. Although they provide no new information on the basic interaction

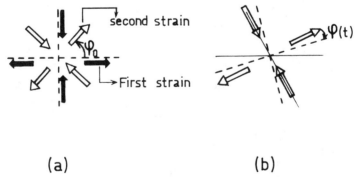

(a)   (b)

Figure 4 (*a*) The two successive plane strains. – – – – represents the principal axes of the Reynolds stress tensor at the inlet of the second strain; (*b*) The evolution of the principal axes of the Reynolds stress tensor under the influence of the second strain.

mechanisms, their results prove useful when testing numerical prediction methods.

4.2   The Pure Rotation

In the case of a pure rotation, the interaction term between the mean gradient and the fluctuating motion, which is underlined in Equation (13) is zero since λ_{ij} and $\langle u_i u_j \rangle$ are respectively antisymmetrical and symmetrical. Accordingly, the rate equation for $\langle q^2/2 \rangle$ reduces to form (4). One of the main questions that arises is whether the mean rotation has any influence on the energy decay and, more generally, on the evolution of the Reynolds stress tensor. This problem is still very controversial, since opposite conclusions depending on the experimental setup have been drawn by Ibbetson & Tritton (1975) and by Wigeland & Nagib (1978), who generalized Traugott's (1958) results.

Ibbetson & Tritton (1975) considered an enclosed volume of fluid consisting of a cylindrical annulus turning around its axis (Figure 5a). The turbulent motion is produced by horizontal grids that rotate with the annulus. The grids are also subjected to a relative vertical motion during a short initial period, at the end of which $(t = 0)$ the decay of the turbulence is studied from $t = 4$ s to $t = 100$ s. Different rotation rates Ω are used, and the maximum value obtained for the product $\tau = \Omega t$ is 640. No information concerning the homogeneity of the Reynolds stress tensor and of the different scales is given. Moreover, the only components of the Reynolds stress tensor that have been measured are $\langle u_2^2 \rangle$ and $\langle u_3^2 \rangle$ because of experimental difficulties associated with the

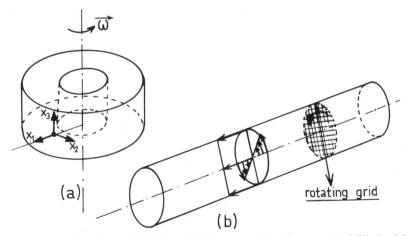

Figure 5   (a) Experiment of Ibbetson & Tritton (1975); (b) Experiment of Wigeland & Nagib (1978).

measurement of $\langle u_1^2 \rangle$. It must be emphasized that the initial turbulence is not isotropic, since the initial ratios $(\langle u_2^2 \rangle / \langle u_3^2 \rangle)^{1/2}$ obtained for the different rotation rates lie between 1.3 and 1.8. The authors observed that the rotation induces an increase of the rate of decay, whose characteristic time varies linearly with $\Omega^{-1/2}$ within the range of rotation rates investigated (from $1\ s^{-1}$ to $6.4\ s^{-1}$). They also showed that this behavior can be predicted by using a model in which the turbulent kinetic energy is transferred to the walls by inertial waves and is dissipated in the boundary layers. Such an agreement, however, supports the view that their flow is not homogeneous, since their interpretation of the dynamics of the decay is strongly influenced by the boundary layers.

Wigeland & Nagib (1978) considered a constant-area duct in which the fluid undergoes a solid rotation around the axis x_1 of the duct and is convected with a constant mean velocity $\langle U_1 \rangle$ along x_1, so that the downstream relative mean flow is the same as in usual grid-generated turbulence (see Figure 5b). The authors checked that for different ranges of rotation rate Ω (6–$180\ s^{-1}$) and convecting velocity $\langle U_1 \rangle$ (6–15 m s^{-1}) and for three different grids, the mean axial profile was uniform while the mean tangential velocity varied linearly within a large domain of each cross section whose radius was of 15 cm. The maximum value of Ωt obtained with such mean velocities was 6.28. In addition $\langle u_\theta^2 \rangle$ and $\langle u_1^2 \rangle$ were shown to be homogeneous inside a circle of radius 3 cm located in the farthest cross section downstream from the grid. In the latter cross section, however, the length scales are of the order of 1 cm, which suggests that the influence of boundary layers is not negligible. The authors measured three components of the Reynolds stress tensor: $\langle u_s^2 \rangle$ in the streamwise direction, $\langle u_r^2 \rangle$ in the radial direction, and $\langle u_n^2 \rangle$ in the direction normal to the two as well as $\langle u_s u_n \rangle$, the latter decreasing rapidly to zero. Their conclusion regarding the rate of decay of $\langle q^2 \rangle$ is opposite from that of Ibbetson & Tritton. Furthermore, using Taylor's assumption, the authors determined L_s, L_n, and L_r, which are defined as the integral length scales in the direction of the mean streamline and are associated with the fluctuations u_s, u_n and u_r. They showed that the scales increase faster than when no rotation takes place, which is consistent with the decrease of the decay rate, since $\langle \varepsilon \rangle$ can always be evaluated from $(\langle q^2 \rangle)^{3/2}/L$, where L is the order of magnitude of these scales.

Although no mean drastic change in the trend indicated by the authors is a priori expected, it is obvious that a larger range of values for Ωt should be investigated in order to confirm the above-described behavior.

In the same experimental context, it would probably be both possible and worthwhile to superimpose a rigid body rotation to a constant strain, thus generating arbitrary plane uniform gradient flows, of which the shear flow is a special case.

4.3   The Uniform Shear Flow

Among all uniform mean gradient flows, the shear flows are the closest to real physical situations such as those that are found in turbulent boundary or mixing layers, jets, and wakes.

The first attempt to generate such a flow was made by Rose (1966). In this experiment, both the mean gradient and the quasi-isotropic turbulent motion were created by a grid composed of horizontal rods whose spacing increased with x_1 (see Figure 6a). The mean gradient ($\lambda_{12} = 13.6$ s^{-1}) is uniform along x_1, as expected, and the Reynolds stress tensor components are homogeneous in planes parallel to the grid. Unfortunately, the integral scales and the Taylor scales depend on x_2 because they have a memory for the nonuniformity of the grid. Therefore, the fluctuating motion cannot be considered as strictly homogeneous in planes perpendicular to x_1.

This difficulty, however, was overcome by Champagne et al. (1970), who used a different setup consisting of a row of equal-width channels having adjustable internal resistances. On the long centerline of each channel's exit plane, a rod was installed in order to reduce the scale of the initial turbulence (Figure 6b). The experiment, in which λ_{12} was equal to 12.9 s^{-1} and the nondimensional product $\lambda_{12} t_{\text{Max}} = \tau$ was about 3, showed that the diagonal components of the Reynolds stress tensor and, of course, $\langle q^2 \rangle$ first slightly decay and then reach an asymptotic value such that

$$\langle u_2^2 \rangle < \langle u_3^2 \rangle < \langle u_1^2 \rangle. \tag{17}$$

Moreover, the correlation coefficient associated with $\langle -u_1 u_2 \rangle$ grows from zero to a constant value of 0.5. It was also observed that the three

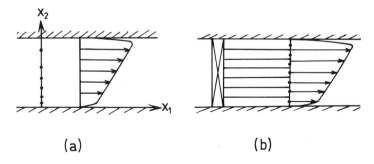

(a)   (b)

Figure 6   (a) Experiment of Rose (1966); (b) experiment of Champagne et al. (1970).

integral and Taylor scales that can be defined from the two-point double correlation of the u_1 fluctuation increase continuously, so that in a framework convected with the axial mean velocity $\langle U_a \rangle$, the fluctuating motion is not statistically stationary. All these results had already been found by Rose (1966) and are quite paradoxical, as underlined by Champagne et al. Indeed, since the time behavior of $\langle q^2 \rangle$ in the convected framework is governed by

$$\frac{d}{dt}\langle q^2 \rangle = \underline{-2\lambda_{12}\langle u_1 u_2 \rangle} - 2\langle \varepsilon \rangle,$$

the constant production of $\langle q^2 \rangle$ given by the underlined term should be balanced by a constant dissipation term if $\langle q^2 \rangle$ is stationary. In fact, $\langle \varepsilon \rangle$ can certainly not be constant, according to the classical evaluation formula

$$\langle \varepsilon \rangle \sim \nu \langle q^2 \rangle / \lambda^2,$$

since Taylor's length scale λ is known to increase with time.

Realizing that no definite statement concerning the constancy of $\langle q^2 \rangle$ could be made unless the range of nondimensional product τ was really significant, Harris et al. (1977) designed a new experiment, with a constant shear of 45 s^{-1} providing a large enough value of τ (about 12). Under these conditions, it was observed that the principal components of the Reynolds stress tensor, while still satisfying inequality (17), kept increasing beyond $\tau = 4$ within that range, which removes the above-mentioned paradox. The correlation coefficient associated with $\langle -u_1 u_2 \rangle$ is equal to 0.47 and the integral scale $L_{11}^{(1)}$ grows. Furthermore, the principal axes of the Reynolds stress tensor were shown not to be aligned with those of the strain tensor and to make angles with x_1 that are respectively equal to 22.3° and 67.7°. Finally, using the equation for the Reynolds stress tensor components and assuming that the viscous term could be approximated by the expression known to be valid in isotropic turbulence, they evaluated the pressure strain correlation tensor and determined its principal axes. These axes were found to be quite close to those of the Reynolds stress tensor, since the angles they made with x_1 were respectively equal to $-27°$ and 63°.

The authors concluded that "because of the unavoidable downstream growth of transverse inhomogeneity it seems unlikely that this tunnel technique for generating nearly homogeneous turbulent shear flows can be extended much further in terms of the total strain."

5.   CONCLUSION

The various experiments described in this review provide fundamental information essential to the understanding of many mechanisms taking place in homogeneous turbulent flows. Two particular situations, however, require a deeper analysis:

1. The return to isotropy, in which the influence of a particular parameter must be investigated in detail.
2. The action of a mean rotation on an initially isotropic turbulence, since the only two experiments devoted to this topic yield opposite results.

It is obvious that the homogeneous turbulence model cannot explain phenomena that are influenced by boundary effects, such as intermittency or the generation of bursts in the vicinity of a wall. Nevertheless, most of the prediction methods used in the calculation of turbulent flows are still derived from the basic properties of homogeneous flows. In this context, the nonhomogeneity effects are considered as mere corrections that are added, in a second step, to a set of equations whose closures are known to describe homogeneous flows adequately.

Finally, although it has become the subject of a review in itself and lies beyond the scope of this paper, it is worth mentioning that the homogeneous turbulence model has been extended to experimental situations in which a passive scalar fluctuating field is superimposed on a homogeneous velocity field.

ACKNOWLEDGMENTS

The author wishes to express his deep appreciation to Professors G. Comte-Bellot and J. Bataille for their numerous and useful comments.

Literature Cited

Batchelor, G. K. 1952. *The Theory of Homogeneous Turbulence.* Cambridge Univ. Press. 197 pp.

Batchelor, G. K., Proudman, I. 1954. The effect of rapid distortion of a fluid in turbulent motion. *J. Mech. Appl. Math.* 7:83–103

Batchelor, G. K., Stewart, R. W. 1950. Anisotropy of the spectrum of turbulence at small wave-numbers. *Q. J. Mech. Appl. Math.* 3:1

Batchelor, G. K., Townsend, A. A. 1948. Decay of turbulence in the final period. *Proc. R. Soc. London Ser. A* 194:527

Bennett, J. C., Corrsin, S. 1978. Small Reynolds number nearly isotropic turbulence in a straight duct and a contraction. *Phys. Fluids* 21:2129–40

Bertoglio, J. P., Charnay, G., Gence, J. N., Mathieu, J. 1978a. Effet d'une rotation en bloc sur les spectres de corrélation double d'une turbulence homogène soumise à des gradients de vitesse. *C. R. Acad. Sci. Ser. A* 286:167–70

Bertoglio, J. P., Charnay, G., Gence, J. N., Mathieu, J. 1978b. Calcul d'une turbulence homogène soumise à une rotation en bloc et à un cisaillement. *C. R. Acad. Sci. Ser. A* 286:957–60

Burgers, J. M., Mitchner, M. 1953. On homogeneous non-isotropic turbulence connected with a mean motion having a constant velocity gradient. *Proc. K. Ned. Akad. Wet. B* 56:228–35, 343–54

Champagne, F. H., Harris, V. G., Corrsin, S. 1970. Experiments on nearly homogeneous turbulent shear flow. *J. Fluid Mech.* 41:81–139

Comte-Bellot, G., Corrsin, S. 1966. The use of a contraction to improve the isotropy of grid generated turbulence. *J. Fluid Mech.* 25:657–82

Comte-Bellot, G., Corrsin, S. 1971. Simple Eulerian time correlation of full and narrow-band velocity signals in grid-generated, "isotropic" turbulence. *J. Fluid. Mech.* 48:273–337

Corrsin, S. 1963. Turbulence: experimental methods. *Handb. Phys.* 8:523–90. Berlin: Springer

Craya, A. 1958. *Contribution à l'analyse de la turbulence associée à des vitesses moyennes.* Thèse. Univ. Grenoble, Fr. *Publ. Sci. Tech.* 345, Ministère de l'air

Deissler, R. G. 1970. Effect of initial conditions on weak homogeneous turbulence with uniform shear. *Phys. Fluids* 3:1868–69

Frenkiel, F., Klebanoff, P. S. 1967a. Higher-order correlations in a turbulent field. *Phys. Fluids* 10:507–20

Frenkiel, F., Klebanoff, P. S. 1967b. Correlation measurement in a turbulent flow using high-speed computing methods. *Phys. Fluids* 10:1737–47

Gence, J. N., Mathieu, J. 1979. On the application of successive plane strains to grid-generated turbulence. *J. Fluid. Mech.* 93:501–13

Gence, J. N., Mathieu, J. 1980. The return to isotropy of an homogeneous turbulence having been submitted to two successive plane strains. *J. Fluid Mech.* 101:555–66

Harris, V. G., Graham, J. A. H., Corrsin, S. 1977. Further experiments in nearly homogeneous turbulent shear flow. *J. Fluid Mech.* 81:657–87

Ibbetson, A., Tritton, D. I. 1975. Experiments on turbulence in a rotating fluid. *J. Fluid Mech.* 62:639–72

Itsweire, E., Chabert, L., Gence, J. N. 1979. Action d'une rotation pure sur une turbulence anisotrope. *C. R. Acad. Sci. Ser. B* 156:197–200

Loiseau, M. 1973. *Evolution d'une turbulence homogène soumise à un cisaillement.* Thèse de docteur Ingenieur. Univ. Lyon I, Fr.

Lumley, J. L., Newman, G. 1977. The return to isotropy of homogeneous turbulence. *J. Fluid Mech.* 82:161–78

Maréchal, J. 1970. *Contribution à l'étude de la déformation plane de la turbulence.* Thèse. Univ. Grenoble, Fr.

Mills, R., Corrsin, S. 1959. Effect of contraction on turbulence and temperature fluctuations generated by a warm grid. *NASA Memo. 5-5-59 W*

Ogura, Y. 1962. A consequence of the zero-fourth-cumulant approximation in the decay of isotropic turbulence. *J. Fluid Mech.* 16:33–40

Orszag, S. A. 1970. Analytical theories of turbulence. *J. Fluid Mech.* 41:363–86

Proudman, I., Reid, W. H. 1954. On the decay of a normally distributed and homogeneous turbulent velocity field. *Philos. Trans. R. Soc. London Ser. A* 247:163–89

Reynolds, A. J., Tucker, H. J. 1975. The distortion of turbulence by general uniform irrotational strain. *J. Fluid Mech.* 68:673–93

Rose, W. G. 1966. Results of an attempt to generate a homogeneous turbulent shear flow. *J. Fluid Mech.* 25:97–120

Townsend, A. A. 1954. The uniform distortion of homogeneous turbulence. *Q. J. Mech. Appl. Math.* 7:104–27

Traugott, S. C. 1958. *NACA Tech. Note 4135*

Tucker, H. J., Reynolds, A. J. 1968. The distortion of turbulence by irrotational plane strain. *J. Fluid Mech.* 32:657–73

Uberoi, M. S. 1957. Equipartition of energy and local isotropy in turbulent flows. *J. Appl. Phys.* 28:1165–70

Uberoi, M. S. 1963. Energy transfer in isotropic turbulence. *Phys. Fluids* 6:1048–56

Van Atta, C. W., Chen, W. Y. 1968. Correlation measurements in grid turbulence using digital harmonic analysis. *J. Fluid Mech.* 34:497–515

Van Atta, C. W., Chen, W. Y. 1969. Measurements of spectral energy transfer in grid turbulence. *J. Fluid Mech.* 38:743–63

Van Atta, C. W., Yeh, T. T. 1970. Some measurements of multi-point time correlations in grid turbulence. *J. Fluid Mech.* 41:169–78

Wigeland, R. A., Nagib, H. M. 1978. Grid generated turbulence with and without rotation about the streamwise direction. *Rep. R78-1,* Ill. Inst. of Technol., Chicago

Ann. Rev. Fluid Mech. 1983. 15:223–39
Copyright © 1983 by Annual Reviews Inc. All rights reserved

LOW-REYNOLDS-NUMBER AIRFOILS

P. B. S. Lissaman

AeroVironment Inc., Pasadena, California 91107

INTRODUCTION

The airfoil section is the quintessence of a wing or lifting surface and, as such, occupies a central position in any design discipline relating to fluid mechanics, from animal flight through marine propellers to aircraft. The proper functioning of the airfoil is the prerequisite to the satisfactory performance of the lifting surface itself, and thus the airfoil is of fundamental technical importance.

Transcending functional considerations, the physical shape of the airfoil—teardrop-like or paisley motif-like—seems to have some universal aesthetic appeal. As a consequence, the development and selection of airfoils has exercised an almost mystical fascination on designers. Since the early work of Eiffel and Joukowsky at the turn of the century, fluid dynamicists have recognized the importance of the airfoil shape and have developed a bewildering plethora of airfoil designs and families, many with almost magical claims of efficaciousness. But the ideal shape of an airfoil depends profoundly upon the size and speed of the wing of which it is the core. This dependence is called *scale effect*.

In the thirties, the significance of scale effect was first recognized. This relates to the phenomenon that an airfoil that has most excellent qualities on an insect or bird may not exhibit these advantages when scaled up for an airplane wing, and vice versa. Different sizes of airfoils require different shapes. This scale effect is characterized by the chord Reynolds number, R, defined by $R = Vc/v$, where V is the flight speed, c is the chord, and v is the kinematic viscosity of the fluid in which the airfoil is operating. The Reynolds number quantifies the relative importance of the inertial (fluid momentum) effects on the airfoil behavior, compared

223

0066-4189/83/0115-0223$02.00

with the viscous (fluid stickiness) effects. It is the latter effects that essentially control the airfoil performance since they dictate the drag or streamwise resistance as well as limiting and controlling the maximum lift of the airfoil. Normally, these qualities are described by the lift and drag coefficients, C_L and C_D, defined as L/qc and D/qc, respectively, where L and D are the lift and drag per unit span, q is the flow dynamic pressure, and c is the airfoil chord. The lift and drag coefficients depend on the Reynolds number as well as on the angle of attack of the airfoil, which represents its geometric inclination to the incoming flow.

It is interesting to describe the different wing systems occurring in the wide range of Reynolds numbers over which airfoils are used. We briefly outline the various flight vehicles that are discussed in more detail in an outstanding paper by Carmichael (1981). We draw special attention to Carmichael's encyclopedic report that for the first time puts together all the known theoretical and experimental results, gives the highlight conclusions, and provides the most exhaustive set of references available.

Figure 1 shows this huge scale range, which spans the Reynolds numbers from 10^2 to 10^9. Below the lower limit, viscous effects are dominant and it is unlikely that any airfoil-like performance can occur. In the next range, up to 10^4, we find the insects and small model airplanes. Here, the flow is characteristically strongly and persistently laminar. At somewhat higher Reynolds numbers, up to 10^5, one enters the range of flying animals and large model airplanes. Airfoil performance is still relatively low in this range, but a significant improvement in performance occurs as we enter the next regime

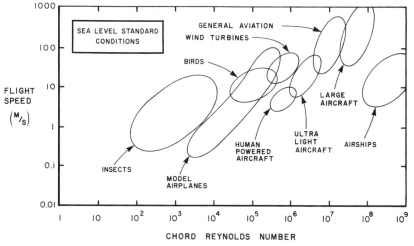

Figure 1   Flight Reynolds-number spectrum.

In this regime, up to about 10^6, we find a major improvement in airfoil performance and the coexistence of a number of fascinating flight systems. We cannot do better here than by quoting Carmichael (1981) directly: "In this regime, we find man and nature together in flight. Large soaring birds of quite remarkable performance, large radio-controlled model aircraft, foot-launched ultralight, man-carrying hang-gliders, and that superb engineering triumph, the human-powered aircraft." Here, we also find the airfoils for smaller modern wind turbines.

Beyond this range, we find sailplanes, light aircraft, and jet transports operating at Reynolds numbers up to and beyond 10^7. In this extensively studied range, some of the highest-performance airfoils have been developed. At high Reynolds numbers, in the neighborhood of 10^8, we enter the regime of large water-immersed vehicles, such as tankers and large nuclear submarines.

The air vehicles described have all been assumed to operate at sea level. Recently, there have been developments in small remotely piloted vehicles (RPVs) used for surveillance, sampling, and monitoring in both military and scientific roles. These vehicles are generally small and often operate at very high altitudes, where the kinematic viscosity is significantly increased by the very low ambient density. These vehicles are not shown in the figure. Because of the extreme operating range, from sea level to 30 km, they span a large Reynolds-number spectrum. Frequently, both the propellers and wings of these RPVs will be required to perform at Reynolds numbers significantly below half a million. This introduces for the first time an aerospace technological requirement for low-Reynolds-number airfoils.

Usually the function of the airfoil is to produce lift, or a force approximately at right angles to its direction of relative motion, while the drag is connected with the forces necessary to propel the lifting surface. Thus, a convenient parameter to measure the effectiveness of an airfoil is its lift-to-drag ratio, C_L/C_D; the maximum value of this quantity gives a good indication of the airfoil effectiveness. For design purposes, it is desirable that this maximum occur at a high lift coefficient so that the physical size of the lifting surface is minimized. An indication of the magnitude of this lift is given by the performance parameter, $C_L^{3/2}/C_D$, which gives somewhat more weighting to the lift coefficient. For detailed design of lifting surfaces, it is necessary to know even more of the performance structure of the airfoil—that is, how the lift and drag vary with angle of attack at a given Reynolds number, the airfoil signature, expressed by its lift-to-drag polar.

We have alluded to the fact that at lower Reynolds numbers the viscous effects are relatively large, causing high drags and limiting the maximum lift, while at the higher values the lift-to-drag ratio improves. There is a critical Reynolds number of about 70,000 at which this performance transition takes place. This dramatic improvement can be seen most vividly in Figure 2, taken from McMasters & Henderson (1980). Here we note the striking change in performance for smooth airfoils near the critical Reynolds number where the lift-to-drag ratio increases more than an order of magnitude. It is of great interest that a rough or turbulated airfoil does not exhibit this abrupt performance change with Reynolds number. We note from Figure 2 that this critical Reynolds number really divides the airfoils of the insect class (less than 10^4) from those of the large airplane class (above 10^6).

Some representative airfoil sections of this transitional range are shown in Figure 3. At the low end, we have the insects, with the interesting feature that it is not necessary to have a smooth surface; in fact, it is likely that the discontinuities are desirable to delay flow separation. For birds, however, smoothness begins to be important, as shown by the pigeon section. In the middle range is the Eppler 193, an airfoil with excellent performance at a Reynolds number of about 100,000, and at the high end, the Lissaman 7769, the airfoil used on the *Gossamer Condor* and *Albatross*, and the Liebeck L 1003, an airfoil of striking performance that provided clues on which the design of the Lissaman 7769 was based.

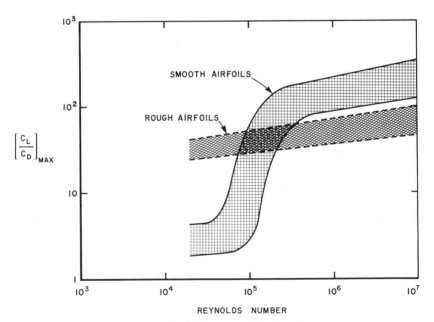

Figure 2 Low-Reynolds-number airfoil performance.

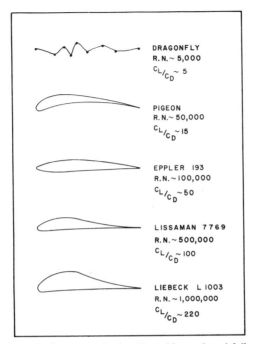

Figure 3   Representative low-Reynolds-number airfoils.

The 7769 airfoil (Lissaman 1980) was developed especially for human-powered airplanes. The design requirements were that the airfoil should have low drag, relatively high lift, a mainly flat undersurface (it is linear for the last 80%), and that it should be tolerant to irregularities due to low-fidelity construction and distortion of covering surface in flight. The latter two requirements, apparently whimsical in the world of high technology, proved to be critically important in the milieu in which the Gossamer aircraft were built, flown, crashed, and repaired.

In the following discussion, we describe the fluid mechanics, performance, and design of low-Reynolds-number airfoils, which we take to include the range between Reynolds numbers of about 10^4 and 10^6. We do not discuss compressibility effects, although these do occur at high altitudes on RPVs, and we confine the discussion to two-dimensional flows, although the third dimension is important on fans, propellers, and low-aspect-ratio wings.

FUNDAMENTAL FLUID MECHANICS

All airfoils have regions of lower-than-static pressure. Normally, these are most pronounced on the suction (or lifting) surface; however, the effects of thickness itself on a symmetrical nonlifting airfoil will introduce a region of accelerated flow and the associated lower pressure. The higher speed flow must then return to approximately free-stream conditions at the trailing edge, experiencing a pressure recovery through an adverse pressure gradient. For airfoils operating in excess of 10^6 Reynolds number, this adverse gradient normally occurs after transition so that it is impressed on a turbulent boundary layer that can negotiate quite severe adverse pressure gradients without separation. However, in lower Reynolds-number ranges, the boundary layer at the onset of the pressure rise may still be laminar, and thus unable to withstand any significant adverse pressure gradients. The performance of low-Reynolds-number airfoils is entirely dictated by the relatively poor separation resistance of the laminar boundary layer.

In the lowest Reynolds-number range (below 30,000), conditions are normally such that the boundary layer is still laminar beyond the point at which pressure recovery commences and, provided the pressure gradient is mild, complete laminar flow can occur for small angles of attack. As the lift is increased, the adverse gradients become more severe and laminar separation occurs, limiting the lift coefficient and significantly increasing the drag. At the lowest Reynolds number, this separation may occur over the entire rear of the airfoil, extending into the wake. However, when a laminar boundary layer separates, the separated layer very rapidly undergoes transition to a turbulent flow, because of the increased transition susceptibility of the separated shear layer. The increased entrainment of this turbulent flow makes it possible for the flow to reattach as a turbulent boundary layer. This forms what is called a *laminar separation bubble*.

Figure 4 shows the general geometric structure of a laminar bubble. After laminar separation, the flow proceeds at an approximately constant separation angle and the processes of transition occur. As turbulence develops, the increased entrainment causes reattachment. After reattachment, the turbulent boundary layer reorganizes itself to form an approximately normal turbulent profile.

These bubbles exhibit a very interesting spectrum of behavior and have been extensively studied, with particular reference to the conditions of reattachment and the length of the bubble. It has been pointed out by Carmichael (1981) that a rough rule is that the distance from separation to reattachment can be expressed as a Reynolds number based on bubble length of approximately 50,000. Thus, for airfoils of chord Reynolds number of about this magnitude, the airfoil is physically too short for reattachment to occur. This accounts for the general observation that the critical Reynolds number of an airfoil is about 70,000. Below this value it is unusual for reattachment to occur.

For airfoils at a Reynolds number higher than 70,000, conditions can exist for reattachment so that a laminar bubble can form. Depending upon the airfoil shape, different types of laminar bubbles occur, characterized by the bubble length as either short or long.

Being a boundary-layer effect, bubble geometry must properly be described in scales of boundary-layer heights, which can be normalized and nondimensionalized by forming a local Reynolds number using length scales from the bubble. The pressure gradients, on the other hand, scale principally with the airfoil shape and chord. However, for the airfoil designer, it is useful to give typical bubble proportions in terms of the airfoil chord. It must be noted that such proportions are strongly Reynolds-number-dependent.

At a Reynolds number of about 10^5, the long bubble generally extends over 20–30% of the airfoil and significantly changes the pressure distribution by effectively altering the shape over which the outer potential stream flows. At higher Reynolds numbers, a short bubble may form. The short bubble is generally of length of the order of a few percent of the airfoil chord and thus does not greatly alter the pressure from its normal attached distribution. So the short bubble, initially, generally represents the transition-forcing mechanism, and as long as it stays short, it does not greatly affect the airfoil performance. However, as the angle of attack increases, requiring a greater pressure recovery in the laminar bubble for reattachment, the short bubble can "burst" to form a long

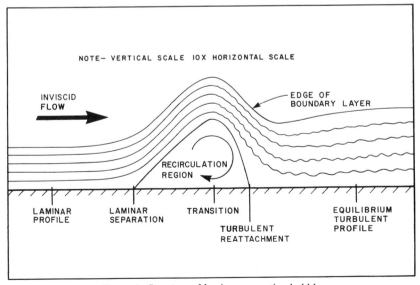

Figure 4   Structure of laminar separation bubble.

bubble. This bursting causes an abrupt stall and sudden severe deterioration in airfoil performance.

As might be expected, the stability of the short bubble is marginal and its bursting can be triggered by many extraneous flow effects. Frequently after bubble burst, reducing the angle of attack will not immediately "unburst" the bubble, so that hysteresis effects occur as the angle of attack is cycled. It is this behavior of short- and long-bubble formation, and bursting with angle of attack and Reynolds number, that causes such striking differences in performance of various airfoil shapes. Carmichael (1981) gives an extensive discussion of the variety of polar shapes that can occur because of these effects and classifies them into five basic characteristic modes.

Figure 5 illustrates the two most distinct of these modes on two different airfoils at a Reynolds number of 50,000. Note that both airfoils have the same minimum drag and $(C_L/C_D)_{max}$. The well-behaved polar is similar to that of conventional airfoils at Reynolds numbers above 1,000,000; the other polar represents the gyrations that can occur in the critical Reynolds-number range. Figure 5 demonstrates the importance of considering the airfoil polar as well as its maxima and minima in lift and drag.

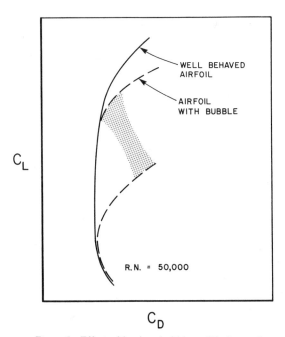

Figure 5   Effect of laminar bubble on lift-drag polar.

As the Reynolds number of the airfoil increases and approaches 200,000, it becomes more likely that the laminar bubble can be avoided, since it is usually possible to design the airfoil so that transition occurs upstream of any severe adverse pressure gradient. Now, the pressure recovery occurs in the turbulent boundary layer, with its much greater resistance to separation. However, it is still apparent that performance is less than would occur at a higher Reynolds number, principally because the separation resistance of the turbulent boundary layer increases as the Reynolds number becomes larger.

Above a Reynolds number of 500,000 there is a further improvement in airfoil characteristics. There is an extensive body of literature available on this regime. The laminar bubble can still occur in injudicious designs but can usually be avoided. It is noted that the laminar bubble is not strictly associated with the chord Reynolds number but, rather, with the local boundary-layer Reynolds number at which pressure recovery first commences. If this recovery is very near the leading edge of the airfoil and the adverse gradients are severe, bubble-type behavior may again develop. Such a situation occurs with thin airfoils of small nose radius at high angle of attack, even at Reynolds numbers exceeding a few million.

Further performance improvements occur for airfoils operating above the 1,000,000 Reynolds number range, although the rate of increase is generally slow and advantages are painfully won. Laminar separation should not be a problem here and the advantages occur through the weak reduction in turbulent skin friction as the Reynolds number (R) increases. The drag coefficient in this case varies approximately proportionally to $R^{-1/5}$. The maximum lift coefficient also increases slowly because of the increased separation resistance of the turbulent boundary layer. Lower limits on drag can readily be estimated; they correspond quite closely to the skin friction of an attached flow with a given transition point. Upper limits on lift have not been as clearly defined; however, a particularly inspired approach was made by Liebeck (1978), who took the point of view that the maximum upper-surface lift would be obtained by developing an upper-surface adverse pressure distribution that was uniformly critically close to separation. On this basis, the entire pressure-recovery region would be operating at its maximum capacity. This involved assuming an incipient separation turbulent profile, calculating the pressure field required to produce this, and deriving the airfoil shape from this critical-velocity distribution. These airfoils caused some surprise by performing in test just as well as, and in some cases better than, the theoretical predictions. It is believed that Liebeck-type airfoils have achieved the best $(C_L/C_D)_{max}$ of any tested in the Reynolds number range from about 500,000 to 2,000,000.

The necessity of eliminating laminar separation at lower Reynolds numbers has led to the development of techniques to artificially accelerate transition, or to "turbulate" the boundary layer. A wide variety of techniques are available to accomplish this and they are discussed in detail by Carmichael (1981). Transition-promoting devices, called *turbulators*, range from simple mechanical roughness elements in the form of serrations, strips, bumps, or ridges near the airfoil leading edge, through transpiration methods using airjets emitted from surface orifices of fractional-percentage chord diameter, to exotic procedures like beaming sound waves of frequencies calculated to cause transition at the wing surface, or mechanically vibrating the wing itself. Transition can also be accelerated by increasing the free-stream turbulence with wires or grids ahead of the airflow. The latter methods sound impractical but are intended to simulate flight conditions in which engine noise, airframe vibration, or strong ambient turbulence introduces such effects.

The design of turbulators is subtle, since the transition-inducing mechanism must be of significant magnitude to produce turbulence and suppress laminar separation without causing the boundary layer to become unnecessarily thick. A thick turbulent boundary layer may again suffer separation, or at least cause an increase in drag. Studies of the effect on $(C_L/C_D)_{max}$ with fixed trip strips, typically of about $1/4\%$ of the chord and located in the first 25% of the chord, have been conducted on a number of airfoils. At a Reynolds number of about 40,000, an increase of about 20% in this parameter is noted, while at 60,000 the increase is only about 10%. At a Reynolds number of 100,000, however, no distinct improvement is seen and some of the airfoils tested experience a reduction in lift-to-drag ratio.

EXPERIMENTAL TESTING OF AIRFOILS

Airfoil testing involves an intrinsic difficulty in that the two quantities to be correlated—the lift and the drag—differ in magnitude by a factor of about 100. In addition, a major region of interest is usually in the vicinity of stall and separation, where small changes can trigger large effects. In wind-tunnel testing, there are frequently difficulties with wall effects. These are both inviscid, where the confined potential flow must be taken into account, and viscous, where boundary layers emanating from the walls or the support system can influence the boundary-layer behavior of the test airfoil section. In addition, it is important to consider the incoming turbulence in the test flow, as well as any perturbations due to mechanical or acoustic disturbances. These disturbances can be partially avoided by free-flight testing (essentially, carefully executed performance measurements of gliding wings, either in the atmosphere or in a large, closed building). The problems here involve the inevitable three-dimensional nature of the flow as well as the difficulty of isolating the airfoil performance from that of the rest of the glider.

For wind-tunnel testing, it is possible to measure the forces directly or to measure the pressures on the airfoil and the velocity and pressure in the wake. Pressure-measuring techniques have the advantage of providing information on the details of the chordwise pressure distribution. For free-flight testing, the principal observable is the flight trajectory, and great pains must be taken to eliminate unsteady effects either from improper launching or from ambient air-mass motions. Flow visualization is applicable to both test methods and will provide valuable qualitative information, as well as quantitative data on the geometrical features of transition and separation.

For all the above reasons, test data in the low-Reynolds-number range have long been regarded with skepticism, especially earlier test results, and there is indeed a substantial record of nonrepeatability of data from tests in different facilities. Sometimes this is attributable simply to inaccurate measurement techniques, but more frequently it can be because the model and environment are actually different from one test to another; the model shape may not be true (the profound effect of turbulator devices of size less than $1/2\%$ of chord illustrates this point), the tunnel turbulence may be different, or the boundary effects may vary.

Even in modern wind-tunnel test facilities with advanced instrumentation and airflows of turbulence levels lower than $1/10\%$, striking differences in airfoil performance are reported, particularly near the critical Reynolds number of about 70,000. Carmichael (1981) reports that for a good standard low-Reynolds-number airfoil, the Eppler 61, tests in two facilities gave a $(C_L/C_D)_{max}$ of about 50, while a third facility reported about 25 for this value. It is disturbing for the practical designer to note that the high data were inferred by integrating wake flow measurements, while the low performance (25) was obtained by measurement of actual forces on the airfoil, a much more convincing observable if properly executed. This discrepancy may be due to spanwise flow variations.

Modern collections of airfoil test data are given by Miley (1982), Althaus & Wortmann (1981), and Althaus (1980).

THEORETICAL DESIGN OF AIRFOILS

The analytical design of airfoils has always occupied an important position in aeronautical research. Many low-Reynolds-number airfoil applications involve flows that are essentially two-dimensional and incompressible, and are thus particularly amenable to analysis. It has already been noted that at high altitudes, low-Reynolds-number airfoils may indeed operate at high Mach numbers and thus experience compressibility effects, while propellers, fans, and low-aspect-ratio wings operate in strongly three-dimensional flow fields. However, here we consider only two-dimensional incompressible flows, a class that covers a large number of applications. We do not discuss airfoils with separate elements, like slats, vanes, or flaps, confining the discourse to single-element airfoils.

Two-dimensional inviscid-flow theory is a very well developed discipline in fluid mechanics and has been a productive research area since the earliest days, when it was first noted that there are a number of simple conformal transformations that convert the circle into airfoil-like shapes with a rounded, bulbous leading edge and a tapered, wedge, or cusp-like trailing edge. This makes it possible to obtain exact analytical formulations of the flow field about airfoils using complex-variable techniques. The availability of numerical computing machines of high speed and capacity has greatly extended the scope and range of flow calculations for airfoils of arbitrary nonanalytical shapes. Thus, the determination of the potential flow about any shape can now be accomplished with dispatch and precision.

The attached boundary layer on the airfoil can also be calculated with good accuracy, providing the location of transition can be reliably estimated. Thus, assuming that separation does not occur, it appears that methods are now available to reliably compute airfoil performance at low Reynolds numbers. An excellent discussion of a modern viscous-inviscid design procedure is given by McMasters & Henderson (1980).

If the flow is fully attached, then the methods described above can be used to predict both the lift and drag with good accuracy and to examine the effects of changes in the airfoil shape.

However, airfoil performance is always limited by separation. In the high-Reynolds-number situation, this usually takes place in the turbulent boundary layer toward the rear of the airfoil. Methods are available to estimate this separation lift coefficient and even the separated drag, although normally this drag is so high that its precise value is not of great interest. Over the last 50 years, an extensive body of research has developed on the stall of airfoils at Reynolds numbers exceeding 1,000,000, and it is correct to say that the field is well understood to the extent that for regular airfoil shapes one can calculate with high precision the lift and drag of such an airfoil up to the separation and also make satisfactory estimates of the poststall performance.

For large Reynolds numbers, above 1,000,000, transition normally occurs near the minimum pressure point, at the first onset of the adverse pressure gradient. This will normally assure that a separation-resistant turbulent boundary layer occurs in the pressure recovery region. However, if the initial adverse gradient is too severe, a laminar separation occurs, as previously described; transition occurs in the separated region, and although reattachment occurs, the drag of the airfoil increases and the maximum lift is reduced. This effect can be eliminated by delicate contouring of the airfoil near the minimum pressure point to create a less severe adverse pressure gradient, called an *instability range*, to accomplish separation-free transition.

These design techniques encompass both the direct procedure, where for a given airfoil shape the pressure fields are determined, and the inverse procedure, where for a given pressure field the airfoil shape is determined. Historically, the direct methods appear to have developed first. Evidently, these methods require an airfoil shape definition that is presumably arbitrary (or even mystical) and not necessarily optimum. A more satisfying and rational procedure is to derive the airfoil shape by using analytical methods from the required pressure characteristics. A most interesting application of this inverse approach is exemplified by the important research of Liebeck, which began as an attempt to develop airfoils of higher lift than had previously been believed possible. These high-lift designs have been proven in test, and also have been shown to have very desirable low-drag characteristics and performance polars. It is believed that these Liebeck airfoils have developed the highest lift-to-drag ratios of any airfoil at any Reynolds number. These design methods assume attached flow and are acceptable down to a Reynolds number of about 300,000.

In the lower Reynolds-number ranges, unusual performance characteristics develop, all caused in one respect or another by different features of transition, laminar separation, and laminar bubble behavior (described in the previous section). Although a considerable amount of theoretical and experimental work has been done here, there is still no generally accepted method of calculating the development of a laminar bubble or even reliably predicting transition at lower Reynolds numbers. However, empirical results are available to the designer, including a number of relatively simple criteria involving relationships between the appropriate boundary-layer parameters in separated flows. Approaching the problem from the pure fluid-mechanical point of view, recent work has involved attempts to calculate the detailed flow structure in the bubble. These use numerical methods to solve fairly complete expressions of the basic flow equations, and correlate well with test data. An excellent summary, including the major results of the various papers, is given by Carmichael (1981).

SPECIAL-PURPOSE AIRFOILS

Airfoil design has progressed through a number of stages. The first consisted of designing families of airfoil shapes on geometrical principles. In some cases, simple analytical shape formulations were intentionally chosen, based on some intuition that such "pure" curves might show magical fluid-mechanical properties. Some members of these families have indeed exhibited admirable characteristics. The next stage involved specifying not the airfoil shape, but the pressure distributions, with various smooth or semi-analytical properties that were generally desirable from lift, transition, and separation aspects. Families of airfoil shapes were then derived to produce these pressure fields. A more sophisticated further stage, exemplified by Liebeck, involved defining the pressure distribution to meet specific high-lift or low-drag requirements while assuring boundary-layer attachment. The airfoil shape was then developed by inverse methods.

The above techniques provide the designer with a compendium of airfoils from which he can choose those which best meet his vehicle requirements. Sometimes the match between the designer specification and the best-fitting performance may leave much to be desired.

Modern computing techniques and an increased understanding and quantification of critical boundary-layer behavior have now made it possible to design airfoils specially tailored for a given flight vehicle, thus assuring the best possible match. Two airfoils designed along these lines are described below, and are shown in Figure 6.

The BoAR 80 (McMasters et al. 1981) was designed for a new ultra-light sailplane of 120-kg all-up weight and 11.5-m span. The high-lift requirements dictated the upper-surface shape and defined the maximum lift that could be carried by the suction (upper surface). Additional lift was required and had to be obtained by undercamber on the lower surface to increase the overpressure on that side. Thus, the pressure distribution required to develop this lift without separation was defined, and the airfoil was designed by inverting this pressure field. Then, using direct methods, the high-speed (low lift and angle of attack) pressure distribution was determined. It was estimated that separation would occur near the leading edge on the lower surface, in this case with undesirably high drag. Thus, the full flight-range requirements could not be met by the highly cambered shape required for high lift.

However, the high-speed requirements could be met by maintaining the upper-surface shape and dropping the lower surface to increase the thickness and reduce the camber. This provided an airfoil with separation-free low-drag performance in the high-speed, low-angle-of-attack flight regime. The airfoil is intended to change configuration during flight, activated by command, as one might use a flap.

The BoAR 80 airfoil has not been tested but the design calculations indicate a minimum drag coefficient in the thin configuration of about 0.0065 with a $(C_L/C_D)_{max}$ in the cambered configuration of 147.

The Lissaman-Hibbs 8025 airfoil (MacCready et al. 1981) was designed for the *Solar Challenger*, a photovoltaic-powered aircraft of 150-kg all-up weight and 14.3-m wingspan. The aerodynamic requirements called for a maximum lift coefficient of about 1.6 and a gentle stall, with low drag at the cruise lift coefficient. These were not severe. The geometrical requirements, however, were draconian—that the upper surface should be flat over its major length. This is a quite traumatic constraint to the designer, who depends upon smooth-flowing curves to achieve good flow characteristics.

The eminently practical reason for this constraint is that the upper surface of the wing and stabilizer are almost entirely covered with rectangular silicon solar cells, 2×6 cm in size. It is of interest that the stabilizer served not in its titular function but principally as a platform for the photovoltaic system. It was decided that the cells should be on the outer surface of the wing to maximize cooling, and that the solar panels

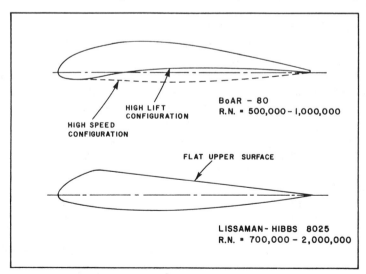

Figure 6   Special purpose low-Reynolds-number airfoils.

should be perfectly flat to simplify installation and also to minimize drag due to unavoidable surface irregularities if a curved surface was approximated by a multifaceted, trapezoidal shape, "tiled" by the cells (each of which was about 1.6% of the airfoil chord). Also, the flat surface exposes all cells to the same sun angle, thus avoiding any voltage unbalances. The airfoil was designed by direct computer methods to cruise with low drag at a lift coefficient of about unity. Desirable features of the pressure distribution on the upper surface could not be obtained by modifying that surface, but had to be achieved by the shape of the nose section and the undersurface.

The flat-topped stabilizer (not shown here) was designed on the same principles, except that since it was mainly a cell platform, its lift coefficient was very low and the design object was to minimize the section drag by achieving the maximum extent of laminar flow on the lower surface. Both airfoils, as finally developed, were perfectly flat for the major portion of the upper surface; the wing for the last 85%, the stabilizer for the last 90%.

No laboratory data are available on these airfoils, which were tested in the ultimate wind tunnel—the sky. In 1981, the *Solar Challenger* flew 350 km from France to England powered only by the sun. Crude flight tests indicate that the minimum drag coefficient of the wing airfoil is about 0.009, while the $(C_L/C_D)_{max}$ is about 160.

PARTING COMMENTS

Low-Reynolds-number airfoils encompass the heart of the lift system of a magnificently wide variety of flying vehicles, from birds and bats to human-powered aircraft and remotely piloted vehicles. There is an extensive body of research on airfoil performance in this range and, at first glance, these data present a bewildering assortment of inconsistencies. However, we now understand the mechanics of the flow well enough to be able to explain this behavior, at least in the sense that for a given performance we can interpret it as a rational consequence of the boundary-layer behavior in the highly sensitive Reynolds-number domain where transition, separation, and reattachment are all occurring in a short length of the airfoil. In other words, if we know the events that happened, we can say why.

The concomitant question of how to make these events happen once it is known why they do is much more difficult and is far from solved. For this reason, the design of low-Reynolds-number airfoils, particularly at Reynolds numbers below 300,000, is still a black art. It is likely, however, that most of the fluid-mechanical processes have been qualitatively defined, and good progress is being made to place these phenomena on a quantitative predictive basis.

The study of low-Reynolds-number airfoils is an intriguing subject. Its fascination can be attested to by the large volume of excellent amateur, unsupported, or minimally funded research that it has engendered. To date, there have been no critical military or commercial requirements for a systematic airfoil technology in this regime. Nonetheless, an extensive and intelligent body of knowledge exists and expands. Apparently, there are people who study this field simply because they find its challenges elegant and attractive. With the appearance of the new small unmanned aircraft, the RPVs, with their military and scientific importance, it is possible that there will be a surge of adequately funded activity. If this research is directed with a sufficiently longsighted vision so that the fundamental issues as well as the immediate practical ones are tackled, we may see a resolution of some of the principal unsolved issues. These issues, relating intimately to the great interconnected puzzle of transition, turbulence, separation, and reattachment, are fundamental to classical fluid mechanics.

Literature Cited

Althaus, D. 1980. *Profilpolaren für den Modellflug.* Karlsruhe: NeckarVerlag

Althaus, D., Wortmann, F. X. 1981. *Stuttgarter Profilkatalog 1.* Braunschweig: Vieweg & Sohn. 320 pp.

Carmichael, B. H. 1981. Low Reynolds number airfoil survey. Vol. 1. *NASA CR 165803*

Liebeck, R. H. 1978. Design of subsonic airfoils for high lift. *J. Aircr.* 15:547–61

Lissaman, P. B. S. 1980. Wings for human-powered flight. *Proc. AIAA Symp. Evol. Aircr. Wing Design,* SP 802, pp. 49–56

MacCready, P. B., Lissaman, P. B. S., Morgan, W. R., Burke, J. D. 1981. Sun-powered aircraft design. *AIAA Repr. 81-0916*

McMasters, J. H., Henderson, M. L. 1980. Low speed single element airfoil synthesis. *Tech. Soaring* 6(2):1–21

McMasters, J. H., Nordvik, R. H., Henderson, M. L., Sandvig, J. H. 1981. Two airfoil sections designed for low Reynolds numbers. *Tech. Soaring* 6(4):2–24

Miley, S. J. 1982. A catalog of low Reynolds number airfoil data for wind turbine applications. Prepared for Rockwell Int. Corp., Energy Syst. Group, Rocky Flats Plant, Wind Syst. Program

Ann. Rev. Fluid Mech. 1983. 15:241–60

NUMERICAL METHODS IN NON-NEWTONIAN FLUID MECHANICS

M. J. Crochet

Université Catholique de Louvain, Louvain-la Neuve, Belgium

K. Walters

University College of Wales, Aberystwyth, Wales

1. INTRODUCTION

The student of Newtonian fluid mechanics has been accustomed for three hundred years to accept the Newtonian constitutive relation as a standard fluid model. The resulting Navier-Stokes equations are noted for their outstanding difficulty due to the presence of inertia terms that are nonlinear in the velocity components; numerical techniques have shed considerable light on this subject over the last 25 years. Yet it is an easy task to generate experiments that could never be simulated even grossly by the Newtonian model. Indeed, Newtonian behavior appears to be the exception rather than the rule, and any fluid that does not conform to it is accordingly called non-Newtonian.

Non-Newtonian behavior has been the object of several books in recent years (Astarita & Marrucci 1974, Bird et al. 1977, Schowalter 1978) and is not reviewed here; let us recall, however, typical manifestations of non-Newtonian behavior. In viscometric flows, which are generalizations of simple-shear flow (see, for example, Walters 1975), one may observe both a viscosity that varies with the shear rate and the presence of normal-stress differences. The former leads, for example, to velocity profiles totally unfamiliar in Newtonian fluid mechanics, while the latter explains the well-known rod-climbing effect. A large amount of experi-

241

0066-4189/83/0115-0241$02.00

mental work has been devoted recently to the peculiar elongational behavior of certain fluids; the ratio of the elongational viscosity to the shear viscosity often differs dramatically from the Newtonian value of three. This elongational behavior is responsible for, among other things, the possibility of spinning polymer fibers. Finally, several polymeric fluids exhibit strong memory effects, in that the stress tensor shows a functional dependence with respect to the strain history.

Polymer solutions and polymer melts, which provide the most common examples of non-Newtonian fluids, are usually quite viscous; it is important to note that observable non-Newtonian effects occur at fairly low Reynolds numbers, where inertia terms have little or no effect upon the flow field. A typical example was shown by Nguyen & Boger (1979), who examined the structure of an 8:1 contraction flow at different rates. The experimental results revealed a dramatic increase of the corner vortex when the memory effects increased, although the Reynolds number was less than 10^{-2}. Similar effects at a slightly higher Reynolds number had already been photographed by Giesekus (1968). No one has yet been able to predict vortex growth on a theoretical basis.

Non-Newtonian fluid mechanics differs from classical Newtonian fluid mechanics in many important respects. For the present review, the most important difference is that the complexity of non-Newtonian response essentially results in governing equations that vary from one flow type to another (see, for example, Astarita 1976), whereas in the classical situation the Navier-Stokes equations are accepted without question as the governing equations.

It is possible to characterize a class of fluids, called simple fluids (Truesdell & Noll 1965), by means of a functional relationship between the stress tensor at time t and the strain history of the material point with respect to its present configuration. Incompressibility is a common assumption in non-Newtonian fluid mechanics, and we may decompose the Cauchy stress tensor P_{ik} (in a rectangular Cartesian coordinate system x_k) as the sum of a pressure contribution and an extra-stress T_{ik}:

$$P_{ik} = -p\delta_{ik} + T_{ik}, \tag{1}$$

where p is the pressure and δ_{ik} is the Kronecker delta. Let $C_{ik}^{-1}(t-s)$ denote the components of the Finger strain tensor of the fluid configuration at time $(t-s)$ with respect to the configuration at time t, which will be given later in terms of the displacement components. A simple fluid is such that, at time t, we have

$$T_{ik} = \mathsf{T}_{ik}\big[C_{mn}^{-1}(t-s), 0 \leqslant s < \infty\big], \tag{2}$$

where T_{ik} is a tensor functional of the deformation history.

One must acknowledge that the general characterization of non-Newtonian behavior by means of a functional constitutive equation like (2) does not lead to tractable equations in any but the simplest of flow problems. Furthermore, even allowing for the significant simplification that often results from a consideration of certain restricted classes of flows, the governing equations can still be significantly more complicated in detail than the Navier-Stokes equations. In the case of highly elastic non-Newtonian fluids flowing in complex geometries (a topic that figures prominently in this review), the problem definition must start, for reasons of tractability, with an approximation of the constitutive equations; even then, the structure of the flow problem can be radically different from that found in Newtonian fluid mechanics. All this means that most flow problems in non-Newtonian fluid mechanics demand numerical solutions, and progress has been largely dependent on the speed and storage capacity of available computers (Walters 1979).

In this review, we find it convenient to introduce a flow classification (Crochet et al. 1983) and to consider briefly the numerical methods required in each flow type. We then concentrate on one subject, namely the flow of highly elastic liquids in complex geometries. The reasons for this emphasis are twofold. First, the related flow problems have a different structure from that found in the corresponding Newtonian situation, and second, much of the worthwhile research effort in non-Newtonian fluid mechanics in the last decade has been expended in this direction.

2. NUMERICAL METHODS EMPLOYED IN RELATIVELY SIMPLE TYPES OF FLOW

It is helpful to consider a simple flow classification with the following headings (cf. Crochet et al. 1983):

1. Flows dominated by the shear viscosity
2. Slow flows (slightly elastic liquids)
3. Nearly viscometric flows
4. Complex flows involving highly elastic liquids

2.1 Flows Dominated by the Shear Viscosity

In many situations of practical importance, the range of flow rates is such that the normal-stress differences are low with respect to the shear stresses. Even when the fluid is highly elastic, the geometry of the flow may be such that fluid elasticity has only a minor effect on the streamlines; this is typical, say, of lubrication analyses. However, shear-thinning effects upon the flow characteristics and the pressure distribution may remain quite important even for low values of the shear rate. The flow problem can then be solved by considering constitutive equations of the generalized Newtonian fluid that have the form

$$P_{ik} = -p\delta_{ik} + 2\eta(I_2)D_{ik},\tag{3}$$

where D_{ik} is the rate-of-deformation tensor, η is the "apparent viscosity," and I_2 is a suitable form of the second invariant of D_{ik}. We note that η is an arbitrary function of I_2 that is selected on experimental grounds; the well-known power-law fluid is just a special case of (3).

The constitutive equations (1) and (3) must be substituted into the stress equations of motion:

$$\frac{\partial P_{ki}}{\partial x_k} + \rho f_i = \rho \frac{Dv_i}{Dt},\tag{4}$$

where f_i is the body-force vector per unit mass, ρ the density, v_i the velocity vector, and D/Dt denotes the material derivative. Since the fluid in general is considered as incompressible, the equation of continuity is given by

$$\frac{\partial v_i}{\partial x_i} = 0.\tag{5}$$

Equations (3)–(5), when combined, have the same general structure as the Navier-Stokes equations; indeed, since the constitutive equations (3) are given explicitly in terms of the velocity gradients, we can write

$$-\frac{\partial p}{\partial x_i} + \frac{\partial}{\partial x_k}[2\eta(I_2)D_{ik}] + \rho f_i = \rho \frac{Dv_i}{Dt}.\tag{6}$$

Numerical methods for calculating Newtonian flow can be adapted without difficulty to meet the variable-viscosity complication (see, for example, Crochet & Walters 1983); a Picard iterative technique may be used, in which the viscosity is updated by means of the previous velocity field.

Early contributions to this area include an influential paper by Duda & Vrentas (1973) on entrance flows and an extension of the Nickell et al. (1974) finite-element algorithm to cover shear-thinning fluids (Tanner et al. 1975). Rheometrical applications based on finite-difference calculations may be found in papers by Paddon & Walters (1979) and Williams (1979), while Halmos et al. (1975) obtain a solution for the contraction flow of a viscoinelastic fluid that is useful in interpreting experimental observations for viscoelastic polymer solutions. A calculation of the end correction for power-law fluids by means of finite elements is provided by Boger et al. (1978). The finite-element technique, which is so convenient for the flow prediction in arbitrary geometries, is exploited by Caswell & Tanner (1978) for the design of wirecoating dies, by Crochet & Keunings (1981) for conical dies, and by Hieber & Shen (1980) for injection molding. A Bingham material with a yield stress is included in the finite-element work of Bercovier & Engelman (1980).

2.2 Slow Flows (Slightly Elastic Liquids)

It may be shown that the constitutive equation of a simple fluid given by (2) implies the existence of a natural time λ that is proper to the fluid. In a given flow, the natural (or relaxation) time λ of the fluid may be compared to a characteristic time L/V of the flow, where V is a typical velocity and L is a characteristic length of the flow geometry. We may define a non-dimensional parameter W through

$$W = \frac{\lambda V}{L}. \tag{7}$$

This is sometimes called the Weissenberg number.

It is well known that if the flow is sufficiently slow (or the viscoelastic fluid is only slightly elastic), the general constitutive equations (2) are significantly simplified by means of a series of approximations in W, which is small under such conditions. This is called the "retarded-motion expansion" (see, for example, Bird 1976) leading to the "hierarchy equations" of B. D. Coleman and W. Noll. For example, the so-called "second-order model" is given by

$$T_{ik} = 2\eta D_{ik} - v_1 \overset{\triangledown}{D}_{ik} + 4v_2 D_{ij} D_{jk}, \tag{8}$$

where η, v_1, and v_2 are material constants and $\triangledown$ denotes the contravariant derivative defined by (Oldroyd 1950)

$$\overset{\triangledown}{D}_{ik} = \frac{\partial D_{ik}}{\partial t} + v_m \frac{\partial D_{ik}}{\partial x_m} - \frac{\partial v_i}{\partial x_m} D_{mk} - \frac{\partial v_k}{\partial x_m} D_{im}. \tag{9}$$

Equation (8) and all higher-order hierarchy equations are explicit in the stress tensor, which means that P_{ik} can be substituted out of the governing equations to yield a flow problem in the variables v_i and p.

To be consistent with the retarded-motion expansion, it is appropriate to seek series solutions to flow problems using W as the small parameter; this leads to a series of partial differential equations having the same structure as the corresponding equations for a Newtonian fluid. They are, however, much more complicated in detail. Standard techniques used in the classical case can be extended without difficulty and no new matters of substance are involved.

Numerical work based on series solutions in W consists mainly of analytical developments leading to ordinary differential equations that must be solved by numerical means. Specific examples of this procedure are provided by the papers of Walters & Waters (1968), Griffiths et al. (1969), Griffiths & Walters (1970), Davies & Walters (1972), and Bhatnagar & Zago (1978). However, the perturbation method may also be carried out entirely with a numerical technique; this is done by Datta & Strauss (1976), who use finite elements for solving the entry flow in a 2:1 contraction of a slightly viscoelastic fluid.

Recently, flow problems for the second-order model per se have been solved without employing a series expansion technique in the parameter W. Finite-difference algorithms to that effect have been developed by Crochet & Pilate (1975, 1976), Pilate & Crochet (1977), and Townsend (1980a); finite-element methods have been proposed by Reddy & Tanner (1977, 1978) and more recently by Mendelson et al. (1982) toward a different goal, which is discussed later.

The use of the hierarchy equations in general flow problems (i.e. in situations for which W is in no sense small) has been rightly criticized by Leal (1979), but it may still be argued that the procedure is not without its benefits in that the solutions are essential (if small) parts of the complete story and their range of applicability is often far in excess of that which might have been anticipated a priori. A good example of this is provided by the calculation of the flow of a viscoelastic fluid around a sphere; the drag calculated for an Oldroyd fluid differs very little from its second-order counterpart up to $W = 1.5$ (Tiefenbruck & Leal 1982, Crochet 1982). However, as is clearly indicated in Section 4, numerical second-order calculations have no future in view of the progress made with more elaborate fluid models.

2.3   Nearly Viscometric Flows

Nearly-viscometric flow is an important category within our overall classification scheme. Such a flow can be regarded as a perturbation about a basic viscometric flow like Poiseuille or Couette flow. Hydrodynamic stability problems provide obvious examples of nearly-viscometric flows, but other examples abound in the non-Newtonian fluid mechanics literature (see, for example, Walters 1979, Crochet et al. 1983).

Usually a suitable perturbation parameter is self-evident in any nearly-viscometric flow and the problem is usually solved by a perturbation expansion in that small parameter, the zero-order solution being the solution appropriate to viscometric flow. Choosing a suitable constitutive equation is a nontrivial process, given that the general characterization of nearly-viscometric flow is ruled out for reasons of tractability. Regardless of the complexity of the ultimate choice of constitutive equation, the resulting differential equations are invariably too complicated to permit analytic solution, and numerical techniques are essential. Solution of the zero-order viscometric flow requires techniques similar to those employed in flow category (1), and the higher-order perturbations result in differential equations that have the same structure as those in the corresponding Newtonian problems, except that they are significantly more complicated in detail.

Examples of the numerical solution of nearly-viscometric pipe-flow problems are to be found in Barnes et al. (1971), Dodson et al. (1974), and Townsend et al. (1976); for the journal-bearing lubrication problem in Davies & Walters (1972); and for rotational flows in Kramer & Johnson (1972) and Williams (1980). Assessment of the worth of these and related solutions requires experience and skill on account of the need to employ approximative constitutive equations.

2.4 Complex Flows Involving Highly Elastic Liquids

We may include in this category all situations which fall outside categories (1)–(3), so that the non-Newtonian fluids must have a finite memory (i.e. the relaxation time λ is not small) and the flow geometry and flow strength are such as to rule out the use of the hierarchy equations and related constitutive expansions.

With the increasing capacity and speed of available computers, attention in non-Newtonian fluid mechanics has naturally turned to complex flows and highly elastic liquids (Walters 1979), not only because of their practical importance in such fields as polymer processing (Crochet & Walters 1983) but also because they have given rise to a number of difficult but stimulating problems of a fundamental nature not encountered in classical fluid mechanics. In addition, the number of workers in this field and the volume of related contributions in the literature has mushroomed in the last decade. For these reasons, we have concentrated on this topic in the present review.

3. BASIC STATEMENT OF THE CURRENT PROBLEM IN COMPUTATIONAL NON-NEWTONIAN FLUID MECHANICS

Having excluded simple models like the generalized Newtonian fluid or the second-order fluid for solving complex flows involving highly elastic liquids, we must resort to constitutive equations that are special cases of the general statement and yet are simple enough to allow numerical calculations. It is generally acknowledged that for reasons of tractability, the complexity of the constitutive model used in non-Newtonian fluid mechanics has to be inversely proportional to the complexity of the flow problem. Our present concern is with very complex flows and, accordingly, the models employed must be relatively simple. At the same time, they must accommodate long-range memory effects.

Two main avenues have been explored for obtaining models that are at least qualitatively meaningful and yet are amenable to computations. The first approach expresses the functional relationship between the stress and the deformation history in terms of a *differential model* that relates algebraically the stress and rate-of-deformation tensors and some of their time derivatives. In the second approach, the functional T in (2) is made explicit by an integral representation, thus giving rise to an *integral model*. Note that kinetic theories of polymer solutions and melts, which serve more and more as a guide for producing constitutive equations, have led to both types of models.

A simple differential model of the Oldroyd type has dominated recent developments in numerical non-Newtonian fluid mechanics. The so-called "upper-convected Maxwell model" has constitutive equations of the form

$$T_{ik} + \lambda \overset{\triangledown}{T}_{ik} = 2\eta D_{ik}, \tag{10}$$

where η and λ are material constants and $\triangledown$ denotes the upper-convected derivative defined in (9). By comparison with (8), we note that (10) is now an implicit expression in the stress tensor.

When (10) is taken in conjunction with the stress equations of motion and the equation of continuity to formulate a flow problem, it is apparent that the stress components have to be considered as dependent variables along with v_i and p, in marked contrast to the classical situation or those discussed in connection with flow categories (1)–(3). To illustrate this point further, consider a two-dimensional steady-state flow problem with nondimensional velocity components $(u, v, 0)$ in the (x, y, z) directions, respectively. The relevant equations for the Maxwell model are (see, for example, Cochrane et al. 1981, Walters & Webster 1982)

$$T_{xx}\left[1 - 2W\frac{\partial u}{\partial x}\right] + W\left[u\frac{\partial T_{xx}}{\partial x} + v\frac{\partial T_{xx}}{\partial y}\right] - 2WT_{xy}\frac{\partial u}{\partial y} = 2\frac{\partial u}{\partial x}, \tag{11}$$

$$T_{yy}\left[1 - 2W\frac{\partial v}{\partial y}\right] + W\left[u\frac{\partial T_{yy}}{\partial x} + v\frac{\partial T_{yy}}{\partial y}\right] - 2WT_{xy}\frac{\partial v}{\partial x} = 2\frac{\partial v}{\partial y}, \tag{12}$$

$$-WT_{xx}\frac{\partial v}{\partial x} - WT_{yy}\frac{\partial u}{\partial y} + W\left[u\frac{\partial T_{xy}}{\partial x} + v\frac{\partial T_{xy}}{\partial y}\right] + T_{xy} = \frac{\partial u}{\partial y} + \frac{\partial v}{\partial x} \tag{13}$$

from the constitutive equation (8);

$$f_x - \frac{\partial p}{\partial x} + \frac{\partial T_{xx}}{\partial x} + \frac{\partial T_{xy}}{\partial y} = R\left[u\frac{\partial u}{\partial x} + v\frac{\partial u}{\partial y}\right], \tag{14}$$

$$f_y - \frac{\partial p}{\partial y} + \frac{\partial T_{xy}}{\partial x} + \frac{\partial T_{yy}}{\partial y} = R\left[u\frac{\partial v}{\partial x} + v\frac{\partial v}{\partial y}\right] \tag{15}$$

from the stress equations of motion; and

$$\frac{\partial u}{\partial x} + \frac{\partial v}{\partial y} = 0 \tag{16}$$

from the equation of continuity. In these equations, R is a Reynolds number defined by $\rho L V / \eta$ and W is the Weissenberg number defined in (7). In the corresponding two-dimensional axisymmetric case the hoop stress $T_{\theta\theta}$ must also be given by a constitutive equation, since the extra-stress tensor T_{ik} in (10) is not traceless (see, for example, Crochet 1982). From (11)–(16), we see that there are six (seven) equations in the six (seven) unknowns T_{xx}, T_{xy}, T_{yy}, u, v, p, ($T_{\theta\theta}$), where the stress components and the pressure are suitably non-dimensionalized. In addition, an inspection of the constitutive relations (11)–(13) clearly shows that the nonlinear character of the constitutive equations is entirely dissociated from the nonlinearity of the inertia terms in (14)–(15).

The upper-convected Maxwell model (10) can also be written in an integral form, which is explicit in the stress tensor. For example, we may write (see, for example, Court et al. 1981)

$$T_{ik}(\mathbf{x}, t) = (\eta/\lambda^2)\int_{-\infty}^{t} \exp[-(t-t')/\lambda][C_{ik}^{-1}(t') - \delta_{ik}]\, dt', \tag{17}$$

where C_{ik}^{-1} is the Finger tensor defined by

$$C_{ik}^{-1} = \frac{\partial x_i}{\partial x_m'}\frac{\partial x_k}{\partial x_m'}, \tag{18}$$

and x_i' is the position at time t' of the element that is instantaneously at the point x_i at time t; the upper-convected Maxwell model in its integral form (17) is a special case of Lodge's rubberlike elastic liquid. In a two-dimensional problem, one of the ways of obtaining the "displacement functions" x_i' is to solve the equations (Oldroyd 1950)

$$\frac{\partial x'}{\partial t} + u\frac{\partial x'}{\partial x} + v\frac{\partial x'}{\partial y} = 0, \tag{19}$$

$$\frac{\partial y'}{\partial t} + u\frac{\partial y'}{\partial x} + v\frac{\partial y'}{\partial y} = 0, \tag{20}$$

in the obvious notation. A two-dimensional steady-state flow problem for the integral Maxwell model essentially involves five unknowns $[u(x, y)$, $v(x, y)$, $p(x, y)$, $x'(x, y, t')$, $y'(x, y, t')]$ and five equations $[(14)–(16)$, after use of (17), and (19)–(20)].

While we have already noted the nonlinear character of the equations, even in the case of creeping flow, and their implicit form, another important difference between the general non-Newtonian flow problem and the corresponding Newtonian equivalent concerns the boundary conditions. The need for a careful consideration of boundary conditions is apparent in view of the convective terms present in the constitutive equations (11)–(13) or the fact that in an actual computation, the infinite domain of the integral in (17) will necessarily be truncated. In the Newtonian case, it is sufficient to specify the velocity or surface force component over the boundary of the domain of interest, and the pressure at one point when no normal surface force has been specified anywhere on the boundary. This is not, in general, sufficient for the flow of a Maxwell fluid on account of fluid memory. If the boundary of the domain contains an entry region, we need to know the strain history of the fluid entering the domain or, equivalently, the knowledge of the stress field on entry to the domain (Crochet & Walters 1983, Crochet et al. 1983). In practice, the boundary condition requirements are often inadvertently satisfied by assuming "fully developed flow conditions" on entry to the domain, an assumption that essentially implies knowledge of the flow field upstream of the domain of interest. In a recent paper, Crochet & Keunings (1982b) have shown that when W is large, the assumption of fully developed flow conditions upstream requires a fairly long entry region for the domain where the flow is being calculated.

The philosophy behind most existing work in this area would appear to be that it is essential to develop consistent numerical algorithms for simple Oldroyd-Maxwell models before proceeding to more realistic and more complicated constitutive models. It is readily acknowledged that the Maxwell model is inadequate for a quantitative characterization of the behavior of real viscoelastic fluids, but analyses for such models can still be qualitatively useful and can provide a launching pad for studies involving more realistic models.

It is arguable if there is yet sufficient confidence in existing algorithms for the Maxwell model for one to think realistically in the short term of extensions to more complicated models. Certainly, the recent work of Mendelson et al. (1982) has highlighted the possibility of obtaining solutions to non-Newtonian flow problems that a cursory glance would indicate to be perfectly respectable but that detailed inspection and appeal to the Tanner-Giesekus theorem (see, for example, Tanner 1966) would show to be spurious; work of this sort is extremely important at this stage in the development of the subject. Another side of the argument is that the simple Maxwell model (10) may bring inherent difficulties to the numerical flow solution. In particular, Crochet & Keunings (1982b) have shown that the use of an Oldroyd-B fluid that contains a retardation time allows one to enlarge by a factor of four the domain of W over which the die swell phenomenon can be calculated with the Maxwell fluid.

Although the present review bears on numerical methods, we emphasize the need for reference analytic solutions. To cite a few, the analytic solutions provided by Hull (1981) for a nontrivial non-Newtonian contraction-flow problem, by Kazakia & Rivlin (1981) for the run-up in a viscoelastic fluid, and by Trogdon & Joseph (1982) for the slow flow over a slot are of utmost importance, not only for the inherent superiority of analytical work (whenever it is feasible) over numerical work, but also for their use as benchmarks in future numerical developments.

4. COMPLETED WORK AND OUTSTANDING PROBLEMS

4.1 Historical

A numerical method for solving time-dependent problems in one space variable for an Oldroyd model was provided by Townsend (1973) and was later used by Akay (1979) and Manero & Walters (1980). We know of no other published work on transient problems that utilizes implicit differential constitutive models.

The first papers on problems involving *two* space variables appeared in 1977. Perera & Walters (1977a,b) solved a number of complex flow problems for a four-constant Oldroyd model using a finite-difference method, while Kawahara & Takeuchi (1977) utilized a finite-element technique in a related context.

Since then there has been a proliferation of papers on the general subject with a strong bias toward finite-element methods (see, for example, Chang et al. 1979a,b, Crochet & Bezy 1979, 1980, Viriyayuthakorn & Caswell 1980, Crochet & Keunings 1980, 1982a,b, Coleman 1980, 1981a, Bernstein et al. 1981, Richards & Townsend 1981, 1982, Mendelson et al. 1982, Jackson & Finlayson 1982). However, further attention has also been given to finite-difference techniques (see, for example, Gatski & Lumley 1978a,b, Perera & Strauss 1979, Davies et al. 1979, Townsend 1980b, Paddon & Holstein 1980, Tiefenbruck & Leal 1982, Cochrane et al. 1982). In addition, there have been isolated attempts to use a collocation method (Chang et al. 1979a, b, Co & Stewart 1980, Malkus 1981) and a suggestion has been made that a boundary-integral approach may be rewarding (Coleman 1981b, Tanner 1981).

In completed work, particular attention has been paid to the following situations:

1. Flow past submerged obstacles (Crochet 1982, Townsend 1980b, Tiefenbruck & Leal 1982).
2. The hole-pressure problem (Crochet & Bezy 1979, Townsend 1980a, Richards & Townsend 1981, Jackson & Finlayson 1982).
3. Free-surface problems, especially die swell (Chang et al. 1979b, Crochet & Keunings 1980, 1981, 1982a, b, Coleman 1981a, Caswell & Viriyayuthakorn 1982). In this connection, the analytic work of Tanner (1980) is of importance in confirming some unexpected numerical results that suggest an initial jet contraction for low elasticity.
4. Flow through a contraction (Gatski & Lumley 1978a, b, Perera & Walters 1977b, Crochet & Bezy 1980, Perera & Strauss 1979, Viriyayuthakorn & Caswell 1980, Bernstein et al. 1981).

4.2 Differential Models

Problems based on differential models have been approached by means of finite differences and finite elements.

Not surprisingly, finite-difference attacks on complex non-Newtonian flow problems have tended to follow and, where necessary, adapt well-established procedures from the classical literature.

Equation (16) is used to introduce a stream function ϕ, defined by

$$u = \frac{\partial \phi}{\partial y}, v = -\frac{\partial \phi}{\partial x}, \tag{21}$$

and the pressure is eliminated between (14) and (15) to yield

$$R\left[\frac{\partial \phi}{\partial x}\frac{\partial \omega}{\partial y} - \frac{\partial \phi}{\partial y}\frac{\partial \omega}{\partial x}\right] = \frac{\partial^2 T_{xx}}{\partial x \partial y} + \frac{\partial^2 T_{xy}}{\partial y^2} - \frac{\partial^2 T_{xy}}{\partial x^2} - \frac{\partial^2 T_{yy}}{\partial x \partial y}, \tag{22}$$

where the vorticity ω is defined by

$$\omega = \frac{\partial v}{\partial x} - \frac{\partial u}{\partial y}, \tag{23}$$

so that

$$\omega = -\nabla^2 \phi. \tag{24}$$

The basic equations are now (11)–(13), (22), and (24) in the five variables T_{xx}, T_{xy}, T_{yy}, ω, and ϕ. Gatski & Lumley (1978a, b) solved (22) as given, but by far the most popular method has been to use the substitution

$$T_{ik} = \overline{T}_{ik} + 2\eta D_{ik} \qquad (25)$$

in dimensionless form, in which case (22) becomes

$$\nabla^2 \omega + R \left[\frac{\partial \phi}{\partial x} \frac{\partial \omega}{\partial y} - \frac{\partial \phi}{\partial y} \frac{\partial \omega}{\partial x} \right] = \frac{\partial^2 \overline{T}_{xx}}{\partial x \partial y} + \frac{\partial^2 \overline{T}_{xy}}{\partial y^2} - \frac{\partial^2 \overline{T}_{xy}}{\partial x^2} - \frac{\partial^2 \overline{T}_{yy}}{\partial x \partial y}, \qquad (26)$$

and the elliptic nature of the equivalent equation in Newtonian fluid mechanics (given by $\overline{T}_{xx} = \overline{T}_{xy} = \overline{T}_{yy} = 0$) is recovered (Perera & Walters 1977a, b, Davies et al. 1979, Townsend 1980b, Cochrane et al. 1981, Walters & Webster 1982).

The Gatski & Lumley (1978a, b) method, which involves a displaced-grid technique, is not applicable to creeping flow ($R = 0$). This method rests heavily on the treatment of the convective terms in (22), and this is presently a subject of relevance and expanding research effort within a much wider context.

Iterative methods are used to solve the set of equations (11)–(13), (22), and (24) (see, for example, Crochet & Walters 1983, Crochet et al. 1983). Heretofore, studies in non-Newtonian fluid mechanics have favored Successive Over Relaxation (SOR) techniques, although a preconditioned conjugate gradient method has also been employed (Court et al. 1981). There is no published finite-difference work on non-Newtonian free-surface problems; recent work by Ryan & Dutta (1981) on Newtonian flow indicates, however, that such an approach is feasible.

The determination of pressure fields in finite-difference approaches to viscoelastic flow involves essentially the same difficulties as those encountered in the corresponding Newtonian situation, and the extra problems are basically related to matters of detail. Existing work contains some examples of the calculation of the pressure variation along a line in the flow (cf. Townsend 1980a, Tiefenbruck & Leal 1982), but we know of no published finite-difference work on the calculation of pressure fields in viscoelastic fluid flow, except for one isolated figure in the Tiefenbruck & Leal (1982) paper.

The first paper on finite-element techniques for solving the flow of a Maxwell fluid was published by Kawahara & Takeuchi (1977). Their method consisted of selecting for the extra-stress components, velocity components, and pressure a finite-element discretization of the form

$$T_{ik} = \sum_m T_{ik}^{(m)} \phi_m, \qquad v_i = \sum_n v_i^{(n)} \psi_n, \qquad p = \sum_q p^{(q)} \pi_q, \qquad (27)$$

where $T_{ik}^{(m)}$, $v_i^{(n)}$, $p^{(q)}$ are nodal values and ϕ_m, ψ_n, π_q are piecewise-polynomial shape functions. The Galerkin method is used for minimizing the residuals of the constitutive equations (10), the momentum equations (4), and the incompressibility equation (5); a weak form is obtained for the momentum equation (see Crochet 1982 or Crochet et al. 1983 for details). The Galerkin method leads to a set of algebraic equations in the nodal values where the number of equations equals the number of unknowns. The central problem is the selection of an element for defining the approximation (27). The paper by Kawahara & Takeuchi (1977) contains an unfortunate misprint: they mention the use of a six-node triangular element, where ψ_n is a second-order polynomial and ϕ_m and π_q are first-order polynomials. This was not in fact the set of shape functions used by Kawahara & Takeuchi (1977), because their use of an order for the stresses less than that for the velocity components, together with their formulation of the Galerkin equations, leads to a singular stiffness matrix (Crochet 1982). In reality, they used second-order polynomials for ψ_n *and* for ϕ_m. The finite-element meshes used by Kawahara & Takeuchi (1977) were too coarse for assessing the potential of their method.

Chang et al. (1979a, b) do not define nodal stresses; an iterative scheme of the Picard type is devised for calculating the extra-stresses in the momentum equations. The procedure diverges for fairly low values of the Weissenberg number, possibly because the iterative process implies the calculation of derivatives of the rate-of-deformation tensor, which is itself discontinuous. Crochet & Bezy (1979) follow the procedure suggested by Kawahara & Takeuchi (1977). They use six-node triangular elements with second-order polynomials for ψ_n, first-order polynomials for ϕ_m, and they obtain a nonsingular stiffness matrix through a discontinuous pressure field that is given by a first-order polynomial over each element. The system is, however, overconstrained by incompressibility and the element has since been forsaken.

The use of piecewise continuous second-order polynomials for the velocity components and the extra-stresses, together with first-order polynomials for the pressure, gives satisfactory results for triangular and Lagrangian elements (Crochet & Bezy 1980, Crochet & Keunings 1980, 1981, 1982a,b, Crochet 1982) and for serendipity elements (Richards & Townsend 1981, 1982). A rather simple quadrilateral with piecewise linear polynomials for ϕ_m and ψ_n and a discontinuous pressure field has been successfully used by Coleman (1980, 1981a). We are not aware of a convergence proof for these various elements, even for the case of Newtonian creeping flow.

A slightly different approach may be used for formulating the Galerkin equations. From (10) we may write

$$T_{ik} = 2\eta D_{ik} - \lambda \overset{\triangledown}{T}_{ik}, \tag{28}$$

and by substituting T_{ik} given by (28) into the momentum equations (5), we obtain

$$-\frac{\partial p}{\partial x_i} + \frac{\partial}{\partial x_k}(2\eta D_{ik}) - \lambda \frac{\partial}{\partial x_k}\left(\overset{\triangledown}{T_{ik}}\right) + \rho f_i = \rho \frac{Dv_i}{Dt}. \tag{29}$$

A finite-element discretization of the form (27) is again adopted, and the Galerkin method is used for obtaining the discretized form of the constitutive equations (10), the momentum equations (29), and the incompressibility equation (6). The main feature of this method is the appearance of the viscous terms in the momentum equation; when λ vanishes, (29) explicitly reduces to the well-known velocity-pressure formulation used by Nickell et al. (1974). The effect of the transformation (28) is twofold: when W is small, the results are smoother than those obtained with the first method, and first-order polynomials may now be used for the extra-stress components because the viscous terms in (29) remove the singularity of the stiffness matrix. The method has been used by Jackson & Finlayson (1982) and Mendelson et al. (1982) with first-order polynomials for the extra-stresses, and by Crochet (1982) and Crochet & Keunings (1982a) with second-order polynomials. Crochet (1982) finds that the second method performs better for low values of W; when W increases, spurious wiggles appear earlier than with the first method. A comparison between the various techniques, based on standard examples run on the same finite-element meshes, is lacking.

A common feature of the papers using finite-element techniques is their failure to provide a solution beyond some relatively low value of W, at a level where the stress distribution differs considerably from its Newtonian counterpart but where the velocity field shows little difference. The lack of convergence of the iterative technique is preceded by the appearance of spurious wiggles in the stress and velocity fields that are typical of the solution of Newtonian flow at high Reynolds numbers; the wiggles may be attributed to steep stress gradients and convected derivatives of the extra-stress tensor, which have not received any special treatment to date.

In a recent paper, Crochet & Keunings (1982b) have extended considerably the range of W for predicting die swell, reaching values of practical interest. This was done by using an Oldroyd-B fluid, which has the main effect of adding a viscous term in the momentum equation. Although the fluid model is still relatively simple, the use of a "retardation time" in the constitutive relation opens up a promising path.

4.3   The Use of Integral Constitutive Models

Existing work on numerical simulation with *integral* constitutive models is limited to one finite-difference attack (Court et al. 1981) and a small but growing number of finite-element approaches (Viriyayuthakorn & Caswell 1980, Malkus 1981, Bernstein et al. 1981, Caswell & Viriyayuthakorn 1982, Winter 1982). The basic problem is that of "tracking" the position of particles, given a known velocity field. From this information the Finger (and related) tensors can be computed, and hence the stress components through an appropriate quadrature procedure. Methods of varying complexity have been proposed; Court et al. (1981) base their technique on the solution of the hyperbolic equations (19) and (20), which appears to be most helpful for future development, and they solve the flow in a contraction for a relatively high value of W. Viriyayuthakorn & Caswell (1980) use an elaborate tracking technique and obtain the best available results for the flow in an axisymmetric contraction.

Convective terms are absent from the integral formulation of the problem; the failure of the algorithm seems due to the lack of compatibility of the Finger strain tensor and the fact that the continuity constraint

$$\det C_{ik}^{-1} = 1 \tag{30}$$

is not satisfied for all past times t' of relevance within the context of the memory of the fluid.

4.4   Boundary Conditions

The specification of boundary conditions, especially near abrupt changes in geometry such as re-entrant corners, has not been given the attention it deserves within the context of flow problems for highly elastic liquids. There is enough evidence of the substantial changes in flow characteristics that can be brought about by small changes in corner geometry (Walters & Webster 1982) to warrant a detailed study. Work to date that deals exclusively with the problem of specifying re-entrant corner conditions is limited to papers by Holstein & Paddon (1981), Cochrane et al. (1982), and Walters & Webster (1982).

The problem of re-entrant corners and stick-slip boundaries deserves special attention in finite-element calculations where all available methods assume continuity of the pressure and the extra-stress components where a singularity is obviously present. There is an urgent need for analytical work identifying the nature of the singularity at re-entrant corners and edges.

4.5   Outstanding Problems

There is no doubt that presently the outstanding problem in the numerical simulation of viscoelastic flow concerns the upper limit on the nondimensional parameter W (found in all published works) above which the numerical algorithms fail to converge (see, for example, Crochet et al. 1983, Mendelson et al. 1982). The limit is relatively low, so low in fact that many of the important and dramatic experimental results fall outside the present scope of numerical simulation. Furthermore, reliable solutions before breakdown are often no more than perturbations about the Newtonian case.

The limit on W is common to all published works. It applies to finite-difference and finite-element techniques, to differential and integral constitutive models, and to flow problems with and without abrupt changes in geometry. Some suggestions for possible causes of the W barrier are the bifurcation phenomena, the unsuitability of the constitutive models, and the failure of the iterative numerical schemes. All workers acknowledge the limit, but the only papers specifically addressing the problem are those by Mendelson et al. (1982) and Tanner (1982).

There have been isolated instances where numerical simulations have been able to cover the range of W over which dramatic changes in flow characteristics take place experimentally. However, the simulations have been singularly unable to predict such changes (see, for example, Crochet & Bezy 1980, Walters & Webster 1982), with the exception of the paper by Crochet & Keunings (1982b) on die swell.

This suggests a further (perhaps related) problem in urgent need of resolution. A study of the appropriateness or otherwise of the chosen constitutive models provides an obvious starting point.

5. SUMMARY

Present work in the numerical simulation of non-Newtonian flow concentrates on the flow of highly elastic liquids in complex geometries. Implicit differential constitutive models (or their integral equivalents) have to be employed, necessitating the consideration of more dependent variables than arise in the classical Newtonian problem. Notwithstanding the need to give further attention to conditions near re-entrant corners, the basic problem is now well posed and resolved—in principle at least. Finite-difference and finite-element techniques have been successfully applied to both differential and integral constitutive models, and for low values of fluid elasticity there is satisfactory agreement between the various treatments.

Most published simulations break down at some finite (and rather small) value of the nondimensional elasticity parameter W. The precise breakdown point may vary from treatment to treatment, but there is a general consensus that this variation is slight and of little consequence. Certainly, the resolution of this numerical breakdown is currently the outstanding problem in this area of research.

Literature Cited

Akay, G. 1979. Non-steady two-phase stratified laminar flow of polymeric liquids in pipes. *Rheol. Acta* 18:256–67

Astarita, G. 1976. Is non-Newtonian fluid mechanics a culturally autonomous subject? *J. Non-Newt. Fluid Mech.* 1:203–6

Astarita, G., Marrucci, G. 1974. *Principles of Non-Newtonian Fluid Mechanics.* London: McGraw-Hill

Barnes, H. A., Townsend, P., Walters, K. 1971. On pulsatile flow of non-Newtonian liquids. *Rheol. Acta* 10:517–27

Bercovier, M., Engelman, M. 1980. A finite element method for incompressible non-Newtonian flows. *J. Comput. Phys.* 36:313–26

Bernstein, B., Kadivar, M. K., Malkus, D. S. 1981. Steady flow of memory fluids with finite elements: Two test problems. *Comput. Methods Appl. Mech. Eng.* 27:279–302

Bhatnagar, R. K., Zago, J. V. 1978. Numerical investigation of flow of a viscoelastic fluid between rotating coaxial disks. *Rheol. Acta* 17:557–67

Bird, R. B. 1976. Useful non-Newtonian models. *Ann. Rev. Fluid Mech.* 8:13–34

Bird, R. B., Armstrong, R. C., Hassager, O. 1977. *Dynamics of Polymeric Liquids*, Vol. 1. New York: Wiley

Boger, D. V., Gupta, R., Tanner, R. I. 1978. The end correction for power-law fluids in the capillary rheometer. *J. Non-Newt. Fluid Mech.* 4:239–48

Caswell, B., Tanner, R. I. 1978. Wirecoating die design using finite element methods. *Polym. Eng. Sci.* 18:416–21

Caswell, B., Viriyayuthakorn, M. 1982. Finite element simulation of die swell for a Maxwell fluid. *J. Non-Newt. Fluid Mech.* In press

Chang, P. -W. Patten, T. W., Finlayson, B. A. 1979a. Collocation and Galerkin finite element methods for viscoelastic fluid flow. I. Description of method and problems with fixed geometries. *Comput. Fluids* 7:267–83

Chang, P. -W., Patten, T. W., Finlayson, B. A. 1979b. Collocation and Galerkin finite element methods for viscoelastic fluid flow. II. Die swell problems with a free surface. *Comput. Fluids* 7:285–93

Co, A., Stewart, W. E. 1980. Viscoelastic radial flow between parallel disks. *Rheol. Res. Cent. Rep. No. 68*, Univ. Wis., Madison, pp. 1–46

Cochrane, T., Walters, K., Webster, M. F. 1981. On Newtonian and non-Newtonian flow in complex geometries. *Philos. Trans. R. Soc. London Ser. A* 301:163–81

Cochrane, T. C., Walters, K., Webster, M. F. 1982. Newtonian and non-Newtonian flow near a re-entrant corner. *J. Non-Newt. Fluid Mech.* 10:95–114

Coleman, C. J. 1980. A finite element routine for analysing non-Newtonian flows. Part 1. Basic method and preliminary results. *J. Non-Newt. Fluid Mech.* 7:289–301

Coleman, C. J. 1981a. A finite element routine for analysing non-Newtonian flows. Part 2. The extrusion of a Maxwell fluid. *J. Non-Newt. Fluid Mech.* 8:261–70

Coleman, C. J. 1981b. A contour integral formulation of plane creeping Newtonian flow. *Q. J. Mech. Appl. Math.* 34:453–64

Court, H., Davies, A. R., Walters, K. 1981. Long-range memory effects in flows involving abrupt changes in geometry. Part 4. Numerical simulation using integral rheological models. *J. Non-Newt. Fluid Mech.* 8:95–117

Crochet, M. J. 1982. The flow of a Maxwell fluid around a sphere. In *Finite Elements in Fluids*, Vol. 4, ed. R. H. Gallagher. New York: Wiley

Crochet, M. J., Bezy, M. 1979. Numerical solution for the flow of viscoelastic fluids. *J. Non-Newt. Fluid Mech.* 5:201–18

Crochet, M. J., Bezy, M. 1980. Elastic effects in die entry flow. In *Rheology, Vol 2: Fluids*, ed. G. Astarita, G. Marrucci, L. Nicolais, pp. 53–58. New York: Plenum

Crochet, M. J., Davies, A. R., Walters, K. 1983. *Numerical Simulation of Non-Newtonian Flow.* Amsterdam: Elsevier. In press

Crochet, M. J., Keunings, R. 1980. Die swell of a Maxwell fluid: numerical prediction. *J. Non-Newt. Fluid Mech.* 7:199–212

Crochet, M. J., Keunings, R. 1981. Numerical simulation of die swell: geometrical effects. *Proc. 2nd World Congr. Chem. Eng., Montreal*, 6:285–88

Crochet, M. J., Keunings, R. 1982a. On numerical die swell calculation. *J. Non-Newt. Fluid Mech.* 10:85–94

Crochet, M. J., Keunings, R. 1982b. Finite element analysis of die swell of a highly elastic fluid. *J. Non-Newt. Fluid Mech.* 10:339–56

Crochet, M. J., Pilate, G. 1975. Numerical study of the flow of a fluid of second grade in a square cavity. *Comput. Fluids* 3:283–91

Crochet, M. J., Pilate, G. 1976. Plane flow of a fluid of second grade through a contraction. *J. Non-Newt. Fluid Mech.* 1:247–58

Crochet, M. J., Walters, K. 1983. Computational techniques for viscoelastic fluid flow. In *Computers and Polymer Processing*, ed. J. R. A. Pearson, S. M. Richardson. Barking, Engl.: Appl. Sci. Publ. In press

Datta, A. B., Strauss, K. 1976. Slow flow of a viscoelastic fluid through a contraction. *Rheol. Acta* 15:403–10

Davies, A. R., Walters, K., Webster, M. F. 1979. Long-range memory effects in flows involving abrupt changes in geometry. Part 3. Moving boundaries. *J. Non-Newt. Fluid Mech.* 4:325–44

Davies, J. M., Walters, K. 1972. The behaviour of non-Newtonian lubricants in journal bearings—a theoretical study. In *Rheology of Lubricants*, ed. T. C. Davenport, pp. 65–80. Barking, Engl.: Appl. Sci. Publ.

Dodson, A. G., Townsend, P., Walters, K. 1974. Non-Newtonian flow in pipes of non-circular cross section. *Comput. Fluids* 2:317–38

Duda, J. L., Vrentas, J. S. 1973. Entrance flows of non-Newtonian fluids. *Trans. Soc. Rheol.* 17:89–108

Gatski, T. B., Lumley, J. L. 1978a. Steady flow of a non-Newtonian fluid through a contraction. *J. Comput. Phys.* 27:42–70

Gatski, T. B., Lumley, J. L. 1978b. Non-Newtonian flow characteristics in a steady two-dimensional flow. *J. Fluid Mech.* 86:623–39

Giesekus, H. 1968. Nicht-lineare Effekte beim Strömen viskoelastischer Flüssigkeiten durch Schlitz- und Lochdüsen. (Non-linear effects in the flow of viscoelastic fluids through slits and circular apertures.) *Rheol. Acta* 7:127–38

Griffiths, D. F., Jones, D. T., Walters, K. 1969. A flow reversal due to edge effects. *J. Fluid Mech.* 36:161–75

Griffiths, D. F., Walters, K. 1970. On edge effects in rheometry. *J. Fluid Mech.* 42:379–99

Halmos, A. L., Boger, D. V., Cabelli, A. 1975. The behaviour of a power-law fluid flowing through a sudden contraction. *AIChE J.* 21:540–49

Hieber, C. A., Shen, S. F. 1980. A finite-element/finite-difference simulation of the injection-molding filling process. *J. Non-Newt. Fluid Mech.* 7:1–32

Holstein, H., Paddon, D. J. 1981. A singular finite difference treatment of re-entrant corner flow. Part I. Newtonian fluids. *J. Non-Newt. Fluid Mech.* 8:81–93

Hull, A. M. 1981. An exact solution for the slow flow of a general linear viscoelastic fluid through a slit. *J. Non-Newt. Fluid Mech.* 8:327–36

Jackson, N. A., Finlayson, B. A. 1982. Calculation of hole pressure. II. Viscoelastic fluids. *J. Non-Newt. Fluid Mech.* 10:71–84

Kawahara, M., Takeuchi, N. 1977. Mixed finite element method for analysis of viscoelastic fluid flow. *Comput. Fluids* 5:33–45

Kazakia, J. K., Rivlin, R. S. 1981. Run-up and spin-up in a viscoelastic fluid. *Rheol. Acta* 20:111–27

Kramer, J. M., Johnson, M. W. Jr. 1972. Nearly viscometric flow in the disk and cylinder system. Part I. Theoretical. *Trans. Soc. Rheol.* 16:197–212

Leal, L. G. 1979. The motion of small particles in non-Newtonian fluids. *J. Non-Newt. Fluid Mech.* 5:33–78

Malkus, D. S. 1981. Functional derivatives and finite elements for the steady spinning of a viscoelastic filament. *J. Non-Newt. Fluid Mech.* 8:223–37

Manero, O., Walters, K. 1980. On elastic effects in unsteady pipe flows. *Rheol. Acta* 19:277–84

Mendelson, M. A., Yeh, P.-W., Brown, R. A., Armstrong, R. C. 1982. Approximation error in finite element calculation of viscoelastic fluid flow. *J. Non-Newt. Fluid Mech.* 10:31–54

Nguyen, H., Boger, D. V. 1979. The kinematics and stability of die entry flows. *J. Non-Newt. Fluid Mech.* 5:353–68

Nickell, R. E., Tanner, R. I., Caswell, B. 1974. The solution of viscous incompressible jet and free surface flows using finite-element methods. *J. Fluid Mech.* 65:189–206

Oldroyd, J. G. 1950. On the formulation of rheological equations of state. *Proc. R. Soc. London Ser. A* 200:523–41

Paddon, D. J., Holstein, H. 1980. A two-dimensional asymmetric flow of a viscoelastic fluid in a T-geometry. In *Rheology, Vol. 2: Fluids*, ed. G. Astarita, G. Marrucci, L. Nicolais, pp. 31–36. New York: Plenum

Paddon, D. J., Walters, K. 1979. On edge effects and related sources of error in rotational rheometry. *Rheol. Acta* 18:565–75

Perera, M. G. N., Strauss, K. 1979. Direct numerical solutions of the equations for viscoelastic fluid flow. *J. Non-Newt. Fluid Mech.* 5:269–83

Perera, M. G. N., Walters, K. 1977a. Long-range memory effects in flows involving abrupt changes in geometry. Part I. Flows associated with L-shaped and T-shaped geometries. *J. Non-Newt. Fluid Mech.* 2:49–81

Perera, M. G. N., Walters, K. 1977b. Long-range memory effects in flows involving abrupt changes in geometry. Part 2. The expansion/contraction/expansion problem. *J. Non-Newt. Fluid Mech.* 2:191–204

Pilate, G., Crochet, M. J. 1977. Plane flow of a second-order fluid past submerged boundaries. *J. Non-Newt. Fluid Mech.* 1:247–58

Reddy, K. R., Tanner, R. I. 1977. Finite element approach to die-swell problem of non-Newtonian fluids. *Australas. Hydraul. Fluid Mech. Conf., 6th, Adelaide*, pp. 431–34

Reddy, K. R., Tanner, R. I. 1978. On the swelling of extruded plane sheets. *Trans. Soc. Rheol.* 22:661–65

Richards, G. D., Townsend, P. 1981. A finite element computer model of the hole pressure problem. *Rheol. Acta* 20:261–69

Richards, G. D., Townsend, P. 1982. Computer modelling of flows of elastic liquids through complex vessels and with forced convection. *J. Non-Newt. Fluid Mech.* 10:175–83

Ryan, M., Dutta, A. 1981. A finite difference simulation of extrudate swell. *Proc. 2nd World Congr. Chem. Eng., Montreal*, 6:277–80.

Schowalter, W. R. 1978. *Mechanics of Non-Newtonian Fluids*. Oxford: Pergamon

Tanner, R. I. 1966. Plane creeping flows of incompressible second-order fluids. *Phys. Fluids* 9:1246–47

Tanner, R. I. 1980. The swelling of plane extrudates at low Weissenberg numbers. *J. Non-Newt. Fluid Mech.* 7:265–67

Tanner, R. I. 1981. Application of boundary element methods to extrusion problems. *Proc. 2nd World Congr. Chem. Eng. Montreal*, 6:281–84

Tanner, R. I. 1982. The stability of some numerical schemes for model viscoelastic fluids. *J. Non-Newt. Fluid Mech.* 10: 169–74

Tanner, R. I., Nickell, R. E., Bilger, R. W. 1975. Finite element methods for the solution of some incompressible non-Newtonian fluid mechanics problems with free surfaces. *Comput. Methods Appl. Mech. Eng.* 6:155–74

Tiefenbruck, G., Leal, L. G. 1982. A numerical study of the motion of a viscoelastic fluid past rigid spheres and spherical bubbles. *J. Non-Newt. Fluid Mech.* 10:115–55

Townsend, P. 1973. Numerical solutions of some unsteady flows of elasticoviscous liquids. *Rheol. Acta* 12:13–18

Townsend, P. 1980a. A computer model of hole-pressure measurement in Poiseuille flow of visco-elastic liquids. *Rheol. Acta* 19:1–11

Townsend, P. 1980b. A numerical simulation of Newtonian and visco-elastic flow past stationary and rotating cylinders. *J. Non-Newt. Fluid Mech.* 6:219–43

Townsend, P., Walters, K., Waterhouse, W. M. 1976. Secondary flows in pipes of square cross-section and the measurement of the second normal stress difference. *J. Non-Newt. Fluid Mech.* 1:107–23

Trogdon, S. A., Joseph, D. D. 1982. Matched eigenfunction expansions for slow flow over a slot. *J. Non-Newt. Fluid Mech.* In press

Truesdell, C., Noll, W. 1965. The non-linear field theories of mechanics. In *Encyclopedia of Physics*, Vol. III, 3, ed. S. Flügge. Berlin/Göttingen/Heidelberg: Springer

Viriyayuthakorn, M., Caswell, B. 1980. Finite element simulation of viscoelastic flow. *J. Non-Newt. Fluid Mech.* 6:245–67

Walters, K. 1975. *Rheometry*. London: Chapman & Hall

Walters, K. 1979. Developments in non-Newtonian fluid mechanics—a personal view. *J. Non-Newt. Fluid Mech.* 5:113–24

Walters, K., Waters, N. D. 1968. On the use of a rheogoniometer. Part I. Steady shear. In *Polymer Systems: Deformation and Flow*, ed. R. E. Wetton, R. W. Whorlow, pp. 212–35. London: Macmillan

Walters, K., Webster, M. F. 1982. On dominating elastico-viscous response in some complex flows. *Philos. Trans. R. Soc. London Ser. A.* In press

Williams, R. W. 1979. Determination of viscometric data from the Brookfield R.V.T. viscometer. *Rheol. Acta* 18:345–59

Williams, R. W. 1980. On the secondary flow induced by spheres and disks rotating in elastico-viscous liquids. *Rheol. Acta* 19:548–73

Winter, H. H. 1982. Modelling of strain histories for memory integral fluids in steady axisymmetric flows. *J. Non-Newt. Fluid Mech.* 10:157–67

Ann. Rev. Fluid Mech. 1983. 15:261–91

MATHEMATICAL MODELING OF TWO-PHASE FLOW

D. A. Drew[1]

Department of Mathematical Sciences, Rensselaer Polytechnic Institute, Troy, New York 12181

INTRODUCTION

Dispersed two-phase flows occur in many natural and technological situations. For example, dust in air and sediment in water contribute to erosion and silting, and can cause problems for machinery such as helicopters and power plants operating in such an environment. Also, many energy-conversion and chemical processes involve two-phase flows. Boiling a fluid such as water or sodium and extracting the heat by condensation at a different location provides a practical energy-flow process. Many kinds of chemical reactants and catalysts are mixed in a dispersion of fine particles in order to expose as much interfacial area as possible.

In such a large class of problems, many diverse mechanisms are important in different situations. The models used for these different situations have many common features, such as interfacial drag.

This paper examines the common features of dispersed two-phase flows from a continuum-mechanical approach. Since it is not universally accepted that such an approach is valid, we discuss some philosophical reasons for taking such an approach. The approach is based on the view that it is sufficient to describe each material as a continuum, occupying the same region in space. This new "material" consists of two interacting materials (called phases, even though they often are not different phases of the same material). The two-phase material is often called the mixture. In analogy with continuum mechanics, we shall have to specify how the

---

[1] On sabbatical leave at Mathematics Research Center, University of Wisconsin, Madison, WI.

0066-4189/83/0115-0261$02.00

mixture interacts with itself. In ordinary continuum mechanics (ignoring thermodynamic considerations), that requires a constitutive equation expressing the stress as a function of various fields. In two-phase flow mechanics, it requires specification of stresses for each phase, plus a relation for the interaction of the two materials.

Researchers who do not subscribe to this view treat the mechanics of the two materials, plus the dynamics of the interface, as fundamental, and use derived results and/or measurements to gain the understanding needed for their particular application. It is instructive to consider the approaches to gas dynamics. Most scientists believe that a gas is a collection of many molecules that move, vibrate, and interact in a complex, but describable way. Indeed, with the aid of large computers, molecular dynamics has made great strides in understanding such phenomena as shocks and phase transition, among others. In spite of the knowledge of the "correctness" of this model, many scientists and engineers use a continuum model for gas dynamics. Indeed, anyone wishing to describe, for example, the flow around an airfoil would be hard-pressed to find a computer big enough to do it as a problem in molecular dynamics. As a problem in continuum gas dynamics, it is still a large problem; however, it is done quite routinely numerically, including shocks. In addition, certain solutions to the continuum equations can be obtained analytically. While these are not always of direct technical interest, they often suggest phenomena or techniques that do have direct bearing on problems of interest. Experiments in gas dynamics can be useful. Without a model for comparison or scaling, however, one is limited to understanding only the particular geometry and scale of the experiment. The ability to infer is lost.

While these points seem obvious for gas dynamics, they nevertheless should be discussed and understood for the analogous situation of two-phase flow mechanics. For a particular flow with only a few particles that interact only slightly, it is best (and perhaps even necessary) to describe them "molecularly," that is, by predicting the trajectories of each one. If many particles are involved, it is better to use a continuum description. As in gas dynamics, both descriptions have their place. Furthermore, it is often instructive to try to ascertain what one model implies about the other, as in using the Boltzmann description of a gas to get continuum properties.

The analogy is apt for a continuum description of two-phase flows. This paper reviews the connection with the exact, or microscopic description, through the application of an averaging process to the continuum-mechanical equations describing the exact motion of each material at each point. If the exact flows were known, the averaged equations would be completely determined, and therefore unnecessary, since desired averaged information (such as the average concentration of particles) could be determined without using the averaged equations. The solutions are not known, nor is it necessary to the continuum approach that they be known. The resulting averaged equations are assumed to describe a material (the mixture) for which the interactions must be specified. This specification is then done according to certain rules that are reasonable and not too limiting. We shall discuss these rules and their implications.

The forms of the resulting equations are determined by the choice of a list of variables that are assumed to influence the interactions. The resulting equations have several unknown coefficients. These coefficients are assumed to be determinable by experiments. For the model we propose, we examine the experimental data and their implications on the coefficients. The result is, in essence, a recommendation for a model that has many known features of two-phase flow dynamics. An enlightened investigator can use the model to make predictions in a situation that falls within the range of assumptions made. The model can also be a starting point to obtain generalizations (such as the inclusion of electromagnetic effects). As with all models, it should never be used blindly, but with caution and careful examination of the results and implications.

Historically, geophysical flows involving sediment and clouds were among the first two-phase flows observed from a scientific point of view. Sediment meant erosion and loss of property, or deposition and loss of navigability. Clouds signaled possible rainfall. The amounts could mean drought, bountiful harvests, or devastating floods. An analytical description awaited the development of the calculus, fluid mechanics, and partial differential equation theory. The industrial revolution spurred the need for an understanding of all of basic science, including two-phase flows, although the correlation between progress and the acquisition of fundamental knowledge is not exact. For example, automobiles have worked reasonably well without a detailed knowledge of the flow and evaporation of fuel droplets; on the other hand, efficient design of fluidized beds relies on the understanding and stabilizing effects obtained from analytical models.

Early work on the form taken by beds of particles subject to forces due to flowing fluid (DuBuat 1786, Helmholtz 1888, Blasius 1910, Exner 1920) led to interesting fundamental results, such as the stability of the interface between two fluids and a practical understanding of the macroscopic properties of sedimentation and bed-form evolution. A more microscopic look (Bagnold 1941) dealt with the mechanics of the interactions between the particles and the fluid, and also the particles and the bed. A recent perspective is given by Engelund & Fredsøe (1982).

Developments in cloud physics occurred in physical chemistry of nucleation and formation. The need for detailed mechanical considerations was low due to the fact that the vapor, nucleation sites, and the formed droplets all flow with the surrounding air (until raindrops form).

Porous-medium theory, a two-phase flow where the solid phase does not move appreciably, developed with some different concerns (Darcy 1856), but Biot (1955) was instrumental in putting the empirical knowledge on a sound continuum basis. The desire to extract hydrocarbons from deep inside the earth's crust has given a great impetus to the study of flow in porous media. See Scheidegger's (1974) book for a discussion of porous-bed-modeling concerns.

The next important milestone in the development of two-phase-flow continuum-mechanics theory was the development and use in the 1950s of equations of dusty gases. Models for dusty gases are summarized by Marble (1970).

Chemical processing in fluidized beds gave an urgency to the development of the theory of particle-fluid systems. Early theoretical papers include those of Jackson (1963), Murray (1965a,b), and Anderson & Jackson (1967). The flow regime involved in fluidization is one of the most difficult for particle-fluid flows. The particle concentrations are high, the dispersed phase is relatively dense, the dispersed phase undergoes a random micromotion, and often it is desirable to have chemical reactions occur in the flow.

In the early 1960s the emergence of commercial atomic energy spurred the study of flows of steam and water. The work of Zuber (1964) is a pioneering landmark. Fluid-fluid flows have a difficulty not encountered in particle-fluid flows, namely that the shape of the dispersed phase can change, resulting in changing interfacial area and consequent interactions between the fluids. In spite of much progress (Lahey & Moody 1977), two-phase flow studies in nuclear reactors are still a concern.

The parallel development of mixture theory had some small influence on the progress of two-phase flows. Indeed, the concept of interpenetrating continua is natural in mixtures where the dispersion occurs on the molecular level. The development of the ideas of diffusion by Fick (1855) and the thermodynamic concepts in mixtures by Duhem (1893), Meixner (1943), and Prigogine & Mazur (1951) lead naturally to the theory of mixtures expounded by Truesdell & Toupin (1960). The theory of mixtures, as applied to the specific mixtures where the two constituents remain unmixed, is the basis of the present work, and (whether explicitly recognized or not) much of the more specific previous work. Kenyon (1976) applies the mixture theory to multiphase flows. See also the review by Bedford & Drumheller (1982).

By the 1960s enough of the common features of the diverse two-phase flows were evident. Several influential books of a general nature appeared, including *Fluid Dynamics of Multiphase Systems* by Soo (1967), *One-Dimensional Two-Phase Flow* by Wallis (1969), and *Flowing Gas-Solid Suspensions* by Boothroyd (1971). Wallis' approach was strongly influenced by gas-liquid flows and dealt with the basic concept of interpenetrability by considering cross-sectionally averaged equations; in addition, he introduced constitutive assumption by quoting appropriate experiments. Soo's work was largely based on particle-fluid flows. He assumed interpenetrability from the start, and included forces in the particular momentum equation known from experiments or inferred from calculations. Boothroyd was interested in particle-gas flows, and contributed ideas on turbulence and drag reduction.

A different, but somewhat related, approach to the problem of a mechanical description of two-phase flows uses the single-particle or several-interacting-particle flow fields, along with an averaging approach to yield rheological or transport properties. The celebrated result in this area is the effective viscosity result due to Einstein (1906), which shows that in slow flow the mixture behaves like a fluid with viscosity increased by a factor of $1 + \frac{5}{2}\alpha$, where α is the volumetric concentration of particles. An approach outlined by Brinkman (1947) has been influential. The philosophical roots of this approach are elegantly discussed in the book *Low Reynolds Number Hydrodynamics*, by Happel & Brenner (1965). The idea is to use solutions of the flow equations in special cases (such as Stokes flow around an array of spheres) to infer information about the flow (such as the total force on the particular phase). Several aspects of this work have been discussed in this Review; see the papers by Brenner (1970), Batchelor (1974), Herczyński & Pieńkowska (1980), Leal (1980), and Russel (1981).

Several authors (Bedford & Drumheller 1978, Drumheller & Bedford 1980) have pursued a variational approach to two-phase flows, generalizing the work of Biot (1977). The variational formulation starts with Hamilton's principle, given by

$$\delta \int_{t_2}^{t_1} (T - V)\, dt + \int_{t_1}^{t_2} \delta W\, dt = 0,$$

where δ represents the variation over an appropriate space of functions, T and V are the kinetic and potential energies, δW is the virtual work, and t_1 and t_2 are two arbitrary times. A variational formulation has an advantage in that if it is desired to include a certain effect (for example, the virtual mass), that effect would be included consistently in the mass, momentum, and kinetic-energy equations. A more concrete and immediate advantage is in formulating numerical techniques, and specifically, finite-element techniques, where variational formulation eases the translation of the partial differential equations into discrete equations.

In order to use a variational formulation, it is necessary to define the variation δ, the kinetic and potential energies T and V, and the virtual work δW. Constraints must be included with Lagrange multipliers. Bedford & Drumheller (1982) note that Hamilton's principle is usually formulated with a "control mass," that is, for material volumes. The problem of having two materials moving at different velocities is dealt with by assuming a control volume that coincides with a rigid surface through which no material of either phase passes.

Bedford & Drumheller use the technique to show how the effect of oscillations in bubble diameter can be taken into account by including the kinetic energy of the liquid due to a change in bubble diameter. They similarly include virtual mass by arguing that relative accelerations increase the kinetic energy. A conceptual difficulty with the approach is that the user must decide which fluctuations contribute to the total kinetic energy and which contribute to virtual work. For example, viscous drag is due to "fluctuations" of the fluid velocity near a particle, but it is included as a virtual-work term. A rule of thumb might be to consider whether the energy is recoverable or not. Virtual work is associated with unrecoverable energy, kinetic and potential energy with recoverable energy. Variational formulations (in general) do not mention the effect of Reynolds stresses, which are one manifestation of fluctuations. We conclude that while variational formulations are useful, it is not always straightforward to formulate them.

EQUATIONS OF MOTION

We start by assuming that each material involved can be described as a continuum, governed by the partial differential equations of continuum mechanics. The materials are separated by an interface, which we assume to be a surface. At the interface, jump conditions express the conditions of conservation of mass and momentum.

The equations of motion for each phase are (Truesdell & Toupin 1960)
1. conservation of mass

$$\frac{\partial \rho}{\partial t} + \nabla \cdot \rho \mathbf{v} = 0, \tag{1}$$

2. conservation of linear momentum

$$\frac{\partial \rho \mathbf{v}}{\partial t} + \nabla \cdot \rho \mathbf{v} \mathbf{v} = \nabla \cdot \mathbf{T} + \rho \mathbf{f}, \tag{2}$$

valid in the interior of each phase. Here ρ denotes the density, $\mathbf{v}$ the velocity, $\mathbf{T}$ the stress tensor, and $\mathbf{f}$ the body force density. Conservation of angular momentum becomes $\mathbf{T} = \mathbf{T}^t$, where t denotes the transpose. At the interface, the jump conditions are

1. jump condition for mass

$$\llbracket \rho(\mathbf{v} - \mathbf{v}_i) \cdot \mathbf{n} \rrbracket = 0, \tag{3}$$

2. jump condition for momentum

$$\llbracket \rho\mathbf{v}(\mathbf{v} - \mathbf{v}_i) \cdot \mathbf{n} - \mathbf{T} \cdot \mathbf{n} \rrbracket = \sigma\kappa\mathbf{n}. \tag{4}$$

Here $\llbracket \ \rrbracket$ denotes the jump across the interface, $\mathbf{v}_i$ is that velocity of the interface, σ is the surface tension, assumed to be a constant, κ is the mean curvature of the interface, and $\mathbf{n}$ is the unit normal (Aris 1962). We shall assume that $\mathbf{n}$ points out of phase k, and that the jump between f in phase k and f in phase l is defined by $\llbracket f \rrbracket = f^l - f^k$, where a superscript k denotes the limiting value from the phase k side. As a sign convention for the curvature, we assume that κ is positive (concave) toward $-\mathbf{n}$. The mass of the interface has been neglected, and the surface stresses have been assumed to be in the form of classic surface tension.

We do not discuss any thermodynamic relations in this paper. Thermodynamic considerations are important for many multiphase flows. Our discussion focuses on the mechanics of two-phase flows; hence we elect to forego a discussion of thermodynamics for the sake of simplicity.

Constitutive equations must be supplied to describe the behavior of each material involved. For example, if one material is an incompressible liquid, then specifying the value of ρ and assuming $\mathbf{T} = -p\mathbf{I} + \mu(\nabla\mathbf{v} + (\nabla\mathbf{v})^t)$ determines the nature of the behavior of the fluid in that phase. Similar considerations are possible for solid particles or a gas. The resulting differential equations, along with the jump conditions, provide a fundamental description of the detailed or exact flow.

Usually, however, the details of the flow are not required. For most purposes of equipment or process design, averaged, or macroscopic flow information is sufficient. Fluctuations, or details in the flow must be resolved only to the extent that they effect the mean flow (like the Reynolds stresses effect the mean flow in a turbulent flow).

Averaging

In order to obtain equations that do not contain the details of the flow, it has become customary to apply some sort of averaging process. It is not essential to do so; indeed, some researchers prefer to postulate macroscopic equations without reference to any microscopic equations. Certainly the necessary terms in the macroscopic equations can be deduced without using an averaging process. One advantage of a postulation approach is obvious—not having to deal with the worries of the averaging process. The advantages of averaging are less obvious. First, the various terms appearing in the macroscopic equations are shown to arise from appropriate microscopic considerations. For example, stress terms arise from microscopic stresses (pressure, for instance) and also from velocity fluctuations (Reynolds stresses). Knowledge of this fact does give a modicum of insight into the formulation of constitutive equations. If a term appeared in the averaging that was not expected in the postulational approach, then it would be obligatory to include it in the postulated model, or else explain why it is superfluous.

An additional advantage of averaging is that the resulting macroscopic variables are related to microscopic variables. If the microscopic problem can be solved for some special situation, the solution can then be used to get values of the macroscopic variables. This cannot replace the need to postulate constitutive equations, but it can give insights into the types of terms expected to be important in the constitutive relations. The philosophy taken here is that any information that can be gained from averaging is worthwhile, and therefore we present a generic averaging method and its results. If the reader feels that averaging is unnecessary, he can skip this section and assume that Equations (40)–(43) have been postulated, along with the interpretations given at the end of this section.

Averaging the equations of motion is suggested by the averaging approach to turbulence (see Hinze 1959). Time and space averages were the first to appear (Frankl 1953, Teletov 1958). These averages have been refined by weighting, by multiple application, and by judicious choice of averaging region. The highlights can be found in the work of Anderson & Jackson (1967), Vernier & Delhaye (1968), Drew (1971), Whitaker (1973), Ishii (1975), Nigmatulin (1979), and Gough (1980). Statistical averages are most convenient for the rheology work; see Batchelor (1974) for a summary. The applications of statistical averages to the equations of motion is straightforward; the paper of Buyevich & Shchelchkova (1978) summarizes this approach nicely.

Let $\langle \; \rangle$ denote an averaging process so that if $f(\mathbf{x}, t)$ is an exact microscopic field, then $\langle f \rangle(\mathbf{x}, t)$ is the corresponding averaged field. We shall specify shortly the requirements that an average should possess; however, for now we merely suggest that it should be smoothing in the sense that no details appear in the averaged variables. Some examples of commonly used averages in multiphase flow are the time average

$$\langle f \rangle_1(\mathbf{x}, t) = \frac{1}{T} \int_{t-T}^{t} f(\mathbf{x}, t') \, dt', \tag{5}$$

where T is an averaging time scale; the space average

$$\langle f \rangle_2(\mathbf{x}, t) = \frac{1}{L^3} \int_{x_1-(1/2)L}^{x_1+(1/2)L} \int_{x_2-(1/2)L}^{x_2+(1/2)L} \int_{x_3-(1/2)L}^{x_3+(1/2)L} f(\mathbf{x}', t) \, dx_3' \, dx_2' \, dx_1',$$

(6)

where L is an averaging length scale; a weighted space average

$$\langle f \rangle_3(\mathbf{x}, t) = \iiint_{R^3} g(\mathbf{x}-\mathbf{x}')f(\mathbf{x}', t) \, d\mathbf{x}',$$

(7)

where $\iiint_{R^3} g(\mathbf{s}) \, d\mathbf{s} = 1$; and various combinations of averages and/or specific types of weightings. Also mentioned in the literature are ensemble averages

$$\langle f \rangle_4(\mathbf{x}, t) = \frac{1}{N} \sum_{n=1}^{N} f_n(\mathbf{x}, t)$$

(8)

or

$$\langle f \rangle_5(\mathbf{x}, t) = \int_\Omega f(\mathbf{x}, t; \omega) \, d\mu(\omega),$$

(9)

where $f_n(\mathbf{x}, t)$ or $f(\mathbf{x}, t; \omega)$ denote a realization of the quantity f over a set of possible "equivalent" realizations Ω. One way in which randomness might be introduced in a particular flow situation is by allowing the particles to have random initial positions.

The averaging process is assumed to satisfy

$$\langle f + g \rangle = \langle f \rangle + \langle g \rangle$$

(10)

$$\langle \langle f \rangle g \rangle = \langle f \rangle \langle g \rangle$$

(11)

$$\langle c \rangle = c$$

$$\left\langle \frac{\partial f}{\partial t} \right\rangle = \frac{\partial}{\partial t} \langle f \rangle$$

(12)

$$\left\langle \frac{\partial f}{\partial x_i} \right\rangle = \frac{\partial}{\partial x_i} \langle f \rangle.$$

(13)

The first three of these relations are called Reynolds' rules, the fourth is called Leibniz' rule, and the fifth is called Gauss' rule.

Some difficulty is encountered when trying to apply the average to the equations of motion for each phase. In order to do this, we introduce the phase function $X_k(\mathbf{x}, t)$, which is defined to be

$$X_k(\mathbf{x}, t) = \begin{cases} 1 & \text{if } \mathbf{x} \text{ is in phase } k \text{ at time } t \\ 0 & \text{otherwise.} \end{cases}$$

(14)

We shall deal with X_k as a generalized function, in particular with regard to differentiating it. Recall that a derivative of a generalized function can be defined in terms of a set of "test functions" ϕ, which are "sufficiently smooth" and have compact support. Then $\partial X_k / \partial t$ and $\partial X_k / \partial x_i$ are defined by

$$\int_{R^3 \times R} \frac{\partial X_k}{\partial t} (\mathbf{x}, t) \phi(\mathbf{x}, t) \, d\mathbf{x} \, dt = - \int_{R^3 \times R} X_k(\mathbf{x}, t) \frac{\partial \phi}{\partial t} (\mathbf{x}, t) \, d\mathbf{x} \, dt,$$

(15)

$$\int_{R^3 \times R} \frac{\partial X_k}{\partial x_i} (\mathbf{x}, t) \phi(\mathbf{x}, t) \, d\mathbf{x} \, dt = - \int_{R^3 \times R} X_k(\mathbf{x}, t) \frac{\partial \phi}{\partial x_i} (\mathbf{x}, t) \, d\mathbf{x} \, dt.$$

(16)

It can be shown that

$$\frac{\partial X_k}{\partial t} + \mathbf{v}_i \cdot \nabla X_k = 0$$

(17)

in the sense of generalized functions. To see this, consider

$$\int_{R^3 \times R} \left(\frac{\partial X_k}{\partial t} + \mathbf{v}_i \cdot \nabla X_k \right) \phi \, d\mathbf{x} \, dt = \int_{R^3 \times R} X_k \left(\frac{\partial \phi}{\partial t} + \nabla \cdot \phi \mathbf{v}_i \right) d\mathbf{x} \, dt$$

$$= - \int_{-\infty}^{\infty} \left(\int_{R_k} \left(\frac{\partial \phi}{\partial t} + \nabla \cdot \phi \mathbf{v}_i \right) d\mathbf{x} \right) dt$$

$$= - \int_{-\infty}^{\infty} \left(\frac{d}{dt} \int_{R_k(t)} \phi \, d\mathbf{x} \right) dt$$

$$= \int_{R_k} \phi \, dx \Big|_{-\infty}^{\infty} = 0,$$

(18)

since ϕ has compact support in t. Here $R_k(t)$ is the region occupied by phase k at time t, and we assume that $\mathbf{v}_i$ is extended smoothly through phase k (in order that the second line makes sense).

If f is smooth except at the interface $\mathfrak{S}$, then $f \nabla X_k$ is defined via

$$\int_{R^3 \times R} f \nabla X_k \phi \, d\mathbf{x} \, dt = - \int_{R^3 \times R} X_k \nabla (f \phi) \, d\mathbf{x} \, dt$$

$$= \int_{-\infty}^{\infty} \int_{R_k(t)} \nabla (f \phi) \, d\mathbf{x} \, dt$$

$$= \int_{-\infty}^{\infty} \int_{\mathfrak{S}} \mathbf{n}_k f^k \phi \, dS \, dt,$$

(19)

where $\mathbf{n}_k$ is the unit normal exterior to phase k, and f^k denotes the limiting value of f on the phase-k side of $\mathcal{S}$.

It is also clear that ∇X_k is zero, except at the interface. Equation (19) describes the behavior of ∇X_k at the interface. Note that it behaves as a "delta-function," picking out the interface $\mathcal{S}$, and has the direction of the normal interior to phase k.

This motivates writing

$$\nabla X_k = \mathbf{n}_k \frac{\partial X}{\partial n}, \tag{20}$$

where $\partial X/\partial n$ is a scalar-valued generalized function such that

$$\int_{R^3 \times R} \left[\frac{\partial X}{\partial n}(\mathbf{x}, t) \right] \phi(\mathbf{x}, t)\, d\mathbf{x}\, dt = -\int_{-\infty}^{\infty} \int_{\mathcal{S}} \phi(\mathbf{x}, t)\, dS\, dt. \tag{21}$$

The quantity $\partial X/\partial n$ then picks out the interface $\mathcal{S}$. We write

$$\left\langle \frac{\partial X}{\partial n} \right\rangle = s, \tag{22}$$

where s is the average interfacial area per unit volume.

Averaged Equations

In order to derive averaged equations for the motion of each phase, we multiply the equation of conservation of mass valid in phase k (1) by X_k and average. Noting that

$$X_k \frac{\partial \rho}{\partial t} = \frac{\partial}{\partial t} X_k \rho - \rho \frac{\partial X_k}{\partial t} = \frac{\partial}{\partial t} X_k \rho + \rho \mathbf{v}_i \cdot \nabla X_k \tag{23}$$

and

$$X_k \nabla \cdot \rho \mathbf{v} = \nabla \cdot X_k \rho \mathbf{v} - \rho \mathbf{v} \cdot \nabla X_k, \tag{24}$$

we have

$$\frac{\partial}{\partial t} \langle X_k \rho \rangle + \nabla \cdot \langle X_k \rho \mathbf{v} \rangle = \left\langle \left[\rho (\mathbf{v} - \mathbf{v}_i) \right]^k \cdot \nabla X_k \right\rangle. \tag{25}$$

Similar considerations for the momentum equations yield

$$\frac{\partial}{\partial t} \langle X_k \rho \mathbf{v} \rangle + \nabla \cdot \langle X_k \rho \mathbf{v} \mathbf{v} \rangle = \nabla \cdot \langle X_k T \rangle + \langle X_k \rho f \rangle$$
$$+ \left\langle \left[\rho \mathbf{v} (\mathbf{v} - \mathbf{v}_i) - \mathbf{T} \right]^k \cdot \nabla X_k \right\rangle. \tag{26}$$

The terms

$$\left\langle \left[\rho (\mathbf{v} - \mathbf{v}_i) \right]^k \cdot \nabla X_k \right\rangle = \Gamma_k \tag{27}$$

and

$$\left\langle \left[\rho \mathbf{v}(\mathbf{v}-\mathbf{v}_i)-\mathbf{T}\right]^k \cdot \nabla X_k \right\rangle = \mathbf{M}_k \tag{28}$$

are the interfacial source terms. As noted, ∇X_k picks out the interface and causes discontinuous quantities multiplying it to be evaluated on the phase-k side of the interface.

The jump conditions are derived by multiplying Equations (27) and (28) by $\partial X/\partial n$, and recognizing that $\mathbf{n}_1 = -\mathbf{n}_2$. We obtain

$$\sum_{k=1}^{2} \left\langle \left[\rho(\mathbf{v}-\mathbf{v}_i)\right]^k \cdot \nabla X_k \right\rangle = \sum_{k=1}^{2} \Gamma_k = 0 \tag{29}$$

$$\sum_{k=1}^{2} \left\langle \left[\rho \mathbf{v}(\mathbf{v}-\mathbf{v}_i)-\mathbf{T}\right]^k \cdot \nabla X_k \right\rangle = \sigma \langle \kappa \nabla X_1 \rangle = \mathbf{M}_m. \tag{30}$$

The term $\mathbf{M}_m$ is the contribution to the total force on the mixture due to the interface, and specifically due to surface tension.

Applying a different averaging process requires a different set of calculations regarding the interfacial source terms (Anderson & Jackson 1967, Drew 1971, Ishii 1975, Delhaye & Achard 1976). Almost all of the derivations for specific averaging processes seem to be more complicated than the above; however, the trade-off for the simple derivation is that all manipulations now involve generalized functions.

The volumetric concentration (or volume fraction, or relative residence time) of phase k is defined by

$$\alpha_k = \langle X_k \rangle. \tag{31}$$

We note that

$$\frac{\partial \alpha_k}{\partial t} = \left\langle \frac{\partial X_k}{\partial t} \right\rangle \tag{32}$$

and

$$\nabla \alpha_k = \langle \nabla X_k \rangle. \tag{33}$$

There are two types of averaged variables that are useful in two-phase mechanics, namely the phasic, or X_k-weighted average, and the mass-weighted average. Which is appropriate is suggested by the appearance of the quantity in the equation of motion. The phasic average of the variable ϕ is defined by

$$\tilde{\phi}_k = \langle X_k \phi \rangle / \alpha_k \tag{34}$$

and the mass-weighted average of the variable ψ is defined by

$$\hat{\psi}_k = \langle X_k \rho \psi \rangle / \alpha_k \tilde{\rho}_k. \tag{35}$$

It is convenient to write the stresses $\tilde{\mathbf{T}}_k$ in terms of pressures plus extra stresses. Thus,

$$\tilde{\mathbf{T}}_k = -\tilde{p}_k \mathbf{I} + \tilde{\boldsymbol{\tau}}_k. \tag{36}$$

It is expected that readers familiar with fluid-dynamical concepts are familiar with the concept of pressure in fluids; in this case, $\tilde{p}_k$ can be thought of as the average of the microscopic pressure. If one of the phases consists of solid particles, the concept is less familiar. In this case, the microscopic stress (involving small elastic deformations, for example) is thought of as made up of a spherical part (acting equally in all directions) plus an extra stress. The spherical part, when averaged, yields the pressure $\tilde{p}_k$ in Equation (36).

It has further become customary to separate various parts of the interfacial momentum-transfer term. This is done by defining the interfacial velocity of the kth phase by

$$\Gamma_k \mathbf{v}_{k,i} = \left\langle \left[\rho \mathbf{v}(\mathbf{v} - \mathbf{v}_i) \right]^k \cdot \nabla X_k \right\rangle, \tag{37}$$

and the interfacial pressure on the kth phase by

$$p_{k,i} |\nabla \alpha_k|^2 = \left\langle p^k \nabla X_k \right\rangle \cdot \nabla \alpha_k. \tag{38}$$

Equation (38) is the dot product of $\nabla \alpha_k$ with the "standard" definition (Ishii 1975) of the interfacial pressure. This definition uses three equations to define one scalar quantity and cannot be a generally valid definition. Here the remaining part of the contribution of the pressure at the interface is lumped with the viscous-stress contribution at the interface and is treated through the use of a constitutive equation. Thus, we write

$$\mathbf{M}^k = \Gamma_k \mathbf{v}_{k,i} - p_{k,i} \nabla \alpha_k + \mathbf{M}_k^d, \tag{39}$$

where $\mathbf{M}_k^d = \left\langle (p - p_{k,i})^k \nabla X_k - \tau_k^k \cdot \nabla X_k \right\rangle$ is referred to as the interfacial force density, although it does not contain the effect of the average force on the interface due to the average interfacial pressure. The term $-p_{k,i} \nabla \alpha_k$, which does contain the force due to the average interfacial pressure, is sometimes referred to as the buoyant force. The reason for this terminology is, of course, that the buoyant force on an object is due to the distribution of the pressure of the surrounding fluid on its boundary.

With Equations (31) and (34)–(39), the equations of motion, (25) and (26), become

$$\frac{\partial \alpha_k \tilde{\rho}_k}{\partial t} + \nabla \cdot \alpha_k \tilde{\rho}_k \hat{\mathbf{v}}_k = \Gamma_k, \tag{40}$$

$$\frac{\partial \alpha_k \tilde{\rho}_k \hat{\mathbf{v}}_k}{\partial t} + \nabla \cdot \alpha_k \tilde{\rho}_k \hat{\mathbf{v}}_k \hat{\mathbf{v}}_k = - \alpha_k \nabla \tilde{p}_k + \nabla \cdot \alpha_k (\tilde{\tau}_k + \tilde{\sigma}_k)$$
$$+ \Gamma_k \mathbf{v}_{k,i} + (p_{k,i} - \tilde{p}_k) \nabla \alpha_k$$
$$+ \mathbf{M}_k^d. \tag{41}$$

The jump conditions, (29) and (30), are

$$\sum_{k=1}^{2} \Gamma_k = 0, \tag{42}$$

$$\sum_{k=1}^{2} \left[\Gamma_k \mathbf{v}_{k,i} + p_{k,i} \nabla \alpha_k + \mathbf{M}_k^d \right] = \mathbf{M}_m. \tag{43}$$

Adequate models for compressibility and phase change require consideration of thermodynamic processes. These are beyond the scope of this paper; therefore we shall restrict our attention to incompressible materials where no phase change occurs. Thus we assume that

$$\tilde{\rho}_k = \text{constant} \tag{44}$$

and

$$\Gamma_k = 0. \tag{45}$$

In order to simplify the notation, we shall *drop all symbols denoting averaging.*

CONSTITUTIVE EQUATIONS

In order to have a usable model for an ordinary single-phase material, relations must be given that specify how the material interacts with itself. These relations specify how the material transmits forces from one part of the body to another. For single-phase materials a relation between the stress and other field variables is usually required. For two-phase materials, where we desire to track both phases, we must specify the stresses for both phases and the way in which the materials interact across the interface. Thus, we require constitutive equations for the stresses ($\tau_k + \sigma_k$), the interfacial force density $\mathbf{M}_k^d$, and the pressure differences $p_k - p_{k,i}$, consistent with the equations of motion and the jump conditions.

The fundamental process consists of proposing forms for the necessary terms within the framework of the principles of constitutive equations, finding solutions of the resulting equations, and verifying against experiments. The process can be iterative, with some experiment indicating that a more involved form is needed for some constituted variable. The ideal end result of the process is a set of equations that could be used to predict the behavior of the two-phase flow, for example with a computer code. With the equations should come a set of conditions for the validity of the values of the constants and other functions used in the constitutive equations.

The stresses $\tau_k + \sigma_k$, the interfacial force density $\mathbf{M}_k^d$, and the pressure differences $p_k - p_{k,i}$ are assumed to be functions of $\alpha_k, \partial\alpha_k/\partial t,$ $\nabla\alpha_k, \mathbf{v}_k, \nabla\mathbf{v}_k, \partial\mathbf{v}_k/\partial t, \ldots$, where … represents the material properties, such as the viscosities and densities of the two materials, and other geometric parameters, such as the average particle size or the interfacial area density.

For concreteness, we shall refer to phase one as the particulate, or dispersed, phase and include in that description solid particles, droplets, or bubbles. Phase two is then the continuous, or carrier phase and can be liquid or gas. We shall denote

$$\alpha = \alpha_1, \tag{46}$$

so that

$$1 - \alpha = \alpha_2. \tag{47}$$

It is evident that both α and $1 - \alpha$ need not be included as independent variables in forming constitutive equations.

Drew & Lahey (1979) consider the general process of constructing constitutive equations. The principle of consequence is that of *material frame indifference*, or *objectivity*. This principle requires that constitutive equations be appropriately invariant under a change of reference frame. The motivation behind this assumption is that we expect the way in which a two-phase material distributes forces to be independent of the coordinate system used to express it. Those who doubt this principle often argue that the momentum equation is not frame indifferent, needing Coriolis forces and centrifugal forces to correct it in noninertial frames. This argument is incorrect, since the appropriate formulation of the momentum equation deals not with "acceleration" but with "acceleration relative to an inertial frame." When formulated with the "acceleration relative to an inertial frame," the momentum equations are frame indifferent (Truesdell 1977). This introduces an interesting question: Can the mixture know whether or not it is being referred to a inertial frame? That is, can the interfacial force and the Reynolds stresses depend on the reference frame? The principle of material frame indifference, as used here, claims that they cannot. Others claim that they must (Ryskin & Rallison 1980).

The principle of objectivity is now examined (Drew & Lahey 1979). The approach we take deals with coordinate changes, but is equivalent to more abstract approaches (Truesdell 1977). Consider a coordinate change from system $\mathbf{x}$ to system $\mathbf{x}^*$, given by

$$\mathbf{x}^* = \mathbf{Q}(t)\cdot\mathbf{x} + \mathbf{b}(t), \tag{48}$$

where $\mathbf{Q}(t)$ is an orthonormal tensor and $\mathbf{b}$ is a vector. A scalar is objective if its value is the same in both the starred and unstarred systems, that is, if

$$\phi^*(\mathbf{x}^*, t) = \phi(\mathbf{x}, t). \tag{49}$$

A vector is objective if it transforms coordinates correctly, that is, if

$$\mathbf{v}^* = \mathbf{Q}\cdot\mathbf{v}. \tag{50}$$

A tensor is objective if

$$\mathbf{T}^* = \mathbf{Q}\cdot\mathbf{T}\cdot\mathbf{Q}^t. \tag{51}$$

The scalar α is objective; however $\partial\alpha/\partial t$ is not. It is straightforward to show that $D_k\alpha/Dt = \partial\alpha/\partial t + \mathbf{v}_k\cdot\nabla\alpha$ is objective.

Let us next consider objective vectors. The volume fraction gradient $\nabla\alpha$ is objective. If we differentiate (48) with respect to t following a material particle of phase k, we see that

$$\mathbf{v}_k^* = \mathbf{Q}\cdot\mathbf{v}_k + (\dot{\mathbf{Q}}\cdot\mathbf{x} + \dot{\mathbf{b}}). \tag{52}$$

Hence velocities are not objective. This is obvious physically, since the velocity of a point depends on the coordinate system. If we take Equation (52) for $k = 1, 2$, and subtract, we have

$$\mathbf{v}_1^* - \mathbf{v}_2^* = \mathbf{Q}\cdot(\mathbf{v}_1 - \mathbf{v}_2). \tag{53}$$

Thus, the relative velocity between the phases is objective.

For accelerations, differentiating (52) following a material particle of phase j yields

$$\frac{D_j\mathbf{v}_k^*}{Dt} = \mathbf{Q}\cdot\frac{D_j\mathbf{v}_k}{Dt} + \ddot{\mathbf{Q}}\cdot\mathbf{x} + \dot{\mathbf{Q}}\cdot\mathbf{v}_j + \dot{\mathbf{Q}}\cdot\mathbf{v}_k + \ddot{\mathbf{b}}. \tag{54}$$

Thus, subtracting the value for $j = 1$, $k = 2$ from the value for $j = 2$, $k = 1$ yields the result that

$$\mathbf{a}_{12} = \frac{D_1\mathbf{v}_2}{Dt} - \frac{D_2\mathbf{v}_1}{Dt} \tag{55}$$

is objective.

The list of tensors that we consider is motivated by the desire to model two-phase flows. Thus, velocity gradients (expressing rate of deformation) are included, but displacement gradients (expressing deformation) are not. It is well known (Truesdell & Toupin 1960) that

$$\mathbf{D}_{k,b} = \tfrac{1}{2}\left(\nabla\mathbf{v}_k + (\nabla\mathbf{v}_k)^t\right) \tag{56}$$

are objective. We further note that $\nabla(\mathbf{v}_1 - \mathbf{v}_2)$ is an objective tensor.

Thus, the problem of forming constitutive equations reduces to expressing

$$\tau_k + \sigma_k, p_k - p_{k,i}, \mathbf{M}_k^d \tag{57}$$

in terms of

$$\alpha, D_k\alpha/Dt, \nabla\alpha, \mathbf{v}_1 - \mathbf{v}_2, D_2\mathbf{v}_1/Dt - D_1\mathbf{v}_2/Dt, \mathbf{D}_{1,b}, \mathbf{D}_{2,b}, \nabla(\mathbf{v}_1 - \mathbf{v}_2). \tag{58}$$

We shall assume that the materials are isotropic. This means that no direction is preferred in either material or in the way they interact. Note that a particular flow may occur in such a way that a preferred direction (down, for example) might emerge. In that case, the *flow* is anisotropic, but the materials and the mixture are isotropic.

If f is a scalar, then f can depend on the scalars and the scalar invariants formed from the vectors and tensors in the list (58). The length of a vector is a scalar invariant, as are the trace, determinant, and Euclidean norm of a tensor.

If $\mathbf{F}$ is a vector, then $\mathbf{F}$ must be linear in all the vectors in the list (58), plus any vectors that can be formed in an invariant way from the vectors and tensors in that list. Thus

$$\mathbf{F} = \sum a_i \mathbf{V}_i, \tag{59}$$

where $\mathbf{V}_i$ are the objective vectors so formed. Each scalar coefficient a_i can be a function of all the scalars and scalar invariants.

If $\mathscr{F}$ is a second-order tensor, then

$$\mathscr{F} = \sum B_j \gamma_j, \tag{60}$$

where γ_j are the objective tensors that can be formed from the list (58). The scalars B_j can be functions of all the scalars. If $\mathscr{F}$ is symmetric, we need only consider symmetric γ_j.

The general approach (Drew & Lahey 1979) gives a problem too big to be practical or meaningful. Thus, we shall consider here forms for the stresses, etc., that we expect to be important in two-phase flows. Obviously, we shall have the constant worry that something has been forgotten; however, the alternative of stopping at this point, awaiting evaluation of over two-thousand scalar coefficients, seems less attractive.

Stresses

Let us now discuss specific considerations for the specific constitutive equations needed. We begin with the stresses. The extra stress (viscous or elastic) and turbulent stresses are combined as $\tau_k + \sigma_k$. Macroscopically, at least in the context of this model, one sees no way to separate them. Microscopically, one is due to the actual stresses and the other is due to velocity fluctuations from the mean. In many cases of interest the extra stress is smaller than the turbulent stress, one reason being that any appreciable slip between the phases generates velocity fluctuations, which appear in this model in the turbulent stresses. These fluctuations can transport momentum. The conceptual situation is analogous to that for single-phase turbulent momentum transport, with "eddies" carrying momentum across planes parallel to the shearing direction and losing it into the mean flow. The fluctuations in single-phase turbulent flow are there because of instabilities in the laminar shear flow; in two-phase flow they also come from fluctuations generated by the flow around individual particles.

Drew & Lahey (1980b) have generated a model for the turbulent stresses in bubbly air-water flow. The model they deduce in this case has the form

$$\sigma_2 = 2\mu_2^T \mathbf{D}_{2,b} + a_2 \mathbf{I} + b_2 (\mathbf{v}_1 - \mathbf{v}_2)(\mathbf{v}_1 - \mathbf{v}_2) \tag{61}$$

for the liquid phases. Here μ_2^T is the eddy viscosity, a_2 is an induced eddy "pressure," and b_2 is associated with bubble passages. The parameters μ_2^T, a_2, and b_2 depend on α, the bubble radius r, the relative velocity $|\mathbf{v}_1 - \mathbf{v}_2|$, and the Euclidean norm of $\mathbf{D}_{2,b}$. Nigmatulin (1979) uses a cell model, assuming inviscid flow around a spherical particle in each cell, to calculate values of a_2 and b_2. He obtains

$$b_2 = -\tfrac{1}{2}\rho_2, \tag{62}$$

$$a_2 = \tfrac{1}{6}\rho_2 |\mathbf{v}_1 - \mathbf{v}_2|^2. \tag{63}$$

He does not obtain a term involving the eddy viscosity because the cell model is unsuited to obtaining shear-flow results. His results are valid for dilute flows. Other effects are present in most turbulent flows.

Drew & Lahey (1980b) use mixing lengths to evaluate the coefficients in Equation (61). The data available thus far (Serizawa 1975) allow only evaluation of the ratios of various Kármán constants; nonetheless, the importance of the problem in nuclear-reactor technology has spurred the acquisition of more direct data in special cases (Lance 1981). These data should be available shortly and should allow the evaluation of a_2 and b_2 for bubbly flows.

Evidence (Serizawa 1975) also indicates that

$$\sigma_1 = O\!\left(\frac{\rho_1}{\rho_2}\sigma_2\right) \tag{64}$$

and that it is negligible in dispersed bubbly flows. A model for σ_1 analogous to Equation (61) is

$$\sigma_1 = 2\mu_1^T \mathbf{D}_{1,b} + a_1 \mathbf{I} + b_1(\mathbf{v}_1 - \mathbf{v}_2)(\mathbf{v}_1 - \mathbf{v}_2). \tag{65}$$

If the particulate phase follows the fluid phase closely, the quantities in Equations (61) and (65) representing both velocity scales and mixing lengths of large eddy processes should be approximately equal. If this is the case, then

$$\mu_1^T \cong \frac{\rho_1}{\rho_2}\mu_2^T. \tag{66}$$

The remaining terms in Equation (65) arise at least in part because of the motion of the particles through the fluid. No constraints are placed on these coefficients.

Ishii (1975) argues that for the viscous-fluid phase, the average of the microscopic viscous stress leads to

$$\tau_2 = 2\mu_2 \mathbf{D}_{2,b} + \mu_2 \frac{b(1-\alpha)}{1-\alpha}\left[\nabla\alpha(\mathbf{v}_1 - \mathbf{v}_2) + (\mathbf{v}_1 - \mathbf{v}_2)\nabla\alpha\right], \tag{67}$$

where μ_2 is the exact viscosity of the fluid and b is called the mobility of phase two. He argues that $b(1-\alpha) \cong 1$ for α near zero. No experimental evidence has been offered to verify (67).

In the case when the particulate phase consists of solid particles, it is usually assumed that $\tau_1 = 0$. If the particulate phase is also a viscous fluid, then it is sometimes assumed that

$$\tau_1 = 2\mu_1 \mathbf{D}_{1,b}. \tag{68}$$

This corresponds to Ishii's result [see Equation (67)] with $b = 0$.

Murray (1965a) offers a model for a particle-fluid flow in which it is assumed that the particles experience a viscous stress with a viscosity coefficient proportional to the fluid viscosity. His argument indicates that the force being calculated is not a stress, but an interfacial force.

Pressure Relations

The pressure differences $p_k - p_{k,i}$ are discussed next. First, we note that with the assumption of local incompressibility of each phase, we must constitute all but one of p_1, p_2, $p_{1,i}$, and $p_{2,i}$. The other will then be a solution of the equations of motion.

The simplest assumption for pressure differences is to assume that there is none. That is, assume $p_k = p_{k,i}$ for $k = 1, 2$. This supposes that there is "instantaneous" microscopic pressure equilibration, which will be the case if the speed of sound in each phase is large compared with velocities of interest. In applications that do not deal with acoustic effects or bubble expansion/contraction, this assumption is adequate. For a discussion of bubble oscillation effects, see the article by van Wijngaarden (1972). Cheng (1982) studies the consequences of several modeling assumptions on wave propagation in a bubble mixture.

In situations where surface tension is important and no contacts occur between the particles, we assume that the jump condition (4) becomes

$$p^1 - p^2 = \sigma \kappa', \tag{69}$$

where κ' is the exact curvature of the interface. Multiplying by ∇X_1 and averaging give

$$(p_{1,i} - p_{2,i}) \nabla \alpha = \sigma \kappa \nabla \alpha, \tag{70}$$

where κ is the average mean curvature. Since this equation holds for all $\nabla \alpha$, we have

$$p_{1,i} - p_{2,i} = \sigma \kappa. \tag{71}$$

If the surface tension is negligible, Equation (71) gives $p_{1,i} = p_{2,i}$.

When contacts occur between particles, these considerations are not always simple. The contacts provide mechanisms for causing the average pressure at the interface in one phase to be higher than the other. Consider solid particles submerged in fluid, with no motion, and with the particles under a compressive force (supporting their own weight, or the weight of an object, for example). The resulting stress will be transmitted from particle to particle through the areas of particle-particle contact. Just on the particle side of the interface where the contact occurs, the stress (pressure) level will be large. At places where no contact occurs, the stress will be equal to the pressure in the surrounding fluid. Under normal circumstances, the contact areas will be a small fraction of the total interfacial area. Thus, the approximation $p_{1,i} \cong p_{2,i} = p_i$ is usually valid. In the more general situation, these contacts may be intermittent, as when the particles are bumping together, or may be constant, as when the particles are at rest. In this case also we expect $p_{1,i} = p_{2,i} = p_i$.

Stuhmiller (1977) argues that

$$p_2 - p_{2,i} = \xi \rho_2 |v_2 - v_1|^2, \tag{72}$$

where ξ is taken to be a constant. The argument leading to (72) is exactly the calculation of form drag, which is usually part of the interfacial drag. It is not clear whether using (72) and a drag law (79) includes the form drag consistently.

We shall assume that for the fluid phase, $p_2 = p_i$. To allow for the possibility of the extra stress due to contacts, in the particulate phase, when it consists of solid particles, we shall write

$$p_1 = p_{1,i} + p_c = p_2 + p_c, \tag{73}$$

where p_c is the pressure in the particles due to contacts. A model is needed for p_c. Several modelers (Gough 1980, Kuo et al. 1976) have used a model with $p_c = p_c(\alpha)$, with $p_c(\alpha) = 0$ for $\alpha < \alpha_c$, where α_c is the packing concentration of the particles and $p_c(\alpha)$ is rapidly increasing for $\alpha > \alpha_c$. This further suggests an "incompressible" model, with $p_c = 0$ in the regions where $\alpha < \alpha_c$, and $\alpha \equiv \alpha_c$ otherwise. In this latter region, p_c is unspecified and the particle momentum equation is meaningless. Such a model has not been exploited.

Interfacial Momentum Transfer

Let us now consider the interfacial-momentum-transfer terms. In a sense, the crucial modeling for two-phase flow is done with these terms, since they are the terms that couple the two phases together.

We now consider $\mathbf{M}_m$. In analogy with Ishii (1975), we write

$$\mathbf{M}_m = \sigma \kappa \nabla \alpha, \tag{74}$$

where κ is the average mean curvature of the interface. In this case, the momentum jump condition becomes

$$\mathbf{M}_1^d + p_{1,i} \nabla \alpha + \mathbf{M}_2^d - p_{2,i} \nabla \alpha = \sigma \kappa \nabla \alpha. \tag{75}$$

Using condition (71), we have

$$\mathbf{M}_1^d + \mathbf{M}_2^d = 0. \tag{76}$$

Therefore, we need only provide a model for $\mathbf{M}_1^d$. The jump condition (76) then determines $\mathbf{M}_2^d$.

The interfacial momentum transfer $\mathbf{M}_1^d$ contains the forces on the particulate phase due to viscous drag, wake and boundary-layer formation, and unbalanced pressure distributions leading to lift or virtual mass effects, except for the mean interfacial pressure. The model that we discuss here should contain, as much as possible, the above forces. Indeed, this motivates the inclusion of certain quantities in the list of variables (58).

We postulate

$$\mathbf{M}_1^d = A_1(\mathbf{v}_2 - \mathbf{v}_1) + A_2\left[\left(\frac{\partial \mathbf{v}_2}{\partial t} + \mathbf{v}_1 \cdot \nabla \mathbf{v}_2\right) - \left(\frac{\partial \mathbf{v}_1}{\partial t} + \mathbf{v}_2 \cdot \nabla \mathbf{v}_1\right)\right]$$
$$+ A_3(\mathbf{v}_2 - \mathbf{v}_1) \cdot \mathbf{D}_{1,b} + A_4(\mathbf{v}_2 - \mathbf{v}_1) \cdot \mathbf{D}_{2,b} + A_5(\mathbf{v}_2 - \mathbf{v}_1) \cdot \nabla(\mathbf{v}_2 - \mathbf{v}_1)$$
$$+ A_6(\mathbf{v}_1 - \mathbf{v}_2) \cdot [\nabla(\mathbf{v}_1 - \mathbf{v}_2)]^t \tag{77}$$

where $A_1, \ldots, A_6$ are scalar functions of the invariants.

The first term represents the classical drag forces. To conform with customary usage, we write

$$A_1 = \frac{3}{8}\alpha\rho_2 \frac{C_D}{r}|\mathbf{v}_1 - \mathbf{v}_2| \tag{78}$$

where r is the effective particle radius and C_D is the drag coefficient. It is usually assumed that $C_D = C_D(\alpha, \mathrm{Re})$, where $\mathrm{Re} = 2\rho_2|\mathbf{v}_1 - \mathbf{v}_2|r/\mu_2$ is the particle Reynolds number. A careful study, including extensive comparisons with data, is given by Ishii & Zuber (1979). Their conclusions for the drag coefficient are summarized in Table 1, which is adapted from their paper.

The combination

$$A_2\left[\left(\frac{\partial \mathbf{v}_2}{\partial t} + \mathbf{v}_1 \cdot \nabla\mathbf{v}_2\right) - \left(\frac{\partial \mathbf{v}_1}{\partial t} + \mathbf{v}_2 \cdot \nabla\mathbf{v}_1\right)\right] + A_5(\mathbf{v}_2 - \mathbf{v}_1) \cdot \nabla(\mathbf{v}_2 - \mathbf{v}_1) \tag{79}$$

is an acceleration term. Drew et al. (1979) write

$$A_2 = \alpha\rho_2 C_{VM}(\alpha), \tag{80}$$

$$A_5 = A_2(1 - \lambda(\alpha)), \tag{81}$$

where C_{VM} is referred to as the virtual volume. If $C_{VM} = 1/2$, and no spatial velocity gradients are present, then (79) reduces to the classically accepted virtual-mass force. It is harder to obtain correlations for C_{VM} than for C_D, since the ratio of virtual-mass forces to drag forces is of order Vt^*/r, where V is a velocity scale and t^* is a time scale. In order for the virtual-mass force to be appreciable, the time scale must be of the order r/V. For most multiphase flow applications, r is small and V is not. Thus, we see virtual-mass effects only at relatively high frequencies.

Zuber (1964) examined finite-concentration effects on virtual mass by considering a sphere moving inside a spherical "cell." He obtained

$$C_{VM}(\alpha) \cong \frac{1}{2}\frac{1 + 2\alpha}{1 - \alpha} \cong \frac{1}{2} + \frac{3}{2}\alpha \tag{82}$$

for small α. Nigmatulin (1979) also calculated the virtual-mass coefficient from a cell model. He obtained $C_{VM}(\alpha) \cong \frac{1}{2}$ to $O(\alpha)$. The form of the acceleration that he derives is not objective. It is possible that this is due to the inability of cell models to deal adequately with velocity-gradient terms. The value of the virtual-mass coefficient derived by van Wijngaarden (1976) is

$$C_{VM}(\alpha) = 0.5 + 1.39\alpha. \tag{83}$$

Mokeyev (1977) obtained an empirical correlation

$$C_{VM}(\alpha) = 0.5 + 2.1\alpha \tag{84}$$

Table 10 Local drag coefficients in multiparticle system

	Fluid particle system			Solid particle system
	Bubble in liquid	Drop in liquid	Drop in gas	
Viscosity model	$\dfrac{\mu_m}{\mu_2} = \left(1 - \dfrac{\alpha}{\alpha_m}\right)^{-2.5\alpha_m\mu^*}$, $\quad\mu^* \equiv \dfrac{\mu_1 + 0.4\mu_2}{\mu_1 + \mu_2}$			
Maximum packing (α_m)	~ 1	~ 1	$0.62 \sim 1$	~ 0.62
μ^*	0.4	~ 0.7	1	1
$\dfrac{\mu_m}{\mu_2}$	$(1-\alpha)^{-1}$	$(1-\alpha)^{-1.75}$	$\sim(1-\alpha)^{-2.5}$	$\left(1-\dfrac{\alpha}{0.62}\right)^{-1.55}$
Stokes' regime (C_D)	$C_D = 24/N_{Re}$ where $N_{Re} = 2r\rho_2|\mathbf{v}_2 - \mathbf{v}_1|/\mu_m$			
Viscous regime (C_D)	$C_D = 24(1 + 0.1 N_{Re}^{3/4})/N_{Re}$			
Newton's regime (C_D) $r\left(\dfrac{\rho_2 g\Delta\rho}{\mu_2}\right)^{1/3} \geq 34.65,\ \Delta\rho = |\rho_2 - \rho_1|$				$C_D = 0.45\left\{\dfrac{1+17.67[f(\alpha)]^{6/7}}{18.67 f(\alpha)}\right\}^2$
Distorted particle regime (C_D) $\dfrac{\mu_2}{\left(\rho_2\alpha\sqrt{\dfrac{\alpha}{g\Delta\rho}}\right)^{1/2}} \geq 0.11(1+\psi)/\psi^{8/3}$	$C_D = \dfrac{4}{3} r_1\sqrt{\dfrac{g\Delta\rho}{\alpha}}\left\{\dfrac{1+17.67[f(\alpha)]^{6/7}}{18.67 f(\alpha)}\right\}^2$ $f(\alpha) = (1-\alpha)^{1.5}$	$(1-\alpha)^{2.25}$	$(1-\alpha)^3$	where $f(\alpha) = \sqrt{1-\alpha}\left(\dfrac{\mu_2}{\mu_m}\right)$

$\psi = 0.55[(1 + 0.08r^3)^{4/7} - 1]^{3/4}$

by using an electrodynamic analog. Thus, there seems to be a consensus that $C_{VM} \to \frac{1}{2}$ as $\alpha \to 0$. There is less agreement on the $O(\alpha)$ correction. In addition, there are several models in the literature (Wallis 1969, Hinze 1972) that use nonobjective forms of the virtual mass. There is no compelling evidence at this point as to whether the objective or nonobjective forms conform more closely to reality.

Cheng (1982) has done extensive predictions using virtual-mass models for accelerating air-water bubbly flows. For low-frequency small disturbances in a one-dimensional flow, he found that the value of C_{VM} had a significant effect on the propagation speed, but that the data were scattered so that a particular value could not be estimated. For higher-frequency waves (sound waves) and for critical flow, compressibility effects were important. For nozzle flow, the virtual-mass effects were insignificant. None of the calculations showed any effect of different values of λ.

The term

$$A_4(\mathbf{v}_2 - \mathbf{v}_1) \cdot \mathbf{D}_{2,b} \tag{85}$$

contains the effect of the lift. We write $A_4 = \alpha L$. On the other hand, the forces represented by $A_3(\mathbf{v}_2 - \mathbf{v}_1) \cdot \mathbf{D}_{1,b}$ and $A_6(\mathbf{v}_1 - \mathbf{v}_2) \cdot [\nabla(\mathbf{v}_1 - \mathbf{v}_2)]'$ have no analogs in single-particle calculations, and no observations confirm their presence. Thus we assume $A_3 = A_6 = 0$. Very little experimental evidence for A_4 exists.

Interfacial Geometry Models

It is evident from Equations (71) and (78) that a relation is needed for expressing the geometry of the interface. Several such relations have been used; however, none is so compelling as to be called general. The one that comes closest is due to Ishii (1975) (derived for the time average; the generalization is obvious). We discuss this later. First, we present the simplest models. If the particles are solid spheres, each of the same radius r, we have

$$\alpha = \tfrac{4}{3}\pi r^3 n, \tag{86}$$

where n is the number density. Other quantities of interest are immediately obtainable from α and r (for example, the average interfacial area per unit volume is $3\alpha/r$). Note that n is conserved because α is; specifically, we have

$$\frac{\partial n}{\partial t} + \nabla \cdot n\mathbf{v}_1 = 0. \tag{87}$$

If the particles are not monodisperse, it may be necessary to derive a model where each size of particle is treated separately. Some recent efforts along these lines show great promise (Greenspan & Ungarish 1982). Particle breaking, agglomeration, and accretion/erosion also necessitate a more complicated model. An effort (without macroscopic mechanical considerations) is given by Seinfeld (1980). Both approaches are essentially geared to an equation analogous to (87) for a number density.

A more general approach is suggested by Ishii (1975) (for time-averaged equations), where he writes

$$\frac{\partial s}{\partial t} + \nabla \cdot \mathbf{v}_i s = \phi_i, \tag{88}$$

where s is the interfacial area per unit volume, which in the present notation is $\langle \partial X / \partial n \rangle$; $\mathbf{v}_i$ is the averaged interfacial velocity; and ϕ_i is a source of interfacial area due to "bulging" of the interface. This approach is probably best suited to flows where the particulate phase (bubbles or drops) can coalesce or break up. It has not been investigated to any degree.

MODEL CONSIDERATIONS

The main aspect of two-phase flows that makes them fundamentally different from a single-phase flow is the interfacial interaction terms, and thus in the absence of interfacial mass transfer the difference is contained in the interfacial force $\mathbf{M}_k^d$. The modeling of this term must proceed carefully, with attention paid to any possible inconsistencies or untoward predictions. A canonical problem involving simplified two-phase flow models has surfaced in the literature; namely, that the one-dimensional, incompressible, inviscid-flow equations without virtual-mass effects are ill-posed as an initial-value problem (Ramshaw & Trapp 1978).

To see this, consider the equations with $C_{\mathrm{VM}} = 0$ and $L = 0$. (This represents the model used in most practical calculations.)

$$\frac{\partial \alpha}{\partial t} + \frac{\partial \alpha v_1}{\partial x} = 0, \tag{89}$$

$$-\frac{\partial \alpha}{\partial t} + \frac{\partial (1 - \alpha) v_2}{\partial x} = 0, \tag{90}$$

$$\alpha \rho_1 \left(\frac{\partial v_1}{\partial t} + v_1 \frac{\partial v_1}{\partial x} \right) = -\alpha \frac{\partial p}{\partial x} + A_1 (v_2 - v_1), \tag{91}$$

$$(1 - \alpha) \rho_2 \left(\frac{\partial v_2}{\partial t} + v_2 \frac{\partial v_2}{\partial x} \right) = -(1 - \alpha) \frac{\partial p}{\partial x} + A_1 (v_1 - v_2). \tag{92}$$

If the pressure is eliminated from system (89)–(92) and the resulting equations are expressed in the form $A \partial u / \partial t + B \partial u / \partial x = C$, where $u =$

$[\alpha, v_1, v_2]'$, A and B are 3×3 matrices and C is a 3-vector, then the characteristics of the system are given by $dz_i/dt = v_i$, $i = 1, 2, 3$, where v_i are the three characteristic values of $\det[Av - B] = 0$. Note that since C contains no derivative terms, it is not involved in determining the characteristics. The characteristic values v_i for the system as given are a pair of complex-conjugate roots $v_{1,2} = \alpha \pm i\beta$ (with $\beta \neq 0$), and $v_3 = \infty$ (resulting from the assumption of incompressibility). This implies (Garabedian 1964) that the initial-value problem is ill-posed.

One implication of the ill-posed nature of the system can be seen as follows. Consider the linear stability of a constant solution. The solution of the linearized equations has the form $u(x, t) = u_0 + u_1 e^{vt + ikx}$, where k is real. Substitution into the linearized system results in the eigenvalue problem

$$[A(u_0)v + B(u_0)ik - C(u_0)]u_1 = 0 \tag{93}$$

for v. The eigenvalue equation is

$$\det\left[A(u_0)\frac{iv}{k} - B(u_0) - \frac{C(u_0)}{k}i\right] = 0. \tag{94}$$

For k large, two roots of (94) are

$$\frac{vi}{k} = \alpha \pm i\beta. \tag{95}$$

Thus,

$$v_{1,2} = k(\pm\beta - \alpha i). \tag{96}$$

As long as $\beta \neq 0$, one of the v's will have a positive real part, indicating linear instability. Note, too, that as k increases, the growth rate $|\beta k|$ increases. This shows that finer disturbances grow faster, implying that at any $t > 0$, $\max|u_1 e^{vt + ikx}|$ can be made as large as desired by taking k sufficiently small. Contrast this with the behavior of an instability of a well-posed system, where the solution $u_1 e^{vt + ikx}$ can be made as small as desired at a finite t by making u_1 small, for any k. The instability in the well-posed system has some realistic physical meaning, while the instability always present in the ill-posed system suggests that the model is not treating small-scale phenomena correctly.

The extension of the argument to three space dimensions is straightforward and shows that the system is ill-posed in three dimensions also.

The ill-posed nature of the model makes it desirable to seek physical modifications of that simple model to find a well-posed model. An effort has been made to determine whether and/or to what extent virtual mass makes the inviscid model well-posed. Lahey et al. (1980) have shown that the model becomes well-posed for certain values of C_{VM}. The value of

C_{VM} that makes the system well-posed is usually large unless α is extremely small, or unless other somewhat nonphysical assumptions are made. Thus, even though it is a possible "fix" to the question of the well- vs. ill-posed nature of the model, virtual mass does not seem to be the total answer.

We note that if the viscous and Reynolds stresses are included in the equation, and if they are assumed to depend on $\nabla \mathbf{v}_k$, the equations become parabolic, and hence are well-posed if enough boundary conditions are used. Also Prosperetti & van Wijngaarden (1976) have shown that using certain compressibility assumptions in gas-liquid systems gives real characteristics. Stuhmiller (1977) shows that his model including the form drag [see Equation (72)] gives real characteristics for sufficiently large ξ.

The question of the role of the inviscid model remains. In single-phase fluid mechanics, the inviscid model is extremely important and governs the flow outside of boundary layers, shear layers, and other singular structures. In two-phase flows, the inviscid equations are ill-posed, possessing the unrealistic feature that small-scale phenomena grow rapidly. This seems to imply that an inviscid model is nowhere valid for two-phase flows, that is, viscous or eddy stresses are important ev- erywhere. An important unanswered question is whether the inviscid limit of the two-phase flow equations is meaningful.

Let us examine another possible hypothesis about the interfacial force that leads to an interesting form of the virtual-mass and lift terms. The hypothesis is that the interfacial force must be dissipative for all con- ceivable flows. This requirement can be derived from considerations involving an entropy inequality. The argument is complicated (Drew & Lahey 1980a) and will not be presented here. It depends on the correct grouping of several terms in the energy and entropy equations.

If we form the total kinetic-energy equation by dotting (42) with $\mathbf{v}_k$ and adding for $k = 1, 2$, we obtain

$$
\frac{\partial}{\partial t} \sum_{k=1}^{2} \alpha_k \rho_k \frac{v_k^2}{2} + \nabla \cdot \sum_{k=1}^{2} \alpha_k \rho_k \mathbf{v}_k \frac{v_k^2}{2}
$$

$$
= - \sum_{k=1}^{2} \alpha_k \mathbf{v}_k \cdot \nabla p_k + \nabla \cdot \sum_{k=1}^{2} \left[\alpha_k (\tau_k + \sigma_k) \cdot \mathbf{v}_k \right]
$$

$$
+ \sum_{k=1}^{2} (p_{k,i} - p_k) \mathbf{v}_k \cdot \nabla \alpha_k
$$

$$
+ \sum_{k=1}^{2} \alpha_k (\tau_k + \sigma_k) : \nabla \mathbf{v}_k + \mathbf{M}_1^d \cdot (\mathbf{v}_1 - \mathbf{v}_2).
$$

$$(97)$$

The terms on the left represent the rate-of-change of the total (mean flow) kinetic energy. The pressure terms contribute to the stored internal energy. The term $\sum_{k=1}^{2}\alpha_k(\tau_k + \sigma_k)\cdot\mathbf{v}_k$ is the working of the stresses. The two terms $-\sum_{k=1}^{2}\alpha_k(\tau_k + \sigma_k):\nabla\mathbf{v}_k$ and $-\mathbf{M}_1^d\cdot(\mathbf{v}_1 - \mathbf{v}_2)$ are the (bulk) dissipation and the interfacial dissipation, respectively. We require the total dissipation to be expressible as

$$\frac{\partial A}{\partial t} + \nabla\cdot\mathbf{B} + C, \tag{98}$$

where $A \geqslant 0$ and $C \geqslant 0$ for all possible flows. The term A represents the extra kinetic energy attributable to interactions, $\mathbf{B}$ represents the extra kinetic-energy flux, and C is the (actual) total dissipation.

We shall present the results of the disspativity requirement for $\tau_k = 0$ and $\sigma_k = 0$. If we then take Equation (78) for $\mathbf{M}_1^d$ and assume that C_{VM}, λ, and L are constants, we have

$$-\mathbf{M}_1^d\cdot(\mathbf{v}_1 - \mathbf{v}_2) = A_1(v_1 - v_2)^2$$
$$+ \frac{\partial}{\partial t}\left[\alpha C_{\mathrm{VM}}\rho_2 \frac{(v_1 - v_2)^2}{2}\right]$$
$$+ \nabla\cdot\left\{\alpha C_{\mathrm{VM}}\rho_2[\mathbf{v}_2 - (1-\lambda)(\mathbf{v}_1 - \mathbf{v}_2)]\frac{(v_1 - v_2)^2}{2}\right\}$$
$$- \frac{(v_1 - v_2)^2}{2}[\alpha C_{\mathrm{VM}}\rho_2(\lambda - 2)\nabla\cdot(\mathbf{v}_1 - \mathbf{v}_2)]$$
$$- (\alpha C_{\mathrm{VM}}\rho_2 + \alpha L)(\mathbf{v}_1 - \mathbf{v}_2)\cdot\nabla\mathbf{v}_2\cdot(\mathbf{v}_1 - \mathbf{v}_2). \tag{99}$$

If we examine (99) for uniform fields, we obtain the first result of the dissipativity requirement, namely that $A_1 \geqslant 0$. Furthermore, we note that since the other terms depend on spatial gradients that can be imposed at will, we have

$$\lambda = 2, \tag{100}$$
$$L = -C_{\mathrm{VM}}\rho_2. \tag{101}$$

Note also that the requirement that $A \geqslant 0$ results in $C_{\mathrm{VM}} \geqslant 0$. The resulting form for $\mathbf{M}_1^d$ is

$$\mathbf{M}_1^d = A_1(\mathbf{v}_2 - \mathbf{v}_1) + \frac{\partial}{\partial t}\alpha C_{\mathrm{VM}}\rho_2(\mathbf{v}_2 - \mathbf{v}_1)$$
$$+ \nabla\cdot[\alpha C_{\mathrm{VM}}\rho_2\mathbf{v}_1(\mathbf{v}_2 - \mathbf{v}_1)]$$
$$+ \alpha C_{\mathrm{VM}}\rho_2(\mathbf{v}_1 - \mathbf{v}_2)\cdot\frac{1}{2}[\nabla\mathbf{v}_2 - (\nabla\mathbf{v}_2)^t]. \tag{102}$$

In this form, the virtual-mass terms are in conservation form with the convective velocity v_1. The nonobjective nature of these terms must be compensated for by the lift term, resulting in the last term in (102). Note that if $\omega_2 = \nabla \times v_2$ is the fluid vorticity, the last term becomes $\alpha C_{VM} \rho_2 (v_1 - v_2) \times \omega_2$. Thus, the remaining part of the lift is always perpendicular to the slip and the fluid vorticity.

Literature Cited

Anderson, T. B., Jackson, R. 1967. A fluid mechanical description of fluidized beds. *Ind. Eng. Chem. Fundam.* 6:527–39

Aris, R. 1962. *Vectors, Tensors and the Basic Equations of Fluid Mechanics.* Englewood Cliffs, N.J.: Prentice Hall

Bagnold, R. A. 1941. *The Physics of Blown Sand and Desert Dunes.* London: Methuen

Batchelor, G. K. 1974. Transport properties of two-phase materials with random structure. *Ann. Rev. Fluid Mech.* 6:227–55

Bedford, A., Drumheller, D. S. 1978. A variational theory of immiscible mixtures. *Arch. Ration. Mech. Anal.* 68:37–51

Bedford, A., Drumheller, D. S. 1982. Theory of immiscible and structured mixtures. *Int. J. Eng. Sci.* In press

Biot, M. A. 1955. Theory of elasticity and consolidation for a porous anisotropic solid. *J. Appl. Phys.* 26:182–85

Biot, M. A. 1977. Variational Lagrangian thermodynamics of nonisothermal finite strain mechanics of porous solids and thermomolecular diffusion. *Int. J. Solids Struct.* 13:579–97

Blasius, H. 1910. Über die Abhängigkeit der Formen der Riffeln und Geschiebebänke vom Gefälle. *Z. Bauwesen.* 60:466–71

Boothroyd, R. G. 1971. *Flowing Gas-Solid Suspensions.* London: Chapman & Hall

Brenner, H. 1970. Rheology of two-phase systems. *Ann. Rev. Fluid Mech.* 2:137–76

Brinkman, H. C. 1947. A calculation of the viscous force exerted by a flowing fluid on a dense swarm of particles. *Appl. Sci. Res. Sect. A* 1:27

Buyevich, Yu. A., Shchelchkova, I. N. 1978. Flow of dense suspensions. *Prog. Aerosp. Sci.* 18:121–50

Cheng, L. Y. 1982. *An analysis of wave dispersion, sonic velocity and critical two-phase flow.* PhD thesis. Rensselaer Polytech. Inst., Troy, N.Y.

Darcy, H. 1856. *Les Fontaines Publiques de la Ville de Dijon.* Paris: Dalmont

Delhaye, J. M., Achard, J. L. 1976. On the averaging operators introduced in two-phase flow modeling. *CSNI Spec. Meet. Transient Two-Phase Flow*, Toronto

Drew, D. A. 1971. Average field equations for two-phase media. *Stud. Appl. Math.* 50:133–66

Drew, D. A., Lahey, R. T. 1979. Application of general constitutive principles to the derivation of multidimensional two-phase flow equations. *Int. J. Multiphase Flow* 5:243–64

Drew, D. A., Lahey, R. T. 1980a. Interfacial dissipation in two-phase flow. *Basic Mechanisms in Two-Phase Flow and Heat Transfer.* ASME

Drew, D. A., Lahey, R. T. 1980b. A mixing length model for fully developed turbulent two-phase flows. *Trans. Am. Nucl. Soc.* 35:624–25

Drew, D. A., Cheng, L. Y., Lahey, R. T. 1979. The analysis of virtual mass effects in two-phase flow. *Int. J. Multiphase Flow* 5:233–42

Drumheller, D. S., Bedford, A. 1980. A thermomechanical theory for reacting immiscible mixtures. *Arch. Ration. Mech. Anal.* 73:257–84

DuBuat, P. 1786. *Principles d'Hydraulique*, Paris: De l'Imprimerie de Monsieur. 2nd ed.

Duhem, P. 1893. Le potential thermodynamique et la pression hydrostatique. *Ann. Ec. Norm.* 10:187–230

Einstein, A. 1906. Eine neue Bestimmung der Moleküldimensionen. *Ann. Phys.* 19:289–306

Engelund, F., Fredsøe, J. 1982. Sediment ripples and dunes. *Ann. Rev. Fluid Mech.* 14:13–37

Exner, F. M. 1920. Zur Physik der Dünen. *Sitzungsber. Akad. Wiss. Wien*, Pt. IIa. 129.

Fick, A. 1855. Über Diffusion. *Ann. Phys. Chem.* 94:59–86

Frankl, F. I. 1953. On the theory of motion of suspended sediments. *Dokl. Akad. Nauk SSSR* 92:247

Garabedian, P. R. 1964. *Partial Differential Equations.* New York: Wiley

Gough, P. S. 1980. On the closure and character of the balance equations for heterogeneous two-phase flow. In *Dynamics and Modelling of Reactive Systems*. New York: Academic

Greenspan, H. P., Ungarish, M. 1982. On hindered settling of particles of different sizes. *Int. J. Multiphase Flow*. In press

Happel, J., Brenner, H. 1965. *Low Reynolds Number Hydrodynamics*. Englewood Cliffs, N.J.: Prentice Hall

Helmholtz, H. von. 1888. Über atmosphaerishe Bewegungen. *Sitzungsber. K. Preuss. Akad. Wiss. Berlin* 1888:647–63

Herczyński, R., Pieńkowska, I. 1980. Toward a statistical theory of suspension. *Ann. Rev. Fluid Mech.* 12:237–69

Hinze, J. O. 1959. *Turbulence*. New York: McGraw-Hill

Hinze, J. O. 1972. Turbulent fluid and particle interaction. *Prog. Heat Mass Transfer* 6:433–52

Ishii, M. 1975. *Thermo-fluid Dynamic Theory of Two-Phase Flow*. Paris: Eyrolles

Ishii, M., Zuber, N. 1979. Drag coefficient and relative velocity in bubbly, droplet or particulate flows. *AIChE J.* 25:843–55

Jackson, R. 1963. The mechanics of fluidized beds. I. *Trans. Inst. Chem. Eng.* 41:13–28

Kenyon, D. E. 1976. The theory of an incompressible solid-fluid mixture. *Arch. Ration. Mech. Anal.* 62:131–47

Kuo, K. K., Koo, J. H., Davis, T. R., Coates, G. J. 1976. Transient combustion in mobile, gas-permeable propellants. *Acta Astron.* 3:575–91

Lahey, R. T., Moody, F. J. 1977. *The Thermal-Hydraulics of a Boiling Water Nuclear Reactor*. Hinsdale, Ill.: Am. Nucl. Soc.

Lahey, R. T., Cheng, L. Y., Drew, D. A., Flaherty, J. E. 1980. The effect of virtual mass on the numerical stability of accelerating two-phase flows. *Int. J. Multiphase Flow* 6:281–94

Lance, M. 1981. Turbulence measurements in two-phase bubbly flows. *18th Ann. Meet. Soc. Eng. Sci., Providence, R.I.*

Leal, L. G. 1980. Particle motions in a viscous fluid. *Ann. Rev. Fluid Mech.* 12:435–76

Marble, F. E. 1970. Dynamics of dusty gases. *Ann. Rev. Fluid Mech.* 2:397–446

Meixner, J. 1943. Zur Thermodynamik der irreversiblen Prozesse. *Z. Phys. Chem. B* 53:235–63

Mokeyev, Yu. G. 1977. Effect of particle concentration on their drag and induced mass. *Fluid. Mech. Sov. Res.* 6:161

Murray, J. D. 1965a. On the mathematics of fluidization. I. Fundamental equations and wave propagation. *J. Fluid Mech.* 21:465–93

Murray, J. D. 1965b. On the mathematics of fluidization. II. Steady motion of fully developed bubbles. *J. Fluid Mech.* 22:57–80

Nigmatulin, R. I. 1979. Spatial averaging in the mechanics of heterogeneous and dispersed systems. *Int. J. Multiphase Flow* 5:353–85

Prigogine, I., Mazur, P. 1951. Sur deux formulations de l'hydrodynamique et le problème de l'hélium liquide. II. *Physica (den Haag)* 17:661–79

Prosperetti, A., van Wijngaarden, L. 1976. On the characteristics of the equations of motion for a bubbly flow and the related problem of critical flow. *J. Eng. Math.* 10:153–62

Ramshaw, J. D., Trapp, J. A. 1978. Characteristics, stability, and short-wavelength phenomena in two-phase flow equation systems. *Nucl. Sci. Eng.* 66:93–102

Russel, W. B. 1981. Brownian motion of small particles suspended in liquids. *Ann. Rev. Fluid Mech.* 13:425–55

Ryskin, G., Rallison, J. M. 1980. On the applicability of the approximation of material frame-indifference in suspension mechanics. *J. Fluid Mech.* 99:525–29

Scheidegger, A. E. 1974. *The Physics of Flow through Porous Media*. Toronto: Univ. Toronto Press. 3rd ed.

Seinfeld, J. H. 1980. Dynamics of aerosols. *Dynamics and Modelling of Reactive Systems*. New York: Academic

Serizawa, A. 1975. *Fluid dynamic characteristics of two-phase flow*. PhD thesis. Kyoto Univ., Jpn.

Soo, S. L. 1967. *Fluid Dynamics of Multiphase Systems*. Waltham, Mass: Blaisdell

Stuhmiller, J. H. 1977. The influence of interfacial pressure forces on the character of two-phase flow model equations. *Int. J. Multiphase Flow* 3:551–60

Teletov, S. G. 1958. Problems of the hydrodynamics of two-phase mixtures. I. *Vestn. Mosk. Gos. Univ., Ser. Mat. Mekh. Astron. Fiz. Khim.* 1958(2):15–27

Truesdell, C. 1977. *A First Course in Rational Continuum Mechanics*. New York: Academic

Truesdell, C., Toupin, R. 1960. The classical field theories. In *Encyclopedia of Physics*, Vol. III/1, pp. 226–793. Berlin, Göttingen, Heidelberg: Springer van Wijngaarden, L. 1972. One-dimensional flow of liquids containing small gas bubbles. *Ann. Rev. Fluid Mech.* 4:369–96

van Wijngaarden, L. 1976. Hydrodynamic interaction between gas bubbles in liquid. *J. Fluid Mech.* 77:27–44

Vernier, P., Delhaye, J. M. 1968. General two-phase flow equations applied to the thermodynamics of boiling water nuclear reactors. *Energ. Primaire* 4:5–46

Wallis, G. B. 1969. *One-Dimensional Two-Phase Flow.* New York: McGraw-Hill

Whitaker, S. 1973. The transport equations for multiphase systems. *Chem. Eng. Sci.* 28:139–47

Zuber, N. 1964. On the dispersed two-phase flow in the laminar flow regime. *Chem. Eng. Sci.* 19:897–903

Ann. Rev. Fluid Mech. 1983. 15:293–319

COMPLEX FREEZING-MELTING INTERFACES IN FLUID FLOW

Michael Epstein

Fauske & Associates, Inc., West Coast Office, Encino, California 91316

F. B. Cheung

Reactor Analysis and Safety Division, Argonne National Laboratory, Argonne, Illinois 60439

INTRODUCTION

Much of the early work on problems involving liquid-solid phase change was confined essentially to the cases in which the liquid melt is stagnant or the motion of the liquid is due entirely to the phase-change process itself. In these transient (conduction) phase-change problems, the unknowns are the interface that separates the regions of liquid and solid and the temperature distributions in both regions. The interface motion is expressed implicitly in an equation for the conservation of thermal energy at the interface. This introduces a nonlinear character to the problem that has intrigued and challenged engineers and mathematicians since Stefan's (1891) pioneering study of the growth of sea ice. Many powerful analytic and numerical techniques have been reported for dealing with heat-diffusion-controlled freezing or melting problems, and references to the mathematical studies on this subject may be found in several previous surveys by Bankoff (1964), Muehlbauer & Sunderland (1965), Rubinshtein (1967), Ockendon & Hodgkins (1975), and Wilson et al. (1978).

In contrast to the earlier studies of phase change in stagnant media, problems in which the melt material is in motion have recived detailed attention only during the past two decades. The stimulus was provided to a large extent by requirements in the nuclear and "alternate" energy

0066-4189/83/0115-0293$02.00

industries (such as solar and waste-heat utilization) and by a desire to obtain a better understanding of naturally occurring ice.

Two distinctive features of the growth or decay of a solid in a flowing melt, which introduce additional complications besides the basic nonlinearity of the transient phase-change problem, are that the shape and the stability of the interface are unknown a priori and that they may be heavily influenced by the variation of the heat transfer and pressure gradient along the solid-liquid interface. These complications can arise at steady state as well as under conditions of phase change in transient flow. In many flow geometries, the interaction of the convective flow and interface is minimized by the fact that the curvature of the interface is small. In some cases of practical importance, however, a strong mutual interaction can occur such that it is difficult to describe the heat-transfer rate across the interface without knowledge of the interface shape and vice versa. In fact, observations have been made where the interaction of melt flow and phase change is so strong as to result in unstable phase-boundary behavior.

The emphasis in this paper is on the class of problems in which the interaction between the flow and the phase-change interface is strong and results in complex interface shapes. No unified theoretical treatment of the effects of strong coupling between flow and solid-liquid phase change is within our grasp, as each flow geometry has its own peculiarities, and in many cases present understanding relies largely on experiment. For this reason, it is convenient to arrange and discuss the problems and phenomena of solid-liquid phase change in a flowing melt in three main groups. As in a text book on heat transfer in fluid flow, these are (a) phase change in external forced flow, (b) phase change in internal forced flow, and (c) phase change in natural-convection flow.

EXTERNAL FORCED FLOW

Simple Phase Change

Suppose a liquid at a fixed bulk temperature T_∞ above its solidification temperature T_{mp} suddenly flows with velocity U_∞ over a cold wall maintained at a uniform temperature $T_0 < T_{mp}$. A solidification front then propagates into the liquid from the surface of the cold wall at a rate determined both by the convective heat flux, q, from the flowing liquid to the phase boundary and the rate at which the latent heat of solidification can be conducted through the frozen layer to the cold wall. In general, both q and the thickness, δ, of the frozen layer are functions of time and the streamwise coordinate, x, and are mutually dependent upon each other. However, two important simplifying assumptions are usually made that eliminate the interaction between the flow and the freezing process. First, the solidified layer is assumed to be thin compared with its extension in the direction of flow so that heat conduction in this direction is small compared with that normal to the flow. Second, it is assumed that convection in the liquid is relatively undisturbed by the moving phase boundary, and the liquid supplies a constant convective heat flux $q = h(T_\infty - T_{mp})$ to the crust surface at all times. The convective heat-transfer coefficient h may then be obtained from the conventional solution of steady thermal convection without phase change; it is regarded as an input rather than an output to the problem. In this way, the problem reduces to one that primarily concentrates on the nonlinear nature of the transient phase-change process. This, of course, greatly facilitates the task of solution.

In many flow situations, experimental observations have confirmed freezing-rate predictions based on negligible interactions between flow and change of phase. The data of Savino & Siegel (1967) for liquid flow over a chilled flat plate, of Petrie et al. (1980) for two-phase gas-liquid flow over a chilled flat plate, and of Savino et al. (1970) for solidification in plane stagnation flow directed against a cold surface, for example, indicate that freezing in a flowing medium can be treated simply as a heat-conduction problem with a moving boundary by using ordinary concepts of convective heat exchange at the solid-liquid interface. Hirata et al. (1979a) studied experimentally and theoretically the phenomena that determine the steady-state shape of an ice layer on a flat plate in Blasius (laminar) flow. The overall conclusions from the measurements and theory are that, for all practical purposes, the flow is unaffected by the shape of the ice layer and that streamwise heat conduction in the ice layer is negligible.

Transition Flow Over a Meltable Surface

A situation in which the interaction between the flow and the phase-change boundary cannot be ignored or predicted is when the transition from laminar to turbulent flow occurs on a frozen surface.

Hirata et al. (1979b) performed an experimental study of a parallel forced-convection flow over a "steady-state" ice layer formed on a constant-temperature plate in the transition and turbulent regimes. The general nature of the flow is shown schematically in Figure 1, indicating the different flow regimes and the ice profile in each regime. At the location where transition occurs, there is a substantial increase in the heat-transfer coefficient. This causes a marked decrease in the ice thickness, which results in a region of unfavorable pressure gradient. In the highly turbulent regime, owing to the gradual decrease in heat-transfer rate along the plate, the ice-layer thickness again increases smoothly. Because the flow and the ice surface could interact in various ways, two distinct modes of transition were observed in the experiment. If the flow remained attached to the ice surface through the region of unfavorable pressure gradient, a "smooth" transition mode was obtained, where the ice thickness decreased gradually through the transition regime. On the other hand, if flow separation occurred at transition, a "step" transition mode was found, where a sudden decrease in ice thickness resulted. Generally, the "smooth" transition was observed to occur on thin ice layers and the "step" transition on thick ice layers. There was, however, a range of flow conditions for which either mode could exist.

The Reynolds numbers at which transition began on the ice surface were found to be much lower than those for a flat plate with no ice. This is believed to be caused by the presence of an adverse pressure gradient in the transition regime that tends to promote the amplification rate of turbulence (see White 1974, p. 441). As might be expected, the peculiar shape of the ice layer in the transition region has a rather profound influence on heat transfer in the turbulent regime downstream of the transition. The "smooth" transition results in a gradual increase in the Nusselt number, whereas the "step" transition results in a peak in the Nusselt number. In both cases, heat-transfer rates are 1.5 to 2.5 times greater than those for conventional turbulent boundary layers without ice formation.

In another experimental study by Cheng et al. (1981), the local and average heat-transfer rates were obtained at the ice-water interface of an ice layer grown on a cooled circular cylinder in a crossflow. The local heat-transfer results were used to determine the approximate locations of flow separation and transition on the ice surface. Consistent with the observations of an ice layer on a flat plate, the laminar-to-turbulent boundary-layer transition was found to occur on the ice-covered cylinder

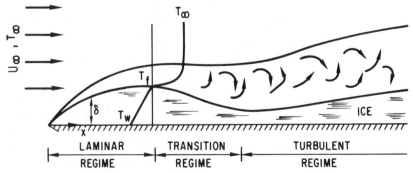

Figure 1   Steady-state ice-layer profile on a flat plate, indicating various flow regimes.

at Reynolds numbers as much as an order of magnitude lower than on a cylinder free of ice. Again, the reason is the sharp change in the ice profile that is produced by the change in local heat flux that occurs in the vicinity of flow separation.

Wavy Interfaces in Turbulent Flow

Were the above complications near the leading edge of a chilled flat plate not enough, it turns out that the phase-boundary behavior can be quite complex far downstream of the transition to turbulent flow. Observations of Carey (1966), Larsen (1969), and Ashton & Kennedy (1972) revealed that the turbulent flow of river water at the underside of an ice cover results in the formation of a wavy pattern called ice ripples. The formation of these ripples on an initially flat ice layer has been studied in some depth in the laboratory by Ashton (1972), Hsu (1973), Gilpin et al. (1980), and Hsu et al. (1979), and their behavior seems to be reasonably well understood.

Clearly, in order for wave patterns to develop, there must be a maximum in the heat-transfer rate in the trough regions between the waves and a minimum in the heat-transfer rate at the crest regions. Furthermore, if the waves are to move downstream as is observed experimentally, the maximum in the heat-transfer rate must occur in the downstream half of the trough. Thorsness & Hanratty (1979a) developed a linear model of the transfer of mass or heat between a turbulent flow and a small-amplitude solid wavy surface and indeed found that the maximum in the heat-transfer rate can exist somewhere in the trough regions. Their model incorporates the experimental observations that turbulence, and therefore heat transfer, is damped in regions of accelerating flow, whereas in a decelerating flow, turbulence and heat transfer are enhanced. Thus, the model predicts that the wave-induced accelerations and decelerations of the flow associated with the variation of the pressure gradient play a principal role in governing turbulent heat transport at wavy surfaces. In a subsequent paper, Thorsness & Hanratty (1979b) used the results of their model to examine the stability of the surface of an initially flat ice slab raised close to a uniform temperature of $0°C$ and then melted in a turbulent flow. Their calculated properties of the ice waves agree favorably with measured values by Gilpin et al. (1980).

Gilpin et al. (1980) extended the stability analysis of Thorsness & Hanratty (1979b) to include both heat flow due to temperature gradients in the ice and convective heat transfer at the surface exposed to turbulent water flow. It was found that the main damping factor for the growth of ice waves is heat conduction in the ice. The parameter that controls the magnitude of the wave growth is G, the ratio of the heat flux conducted away from the ice surface through the ice slab to the convective heat flux from the flowing water to the ice. From the measurements of the properties of small-amplitude waves at steady-state conditions ($G = 1$), it was predicted that the maximum value of G for which unstable behavior is possible is $G = 2.3$. Thus wave development can occur under conditions of ice growth, namely $1 < G < 2.3$. However, an initially planar ice surface was predicted to display its greatest instability for values of $G < 1$, i.e. for a melting ice surface. For the more unstable conditions, the waves grow in size until a "ripple" pattern develops and are large enough that separations occur on each wave. Eventually, a steady state is reached in which the waves can have an asymmetric form. The measured heat-transfer rates for flow over the rippled surface are 30 to 60 percent larger than those for flow over a flat or small-amplitude wavy ice surface.

Impingement Melting

The melting erosion of frozen material by the action of a warm fluid jet presents another complex heat-transfer problem involving a strong interaction between flow and change of phase. The main feature of the melting problem that makes it particularly difficult is the fact that the shape of the jet impingement surface is not fixed but rather is determined by the flow pattern within the jet.

Yen & Zehnder (1973) and Yen (1976) reported two experimental studies on the ablation of ice by a water jet emanating from a nozzle and directed against an ice block to simulate the application of using water jets in excavating high-ice-content soil. It was found that the jet rapidly forms a cavity in the ice and, therefore, a rather complex flow geometry. Similar experiments were carried out by Gilpin (1973).

The data of Yen & Zehnder showed that the rate of penetration of a free water jet into a solid block of ice was essentially constant. The penetration rate was found to be linearly dependent on the bulk water temperature, indicating that convective heat transfer was controlling the melting process. Based upon the physical properties of water taken at $0°C$, a least-squares analysis of the data obtained by Yen & Zehnder was performed by Yen (1976). The results were represented by an expression typical of heat transfer at a stagnation point with high-level turbulence in the flow. The ice cavities observed were almost cylindrical in shape, similar to that of a drill bit in metal with the exception that the ice surface was rough rather than smooth. The diameters of the cavities were found to be dependent only on the size of the nozzle, not on the jet velocity or bulk water temperature.

Different phenomena were observed in the experiments of Gilpin (1973), where the melting rate and the shape of the ice cavity were found to depend heavily on the mode of melting. Two distinctly different ablation modes were found to exist for any given set of externally controllable parameters. In the so-called slow mode, relatively larger cavities were produced with smooth rounded shapes at the stagnation point. The variation of melting rate with jet Reynolds number indicated that the mechanism of heat transfer in the ice cavity was predominantly laminar with some degree of turbulence or unsteadiness in it caused by the incoming jet. In the so-called fast mode, the cavities were found to have smaller diameters and rough cylindrical surfaces, similar to those observed by Yen & Zehnder. The heat-transfer correlation in this mode is similar to the one reported by Yen and is also typical of turbulent flow. The fast mode was preferentially produced at higher Reynolds numbers; however, the main factor controlling the occurrence of either mode was found to be the starting condition. If initially the ice surface was smooth and flat, the slow mode would generally be developed. On the other hand, if the ice surface was initially indented, the fast mode would more likely prevail. The mutual interaction between the flow and the melting process (i.e. the resultant cavity shape) leads to a situation in which more than one steady-state configuration may exist. In Gilpin's free-jet experiment, the nozzle tips were in contact with the ice surfaces at the commencement of the test; the nozzle tips in Yen & Zehnder's experiment were set at a desired distance from the initial ice surfaces. This difference in starting points may be responsible for the absence of the slow mode of melting in the experiments of Yen & Zehnder.

Simultaneous Melting and Freezing

An interesting dual phase-change process can arise in a two-component liquid-solid system when the fusion temperature of the hot flowing material exceeds that of a solid body immersed in the flow. The special circumstance here is that the interfacial contact temperature of the two materials, calculated from classical conduction theory, lies between the fusion temperatures of the two materials so that melting and freezing take place simultaneously. An example, which is important to nuclear-reactor safety research, is the melting of solid steel [melting point (mp) 1410°C] immersed in a flow of molten ceramic UO_2 (mp 2850°C). The steel melting rate in this situation depends on the behavior of the solid UO_2 layer that forms when the flowing UO_2 melt comes into contact with the solid steel surface. Since the frozen UO_2 layer does not have a solid surface to "stick" to, mechanical processes for crust removal may prevent the UO_2 crust from insulating the melting steel surface. The rate of melting of the steel will vary by an order of magnitude, depending upon whether the crust is stable or tends to be broken up and disappear. Thus it would be highly desirable to characterize the stability of the frozen UO_2 layer in a given flow geometry.

An experimental and analytical investigation of the stability of an ice layer formed in a laminar flow of water over a flat plate of melting decane (mp $-30°C$) was conducted by Ganguli & Bankoff (1979). The experimental apparatus consisted of a once-through flow system for suddenly passing water over a frozen, thick decane slab in a rectangular test section. As soon as the water contacted the solid decane, the decane melted while the water froze, forming a thin protective ice crust over the decane melt. The ice crust was then observed to remelt, forming a leading edge that started from the upstream end of the decane slab and proceeded down the length of the melting slab. Severe ablation of the decane block was observed in front of the leading edge, where there was no crust to shield it from contact with the relatively warm water. The observed crust-removal process can be explained in terms of one-dimensional conduction theory. Solidification of the water proceeds as long as the conduction heat flux through the ice layer to the decane slab exceeds that to the ice surface from the flowing water. Once the thermal boundary layer in the decane slab and/or the decane melt layer is sufficiently large, convective heat transfer becomes more than can be conducted away, resulting in reduction in thickness through melting. The ice crust must ultimately disappear by melting to maintain the heat balance. An approximate numerical method, using polynomial temperature profiles with time-dependent coefficients, was used by Ganguli & Bankoff to solve this one-dimensional heat-conduction problem. Fair agreement between the predicted and observed motions of the leading edge of the ice sheet was reported.

It is important to note that no secondary freezing of the water was observed by Ganguli & Bankoff after removal of the primary ice crust. This was explained in terms of the lower thermal conductivity of the decane compared with that of the water. Refreezing might possibly occur in a system where the thermal conductivity of the meltable substrate is much higher than that of the freezing liquid flow, so that the driving "force" for refreezing is sufficiently strong to overcome any mechanical or thermal crust-destabilization effects. Indeed, this was confirmed by the experiments of Yim et al. (1978), who investigated a similar process with Freon flowing through a melting ice pipe. This study is discussed at the end of the next section on internal flow.

Epstein et al. (1980b) conducted an experimental investigation of the behavior of a freeze layer on a melting surface in a stagnation-region flow geometry under steady-state conditions. The work was directed toward a study of crust stability in the presence of a well-defined flow geometry.

In the experiment, a water jet was directed upward against the lower end of a meltable rod of solid octane (mp $-56.6°C$) or solid mercury (mp $-38.9°C$). To eliminate the geometry-variation effect, which occurred in the melting impingement studies of Yen & Zehnder (1973) and Gilpin (1973), the selection of the diameters of the water jet and the impingement surface was made to ensure that a flat melting surface was maintained; that is, melting-rate measurements were restricted to the stagnation region. Also developed in this study was a dual-phase-conversion model describing melting heat transfer in the presence of jet solidification and predicting the threshold jet temperature for incipient jet solidification.

For a given jet velocity, two sharply defined melting regimes for the water-jet/octane-rod system were observed by Epstein et al. For jet temperatures in the range $0 < T_j < 13°C$, the melting velocity was found to be dependent on the liquid jet temperature relative to its freezing point, indicating that a protective ice crust on the octane melt layer controls the melting rate. On the other hand, when $T_j > 13°C$, the thermally protective ice crust is absent, as the relevant temperature difference for heat convection was observed to be the jet temperature minus the octane melting temperature ($-56.5°C$). Over the range of jet temperatures studied (1.0–$80°C$), an ice crust was always present on melting mercury. The measured octane and mercury melting rates and the temperature threshold $T_j = 13°C$ for jet solidification on melting octane were found to agree quite well with the values predicted with the dual-phase-conversion model. Melting-rate predictions based on a two-material axisymmetric flow model that accurately predicts melting rates in situations in which the impinging liquid does not freeze on the melting solid (Swedish et al. 1979) were found to overestimate the mercury melting rate by as much as an order of magnitude. Hence it is important to determine the conditions for stable crust formation before calculating the jet penetration rate.

INTERNAL FORCED FLOW

Crust Profiles and Pressure Drops in Steady-State Pipe Flow

While the problem of the formation of ice in flow through pipes has been with us for centuries, the productive period of research in this area did not begin until the late-1960s. Figure 2 describes a practical situation in which liquid flowing in steady state in a pipe is being frozen along the cold pipe wall. The flow first passes through a solidification-free zone,

$z < 0$, where the pipe-wall temperature is maintained constant and above the liquid freezing temperature T_{mp}. In this region the flow is at a uniform temperature $T_0 > T_{mp}$ and, assuming the flow is laminar, the flow velocity profile is parabolic. At $z = 0$, the wall temperature undergoes a step reduction in temperature to a value T_w that is below the freezing point of the liquid ($T_w < T_{mp}$). Downstream of this location the situation is complicated by the constriction produced through solid phase growth. This poses two problems: what will be the shape of the frozen layer; and what is the pressure in the pipe due to solidification?

To make this problem readily amenable to solution in a straightforward manner, Zerkle & Sunderland (1968) assumed that the axial velocity profile remains parabolic in form even in the presence of an axially increasing frozen shell. They argued that the "flattening" of the velocity profile due to the converging-channel effect is compensated by the increased viscosity of the liquid (water) in the colder region near the wall. In this manner the problem is reduced to the classical Graetz-Nusselt problem, which describes the temperature profiles in pipe flow in the absence of a frozen shell along the pipe wall. Once the interface temperature gradient in the liquid is determined, the radial location of the frozen layer is obtained from the one-dimensional temperature distribution in the frozen crust combined with an energy balance at the liquid-solid interface. A typical predicted interface radius, $R(z)$, is a slowly varying function of axial distance, which is consistent with the assumptions in the theoretical model. The pressure drop is then obtained from an integrated form of the axial momentum equation.

Zerkle & Sunderland's theoretical work was accompanied by an experimental study of ice formation in water flowing through a circular tube. Observations of the ice crust that formed along the wall confirmed the validity of the physical picture shown in Figure 2. The ice crust gradually increased in thickness with axial distance, starting from zero thickness at

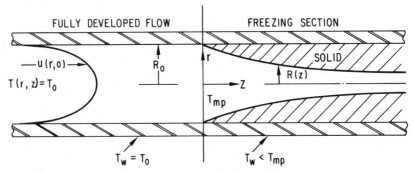

Figure 2   Thermal entrance of a tube with frozen layer.

the freeze-section inlet. Their experimental data on the average heat loss rate from the solidifying water flow indicated a very strong natural-convection effect that increased the heat transfer from 25 to 100% higher than the classical Graetz solution, depending on the liquid flow rate and the test-section length-to-diameter ratio. Experimental data on the pressure drop were presented for two runs with different inlet Reynolds numbers but essentially identical temperature conditions. The experimentally determined pressure-drop values were found to exceed the theoretical predictions. Zerkle & Sunderland attributed these differences to natural convection. This reasoning seems consistent with the fact that the discrepancy between theory and experiment is much greater for the case where the inlet Reynolds number is low. However, natural-convection effects cannot explain why the actual pressure drop exceeds the theoretical results. Enhanced heat transport to the liquid-solid interface due to natural convection tends to reduce the frozen-layer thickness and thereby reduces the pressure drop for all but very long test sections.

Experimental studies of freezing in thermal entrance-region laminar (water) flow were also carried out by Depew & Zenter (1969), Mulligan & Jones (1976), and Hwang & Sheu (1976). The experimental systems used by these workers were similar to that of Zerkle & Sunderland, although the experiments were conducted under conditions of higher tube-wall temperatures and with tubes of smaller inside diameter. The experimental study of Hwang & Sheu yielded heat-transfer data in close accord with the Graetz theory. A much smaller diameter tube was used in these experiments and the effects of free convection were suppressed. Such a test allows a comparison with the predicted pressure drop in the absence of natural convection. Unfortunately, their experimental program did not include measurements of the pressure drop. Interestingly enough, except for the one data point reported by Zerkle & Sunderland that we have already discussed, all the pressure-drop data reported in the literature are in reasonable agreement with the theory of Zerkle & Sunderland wherein the effects of natural convection are ignored. Mulligan & Jones argued that this agreement is fortuitous: the decrease in pressure drop due to natural convection thinning the frozen layer is offset by the increase in pressure drop caused by the flattening of the velocity profile by the converging flow. The higher pressure drops obtained from a combined hydrodynamic and thermal entrance-region theory (Hwang & Sheu 1976) of liquid solidification compared with the Graetz theory lends support to this conjecture. The extent to which these compensating phenomena can be exploited in making pressure-drop calculations remains an open question that can only be answered after a much more extensive set of data is obtained. However, it is already known from the results of an experimental study of the conditions that lead to freeze blockage in a tube that the Graetz solution for predictions of pressure drop can be greatly in error (DesRuisseaux & Zerkle 1969). The theoretical solution of the combined-convection problem, together with the variable cross section, has yet to appear.

An analysis and measurements of the steady-state heat transfer and pressure drop in turbulent freezing flow in a tube were reported by Thomason et al. (1978). Their mathematical model is similar to that for laminar flow in that a fully developed (turbulent) flow profile is assumed in both the solidification-free zone and in the freeze zone (see Figure 2). Implicit in the analysis is the assumption that the turbulent fluid motion and heat transport are obtainable directly from experiments or from independent analyses of turbulent tube flow in the absence of solidification. The Sparrow-Hallman-Siegel (Sparrow et al. 1959) solution for the heat-transfer coefficient in turbulent, thermal entrance-region flow and the Blasius pressure-drop expression were employed. The expression for the heat-transfer coefficient is an infinite series solution to a rather complex eigenvalue problem where the eigenvalues of the solution are functions of local Reynolds number and Prandtl number. Accordingly, numerical solution of the governing equations for the solid-liquid interface profile and the turbulent-flow pressure drop was required. [A simple, asymptotic closed-form solution to this problem has been reported by Epstein & Cheung (1982).] The pressure-drop predictions are in favorable agreement with the experimental data. An interesting result of the analytical work is that a rather peculiar crust profile having a "concave curvature" over a portion of the freeze length is predicted for freezing in turbulent pipe flow, as opposed to the strictly convex shape predicted and observed in laminar pipe flow (see Figure 2). It is important to note, however, that no observations of the crust profile were made and that the pressure drops measured by Thomason et al. were not greatly in excess of predicted pressure drops for the case of no solidification. Thus the ice-layer thickness must have been a small fraction of the tube radius and, therefore, the experimental results do not really constitute a stringent test of the theory. Pressure-drop measurements under more severe freezing conditions are needed.

Wavy Interfaces in Internal Solidification

A common assumption made in most studies of solidification in internal flow is that the solid-liquid interface is smooth and the flow cavity has a monotonically decreasing diameter beginning at the freeze-section entrance. In the theoretical treatment of internal flows involving phase change, a certain amount of caution must be exercised to ensure the validity of this assumption. We learned in the previous section on external forced flow that a planar solid-liquid interface may not be stable in a turbulent flow. Two recent studies by Gilpin (1979, 1981) of the shape and stability of a growing ice layer show this conclusion to also be true for ice growth in a laminar or turbulent tube flow.

The first experimental study by Gilpin (1979) of visual observations of ice growth in a tube flow was carried out at laminar or transition Reynolds numbers. The final steady-state ice profile observed by Gilpin was not the solid-phase shell of gradually increasing thickness observed or predicted in earlier studies of solidification in internal flow. Instead, a flow cavity with a wavelike or cyclic variation in cross section along the axial length of the tube was observed. Early in the experiment, after the tube-wall temperature was lowered below 0°C, the anticipated ice shell formed along the tube wall; the thickness of the ice increased in the direction of flow. Several hours after the initiation of ice growth, however, a most interesting phenomenon was observed. At the exit of the test section, where the ice layer abruptly ends and the flow cross section suddenly expands, the downstream face of the ice layer begins to melt away. This melt (or expansion) "wave" propagates slowly upstream, leaving behind a flow passage free of ice, until it reaches some equilibrium position and stops. Depending on the values of the water and coolant temperatures, new ice growth may begin in the region between the melt wave and the test-section exit. This could be followed by the appearance of another melt wave that propagates upstream and comes to rest at some equilibrium position downstream of the first melt wave. The final steady-state ice profile then is best described as a series of regularly narrowed ice cavities (or bands) of length L_c, each separated by a sudden ice-free expansion zone (see Figure 3).

Gilpin attributed this unanticipated behavior to a laminar-to-turbulent transition at the sudden enlargement that exists at the end of the ice layer and causes the enlargement zone to move upstream as a melt wave that cuts into thinner ice layers as it moves. He postulated that this upstream migration stops when the turbulent heat flux impinging on the ice face

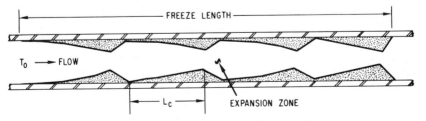

Figure 3   Pictorial representation of the ice-band structure.

balances the conduction heat flux through the ice thickness. Moreover, the second and subsequent tapered ice cavities are maintained by the relaminarization of the flow and concomitant reduction in heat-transfer rates in these tapered sections, followed by a flow separation and transition back to turbulent-flow heat-transfer rates in the expansion zones. This relaminarization and separation instability may even be responsible for the unexpected periodic flow and pressure-drop response observed by Thomason & Mulligan (1980) during turbulent-flow freezing experiments in a copper tube.

The purposes of the second pipe experiment performed by Gilpin (1981) were to investigate the form of the ice growth in a pipe containing flows that are initially turbulent and to determine the effect of ice-band structure on pressure drop. The transient development of the ice structure in this case was quite different from the behavior observed at low Reynolds numbers. The ice-band structure did not begin at the exit of the freeze section, but, instead, each band grew from small undulations in the flow cavity along the length of the freeze section to its final steady-state size. The development of the ice-band structure in a turbulent pipe flow is most likely caused by the same mechanism that produces ice waves on an initially flat ice-water interface melted by a turbulent flow (Thorsness & Hanratty 1979b, Gilpin et al. 1980). Recall that the mechanism of the instability involves the effects of flow acceleration and deceleration on turbulence and thus on heat transfer. While the ice-band development phase was observed to be quite different at high Reynolds numbers, the final steady-state ice-band structure is very similar in shape to that observed during laminar or transition flow.

Gilpin found that the length of each tapered section (or ice band) normalized by the tube diameter, namely L_c/D, is largely independent of the flow Reynolds number. The length of the tapered cavity that begins at the tube inlet is well correlated for laminar or turbulent flow by the freezing parameter, $(T_0 - T_{mp})/(T_{mp} - T_c)$, where T_c is the coolant temperature. The main difference between Gilpin's study and previous pipe-freezing studies in which the ice-band structure did not materialize is the value of this temperature ratio. For example, in the work of Zerkle & Sunderland (1968) the smallest value of this ratio was approximately 2.5, corresponding to an ice band $L_c/D = 500$. Thus the ice-band length is very large compared with that of the test sections employed by Zerkle & Sunderland (length-to-diameter ratios of 19 and 53.7). Unless the ratio $(T_0 - T_{mp})/(T_{mp} - T_c)$ is less than unity, a smooth freeze cavity is likely to be the more common ice configuration in laboratory scale experiments. Alternatively, in many industrial and commercial piping applications the ice band is likely to be the common freeze-layer configuration.

Epstein & Cheung (1982) recently constructed a flow-regime map that separates the smooth-crust regime from the ice-band regime on the basis of the pressure drop required to prevent the pipe flow from freezing shut.

The consequence of the ice-band structure of most interest is its effect on the axial pressure drop. Gilpin (1981) measured the pressure drop under conditions in which the pipe contained many ice bands. By ignoring viscous drag and assuming that each ice band acts as a nozzle constriction in the flow, he was able to predict with fair accuracy the measured pressure-drop values.

Freezing of an Advancing Pipe Flow

The transient solidification of a liquid along a specified freezing length in pipe flow has received considerable attention. In early theoretical studies of this problem (Ozisik & Mulligan 1969, Bilenas & Jiji 1970, Martinez & Beaubouef 1972), the transient solidification process was initiated by considering an instantaneously cooled tube wall surrounding an otherwise isothermal steady (internal) flow of liquid. In many physically realistic situations, it is difficult to imagine that freezing would begin in this manner, especially when the freeze section is very long. In casting technology, for example, the liquid to be solidified freezes on the walls of an initially empty channel (or mold cavity) while flowing into the channel. The salient feature of this advancing-flow problem is that the freezing length increases with time. The analysis of the problem is even more difficult than that for transient solidification along a fixed freezing length, because the crust thickness depends not only on the local convective heat exchange but on the time of arrival of the flow front as well. Moreover, the flow-front velocity is dependent both on its position and the instantaneous crust shape. For a liquid initially at its fusion temperature, convective heat exchange at the solid-liquid interface is absent and the problem is simplified, since the growth of the solid layer is not influenced by the flow. However, even in this limiting case, the coupling between fluid flow and crust growth results in a rather formidable second-order nonlinear integro-differential (momentum) equation for the position of the flow front.

The first study of freezing of a penetrating tube flow was performed by Cheung & Baker (1976). They conducted a series of tests in which various liquids were allowed to flow under gravity into long copper tubes cooled by liquid nitrogen or by a bath of dry ice and acetone. Different constant driving heads in the liquid reservoir were used and the penetration distances were measured. An empirical correlation of the penetration length in terms of the measured average flow velocity was found. Madejski (1976) obtained simple solutions to the penetration and freezing of a liquid at its fusion temperature into cold cavities and channels. His flow-freezing model was based on the assumption that the pressure drop over the instantaneous freezing length is the same as in channels of constant cross section. The penetration of a liquid at its freezing temperature into a cold tube was treated theoretically and experimentally by Epstein et al. (1977). They obtained an approximate closed-form solution for the flow-penetration length before solidification is complete (in the tube-inlet region) by postulating a reasonable functional form for the instantaneous shape of the frozen layer ("crust profile model"). In a later paper, Epstein & Hauser (1977) verified the assumption on the frozen-layer shape by a numerical solution of the full integro-differential equation for the liquid motion. To test the validity of the crust-profile model, the experimental studies on liquid penetration and freezing were carried out in a copper coil immersed in a bath of liquid coolant. The principal working fluid used was Freon 112A, which melts at 40.5°C. The penetration data for Freon 112A compare well with the model. Experiments were also carried out with both water and benzene flows. The predicted penetration lengths fall below the experimental data because of the difficulty in maintaining a constant tube-wall temperature with these materials. Good agreement between theory and experiment was obtained for all freezing materials with a more elaborate crust-profile model that accounts for resistance to heat transfer in the liquid coolant (Epstein & Hauser 1978). The crust-profile model has been extended by Epstein et al. (1980a) to consider the amount of liquid displaced into or through a rod bundle before freezing shut.

It should be recognized that if the liquid entering the tube is at a temperature higher than the freezing point (superheated), complete occlusion of the tube with frozen material ultimately occurs at some location between the inlet and the leading edge of the liquid flow. Calculation of the penetration distance in this most general situation is quite difficult, yet a numerical solution for the instantaneous shape of a solidified layer in an advancing tube flow of initially superheated liquid was recently obtained by Kolling & Grigull (1980). Penetration experiments were also conducted by the authors by pouring molten Sn, Pb, Zn, or Al into a thick-walled, horizontal casting tube. The observed metal-penetration lengths are in good agreement with the predicted values. It should be noted, however, that the calculated values are based on the assumption that perfect thermal contact is not achieved at the interface between the frozen crust and the tube wall. A concentric air gap was postulated to exist between the crust and the tube wall, the thickness of which is determined by the thermal contraction of the solidified crust shell and tracked by the numerical solution procedure. Unfortunately, it was not possible to confirm the development of such an air gap with available measurement techniques.

Simultaneous Melting and Freezing

The only experimental study of the behavior of a freeze layer on a melting surface in an internal-flow configuration is that reported by Yim et al. (1978). In these experiments, hot Freon 112A (mp 40.5°C) was injected into an ice pipe in turbulent flow, resulting in the formation of a Freon crust with the simultaneous melting of the ice wall. The effects of Freon injection pressure and Freon temperature were investigated on the amount of melted ice collected at the pipe exit over a fixed injection period. The transient shape of the melted ice cavity was also determined. The Freon crust behavior was visible through the ice-pipe wall and sections of Freon crust were clearly seen to move or slide along the ice-pipe wall in the direction of flow with a jerking motion. At times, the crust sections were observed to accumulate in the exit region of the ice pipe right in front of the exit tube, forming a Freon "crust jam." For some runs performed at low initial Freon temperatures, tiny solid Freon particles were seen flowing into the receiving vessel, indicating "bulk freezing" of Freon around ablated water drops. Numerical results based on a stable Freon-crust/ice-melt film model were compared with the experimentally determined results for ice melting. Despite the complex crust motion, the model was found to agree with the data for relatively warm Freon. At low Freon injection temperatures, the data for the mass of melted ice collected fall below the predicted values. This difference between theory and experiment was attributed to the rapid accumulation of Freon crust in the ice-pipe exit region. Apparently, if the flowing material has a large potential for solidification on the wall material, as in the case of Freon flowing over ice, crust removal leading to exposed wall-melt layers is difficult to attain. Refreezing of the flow will quickly "cover" exposed regions of wall material.

NATURAL CONVECTION

Simple Phase Change

As in the case of forced convection with phase change, in many problems of practical interest the interaction between natural convection and the phase-transformation process can be ignored. Tien & Yen (1966) performed an analytical study of the motion of a solid-liquid interface above a horizontal liquid melt layer created by subjecting its lower solid boundary to a step increase in temperature. The correlation by O'Toole & Silveston (1961) of non-phase-change, natural-convection heat transfer for fluid confined between two parallel horizontal plates was adopted to model the heat transfer to the melting interface. The same approach was employed in the model by Heitz & Westwater (1970) and experimental certification was obtained with their data for water. An experimental and analytical investigation of the motion of a phase boundary above a liquid melt layer during freezing and melting was performed by Hale & Viskanta (1980) for several different materials. Again, the assumption of negligible phase-change effect on convective heat transfer in the liquid layer was made for modeling purposes and the correlation by O'Toole & Silveston was employed. The data for freezing agreed nicely with the predictions of the model. The model, although in reasonable agreement with the melting-rate data, overpredicted the melting rate. A refined model that accounts for "blowing" (injection of melt) at the moving phase boundary would bring the data and theory closer together (Epstein 1975, Epstein & Cho 1976). Lapadula & Mueller (1970) analyzed the growth rate and the spatial distribution of a solid deposit freezing onto a vertical surface for the case in which there is heat transfer at the moving phase boundary due to laminar natural convection in the liquid. The analysis shows that, for purposes of engineering estimates, the convective heating rate at the phase interface can be assumed equal to the steady-state value for a nonmelting vertical flat plate.

Water - Ice System

The simple approach outlined above for a normal liquid is not valid for the case of melting or freezing of water at temperatures near 4°C. This is not because of a strong interaction between convection and phase change unique to water, but, instead, is due to the fact that thermal convection in water near its freezing point is distinguished from more common convective systems, since the density of water attains a maximum value at about 4°C. Obviously, the conventional correlations such as the one by O'Toole & Silveston are not applicable to the case of water. Some detailed theoretical and experimental studies of free-convection heat transfer in fresh water, which may be useful in making ice-melting or freezing-rate calculations, were reported by Saitoh (1976), Bendell & Gebhart (1976), and Yen (1980) for flows over a horizontal cylinder and a vertical ice slab, and for flow in a horizontal water layer, respectively. More complex flow-phase change situations, however, may occur in systems of salt water and ice, as is discussed in a subsequent section on melting in a density gradient.

Embedded Heat Source

The need to gain an understanding of heat transport in latent-heat-of-fusion thermal-energy storage systems has motivated a rather large number of studies of melting from vertical or horizontal cylindrical heat sources embedded in a phase-change material. The difficulty associated with this type of phase-change problem is that the location of the phase boundary is dependent on the natural-convection flow pattern and, therefore, is not known a priori. This precludes the use of non-phase-change, natural-convection heat-transfer results.

The variation with time of the size and shape of the melt region around a heated horizontal cylinder embedded in a meltable solid was observed photographically by White et al. (1977). For a short period after the onset of melting, heat transfer is primarily by conduction and the moving interface is represented by a succession of concentric circles. As natural convection develops in the melt, the solid-liquid interface becomes asymmetrical about the heated cylinder. Thermal plumes formed above the heater convey the hot liquid to the upper portion of the melt region and cause further melting to occur there. The rapid cooling of the plume as it impinges at the top of the melt cavity makes it a less effective melting medium as it recirculates around the sides of the cavity. As a result, melting occurs mainly in a region above the cylinder, with relatively little melting below, and the solid-liquid interface becomes somewhat pear-shaped. It is clear from the foregoing description that melting from an embedded heater belongs to the class of problems on natural convection in enclosures, but with the shape and volume of the enclosure changing with time.

Similar experimental studies with horizontal embedded heaters were performed by Sparrow et al. (1978) and Abdel-Wahed et al. (1979). The variation of the local heat-transfer coefficient around an electrically heated horizontal cylinder embedded in paraffin was measured by Bathelt et al. (1979a). Goldstein & Ramsey (1979) and Bathelt & Viskanta (1980) measured local heat-transfer rates at the moving solid-liquid interface during melting from a horizontal cylindrical heat source. Bain et al. (1974) and Sparrow et al. (1977) employed an implicit finite-difference scheme to study the effect of natural convection on heat transfer in the melt region created by a heated vertical tube embedded in a solid. The experimental observations of Bain et al. and Ramsey & Sparrow (1978) generally confirmed the numerical results. As with the horizontal heater, a recirculating natural-convection flow was found to be responsible for a substantially higher melting rate adjacent to the upper part of the vertical heater. Similar buoyancy-induced, recirculating flows were observed by

Hale & Viskanta (1978) during melting of a solid from a vertical isothermal wall.

The aforementioned studies focus on melting about a single heating cylinder. In practical applications, an array of cylinders is used. A cluster of three horizontal cylinders was used in the experiments of Bathelt et al. (1979b) to model staggered rows. Two different arrangements of the heating cylinders were used. One had a tighter spacing and the other a wider spacing. In both cases, free-convection flow around each cylinder was found to interact strongly with the other two cylinders once a common solid-liquid interface was formed. Interestingly enough, the circumferentially averaged heat-transfer coefficients were found to differ only by about 10% for the two cases. In the experimental study of Ramsey et al. (1979), an array of four parallel, equally spaced cylinders, situated in a common horizontal plane, was used. The main objective of this work was to compare the melting coefficients for the multicylinder array and the single cylinder. It was found that the heat-transfer coefficients for the horizontal array could be regarded as approximately the same as those for the single horizontal cylinder.

Melting in a Density Gradient

If the melting solid and the liquid are of the same material, the motion-driving buoyancy force is simply due to the variation of liquid density with temperature. If the solid and liquid are of different materials, however, the melting solid produces a sharp density interface or gradient that may provide the controlling mechanism for natural-convective movement of the melt layer and liquid adjacent to the solid surface. Melting of this type may occur near the surfaces of glacial ice in sea water and is of interest in the design of postaccident core-retention systems for nuclear reactors.

It would seem that sideward melting of a vertical surface immiscible with the warm liquid can be adequately calculated using a two-component side-by-side boundary-layer treatment, much as has been done in the theory of film condensation or film boiling. Such a model would even apply to most miscible solid-liquid pairs, since the diffusion coefficient is usually too low to cause significant mixing between the warm liquid and the melt layer. For the case of ice melting in sea or "salt" water, however, the flow and heat-transfer behavior near the phase-change interface cannot be represented by this simple description. The work of Josberger & Martin (1981) on melting of a vertical ice wall in salt water indicates that there could be three distinct boundary-layer flow regimes on the ice surface depending on the far-field water temperature, T_∞ (downward buoyancy force), and salinity, S_∞ (upward buoyancy force). The simplest regime was observed when T_∞ and S_∞ fall between the maximum density curve and the freezing curve, where the flow was unidirectional and upward. For $30\% \leqslant S_\infty \leqslant 35\%$ (a typical oceanic salinity) and $T_\infty < 20°C$, there were a laminar bidirectional flow at the bottom of the ice, consisting of a narrow upward inner flow inside of a wider downward outer flow, and a turbulent upward flow along the upper ice wall. For $T_\infty > 20°C$ and for the same range of S_∞, however, the flow reversed; it consisted of a laminar bidirectional boundary layer at the top of the ice and a turbulent downward boundary layer in the lower region. The ice was found to ablate smoothly in the first two cases, with the exception that a notch was observed because of the greater ablation rate near the flow-reversal point in the second case. A significantly different ice ablation was observed in the third case, where the ice surface was found to be smooth only in the laminar region. In the turbulent region the ice was found to melt irregularly and the interface was cusped, as illustrated in Figure 4. Dye injections showed that there were "back eddies" in the cusps. In general, the eddies were found to flow upward and then outward at the crests of the cusps before they were finally carried downward by the main stream. This cusped interface, mentioned only briefly by the authors, appears to be quite similar to the wavy interfaces observed in forced-convection flow as discussed in previous sections, although it is not clear that the mechanism for ice-wave (cusp) formation is the same in both cases.

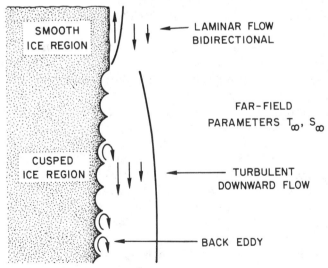

Figure 4 Ice interface during free-convection melting in salt water; $T_\infty > 20°C$, $30\% \leqslant S_\infty \leqslant 35\%$.

Several investigations have been made of the melting of a solid by a heavier overlying hot liquid pool immiscible with the molten phase of the solid. A qualitative experimental study of the melting of frozen benzene and frozen o-xylene under warm water was made by Alsmeyer & Reimann (1977). Taghavi-Tafreshi et al. (1979) obtained detailed data on the melting of a slab of frozen olive oil placed beneath a pool of water and concluded that the melting process is best described as a Taylor instability. The melt layer consists of an array of cosine-shaped melt drops. The flow within the melt layer initiates from the trough region between drops, where the thickness of the melt layer is minimal, and is directed inward toward the center of the drop. The continuous influx of the melt material into the drop increases its amplitude, which ultimately results in the release of the drop through a necking and pinching-off process. The melt-film behavior is thus quite similar to the hanging and dripping of water droplets on the underside of a wet, horizontal surface. In fact, Farhadieh & Epstein (1982) found their measurements of downward melting rates for a wide range of material pairs to be in good agreement with a theory by Gerstmann & Griffith (1967), originally developed for film-condensation heat transfer on the underside of a horizontal surface.

Farhadieh & Baker (1978) have used a polyethylene glycol/aqueous salt solution to make exploratory studies of melting rates and convective motions that can arise when the overlying hot liquid pool (salt solution) is miscible with the molten phase of the solid substrate (glycol). Many different effects were observed with this system, the most important of which are the following. A well-defined melt layer exists between the melting solid and the liquid pool. The interface between the melt and the liquid pool is sharp on the scale of the thickness of the melt layer, indicating that little or no mixing takes place adjacent to the melting surface. Narrow columns or fingers of melt material extend into the overlying heavier solution from discrete random sites at the interface. These fingers are essentially buoyant plumes that inject melt material into the liquid pool. Large and small vortices can be seen at the outer edge of the fingers, which produce vigorous mixing between the melt and pool materials. The downward melting rate is quite sensitive to the density ratio of liquid and melted solid, which was varied in the experiment by changing the salt concentration in the liquid pool. At low density ratios, between 1.09 and 1.25, the flow regime appears to be that of ordinary turbulent convection, as the measured melting rate or transport of heat across the melt film depends on the 1/3 power of the Rayleigh number. When the density ratio exceeds about 1.25, there is a sudden transition to a very vigorous turbulent regime characterized by a nearly linear relation between melting rate and Rayleigh number. Clearly, there are many questions raised by this seemingly simple experiment. Eventually one might hope to understand how the flow within the melt layer and within the needlelike fingers depends on system parameters, such as the density ratio, and to evaluate the effect of the fingers on the level of turbulent agitation in the overlying pool.

Frozen-Crust Stability in Liquid Layers

A description of the heat-transfer characteristics between a molten ceramic pool and its lower solid-steel boundary is of interest in fast breeder-reactor accident analysis. In this situation a thin ceramic crust will form above a steel melt layer as a result of the dual-phase-conversion process discussed in the sections on forced convection and internal flows. It seems likely that the fate of the ceramic crust depends on a race between the crust-growth process and the rate of development of buoyancy forces due to the presence of the underlying lighter steel melt. Slow crust growth should lead to a thin, weak crust that becomes laterally unstable through sidewise buckling (oscillation), whereas rapid crust freezing will lead to a stable crust. An experimental and theoretical study of the mechanical stability of a submerged frozen crust was reported by Epstein (1977). His work parallels that of Rayleigh and Kelvin for the generation of waves at the plane interface between two different fluids, but with both surface tension at the crust surface and elastic forces within the crust acting to stabilize the motion. Equations were derived for the breakup time of the crust and an approximate crust-breakup criterion was proposed, namely that the crust growth (or crystallization) time exceed the crust-breakup time. The criterion shows reasonable agreement with the experimental observations of the stability of a growing frozen crust in a gravity field alone (Rayleigh-Taylor instability). A notable result of this work was that a ceramic UO_2 crust growing on melting steel was predicted to be mechanically stable.

Under certain circumstances, the process of crust formation at the surface of a heat-generating liquid layer may not be "thermally" stable. Laboratory observations by Welander (1977a) of a water layer heated internally at a constant rate while it was cooled at its free surface with air kept at a fixed subzero temperature revealed periodic ice growth and decay at the top of the layer. For large internal heating and negative air temperature, the water temperature with ice was found to be too high and the water temperature without ice too low to allow any steady-state ice layer to exist. Based upon a simple mechanistic model, Welander (1977b) has also presented a theoretical discussion of the temperature transients and thermal oscillations in the system. A more detailed theoretical investigation of the process has been reported by Cheung (1981).

SUMMARY

The intent herein has been to present an overall view of the subject of solid-liquid phase change in fluid flow, with major emphasis on situations in which there are strong interactions among the flow and the motion as well as the shape of the solid-liquid interface. The engineering applications of this subject are ubiquitous; they range from drilling through frozen soil to safety assessments of nuclear reactors. In fact, research activities in the last ten years have become so widespread, with so many interesting phenomena and diverse applications reported on, that it is impossible to discuss the current status of the research on complex freezing-melting interfaces within this short section. Suffice it to say that there is now sufficient information on the subject to permit application to various engineering design problems. This is especially true, for example, in the application of measured heat-transfer coefficients for melting from embedded heaters to the design of thermal-energy storage systems. Also, the work reviewed suggests that it is now possible to make a reasonably accurate prediction of the pressure drop in turbulent flow through partially frozen pipes. In some areas, however, while the research work can be applied, there is less assurance of accuracy and/or understanding. These areas in which room remains for improvement have been pointed out in various places throughout the paper.

Literature Cited

Abdel-Wahed, R. M., Ramsey, J. W., Sparrow, E. M. 1979. Photographic study of melting about an embedded horizontal heating cylinder. *Int. J. Heat Mass Transfer* 22:171–73

Alsmeyer, H., Reimann, M. 1977. *On the heat and mass transport processes of a horizontal melting or decomposing layer under a molten pool.* Presented at ASME Winter Ann. Meet., Atlanta

Ashton, G. D. 1972. Turbulent heat transfer to wavy boundaries. *Proc. 1972 Heat Transfer Fluid Mech. Inst.*, pp. 200–13

Ashton, G. D., Kennedy, J. F. 1972. Ripples on underside of river ice covers. *J. Hydraul. Div. ASCE* 98:1603–24

Bain, R. L., Stermole, F. J., Golden, J. O. 1974. Liquefaction dynamics of n-octadecane in cylindrical coordinates. *J. Spacecr.* 11:335–39

Bankoff, S. G. 1964. Heat conduction or diffusion with change of phase. *Adv. Chem. Eng.* 5:75–150

Bathelt, A. G., Viskanta, R. 1980. Heat transfer at the solid-liquid interface during melting from a horizontal cylinder. *Int. J. Heat Mass Transfer* 23:1493–503

Bathelt, A. G., Viskanta, R., Leidenfrost, W. 1979a. An experimental investigation of natural convection in the melted region around a heated horizontal cylinder. *J. Fluid Mech.* 90:227–39

Bathelt, A. G., Viskanta, R., Leidenfrost, W. 1979b. Latent heat-of-fusion energy storage: experiments on heat transfer from cylinders during melting. *J. Heat Transfer* 101:453–58

Bendell, M. S., Gebhart, B. 1976. Heat transfer and ice-melting in ambient water near its density extremum. *Int. J. Heat Mass Transfer* 19:1081–87

Bilenas, J. A., Jiji, L. M. 1970. Variational solution of axisymmetric fluid flow in tubes with surface solidification. *J. Franklin Inst.* 289:267–79

Carey, K. L. 1966. Observed configuration and computed roughness of the underside of river ice, St. Croix River, Wisconsin. *US Geol. Surv. Prof. Pap. 550-B*, pp. B192–98

Cheng, K. C., Inaba, H., Gilpin, R. R. 1981. An experimental investigation of ice formation around an isothermally cooled cylinder in crossflow. *J. Heat Transfer* 103:733–38

Cheung, F. B. 1981. Periodic growth and decay of a frozen crust over a heat generating liquid layer. *J. Heat Transfer* 103:369–75

Cheung, F. B., Baker, L. Jr. 1976. Transient freezing of liquids in tube flow. *Nucl. Sci. Eng.* 60:1–9

Depew, C. A., Zenter, R. C. 1969. Laminar flow heat transfer and pressure drop with freezing at the wall. *Int. J. Heat Mass Transfer* 12:1710–14

DesRuisseaux, N., Zerkle, R. D. 1969. Freezing of hydraulic systems. *Can. J. Chem. Eng.* 47:233–37

Epstein, M. 1975. The effect of melting on heat transfer to submerged bodies. *Lett. Heat Mass Transfer* 2:97–104

Epstein, M. 1977. Stability of a submerged frozen crust. *J. Heat Transfer* 99:527–32

Epstein, M., Cheung, F. B. 1982. On the prediction of pipe freeze-shut in turbulent flow. *J. Heat Transfer* 104:381–84

Epstein, M., Cho, D. H. 1976. Melting heat transfer in steady laminar flow over a flat plate. *J. Heat Transfer* 98:531–33

Epstein, M., Hauser, G. M. 1977. Freezing of an advancing tube flow. *J. Heat Transfer* 99:687–89

Epstein, M., Hauser, G. M. 1978. Solidification of a liquid penetrating into a convectively cooled tube. *Lett. Heat Mass Transfer* 5:19–28

Epstein, M., Stachyra, L. J., Lambert, G. A. 1980a. Transient solidification in flow into a rod bundle. *J. Heat Transfer* 102:330–34

Epstein, M., Swedish, M. J., Linehan, J. H., Lambert, G. A., Hauser, G. M., Stachyra, L. J. 1980b. Simultaneous melting and freezing in the impingement region of a liquid jet. *AIChE J.* 26:743–51

Epstein, M., Yim, A., Cheung, F. B. 1977. Freezing-controlled penetration of a saturated liquid into a cold tube. *J. Heat Transfer* 99:233–38

Farhadieh, R., Baker, L. Jr. 1978. Heat transfer phenomenology of a hydrodynamically unstable melting system. *J. Heat Transfer* 100:305–10

Farhadieh, R., Epstein, M. 1982. Downward penetration of a hot liquid pool into the horizontal surface of a solid. *J. Heat Transfer* 104:199–201

Ganguli, A., Bankoff, S. G. 1979. Crust behavior in simultaneous melting and freezing on a submerged flat plate. *AIChE Symp. Ser.* 189:40–53

Gerstmann, J., Griffith, P. 1967. Laminar film condensation on the underside of horizontal and inclined surfaces. *Int. J. Heat Mass Transfer* 10:567–80

Gilpin, R. R. 1973. The ablation of ice by a water jet. *Trans. Can. Soc. Mech. Engrs.* 2:91–95

Gilpin, R. R. 1979. The morphology of ice structure in a pipe at or near transition Reynolds numbers. *AIChE Symp. Ser.* 189:89–94

Gilpin, R. R. 1981. Ice formation in a pipe containing flows in the transition and turbulent regimes. *J. Heat Transfer* 103:363–68

Gilpin, R. R., Hirata, T., Cheng, K. C. 1980. Wave formation and heat transfer at an ice-water interface in the presence of a turbulent flow. *J. Fluid Mech.* 99:619–40

Goldstein, R. J., Ramsey, J. W. 1979. Heat transfer to a melting solid with application to thermal energy storage systems. In *Studies in Heat Transfer*, pp. 199–208. Washington DC: Hemisphere Publ. Co.

Hale, N. W. Jr., Viskanta, R. 1978. Photographic observation of the solid-liquid interface motion during melting of a solid heated from an isothermal vertical wall. *Lett. Heat Mass Transfer* 5:329–37

Hale, N. W. Jr., Viskanta, R. 1980. Solid-liquid phase-change heat transfer and interface motion in materials cooled or heated from above or below. *Int. J. Heat Mass Transfer* 23:283–92

Heitz, W. L., Westwater, J. W. 1970. Extension of the numerical method for melting and freezing problems. *Int. J. Heat Mass Transfer* 13:1371–10

Hirata, T., Gilpin, R. R., Cheng, K. C., Gates, E. M. 1979a. The steady state ice layer profile on a constant temperature plate in a forced convection flow. I. Laminar regime. *Int. J. Heat Mass Transfer* 22:1425–33

Hirata, T., Gilpin, R. R., Cheng, K. C. 1979b. The steady state ice layer profile on a constant temperature plate in a forced convection flow. II. The transition and turbulent regimes. *Int. J. Heat Mass Transfer* 22:1435–43

Hsu, K. S. 1973. *Spectral evolution of ice ripples.* PhD thesis. Univ. Iowa. 147 pp.

Hsu, K. S., Locher, F. A., Kennedy, J. F. 1979. Forced-convection heat transfer from irregular melting wavy boundaries. *J. Heat Transfer* 101:598–602

Hwang, G. J., Sheu, J. 1976. Liquid solidification in combined hydrodynamic and thermal entrance region of a circular tube. *Can. J. Chem. Eng.* 54:66–71

Josberger, E. G., Martin, S. 1981. A laboratory and theoretical study of the boundary layer adjacent to a vertical melting ice wall in salt water. *J. Fluid Mech.* 111:439–73

Kolling, M., Grigull, U. 1980. Unsteady heat transfer with solidification applied to the fluidity of pure metals. *Waerme Stoffubertrag.* 14:231–44

Lapadula, C. A., Mueller, W. K. 1970. The effect of buoyancy on the formation of a solid deposit freezing onto a vertical surface. *Int. J. Heat Mass Transfer* 13:13–25

Larsen, P. A. 1969. Head losses caused by an ice cover on open channels. *J. Boston Soc. Div. Eng.* 56:45–67

Madejski, J. 1976. Solidification in flow through channels and into cavities. *Int. J. Heat Mass Transfer* 19:1351–56

Martinez, E. P., Beaubouef, R. T. 1972. Transient freezing in laminar tube flow. *Can. J. Chem. Eng.* 50:445–49

Muehlbauer, J. C., Sunderland, J. E. 1965. Heat conduction with freezing or melting. *Appl. Mech. Rev.* 18:951–59

Mulligan, J. C., Jones, D. D. 1976. Experiments on heat transfer and pressure drop in a horizontal tube with internal solidification. *Int. J. Heat Mass Transfer* 19:213–18

Ockendon, J. R., Hodgkins, W. R. 1975. *Moving Boundary Problems in Heat Flow and Diffusion: Proceedings.* London: Oxford Press. 300 pp.

O'Toole, J. L., Silveston, P. L. 1961. Correlations of convective heat transfer confined in horizontal layers. *AIChE Chem. Eng. Prog. Symp. Ser.* 32:81–86

Ozisik, M. N., Mulligan, J. C. 1969. Transient freezing of liquids in forced flow inside circular tubes. *J. Heat Transfer* 91:385–91

Petrie, D. J., Linehan, J. H., Epstein, M., Lambert, G. A., Stachyra, L. J. 1980. Solidification in two-phase flow. *J. Heat Transfer* 102:784–86

Ramsey, J. W., Sparrow, E. M. 1978. Melting and natural convection due to a vertical embedded heater. *J. Heat Transfer* 100:368–70

Ramsey, J. W., Sparrow, E. M., Varejao, L. M. C. 1979. Melting about a horizontal row of heating cylinders. *J. Heat Transfer* 101:732–33

Rubinshtein, L. I. 1967. *Stefan Problem; Translations of Mathematical Monographs, No. 27.* Providence, R.I: Am. Math. Soc. 419 pp.

Saitoh, T. 1976. Natural convection heat transfer from a horizontal ice cylinder. *Appl. Sci. Res.* 32:429–51

Savino, J. M., Siegel, R. 1967. Experimental and analytical study of the transient solidification of a warm liquid flowing over a chilled flat plate. *NASA TND-4015*

Savino, J. M., Zumdieck, J. F., Siegel, R. 1970. Experimental study of freezing and melting of flowing warm water at a stagnation point on a cold plate. *Proc. Int. Heat Transfer Conf., 4th,* Cu 2.10, pp. 1–8

Sparrow, E. M., Hallman, T. M., Siegel, R. 1959. Turbulent heat transfer in the thermal entrance region of a pipe with a uniform heat flux. *Appl. Sci. Res.* 7:37–52

Sparrow, E. M., Patankar, S. V., Ramadhyani, S. 1977. Analysis of melting in the presence of natural convection in the melt region. *J. Heat Transfer* 99:520–26

Sparrow, E. M., Schmidt, R. R., Ramsey, J. W. 1978. Experiments on the role of natural convection in the melting of solids. *J. Heat Transfer* 100:11–16

Stefan, J. 1891. Über die Theorie der Eisbildung, insbesondere über die Eisbildung in Polarmaere. *Ann. Phys. Chem.* 42:269–86

Swedish, M. J., Epstein, M., Linehan, J. H., Lambert, G. A., Hauser, G. M., Stachyra, L. J. 1979. Surface ablation in the impingement region of a liquid jet. *AIChE J.* 25:630–38

Taghavi-Tafreshi, K., Dhir, V. K., Catton, I. 1979. Thermal and hydrodynamic phenomena associated with melting of a horizontal substrate placed beneath a heavier immiscible liquid. *J. Heat Transfer* 101:318–25

Thomason, S. B., Mulligan, J. C. 1980. Experimental observations of flow instability during turbulent flow freezing in a horizontal tube. *J. Heat Transfer* 102: 782–84

Thomason, S. B., Mulligan, J. C., Everhart, J. 1978. The effect of internal solidification on turbulent flow heat transfer and pressure drop in a horizontal tube. *J. Heat Transfer* 100:387–94

Thorsness, C. B., Hanratty, T. J. 1979a. Mass transfer between a flowing fluid and a solid wavy surface. *AIChE J.* 25:686–97

Thorsness, C. B., Hanratty, T. J. 1979b. Stability of dissolving or depositing surfaces. *AIChE J.* 25:697–701

Tien, C., Yen, Y. C. 1966. Approximate solution of a melting problem with natural convection. *AIChE Chem. Eng. Prog. Symp. Ser.* 62:166–72

Welander, P. 1977a. Observation of oscillatory ice states in a simple convection experiment. *J. Geophys. Res.* 82:2591–92

Welander, P. 1977b. Thermal oscillations in a fluid heated from below and cooled to freezing from above. *Dyn. Atmos. Oceans* 1:215–23

White, F. M. 1974. *Viscous Fluid Flow.* New York: McGraw-Hill. 725 pp.

White, R. D., Bathelt, A. G., Leidenfrost, W., Viskanta, R. 1977. Study of heat transfer and melting front from a cylinder imbedded in a phase change material. *17th Natl. Heat Transfer Conf., Salt Lake City,* Pap. No. 77-HT-42

Wilson, D. G., Solomon, A. D., Boggs, P. T. 1978. *Moving Boundary Problems,* New York: Academic. 329 pp.

Yen, Y. C. 1976. Heat transfer between a free water jet and an ice block held normal to it. *Lett. Heat Mass Transfer* 3:299–308

Yen, Y. C. 1980. Free convection heat transfer characteristics in a melt water layer. *J. Heat Transfer* 102:550–56

Yen, Y. C., Zehnder, A. 1973. Melting heat transfer with water jet. *Int. J. Heat Mass Transfer* 16:219–23

Yim, A., Epstein, M., Bankoff, S. G., Lambert, G. A., Hauser, G. M. 1978. Freezing-melting heat transfer in a tube flow. *Int. J. Heat Mass Transfer* 21:1185–96

Zerkle, R. D., Sunderland, J. E. 1968. The effect of liquid solidification in a tube upon laminar-flow heat transfer and pressure drop. *J. Heat Transfer* 90:183–90

Ann. Rev. Fluid Mech. 1983. 15:321–43

MAGNETO-ATMOSPHERIC WAVES

John H. Thomas

Department of Mechanical Engineering and C. E. Kenneth Mees Observatory, University of Rochester, Rochester, New York 14627

1. INTRODUCTION

Magneto-atmospheric waves, or magneto-acoustic-gravity waves as they are also known, occur in a compressible, stratified, electrically conducting atmosphere under gravity, permeated by a magnetic field. These waves are supported by the combined restoring forces due to compression, buoyancy, and distortion of the magnetic field. They may be thought of as acoustic-gravity waves modified by the magnetic field or as magneto-acoustic waves modified by the stratification and buoyancy. Magneto-atmospheric waves have been studied fairly extensively in recent years, primarily because of their importance in the Sun's atmosphere and the Earth's upper atmosphere. Much of this work appears in the astrophysical literature, and one purpose of this review is to bring this work to the attention of fluid dynamicists.

We concentrate here on cases where all three restoring forces contribute to the wave motion. Acoustic-gravity waves, supported by compression and buoyancy, have an extensive literature of their own (see Eckart 1960, Tolstoy 1963, Gossard & Hooke 1975). Magneto-acoustic waves, supported by compression and magnetic forces, have also been studied extensively (see Ferraro & Plumpton 1966, Jeffrey 1966). Magneto-gravity waves, supported by buoyant and magnetic forces, occur in a conducting Boussinesq fluid permeated by a magnetic field. This case has received only limited attention (Howe 1969).

Almost all studies of magneto-atmospheric waves have dealt with linearized waves in a nondissipative medium. (An exception is the rather formal treatment of the nonlinear case given by Chiu 1971.) Even in the linearized formulation, the problem is complicated because the medium

321

0066-4189/83/0115-0321$02.00

is both inhomogeneous and anisotropic. Gravity and the magnetic field each introduce a preferred direction. For a uniform or slowly varying magnetic field, the inhomogeneity influences wave propagation most strongly through the rapid increase of the Alfvén speed with height.

The first paper on magneto-atmospheric waves was that of Ferraro & Plumpton (1958), who considered the case of a uniform vertical magnetic field in an isothermal atmosphere. Their study was motivated by applications to waves in the solar atmosphere, especially in sunspots, and these same applications have motivated much of the ensuing work. The term *magneto-atmospheric wave* was coined by Yu (1965) in an important paper that develops the dispersion relation in a special case where all three of the wave parameters (the sound speed, the Alfvén speed, and the local density scale height) are constant. In general, at least one of these parameters varies with height, and a dispersion relation for plane waves has only local significance and is restricted to waves of short vertical wavelength. Studies of magneto-atmospheric waves either have been based upon a local dispersion relation or have sought analytical or numerical solutions of the linearized wave equation for particular model atmospheres.

Past studies generally have been restricted to the case of a plane-stratified atmosphere in which all unperturbed quantities vary with height only. This requires a rather simple magnetic field configuration—either uniform, or horizontal with vertical shear. Waves in the case of more structured magnetic fields have recently received attention, but these investigations generally omit the effects of stratification and gravity. The wave modes of an isolated magnetic flux tube in field-free surroundings are of particular interest (Cram & Wilson 1975, Defouw 1976, Roberts & Webb 1978, Spruit 1981). A recent review of waves in magnetic structures, which includes a discussion of the effects of stratification and gravity, has been given by Roberts (1981).

2. MATHEMATICAL FORMULATION AND MODEL ATMOSPHERES

In this section we give a brief summary of the mathematical formulation leading to the linearized wave equation for magneto-atmospheric waves, and we discuss the class of plane-stratified atmospheres used as models in various studies of these waves.

2.1 Basic Linearized Equations

Consider an inviscid atmosphere at rest, having infinite electrical conductivity and permeated by a magnetic field $\mathbf{B}_0$ (not necessarily uniform).

Let $\mathbf{g}$ denote the acceleration of gravity (not necessarily uniform). The undisturbed pressure, density, and temperature are denoted by p_0, ρ_0, and T_0 respectively. Magnetic flux must be conserved,

$$\nabla \cdot \mathbf{B}_0 = 0, \tag{2.1}$$

and the atmosphere must be in static equilibrium,

$$0 = -\nabla\left(p_0 + \frac{B_0^2}{8\pi}\right) + \rho_0 \mathbf{g} + \frac{1}{4\pi} \mathbf{B}_0 \cdot \nabla \mathbf{B}_0. \tag{2.2}$$

Now consider small adiabatic perturbations of this equilibrium atmosphere, letting $\mathbf{u}$, $\mathbf{B}$, p, ρ, and T denote the perturbations in velocity, magnetic field, pressure, density, and temperature, respectively. The linearized equations of continuity, momentum, induction, and energy are then

$$\frac{\partial \rho}{\partial t} + \nabla \cdot (\rho_0 \mathbf{u}) = 0, \tag{2.3}$$

$$\rho_0 \frac{\partial \mathbf{u}}{\partial t} = -\nabla p + \rho \mathbf{g} + \frac{1}{4\pi}(\nabla \times \mathbf{B}) \times \mathbf{B}_0 + \frac{1}{4\pi}(\nabla \times \mathbf{B}_0) \times \mathbf{B}, \tag{2.4}$$

$$\frac{\partial \mathbf{B}}{\partial t} = \nabla \times (\mathbf{u} \times \mathbf{B}_0), \tag{2.5}$$

and

$$\frac{\partial p}{\partial t} + \mathbf{u} \cdot \nabla p_0 = c^2 \left[\frac{\partial \rho}{\partial t} + \mathbf{u} \cdot \nabla \rho_0\right], \tag{2.6}$$

where c is the adiabatic sound speed.

If we take the partial time derivative of the momentum equation (2.4) and then eliminate p, ρ, and $\mathbf{B}$ through the use of (2.3), (2.5), and (2.6), we arrive at the following governing equation for the perturbation velocity $\mathbf{u} = (u, v, w)$ (cf. Ferraro & Plumpton 1958):

$$\begin{aligned}
\frac{\partial^2 \mathbf{u}}{\partial t^2} = {} & \frac{1}{\rho_0} \nabla\left(\rho_0 c^2 \nabla \cdot \mathbf{u}\right) + \nabla(\mathbf{u} \cdot \mathbf{g}) - (\nabla \cdot \mathbf{u})\mathbf{g} \\
& - \frac{1}{\rho_0} \nabla\left[\mathbf{u} \cdot \left\{\nabla\left(\frac{B_0^2}{8\pi}\right) - \frac{1}{4\pi}\mathbf{B}_0 \cdot \nabla \mathbf{B}_0\right\}\right] \\
& + \frac{1}{4\pi\rho_0}\left[\{\nabla \times \nabla \times (\mathbf{u} \times \mathbf{B}_0)\} \times \mathbf{B}_0 + (\nabla \times \mathbf{B}_0) \times \{\nabla \times (\mathbf{u} \times \mathbf{B}_0)\}\right].
\end{aligned} \tag{2.7}$$

This linearized wave equation serves as the basis for most studies of magneto-atmospheric waves. Once the perturbation velocity $\mathbf{u}$ is known, we can return to the first-order system (2.3)–(2.6) to determine the other perturbation quantities. Note that in deriving the wave equation (2.7) the unperturbed atmosphere is assumed to be in static equilibrium, but no additional assumptions are made concerning the distribution of temperature or magnetic field, and no particular equation of state is assumed.

2.2   Plane-Stratified Atmospheres

Almost all work on magneto-atmospheric waves has dealt with plane-stratified atmospheres. In this case the acceleration of gravity is uniform, with $\mathbf{g} = (0, 0, -g)$ in Cartesian coordinates (x, y, z), and all of the unperturbed quantities in the atmosphere are functions of the vertical coordinate z only. We can then Fourier-analyze the perturbation in the horizontal directions as well as in time, thereby reducing the wave equation (2.7) to an ordinary vector differential equation in z.

In a plane-stratified atmosphere we must have $\mathbf{B}_0 = \mathbf{B}_0(z)$ only, which, in view of Equations (2.1) and (2.2), requires that the magnetic field be either uniform, $\mathbf{B}_0 = $ constant, or purely horizontal but varying with height, $\mathbf{B}_0 = [B_{0x}(z), B_{0y}(z), 0]$. For a plane-stratified atmosphere, the magneto-hydrostatic equation (2.2) reduces to

$$\frac{d}{dz}\left(p_0 + \frac{B_0^2}{8\pi} \right) = -\rho_0 g. \tag{2.8}$$

A uniform magnetic field has no effect on the structure of the unperturbed atmosphere, whereas a vertically varying horizontal magnetic field contributes to the static balance (2.8) through the vertical gradient of the magnetic pressure $B_0^2/8\pi$. The first term on the right-hand side of (2.7) can be written as

$$\frac{1}{\rho_0}\nabla\left(\rho_0 c^2 \nabla\cdot\mathbf{u} \right) = \left(\frac{dc^2}{dz} - \frac{c^2}{H} \right)(\nabla\cdot\mathbf{u})\mathbf{e}_z + c^2\nabla(\nabla\cdot\mathbf{u}), \tag{2.9}$$

where $\mathbf{e}_z$ is the unit vector in the z-direction and $H = [-(d\rho_0/dz)/\rho_0]^{-1}$ is the local density scale height. Also, in the wave equation (2.7) it is usually convenient to express the undisturbed magnetic field in terms of the Alfvén speed $v_A = (B_0^2/4\pi\rho_0)^{1/2}$. All of the coefficients in the wave equation (2.7) can then be conveniently expressed in terms of three parameters: the sound speed c, the Alfvén speed v_A, and the local density scale height H. Each of these parameters may vary with height z in an arbitrary plane-stratified atmosphere.

From here on we assume that the atmosphere is a perfect gas, so that $p_0 = \rho_0 R T_0$ and $c^2 = \gamma R T_0$. (We also assume that γ, c_p, c_v, and R are all constant. In a region of partial ionization or variable composition they all vary.) Then the hydrostatic equation (2.8) gives the relation

$$\frac{dc^2}{dz} - \frac{c^2}{H} + \gamma g = -\frac{\gamma}{\rho_0}\frac{d}{dz}\left(\frac{B_0^2}{8\pi} \right) = -\frac{\gamma}{2}\left(\frac{dv_A^2}{dz} - \frac{v_A^2}{H} \right) \tag{2.10}$$

among the three parameters c, v_A, and H. In order to study magneto-atmospheric waves in any detail, we must specify a model atmosphere by giving the distribution of undisturbed magnetic field $\mathbf{B}_0(z)$ and temperature $T_0(z)$. Perhaps the simplest model atmosphere is an isothermal atmosphere (T_0 = constant) with a uniform magnetic field ($\mathbf{B}_0$ = constant). In this case the sound speed $c = (\gamma R T_0)^{1/2}$ and density scale height $H = RT_0/g = c^2/\gamma g$ are both constant, the density decreases exponentially with height, $\rho_0 = \rho_{00}\exp(-z/H)$, and the Alfvén speed increases exponentially with height, $v_A = v_{A0}\exp(z/2H)$, where $v_{A0} = (B_0^2/4\pi\rho_{00})^{1/2}$. Here we see one of the complicating factors in the analysis of magneto-atmospheric waves: even in the simplest model atmospheres the parameters c, v_A, and H are not all constant, so that the wave equation does not have plane-wave solutions. (The sole exception is discussed in Section 3.2).

2.3   Stability of the Basic Atmosphere

A plane-stratified atmosphere may be subject to convective instability, and the criterion for this instability may be modified by the magnetic field. The stability criterion will, in general, emerge naturally from the linearized wave analysis, according to linear stability theory. Here we give a brief summary of results for the stability of plane-stratified magneto-atmospheres.

A uniform magnetic field with a nonzero vertical component has a stabilizing influence on the atmosphere and can allow a superadiabatic temperature gradient in the atmosphere. With dissipation included (thermal conductivity and finite electrical conductivity), the atmosphere may be subject to overstability (instability in the form of a growing oscillation) when the temperature gradient is superadiabatic. (See Chandrasekhar (1961) for a detailed discussion.)

Convective instability in the case of a horizontal magnetic field $\mathbf{B}_0 = (B_0(z),0,0)$ was considered by Newcomb (1961) and later by Thomas & Nye (1975), who showed that a necessary and sufficient condition for stability is

$$\frac{dT_0}{dz} - \left(\frac{dT_0}{dz}\right)_{ad} > -\frac{1}{\rho_0 R}\frac{d}{dz}\left(\frac{B_0^2}{8\pi}\right), \tag{2.11}$$

where $(dT_0/dz)_{ad} = -g(\gamma-1)/\gamma R$ is the adiabatic temperature gradient. For a uniform horizontal magnetic field (B_0 = constant), (2.11) reduces to the Schwarzschild criterion in the absence of a magnetic field. Thus, a uniform horizontal magnetic field has no effect on the onset of convective instability, although it does affect the growth rates of unstable modes, which involve the interchange of long horizontal magnetic flux tubes. A nonuniform horizontal magnetic field can be either stabilizing $(dB_0/dz > 0)$ or destabilizing $(dB_0/dz < 0)$, according to (2.11). For example, in an isothermal atmosphere the stability criterion (2.11) becomes

$$\frac{d}{dz}\left(\frac{B_0^2}{8\pi}\right) > -\rho_0 g\left(\frac{\gamma-1}{\gamma}\right). \tag{2.12}$$

Thus, an isothermal atmosphere, which is convectively stable in the absence of a magnetic field, can be rendered unstable by a horizontal magnetic field whose magnetic pressure decreases more rapidly than the critical rate $-\rho_0 g(\gamma-1)/\gamma$ over some range of heights (Thomas & Nye 1975).

3.   DISPERSION RELATIONS

In a general plane-stratified atmosphere the coefficients in the wave equation (2.7) vary with height z and there are no plane-wave solutions. However, because it is so convenient to study wave propagation in terms of a dispersion relation for plane waves, it is useful to consider cases where a plane-wave solution is at least approximately valid. Such is the case when the vertical wavelength is much smaller than the scale height of variation of any of the atmospheric parameters. By assuming that the coefficients in the wave equation (2.7) are constant locally we are led to an approximate, local dispersion relation for quasi-plane waves. The local dispersion relation and its limitations are discussed in Section 3.1. The one particular case for which the coefficients in the wave equation (2.7) are all strictly constant yields a global dispersion relation for true plane waves. This case is discussed in Section 3.2.

3.1   The Local Dispersion Relation

Consider an isothermal atmosphere at temperature T_0 with a uniform magnetic field $\mathbf{B}_0$ of arbitrary direction. In this case the wave equation (2.7) reduces to [using (2.9) and (2.10)]

$$\frac{\partial^2 \mathbf{u}}{\partial t^2} = c^2 \nabla(\nabla \cdot \mathbf{u}) + \nabla(\mathbf{g} \cdot \mathbf{u}) + \mathbf{g}(\gamma-1)\nabla \cdot \mathbf{u}$$

$$-\frac{B_0^2}{4\pi\rho_0}\left[\mathbf{b} \times \nabla \times \{\nabla \times (\mathbf{u} \times \mathbf{b})\}\right], \tag{3.1}$$

where $\mathbf{b}$ is a unit vector in the direction of $\mathbf{B}_0$ and $c^2 = \gamma R T_0$ is constant. We assume without loss of generality that $\mathbf{B}_0$ has no y-component. The undisturbed density distribution is unaffected by the uniform magnetic field and has the form $\rho_0 = \rho_{00}\exp(-z/H)$, where $\rho_{00} = \text{constant}$ and $H = RT_0/g$ is the density scale height. All of the coefficients in (3.1) are constant except for ρ_0^{-1} in the last term. If we assume that ρ_0 is "locally" constant, then (3.1) has all constant coefficients and hence a plane-wave solution of the form $\mathbf{u}(x, y, z, t) = \mathbf{u}_0\exp[i(k_x x + k_y y + k_z z - \omega t)]$. The resulting dispersion relation is a cubic equation in ω^2 that can be factored, so that either

$$\omega^2 - v_A^2(\mathbf{b}\cdot\mathbf{k})^2 = 0 \tag{3.2}$$

or

$$\omega^4 - \left[\left(c^2 + v_A^2\right)\left(k_x^2 + k_y^2 + k_z^2\right) + i\frac{c^2}{H}k_z\right]\omega^2$$
$$+ g^2(\gamma - 1)\left(k_x^2 + k_y^2\right)$$
$$+ \left[c^2(\mathbf{b}\cdot\mathbf{k})^2 + ig\gamma(\mathbf{b}\cdot\mathbf{e}_z)(\mathbf{b}\cdot\mathbf{k})\right]v_A^2\left(k_x^2 + k_y^2 + k_z^2\right) = 0, \tag{3.3}$$

where $\mathbf{k} = (k_x, k_y, k_z)$ is the wave-number vector and $\mathbf{e}_z$ is the unit vector in the z-direction. The local dispersion relation in this form was first given by McLellan & Winterberg (1968). However, particular forms of this relation were given earlier for the cases where $\mathbf{B}_0$ is either vertical (Deutsch 1967, Stepien 1967) or horizontal (Deutsch 1967). A local dispersion relation in a different form was derived earlier by MacDonald (1961), who applied the local approximation directly to the system of first-order perturbation equations (2.3)–(2.6). This somewhat trouble-some nonuniqueness of the local dispersion relation is discussed later in this section.

The first relation, (3.2), represents a pure transverse Alfvén wave with a purely horizontal, nondivergent motion ($u = w = 0$, $v \neq 0$). Only mag-netic forces come into play in this case. The second relation, (3.3), represents the two magneto-atmospheric wave modes in which compres-sive, buoyant, and magnetic forces all play a role. The local dispersion relation (3.3) is quadratic in ω^2 and has two pairs of real roots $\omega = \pm\omega_1, \pm\omega_2$, is given by

$$\omega_{1,2}^2 = \left[\mathcal{B} \pm (\mathcal{B}^2 - 4\mathcal{C})^{1/2}\right]/2,$$

with (3.3) expressed as $\omega^4 - \mathcal{B}\omega^2 + \mathcal{C} = 0$. For a fixed wave-number vector $\mathbf{k}$ there are two magneto-atmospheric wave modes of frequency ω_1 and ω_2. These are called the *plus* and *minus* modes, respectively (McLellan & Winterberg 1968), or alternatively, the *fast* and *slow* modes, in analogy to magneto-acoustic waves. In the absence of gravity and stratification ($g = 0$, $H \to \infty$), these two modes become the fast and slow magneto-acoustic waves in a uniform medium, governed by the dispersion relation

$$\omega^4 - \left(c^2 + v_A^2\right)\left(k_x^2 + k_y^2 + k_z^2\right)\omega^2$$
$$+ c^2 v_A^2 (\mathbf{b \cdot k})^2\left(k_x^2 + k_y^2 + k_z^2\right) = 0. \tag{3.4}$$

In the absence of a magnetic field ($v_A = 0$) these modes become the acoustic and gravity modes of acoustic-gravity waves in an isothermal atmosphere, governed by the dispersion relation

$$\omega^4 - c^2\left(k_x^2 + k_y^2 + k_z^2 + \frac{i}{H}k_z\right)\omega^2 + g^2(\gamma - 1)\left(k_x^2 + k_y^2\right) = 0. \tag{3.5}$$

For real frequencies ω, the vertical wavenumber k_z given by (3.3) is, in general, complex, $k_z = \alpha + i\beta$, so that the wave amplitude has an exponential as well as oscillatory variation with height. The validity of the local dispersion relation is restricted to the range $|k_z|H \gg 1$. A number of authors have studied the local dispersion relation (3.3) in detail for special cases. Yeh (1974) has studied the wave-number surfaces generated by (3.3) for horizontal and vertical magnetic fields. Bel & Mein (1971) and Bel & Leroy (1977) studied the case of vertical propagation ($k_x = k_y = 0$) and Nakagawa et al. (1973) and Michalitsanos (1973) studied the case of vertically evanescent waves ($\alpha = 0$). However, it has been pointed out recently (Thomas 1982) that most of the numerical results of Bel & Mein (1971), Nakagawa et al. (1973), and Bel & Leroy (1977) lie outside the range of validity of the local approximation (i.e. have $|k_z|H \lesssim 1$) and can be misleading.

The problems encountered when the local dispersion relation is applied outside of its range of validity $|k_z|H \gg 1$ have been discussed recently (Thomas 1982). As a particular example, consider the effect of a uniform horizontal magnetic field on the cutoff frequency for fast modes in an isothermal atmosphere. In the absence of the magnetic field the fast modes reduce to gravity-modified acoustic modes, and the cutoff frequency is the well-known acoustic cutoff frequency $\omega_0 = c/2H$. The local dispersion relation (3.3) predicts that the cutoff frequency *decreases* with increasing magnetic-field strength (Bel & Leroy 1977). [The cutoff frequency corresponds to $\alpha = 0$ and $|\beta|H < 2$, and so is outside the range of validity of (3.3).] A more general WKB approximation predicts that the cutoff frequency *increases* with increasing magnetic-field strength. Finally, the exact solution in this case (Nye & Thomas 1976a, Summers 1976) shows that the cutoff frequency is *unaffected* by the magnetic field. This correct result is easily understood if we recall that the cutoff frequency corresponds to the limit $k_x = k_y = \alpha = 0$, which means that the entire atmosphere moves up and down in phase. There is no distortion of the magnetic field caused by this motion and hence no effect on the cutoff frequency.

Another problem with the local dispersion relation (3.3) is that it is not the only possible form. A different form results if the local approximation is applied to the system of first-order perturbation equations (2.3)–(2.6) instead of the single second-order equation (2.7). The former approach was used by MacDonald (1961) to obtain a local dispersion relation that is different from (3.3). [The two terms containing i in (3.3) are replaced by different terms in MacDonald's relation.] The terms that differ in the two forms of the local dispersion relation are of order $(k_z H)^{-1}$ and thus are negligibly small in the range of validity of either relation. These small terms arise from gradients in the undisturbed atmosphere, which cannot be represented consistently in the local approximation. Thomas (1982) has suggested that all such terms of order one or higher in $(k_z H)^{-1}$ be neglected in any derivation of a local dispersion relation. This approach leads to a unique, unambiguous form for the local dispersion relation consisting of Equation (3.2) and the equation

$$\omega^4 - \left(c^2 + v_A^2\right)\left(k_x^2 + k_y^2 + k_z^2\right)\omega^2 + g^2(\gamma - 1)\left(k_x^2 + k_y^2\right)$$

$$+ (\mathbf{b}\cdot\mathbf{k})^2 c^2 v_A^2\left(k_x^2 + k_y^2 + k_z^2\right) = 0 \tag{3.6}$$

replacing (3.3). This may be taken as a standard form for the local dispersion relation for magneto-atmospheric waves.

3.2   A Global Dispersion Relation for a Special Case

There is one particular case, first studied by Yu (1965), in which the wave equation leads to a global dispersion relation that is not restricted to short vertical wavelengths (see also Deutsch 1967, Chen & Lykoudis 1972, Nye & Thomas 1974, Rudraiah et al. 1977). This is the case of an isothermal atmosphere with a horizontal magnetic field $\mathbf{B}_0 = (B_0(z), 0, 0)$ that decreases exponentially with height according to $B_0(z) = B_{00}\exp(-z/2H)$, where the density scale height H is given by $H = [(c^2/\gamma) + (v_A^2/2)]/g$. In this case the three wave parameters c, v_A, and H are each constant. The magnetic pressure is proportional to the gas pressure, and the vertical gradient of magnetic pressure helps to support the atmosphere against gravity. All the coefficients of the wave equation (2.7) are constant and the equation has an exact plane-wave solution

$\mathbf{u} = \mathbf{u}_0 \exp[i(k_x x + k_y y + k_z z - \omega t)]$ for which the dispersion relation is

$$\omega^6 - \left[(c^2 + v_A^2)\left(k_x^2 + k_y^2 + \alpha^2 + \frac{1}{4H^2}\right) + v_A^2 k_x^2 \right]\omega^4$$

$$+ \left[v_A^2(2c^2 + v_A^2)k_x^2\left(k_x^2 + k_y^2 + \alpha^2 + \frac{1}{4H^2}\right) - g\left(g - \frac{c^2}{H}\right)(k_x^2 + k_y^2) \right.$$

$$\left. + k_y^2 g \frac{v_A^2}{H} \right]\omega^2$$

$$- v_A^2 k_x^2 \left[c^2 v_A^2 k_x^2\left(k_x^2 + k_y^2 + \alpha^2 + \frac{1}{4H^2}\right) - g\left(g - \frac{c^2}{H}\right)(k_x^2 + k_y^2) \right] = 0,$$

$$(3.7)$$

where $k_z = \alpha + i\beta$ with $\beta = -(2H)^{-1}$. In this case the imaginary part β of k_z is independent of frequency and horizontal wavelength, and the exponential factor $\exp(-\beta z) = \exp(z/2H)$ in the wave amplitude gives exact energy conservation for a vertically propagating wave. The dispersion relation (3.7) is *global*, in the sense that it is not restricted to short vertical wavelengths, as is the local dispersion relation (3.3) or (3.6). Because it is not subject to the difficulties of the local approximation, this is a useful example, even though it is a very special case.

Following Yu (1965), we can express the real part of the wave-number vector, $\mathbf{k}_R = (k_x, k_y, \alpha)$, in terms of spherical coordinates in wave-number space, with zenith angle θ and azimuth angle ϕ, in the form $k_x = k \sin\theta \cos\phi$, $k_y = k\sin\theta\sin\phi$, and $\alpha = k\cos\theta$, where $k = |\mathbf{k}_R|$. We also introduce the nondimensional wave number $K = kH$, the nondimensional frequency $\Omega = \omega H/c$, and the nondimensional ratios $M^2 = v_A^2/c^2$ and $G^2 = gH/c^2 = \gamma^{-1} + M^2/2$. Then the dispersion relation (3.7) can be rewritten in the nondimensional form

$$\Omega^6 - \left[(1 + M^2)(K^2 + \tfrac{1}{4}) + M^2 K^2 \sin^2\theta\cos^2\phi\right]\Omega^4$$

$$+ \left[M^2(2 + M^2)K^2(K^2 + \tfrac{1}{4})\sin^2\theta\cos^2\phi + G^2(1 - G^2)K^2\sin^2\theta \right.$$

$$\left. + M^2 G^2 K^2 \sin^2\theta\sin^2\phi\right]\Omega^2$$

$$- M^2 K^4 \sin^4\theta\cos^2\phi\left[M^2(K^2 + \tfrac{1}{4})\cos^2\phi + G^2(1 - G^2)\right] = 0. \quad (3.8)$$

The dispersion relation (3.8) has three real roots of Ω^2, and it can be shown that these roots are all positive provided $M^2 < 2(\gamma - 1)/\gamma$. The critical value $M^2 = 2(\gamma - 1)/\gamma$ agrees with the value given by the stability criterion (2.12) for this case. For $\gamma = 5/3$, stability requires $M^2 < 0.8$, and so the magnetic field can never be dominant in this model. For this reason, it is best to think of the fast and slow modes in this case as acoustic and gravity modes modified by the magnetic field.

Figure 1 shows the dispersion curves given by (3.8) on a K, Ω diagram for different directions (θ, ϕ) of the wave-number vector, using the illustrative values $\gamma = 5/3$ and $M^2 = 0.1$. There are three modes for each value of K: an acoustic (fast) mode, modified by the magnetic field; a gravity (slow) mode, modified by the magnetic field; and a hydromagnetic mode, modified only slightly by compression and gravity. For $K \to 0$ the hydromagnetic and gravity modes have $\Omega \to 0$, while the acoustic modes have a cutoff frequency $\Omega_c = (1 + M^2)^{1/2}/2$, corresponding to a dimensional cutoff frequency $\omega_c = c(1 + M^2)^{1/2}/2H = \omega_0(1 + M^2)^{1/2}/(1 + \gamma M^2/2)$, where $\omega_0 = \gamma g/2c$ is the acoustic cutoff frequency in the absence of a magnetic field. Here the cutoff frequency ω_c decreases with increasing magnetic field strength because of the effect of the nonuniform magnetic field on the structure of the undisturbed atmosphere. (This does not contradict the result in Section 3.1 for a uniform magnetic field, which has no effect on the structure of the undisturbed atmosphere.)

For large K the acoustic mode is similar to a pure sound wave $\Omega = K$ ($\omega = ck$), whereas the gravity mode is considerably different from the nonmagnetic case. Unless $\phi = \pi/2$ ($k_x = 0$), the gravity mode does not have a limiting frequency for $K \to \infty$; rather, it behaves like a slow magneto-acoustic mode in a homogeneous medium. For $\phi = \pi/2$ ($k_x = 0$) the gravity-mode frequency approaches an upper limit $\Omega^2 = [G^2(1 + M^2 - G^2)/(1 + M^2)]\sin^2\theta$ as $K \to \infty$. If we further take $\theta = \pi/2$ ($\alpha = 0$), this limiting frequency in dimensional form is

$$\omega_g^2 = (\gamma - 1)\frac{g^2}{c^2}\left[\frac{1 + \dfrac{\gamma}{2(\gamma - 1)}M^2}{\left(1 + \dfrac{\gamma}{2}M^2\right)(1 + M^2)}\right].$$

The motion is purely vertical in this case, and ω_g may be thought of as a modified Brunt-Väisälä frequency. For $M^2 = 0$, it reduces to the Brunt-Väisälä frequency ω_{BV} in a nonmagnetic isothermal atmosphere, $\omega_{BV}^2 = (\gamma - 1)g^2/c^2$. For $M^2 > 0$, the modified Brunt-Väisälä frequency is less than ω_{BV}.

The mode that becomes unstable for $M^2 > 2(\gamma - 1)/\gamma$ is the hydromagnetic mode. The perturbation in this case consists of an interchange of long tubes of magnetic lines, with motion along the magnetic field lines.

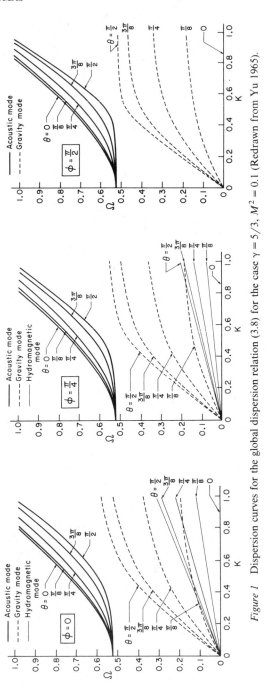

Figure 1 Dispersion curves for the global dispersion relation (3.8) for the case $\gamma = 5/3$, $M^2 = 0.1$ (Redrawn from Yu 1965).

Antia & Chitre (1980) have generalized the dispersion relation (3.7) to include thermal dissipation and temperature stratification. They find that overstable modes can occur when the temperature gradient is superadiabatic.

4. SOLUTIONS OF THE FULL WAVE EQUATION

Because of the limitations of the local dispersion relation, it is useful to have exact solutions of the wave equation (2.7) for special cases. For a plane-stratified atmosphere, the perturbation velocity can be represented in the form $\mathbf{u}(x, y, z, t) = \hat{\mathbf{u}}(z)\exp[i(k_x x + k_y y - \omega t)]$ and then the wave equation (2.7) reduces to an ordinary differential equation for the vertical (z) dependence of the perturbation velocity $\hat{\mathbf{u}}(z) = [\hat{u}(z), \hat{v}(z), \hat{w}(z)]$. The coefficients of the differential equation are all expressed in terms of the three parameters $c(z)$, $v_A(z)$, and $H(z)$. For a few particular choices of the distribution of these parameters an exact solution to the wave equation has been found. For more complicated distributions, numerical solutions may be obtained; this has been done for some rather detailed model atmospheres in connection with oscillations in sunspots (see Section 5.1). All of the exact solutions involve the assumption of either a purely vertical or a purely horizontal magnetic field, and it is convenient to divide our discussion in the same manner.

4.1 Vertical Magnetic Field

PURE ALFVÉN WAVES   For a uniform vertical magnetic field $\mathbf{B}_0 = (0, 0, B_0)$ and a purely horizontal, incompressible motion $\mathbf{u} = (0, v(z, t), 0)$, the wave equation (2.7) reduces to

$$\frac{\partial^2 v}{\partial t^2} = v_A^2(z)\frac{\partial^2 v}{\partial z^2}. \tag{4.1}$$

This equation describes pure Alfvén waves in which compression and buoyancy play no role. Stratification, however, is important because the Alfvén speed $v_A = (B_0^2/4\pi\rho_0(z))^{1/2}$ varies with height as a result of the varying density. If we assume an oscillatory solution $v(z, t) = \hat{v}(z)\exp(i\omega t)$, Equation (4.1) becomes

$$\frac{d^2\hat{v}}{dz^2} + \frac{\omega^2}{v_A^2(z)}\hat{v} = 0. \tag{4.2}$$

This equation can be solved for a specified density distribution $\rho_0(z)$ to determine the variation of wave amplitude with height. Variations in Alfvén speed lead to reflection of Alfvén waves, and solutions of (4.2) can be used to determine a reflection coefficient for a region of varying density. This problem is mathematically analogous to that of determining the propagation of light in a medium in which the index of refraction varies in the direction of propagation. It is a fundamental problem in the theory of waves in layered media, and many analytical solutions may be borrowed from the literature on related problems (see Brekhovskikh 1960).

The problem of reflection and refraction of Alfvén waves at a plane interface separating layers of different uniform density ρ_1 and ρ_2, for an arbitrary angle of incidence and a uniform magnetic field of arbitrary direction, has been solved by Ferraro (1954). For a vertical magnetic field and normal incidence, the amplitudes of the magnetic field perturbations of the incident, reflected, and transmitted waves are in the ratio $(\sqrt{\rho_2} + \sqrt{\rho_1}):(\sqrt{\rho_2} - \sqrt{\rho_1}):2\sqrt{\rho_2}$. Solutions to (4.2) have been given for several cases of smoothly varying density distributions $\rho_0(z)$: an exponential variation, corresponding to an isothermal atmosphere (Ferraro 1954, Thomas 1978); a hyperbolic-tangent distribution (Adam 1976, Geronicolas 1977); and a power-law variation (Thomas 1978). Much of this work has been motivated by the problem of Alfvén-wave propagation in sunspots. Nye & Hollweg (1980) solved (4.2) numerically for a detailed density distribution representing a sunspot umbra.

FULLY COUPLED WAVES   Here we consider a more general disturbance for which compressive, buoyant, and magnetic restoring forces all contribute. For a vertical magnetic field there is no preferred horizontal direction, so we can take the horizontal wave number in the x-direction with no loss of generality. We assume a perturbation velocity $\mathbf{u} = (u, v, w)$ of the form $\mathbf{u} = \hat{\mathbf{u}}(z)\exp[i(k_x x - \omega t)]$, with $\hat{\mathbf{u}}(z) = [\hat{u}(z), \hat{v}(z), \hat{w}(z)]$. The y-component of the wave equation (2.7) is then identical to Equation (4.2) and is decoupled from the other two component equations. It describes pure, transverse Alfvén waves. The x- and z-components of the wave equation (2.7) are

$$\left[v_A^2\left(\frac{d^2}{dz^2} - k_x^2\right) - c^2 k_x^2 + \omega^2\right]\hat{u} + ik_x\left[c^2\frac{d}{dz} - g\right]\hat{w} = 0, \qquad (4.3)$$

$$ik_x\left[c^2\frac{d}{dz} - (\gamma - 1)g\right]\hat{u} + \left[c^2\frac{d^2}{dz^2} - \gamma g\frac{d}{dz} + \omega^2\right]\hat{w} = 0. \qquad (4.4)$$

The coupled equations (4.3) and (4.4) describe magneto-atmospheric waves in which all three restoring forces are involved.

For an isothermal atmosphere the sound speed c is uniform and the Alfvén speed increases exponentially with height, $v_A(z) = v_{A0}\exp(z/2H)$.

Ferraro & Plumpton (1958) found analytical power-series solutions to (4.3) and (4.4) in this case. These solutions are not expressible in terms of tabulated functions, however, so that direct numerical integration is just as useful in some applications. Numerical solutions of (4.3) and (4.4) for an isothermal atmosphere have been calculated by Weymann & Howard (1958), Lüst & Scholer (1966), and Scheuer & Thomas (1981). An important feature of these solutions is the way in which the character of the wave motion changes with height in the atmosphere. Low down in the atmosphere, where the Alfvén speed is less than or comparable to the sound speed, the wave can have horizontal motions comparable to or greater than the vertical motions. High up in the atmosphere, where the Alfvén speed is much greater than the sound speed, the magnetic field is effectively rigid against the reduced density of the gas, and the motions are constrained to be along the vertical magnetic field lines. A fast-mode wave that is forced by horizontal compressive motions low in the atmosphere changes to what is essentially an acoustic wave along the magnetic field lines high in the atmosphere. There is considerable downward reflection, and only a small fraction of the wave energy reaches the upper atmosphere; nevertheless, the amplitude can be large in the upper atmosphere because the density is greatly reduced. This basic mechanism of generation of sound waves in a vertical magnetic field has also been discussed by Parker (1964) in terms of an approximate wave equation and a simple model atmosphere.

4.2   Horizontal Magnetic Field

An exact solution of the wave equation for the case of a uniform horizontal magnetic field $\mathbf{B}_0 = (B_0, 0, 0)$ in an isothermal atmosphere was found by Nye & Thomas (1976a). If we restrict horizontal propagation to the magnetic field direction ($k_y = 0$), then the wave equation (2.7) has components

$$\left(\omega^2 - c^2 k_x^2 \right) \hat{u} + ik_x \left(c^2 \frac{d}{dz} - g \right) \hat{w} = 0, \tag{4.5}$$

$$\left(\omega^2 - v_A^2 k_x^2 \right) \hat{v} = 0, \tag{4.6}$$

$$ik_x \left[c^2 \frac{d}{dz} - (\gamma - 1) g \right] \hat{u}$$

$$+ \left[\left(c^2 + v_A^2 \right) \frac{d^2}{dz^2} - \gamma g \frac{d}{dz} + \left(\omega^2 - v_A^2 k_x^2 \right) \right] \hat{w} = 0. \tag{4.7}$$

The y-component equation (4.6) is decoupled from the other components and describes pure, transverse Alfvén waves. The coupled equations (4.5) and (4.7) describe the coupled magneto-atmospheric waves. We can eliminate $\hat{u}(z)$ to obtain a single second-order equation in $\hat{w}(z)$. This equation can then be transformed into the hypergeometric equation

$$\zeta(1-\zeta)\frac{d^2W}{d\zeta^2} + [C - (A + B + 1)\zeta]\frac{dW}{d\zeta} - ABW = 0 \tag{4.8}$$

through the transformations $W = \hat{w}\exp(Kz/H)$ and $\zeta = \Omega^2 v_{A0}^2 \cdot \exp(-z/H)/c^2(K^2 - \Omega^2)$, where $\Omega = H\omega/c$ is a nondimensional frequency, $K = Hk_x$ is a nondimensional horizontal wave number, and

$$A + B = C = 2K + 1, \qquad AB = \Omega^2 + K + \left(\frac{\gamma - 1}{\gamma^2}\right)\frac{K^2}{\Omega^2}.$$

Solutions to (4.8) can be expressed in terms of hypergeometric functions.

The hypergeometric equation (4.8) has a singular point at $\zeta = 1$, which corresponds to

$$\frac{\omega}{k_x} = \left[\frac{c^2 v_A^2}{c^2 + v_A^2}\right]^{1/2} \tag{4.9}$$

Since v_A^2 varies exponentially with height, for a wave of fixed frequency and horizontal wave number there may be some height z for which (4.9) is satisfied; this height z_c is a *critical layer* for the wave (Thomas 1976). The critical layer occurs where the horizontal phase speed of the wave equals the critical speed $v_c = [c^2 v_A^2/(c^2 + v_A^2)]^{1/2}$. [Roberts (1981) refers to the critical speed v_c as the *cusp speed*.] Note that $v_c < \min(c, v_A)$ for all values of c and v_A. Thus, for fast modes, for which $\omega/k_x > \min(c, v_A)$, a critical layer does not occur. The critical layer occurs only for slow modes. It is somewhat analogous to the critical layer for internal gravity waves in a shear flow. The case $A + B = C$ is an exceptional case for the hypergeometric equation, and the singularity at $\zeta = 1$ is logarithmic rather than algebraic. Thus, the critical layer in this case is significantly different from the internal gravity wave case. An analysis of the present case similar to that of Booker & Bretherton (1967) for the internal gravity wave case shows that waves are partially absorbed at the critical layer (Thomas 1976). The attenuation across the critical layer is algebraic, rather than exponential as in the gravity wave case. A more detailed discussion of the critical layer for magneto-atmospheric waves has been given by Adam (1977, 1981) and El Mekki et al. (1978). Adam (1981) has considered the case of a horizontal magnetic field with the addition of an aligned shear flow $(U_0(z), 0, 0)$.

5. APPLICATIONS TO WAVES IN THE SOLAR ATMOSPHERE

The most frequent application of the theory of magneto-atmospheric waves has been to explain various observed wave motions in the Sun's atmosphere. These include regular oscillations in sunspots, large-scale disturbances in the corona induced by solar flares, and small-scale waves that may be responsible for heating the chromosphere and corona. For large-scale oscillations the motions are observed as a variation of intensity or a Doppler shift of a spectral absorption line, whereas the small-scale waves are observed more indirectly as a Doppler broadening of a spectral line.

5.1   Waves in Sunspots

A sunspot, with its strong magnetic field and strong density stratification, is an ideal site for magneto-atmospheric waves. In the layers of a sunspot just below the visible solar surface the gas is convectively unstable, and in the presence of the sunspot magnetic field this takes the form of an overstability, or growing oscillation. In the steady state this leads to finite-amplitude oscillatory convection, which serves as a continuous source of generation of magneto-atmospheric waves in the stable, visible layers of the sunspot. Observed wave motions in sunspots range from large-scale oscillations of the sunspot as a whole (the umbral oscillations and running penumbral waves) down to small-scale, unresolved waves which may be important in the overall energy balance of the spot. A recent review of observations of waves and other dynamical phenomena in sunspots has been given by Moore (1981), and the related theoretical work has been reviewed by Thomas (1981). Here we discuss two of the most important wave phenomena. Figure 2 provides a schematic guide to waves in sunspots.

UMBRAL OSCILLATIONS   Well-defined, large-scale velocity oscillations are observed in the visible layers of sunspot umbras. These umbral oscillations were first observed by Beckers (Beckers & Tallant 1969, Beckers & Schultz 1972). Oscillations within a particular sunspot have a fairly well defined period, and periods for different sunspots all lie in a fairly narrow range of 145–190 s. The velocity is predominantly vertical with speeds of 1 to 5 km s^{-1} in the chromosphere and much lower speeds in the photosphere. The observations suggest that the umbral oscillations are a resonant mode of the entire sunspot umbra, and this interpretation has served as a basis for the theoretical work.

The exciting mechanism for umbral oscillations is assumed to be overstable, oscillatory convection in a shallow layer (roughly 2000 km thick) just below the visible surface of the umbra. Theoretical models of this driving layer have been given by Moore (1973), Mullan & Yun (1973), and Roberts (1976). These models produce driving oscillations with a range of periods that includes the observed periods of umbral oscillation.

The well-defined observed period in the visible layers represents a sharp resonant response of the umbral atmosphere to the broader-band forcing in the driving layer, in the form of a trapped magneto-atmospheric wave. Uchida & Sakurai (1975) found resonant modes of appropriate period in a simple two-layer model of the umbral atmosphere. Their analysis assumes a uniform vertical magnetic field in the umbra and is based on an approximate form of the wave equation for the case where $v_A^2 \gg c^2$ (the "quasi-Alfvén" approximation). In this approximation, the vertical component of the wave equation decouples from the horizontal components. The resulting waves have very nearly the form of a transverse Alfvén wave, but with some compression which produces passive vertical motions. The approximation is justified in the umbral chromosphere but not in the photosphere, where $v_A^2 \sim c^2$. Uchida & Sakurai calculate resonant modes that are vertically trapped by means of downward reflection due to increasing Alfvén speed and upward reflection due to a rather artificial lower rigid boundary.

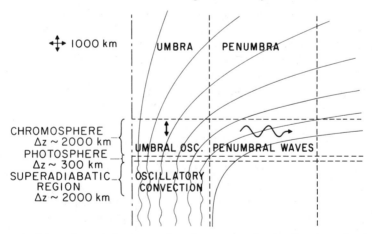

Figure 2 Schematic vertical cross section of a sunspot showing the wave phenomena discussed in Section 5.1. The thin solid lines are representative magnetic field lines, and the dash-dot line is the central axis of the sunspot.

Scheuer & Thomas (1981) studied umbral oscillations in a model atmosphere similar to Uchida & Sakurai's, but without making the quasi-Alfvén approximation. Using a three-layer model atmosphere with a uniform vertical magnetic field, they find resonant modes of appropriate period in the form of trapped fast magneto-atmospheric waves. The photosphere and chromosphere are represented by a single isothermal layer, the chromosphere is represented by another isothermal layer at higher temperature, and the convection zone is represented by a layer with temperature increasing linearly with depth. The upward reflection in this case is due to the increasing sound speed with depth into the convection zone; the compressive nature of the wave is crucial here. Downward reflection is strong, occurring mostly well below the chromosphere-corona transition region, but a small amount of wave energy leaks into the corona in the form of a pure acoustic wave along the vertical magnetic field lines. Subsequently, Thomas & Scheuer (1982) calculated the resonant modes of oscillation in a much more detailed model of the umbral atmosphere, with essentially the same conclusions. In related work, Antia & Chitre (1979) considered less realistic models of the umbral atmosphere, but included the overstable layer just below the surface as part of the eigenmode calculation rather than as an external forcing. In this picture the eigenmodes themselves are overstable and hence self-exciting.

PENUMBRAL WAVES   Running penumbral waves in the sunspot chromosphere were discovered independently by Zirin & Stein (1972) and Giovanelli (1972). They appear as wavefronts in the form of concentric circular arcs which propagate radially outward from the edge of the umbra across the penumbra, with typical horizontal phase speeds of $15-20$ km s^{-1} and periods in the range 180–250 s. A fairly detailed theory of these penumbral waves in terms of magneto-atmospheric waves has been given by Nye & Thomas (1974, 1976b). Their theory is based on simple models of the penumbral atmosphere with a purely horizontal magnetic field. The penumbral waves are identified as ducted fast magneto-atmospheric waves. The vertical trapping is due to increasing Alfvén speed up into the chromosphere, and increasing sound speed down into the convection zone (which is assumed to be free of magnetic field below the penumbra). The downward reflection is strong and the wave is actually vertically evanescent at the levels in the chromosphere where it is most easily observed. The wave amplitude is greatest in the chromosphere, but the density is very low there and most of the wave energy actually lies in the photosphere, where the wave amplitude is lower but the density is much higher.

5.2   Flare-Induced Coronal Disturbances

A large solar flare may generate a very large scale disturbance in the chromosphere and low corona which propagates at a nearly constant speed (of order 1000 km s^{-1}) over great distances across the solar disk (several hundred thousand kilometers). These flare-induced waves were first reported by Moreton and sometimes are called Moreton waves (see Moreton 1960). The fact that the disturbance propagates over a significant fraction of the solar surface indicates that horizontal ducting of the wave occurs. Also, the very high propagation speed indicates that the wave propagates in the lower corona; in the quiet chromosphere, the sound speed is of order 20 km s^{-1} and the Alfvén speed is generally less than 50 km s^{-1}. In the corona the Alfvén speed becomes as large as 1000 km s^{-1} because of the reduced density.

Meyer (1968) and Nye & Thomas (1976b) have presented similar models of this phenomenon in which the disturbance is considered to be a fast magneto-atmospheric wave trapped in the low corona. Meyer assumes a vertical magnetic field and uses an approximate form of the wave equations (the quasi-Alfvén approximation discussed above). Nye & Thomas assume a horizontal magnetic field and use the full magneto-atmospheric wave equations. The basic mechanism clearly works for an inclined field as well. The ducting of the wave is due to upward reflection resulting from the sharp density change across the chromosphere-corona transition region and downward reflection resulting from the increase in Alfvén speed up into the corona. The lowest mode is trapped within the first two density scale heights above the chromosphere-corona transition. Nye & Thomas also show that the horizontal group velocity is nearly constant for horizontal wavelengths of 10^5 km or less, so that a wave packet propagates horizontally with very little dispersion.

An alternative approach was taken by Uchida (1968, 1974), who studied the propagation of disturbances in various coronal models by means of a ray-tracing technique. He finds horizontal ducting of disturbances that is essentially in agreement with the mode calculations of Meyer and of Nye & Thomas. However, Uchida's coronal models are generally more detailed, and he even presents a calculation of ray paths for a specific observed disturbance using a detailed map of the coronal magnetic field for that day (Uchida et al. 1973).

6. CONCLUSIONS

The theory of magneto-atmospheric waves has progressed to the point where we have a good understanding of linear waves in nondissipative, plane-stratified atmospheres. The local dispersion relation is useful but limited and must be applied with care. Exact analytical and numerical solutions of the linearized wave equation have disclosed many important features of the fully coupled waves. Applications of the linear theory have led to a satisfactory understanding of some of the more significant dynamical phenomena in the solar atmosphere, especially the observed oscillations in sunspots.

Future work should include the extension of the linear theory to cases of more structured magnetic fields. For example, a study of waves in a vertical, axially symmetric, spreading magnetic flux tube under gravity would be quite relevant to the problem of sunspot oscillations. Improved observations of waves in the solar atmosphere will encourage theoretical calculations based on more detailed model atmospheres. Work on the nonlinear theory and the effects of dissipation is also needed, especially concerning the steepening and shock formation of vertically propagating waves, in order to understand the heating of the solar chromosphere and corona.

ACKNOWLEDGMENTS

This paper was prepared with the support of the National Aeronautics and Space Administration, NASA Grant NSG-7562, and the National Science Foundation, NSF Grant ATM-8021305. Professor Alfred Clark, Jr., read this article in manuscript form and suggested several improvements.

Literature Cited

Adam, J. A. 1976. Alfvén wave reflection at a density transition region. *J. Phys. A* 9:L193–95

Adam, J. A. 1977. On the occurrence of critical levels in solar magnetohydrodynamics. *Sol. Phys.* 52:293–307

Adam, J. A. 1981. Mechanical wave-energy flux in magneto-atmospheres: discrete and continuous spectra. *Astrophys. Space Sci.* 78:293–350

Antia, H. M., Chitre, S. M. 1979. Waves in the sunspot umbra. *Sol. Phys.* 63:67–78

Antia, H. M., Chitre, S. M. 1980. Stability of magneto-acoustic waves in a thermally conducting compressible fluid. *Astrophys. Space. Sci.* 68:183–200

Beckers, J. M., Schultz, R. B. 1972. Oscillatory motions in sunspots. *Sol. Phys.* 27:61–70

Beckers, J. M., Tallant, P. E. 1969. Chromospheric inhomogeneities in sunspot umbrae. *Sol. Phys.* 7:351–65

Bel, N., Leroy, B. 1977. Analytical study of magneto-acoustic gravity waves. *Astron. Astrophys.* 55:239–43

Bel, N., Mein, P. 1971. Propagation of magneto-acoustic waves along the gravitational field in an isothermal atmosphere. *Astron. Astrophys.* 11:234–40

Booker, J. R., Bretherton, F. P. 1967. The critical layer for internal gravity waves in a shear flow. *J. Fluid Mech.* 27:513–39

Brekhovskikh, L. M. 1960. *Waves in Layered Media.* New York: Academic. 561 pp.

Chandrasekhar, S. 1961. *Hydrodynamic and Hydromagnetic Stability.* Oxford Univ. Press. 652 pp.

Chen, C.-J., Lykoudis, P. S. 1972. Velocity oscillations in solar plage regions. *Sol. Phys.* 25:380–401

Chiu, Y. T. 1971. Arbitrary amplitude magnetoacoustic waves under gravity: An exact solution. *Phys. Fluids* 14:1717–24

Cram, L. E., Wilson, P. R. 1975. Hydromagnetic waves in structured magnetic fields. *Sol. Phys.* 41:313–27

Defouw, R. J. 1976. Wave propagation along a magnetic tube. *Ap. J.* 209:266–69

Deutsch, R. V. 1967. Unde magnetoatmospherice intr-un mediu stratificat orizontal. *Stud. Cercet. Fiz.* 19:807–21

Eckart, C. 1960. *Hydrodynamics of Oceans and Atmospheres.* New York: Pergamon. 290 pp.

El Mekki, O., Eltayeb, I. A., McKenzie, J. F. 1978. Hydromagnetic-gravity wave critical levels in the solar atmosphere. *Sol. Phys.* 57:261–66

Ferraro, V. C. A. 1954. On the reflection and refraction of Alfvén waves. *Ap. J.* 119:393–406

Ferraro, V. C. A., Plumpton, C. 1958. Hydromagnetic waves in a horizontally stratified atmosphere. *Ap. J.* 127:459–76

Ferraro, V. C. A., Plumpton, C. 1966. *An Introduction to Magneto-Fluid Mechanics.* Oxford Univ. Press. 254 pp. 2nd ed.

Geronicolas, E. A. 1977. Alfvén wave propagation in a density gradient in sunspots. *Ap. J.* 211:966–71

Giovanelli, R. G. 1972. Oscillations and waves in sunspots. *Sol. Phys.* 27:71–79

Gossard, E. E., Hooke, W. H. 1975. *Waves in the Atmosphere.* Amsterdam: Elsevier. 456 pp.

Howe, M. S. 1969. On gravity-coupled magnetohydrodynamic waves in the Sun's atmosphere. *Ap. J.* 156:27–47

Jeffrey, A. 1966. *Magnetohydrodynamics.* Edinburgh/London: Oliver & Boyd. 252 pp.

Lüst, R., Scholer, M. 1966. Kompressionswellen in einer isothermen Atmosphäre mit vertikalem Magnetfeld. *Z. Naturforsch. Teil A* 21:1098–1106

MacDonald, G. J. F. 1961. Spectrum of hydromagnetic waves in the exosphere. *J. Geophys. Res.* 66:3639–70

McLellan, A., Winterberg, F. 1968. Magneto-gravity waves and the heating of the solar corona. *Sol. Phys.* 4:401–8

Meyer, F. 1968. Flare-produced coronal waves. In *Structure and Development of Solar Active Regions*, ed. K. Kiepenheuer, pp. 485–89. Dordrecht: Reidel. 608 pp.

Michalitsanos, A. G. 1973. The five-minute period oscillations in magnetically active regions. *Sol. Phys.* 30:47–61

Moore, R. L. 1973. On the generation of umbral flashes and running penumbral waves. *Sol. Phys.* 30:403–19

Moore, R. L. 1981. Dynamic phenomena in sunspots. In *The Physics of Sunspots*, ed. L. E. Cram, J. H. Thomas, pp. 259–311. Sunspot, N. Mex: Sacramento Peak Obs. 495 pp.

Moreton, G. E. 1960. H_α observations of flare-induced disturbances with velocities ~ 1000 km/sec. *Astron. J.* 65:494–95

Mullan, D. J., Yun, H. S. 1973. Can oscillations grow in a sunspot umbra? *Sol. Phys.* 30:83–91

Nakagawa, Y., Priest, E. R., Wellck, R. E. 1973. The trapped magnetoatmosphere waves. *Ap. J.* 184:931–53

Newcomb, W. A. 1961. Convective instability induced by gravity in a plasma with a frozen-in magnetic field. *Phys. Fluids* 4:391–96

Nye, A. H., Hollweg, J. V. 1980. Alfvén waves in sunspots. *Sol. Phys.* 68:279–95

Nye, A. H., Thomas, J. H. 1974. The nature of running penumbral waves. *Sol. Phys.* 38:399–413

Nye, A. H., Thomas, J. H. 1976a. Solar magneto-atmospheric waves. I. An exact solution for a horizontal magnetic field. *Ap. J.* 204:573–81

Nye, A. H., Thomas, J. H. 1976b. Solar magneto-atmospheric waves. II. A model for running penumbral waves. *Ap. J.* 204:582–88

Parker, E. N. 1964. A mechanism for magnetic enhancement of sound-wave generation and the dynamical origin of spicules. *Ap. J.* 140:1170–73

Roberts, B. 1976. Overstability and cooling in sunspots. *Ap. J.* 204:268–80

Roberts, B. 1981. Waves in magnetic structures. In *The Physics of Sunspots*, ed. L. E. Cram, J. H. Thomas, pp. 369–83. Sunspot, N. Mex: Sacramento Peak Obs. 495 pp.

Roberts, B., Webb, A. R. 1978. Vertical motions in an intense magnetic flux tube. *Sol. Phys.* 56:5–35

Rudraiah, N., Venkatachalappa, M., Kandaswamy, P. 1977. Propagation and reflection of Alfvén-acoustic-gravity waves in an isothermal compressible fluid. *J. Fluid Mech.* 80:223–36

Scheuer, M. A., Thomas, J. H. 1981. Umbral oscillations as resonant modes of magneto-atmospheric waves. *Sol. Phys.* 71:21–38

Spruit, H. C. 1981. Motion of magnetic flux tubes in the solar convection zone and chromosphere. *Astron. Astrophys.* 98:155–60

Stepien, K. 1967. Hydromagnetic oscillations in a stratified atmosphere. *Acta Astron.* 17:31–54

Summers, D. 1976. Gravity modified sound waves in a conducting stratified atmosphere. *Q. J. Mech. Appl. Math.* 29:117–26

Thomas, J. H. 1976. Comments on magneto-atmospheric waves. In *The Energy Balance and Hydrodynamics of the Solar Chromosphere and Corona*, *Proc. IAU Colloq. No. 36*, ed. R. M. Bonnet, P. Delache, pp. 134–38. Clermont-Ferrand, Fr.: G. de Bussac. 504 pp.

Thomas, J. H. 1978. The reflection of Alfvén waves and the cooling of sunspots. *Ap. J.* 225:275–80

Thomas, J. H. 1981. Theories of dynamical phenomena in sunspots. In *The Physics of Sunspots*, ed. L. E. Cram, J. H. Thomas, pp. 345–58. Sunspot, N. Mex: Sacramento Peak Obs. 495 pp.

Thomas, J. H. 1982. The local dispersion relation for magneto-atmospheric waves. *Ap. J.* In press

Thomas, J. H., Nye, A. H. 1975. Convective instability in the presence of a nonuniform horizontal magnetic field. *Phys. Fluids* 18:490–91

Thomas, J. H., Scheuer, M. A. 1982. Umbral oscillations in a detailed model umbra. *Sol. Phys.* 79:19–29

Tolstoy, I. 1963. The theory of waves in stratified fluids including the effects of gravity and rotation. *Rev. Mod. Phys.* 35:207–30

Uchida, Y. 1968. Propagation of hydromagnetic disturbances in the solar corona and Moreton's wave phenomenon. *Sol. Phys.* 4:30–44

Uchida, Y. 1974. Behavior of the flare-produced coronal MHD wavefront and the occurrence of type II radio bursts. *Sol. Phys.* 39:431–49

Uchida, Y., Altschuler, M. D., Newkirk, G. 1973. Flare-produced coronal MHD-fast-mode wavefronts and Moreton's wave phenomenon. *Sol. Phys.* 28:495–516

Uchida, Y., Sakurai, T. 1975. Oscillations in sunspot umbras due to trapped Alfvén waves excited by overstability. *Publ. Astron. Soc. Jpn.* 27:259–74

Weymann, R., Howard, R. 1958. Note on hydromagnetic waves passing through an atmosphere with a density gradient. *Ap. J.* 128:142–45

Yeh, T. 1974. Wavenumber surfaces of magnetoatmospheric waves. *Phys. Fluids* 17:2282–89

Yu, C. P. 1965. Magneto-atmospheric waves in a horizontally stratified conducting medium. *Phys. Fluids* 8:650–56

Zirin, H., Stein, A. 1972. Observations of running penumbral waves. *Ap. J.* 178:L85–87

Ann. Rev. Fluid Mech. 1983. 15:345–89

INTEGRABLE, CHAOTIC, AND TURBULENT VORTEX MOTION IN TWO-DIMENSIONAL FLOWS

Hassan Aref

Division of Engineering, Brown University, Providence, Rhode Island 02912

It is indeed rather astonishing how little practical value scientific knowledge has for ordinary men, how dull and commonplace such of it as has value is, and how its value seems almost to vary inversely to its reputed utility.

G. H. Hardy, *A Mathematician's Apology*

INTRODUCTORY REMARKS

Vortex dynamics would appear to be exempt from Hardy's pessimistic verdict. On one hand, the evolution of vorticity, and thus the motions of vortices, are essential ingredients of virtually any real flow. Hence vortex dynamics is of profound practical importance. On the other hand, vortex motion has always constituted a mathematically sophisticated branch of fluid mechanics that continues to invite the application of novel analytical techniques. Indeed it is neither dull nor commonplace.

This central role of vorticity in fluid mechanics is not difficult to understand. As we know, any velocity field, v, can be split into a sum of two fields, one with the same divergence as v, and no curl, and one with the same curl as v and vanishing divergence. This important result is due to Stokes and to Helmholtz (1858; see Sommerfeld 1964). In incompressible flow, as we deal with exclusively here, the first part is irrotational and divergence-free and thus leads to the linear problem of potential flow. The second part, however, derives directly from the vorticity of the field v. In the dynamics of this part lies the essence of the problem.

345

0066-4189/83/0115-0345$02.00

Vorticists, as those who pursue the topic of vortex dynamics may be called, have been frequent reviewers of their subject. Hall (1972), Widnall (1975), Leibovich (1978), and Saffman & Baker (1979) have reviewed aspects of the field in earlier volumes of this publication. Notable reviews appearing elsewhere include the articles by Clements & Maull (1975), Chorin (1980), Leonard (1980), and Zabusky (1977, 1981b). The reader should also consult the recent impromptu remarks of Saffman (1981). With this wealth of material already available, the purpose of yet another review on vortex motion may well be questioned. The intention in this article is to discuss the restricted subject of two-dimensional vortex motion according to the program indicated by the title. As we shall see, even with this restriction the spectrum of necessary analytical techniques and possible physical applications is remarkably broad.

Our discussion, which is almost wholly *mechanistic*, proceeds mainly along two tracks. On one hand, the dynamics of a system of *point vortices* will occupy us extensively; on the other, the effects of finite-area vortices will frequently be addressed using uniform vortices, so-called *vortex patches*. Neither system is entirely satisfactory. Point vortices of course exist in superfluid helium (cf. Roberts & Donnelly 1974), and uniform vortices will prevail whenever the requirements of the "Batchelor-Prandtl theorem" (Batchelor 1956) are met, but we desire greater flexibility than that! Our point of view is that vortex patches provide a convenient mathematical model with which to study the effects of finite vortex cores. Point vortices offer a simplified description, valid when the vortices are concentrated and well separated, wherein just one degree of freedom is required for each vortex structure. Astrophysics, for example, employs a similar double standard, using point masses for discussions of celestial mechanics but extended fluid globes when modeling stellar structure. *Vortex sheets*, another favorite of the theorists' arsenal, only receive scant attention here (see Saffman & Baker 1979 for a detailed account). Note incidentally that all three—points, sheets, and patches—are singular solutions of the *two-dimensional Euler equation* (henceforth abbreviated as 2DEE), which is the basic governing dynamical law for most of our discussion.

PARTICLES AND FIELDS

We start by quoting a recent formal result. Consider *Burgers' equation* (cf. Whitham 1974),

$$\frac{\partial u}{\partial t} + u\frac{\partial u}{\partial x} = v\frac{\partial^2 u}{\partial x^2}, \tag{1}$$

for a complex-valued field $u(x, t)$. This equation admits solutions of the form

$$u(x, t) = -2v \sum_{\alpha=1}^{N} (x - z_\alpha(t))^{-1} \tag{2}$$

if the complex poles $z_\alpha(t)$ evolve according to[1]

$$\dot{z}_\alpha = -2v \sum_{\beta=1}^{N}{}' (z_\alpha - z_\beta)^{-1} \tag{3}$$

(Choodnovsky & Choodnovsky 1977), where N can be any integer. This remarkable result shows that the field equation (1) can be transcribed without approximation into a many-particle problem [Equations (3)] for a (wide) class of initial conditions. If $u(x, t)$ is to be real, as is usually assumed, N must be even and the z_α must constitute a set of $N/2$ complex-conjugate pairs. Since the z_α show up as simple poles if u is analytically continued to complex x, Equation (2) is usually referred to as a *pole decomposition*.

We now return to two-dimensional, incompressible flow, which reduces to a scalar field theory for the *stream function* $\psi(x, y, t)$. The governing dynamical equation is

$$\frac{\partial}{\partial t}\Delta\psi + \frac{\partial\psi}{\partial y}\frac{\partial}{\partial x}\Delta\psi - \frac{\partial\psi}{\partial x}\frac{\partial}{\partial y}\Delta\psi = v\Delta\Delta\psi, \tag{4}$$

where v is again a kinematic viscosity and Δ denotes Laplace's operator. We inquire into the possibility of analogous decompositions for this equation, motivated by the observation that flows governed by Equation (4) frequently are dominated by concentrated regions of vorticity with particlelike attributes. So far as we know, we are now much less fortunate. For Equation (4), the only analog of a pole-decomposition scheme is the classical decomposition into a system of point vortices for the inviscid case. Here the *vorticity*,

$$\zeta \equiv -\Delta\psi = \sum_{\alpha=1}^{N} \kappa_\alpha \delta(x - x_\alpha(t)) \delta(y - y_\alpha(t)), \tag{5}$$

is singular, a sum of δ-functions, one from each vortex, with the one at (x_α, y_α) having circulation κ_α. The equations of motion for these point vortices, the counterparts of Equations (3), are

$$\dot{z}_\alpha^* = (2\pi i)^{-1} \sum_{\beta=1}^{N}{}' \kappa_\beta (z_\alpha - z_\beta)^{-1}, \tag{6}$$

---

[1] A prime on Σ will mean omission of singular term(s), here $\beta \neq \alpha$.

where the asterisk denotes complex conjugation, and $z_\alpha \equiv x_\alpha + iy_\alpha$, $\alpha = 1, \ldots, N$. Equations (6) embody *Helmholtz's* (1858) *vorticity theorems* for two-dimensional flow. Again an equivalence is established between a field equation, now in two space dimensions, and a system of ordinary differential equations of arbitrary, finite size. Several derivations and discussions of Equations (6) exist in the literature. The one by Sommerfeld (1964) is particularly nice.

Notice that the singularities z_α are now imbedded in the flow field itself, whereas for Burgers' equation the poles were situated away from the x-axis (where the field u is defined). Consequently, the solutions produced by the point-vortex scheme are "weak" or "generalized," singular solutions of the 2DEE. There is apparently no reason that schemes providing a nonsingular decomposition should not exist, but so far none has been advanced. (Various approximate decompositions employing nonsingular vortices with finite cores have been used extensively for numerical computations. We return to this topic later.)

Notice furthermore that Equations (3) are *integrable*, since Burgers' equation can be solved by the Cole-Hopf transformation (see Whitham 1974). As we discuss below, Equations (6) are nonintegrable for $N \geqslant 4$. It is thus all the more remarkable that the two systems are so similar. In particular, for identical vortices the prefactors in (3) and (6) can be scaled out and the only remaining difference is the complex conjugation on the left-hand side of Equations (6)!

The main virtue of a vortex decomposition is that it yields a system with the same integrals of the motion, in particular the vorticities of individual fluid elements, as the 2DEE while employing only a finite number of degrees of freedom. In numerical simulations, this property is likely to be of considerable importance in producing reliable results.

Hamiltonian Dynamics of Point Vortices

It is a classical result, due to Kirchhoff (1876), that the equations of point-vortex motion in the unbounded plane define a *Hamiltonian dynamical system*. Indeed, Equations (6) can be written

$$\kappa_\alpha \dot{x}_\alpha = \frac{\partial H}{\partial y_\alpha}, \qquad \kappa_\alpha \dot{y}_\alpha = -\frac{\partial H}{\partial x_\alpha}, \tag{7}$$

where

$$H = -\frac{1}{4\pi} \sum_{\alpha, \beta = 1}^{N}{}' \kappa_\alpha \kappa_\beta \log|z_\alpha - z_\beta| \tag{8}$$

has been aptly called the "kinetic energy of interaction" by Kraichnan. The quantities $\kappa_\alpha x_\alpha$ and y_α are canonically conjugate coordinates and momenta. It is also known that the equations of motion for point vortices in an arbitrary domain are Hamiltonian. This more comprehensive theory, which embraces both Kirchhoff's result just quoted and a Hamiltonian theory for the motion of one vortex inside a bounded domain due to Routh, was developed many years ago by Lin (1941, 1943). Generalizations to flows on curved surfaces have recently been given by Hally (1980). Incorporating the effects of rotation leads after certain approximations to the concept of *geostrophic point vortices* (Morikawa 1960). In all cases the number of degrees of freedom equals the number of vortices and the phase space of the vortex system is simply its configuration space (up to a change of scale) with the Cartesian coordinates of each vortex being canonically conjugate. The profound implications of this fact for the statistical mechanics of a system of point vortices in a bounded container were outlined by Onsager (1949) in a remarkable paper. Subsequent developments are reviewed by Kraichnan & Montgomery (1980).

The Hamiltonian formalism may be carried further by defining a *Poisson bracket*,

$$[f, g] \equiv \sum_{\alpha=1}^{N} \frac{1}{\kappa_\alpha} \left(\frac{\partial f}{\partial x_\alpha} \frac{\partial g}{\partial y_\alpha} - \frac{\partial f}{\partial y_\alpha} \frac{\partial g}{\partial x_\alpha} \right), \tag{9}$$

for any two quantities f and g depending on the vortex positions. The equations of motion (7) then become

$$\dot{x}_\alpha = [x_\alpha, H], \qquad \dot{y}_\alpha = [y_\alpha, H], \tag{10}$$

in the usual way.

There are three well-known integrals of the motion, apart from H itself. These are

$$Q = \sum_{\alpha=1}^{N} \kappa_\alpha x_\alpha, \tag{11a}$$

$$P = \sum_{\alpha=1}^{N} \kappa_\alpha y_\alpha, \tag{11b}$$

and

$$I = \sum_{\alpha=1}^{N} \kappa_\alpha \left(x_\alpha^2 + y_\alpha^2 \right), \tag{12}$$

which arise from the invariance of H [Equation (8)] to translation and rotation of the coordinates. The Poisson bracket between H and either of Q, P, or I vanishes. Calculating the Poisson brackets between Q, P, and I

pairwise, one obtains the important result that

$$[I, P^2 + Q^2] = 0 \tag{13}$$

for arbitrary values of the vortex strengths (Aref & Pomphrey 1982; see also Novikov & Sedov 1978). A general result from Hamiltonian mechanics then immediately provides the corollary that a system of three vortices is integrable. This result is dealt with in more detail below. A system of four (and thus more than four) vortices, however, is nonintegrable. This result, which is also elaborated on later, vindicates the application of statistical mechanics to point-vortex systems.

We thus arrive at the general notion of a *threshold for chaotic behavior* in point-vortex systems. For a given flow geometry, there will be a maximum number of vortices, N_*, that can have integrable dynamics. For $N \leqslant N_*$, the vortex motion is at worst quasi-periodic in time. For $N > N_*$, the dynamics is chaotic with aperiodic behavior and extreme sensitivity to initial conditions. Examples and applications of this idea follow in subsequent sections.

For unbounded flow, $N_* = 3$, but the introduction of boundaries or the imposition of a background potential flow must in general reduce N_*. For point vortices in a half-plane or inside a circular boundary, $N_* = 2$ (Murty & Rao 1970). For a general enclosure, one expects $N_* = 1$ (Novikov & Sedov 1979a, Novikov 1980). In the following subsections, we gradually increase the number of vortices starting with $N \leqslant N_*$, then consider in detail the qualitative changes as N increases beyond N_*, and finally go to the limit $N \gg N_*$, where a variety of collective modes involving many elemental vortices merit attention. Although flows with a small number of discrete vortices ($N > N_*$) may be chaotic and thus in some sense turbulent, it seems better to reserve the term *turbulence* for chaotic flows with many degrees of freedom.

INTEGRABLE FEW-VORTEX SYSTEMS

One Vortex

A single point vortex in the unbounded plane is completely stationary. If enclosed in a boundary its motion defines a one degree of freedom Hamiltonian system that is always integrable. And although the actual trajectories followed for certain geometries of the boundaries may be of use (see references given by Lin 1943, Oberhettinger & Magnus 1949, Singh 1954, Peskin & Wolfe 1978), we must consider vortices of finite spatial extent in order to arrive at a one-vortex problem of substance. Returning to the case of unbounded flow, an infinite family of steady states presents itself: If the vorticity, ζ, is any function of the stream-function, ψ, a steady distribution of vorticity is obtained (cf. Batchelor 1967). Certain very interesting steady states have been found by this method. We note the shear-layer solution found by Stuart (1967) and for later reference the Bessel-function, dipole vortices discussed by Lamb (1932) and Batchelor (1967). Kida (1975) and Lundgren & Pointin (1977) have found a set of steady states by seeking the statistical mechanical equilibrium for a system of identical point vortices. The density of the point vortices gives a smooth vorticity profile [dependent on I; see Equation (12)] that one may expect to have special significance.

The problem of steady motion (rotation) of a vortex patch goes back to Rankine, for whom the uniform circular disk vortex is named, and to Kirchhoff (1876; also Hill 1884, Basset 1888, Lamb 1932), who discovered that a vortex patch of elliptical shape will precess with constant angular velocity. In this problem, the full two-dimensional dynamics of the vortex is reduced to the motion of the one-dimensional boundary curve, and the tremendous degeneracy due to the freedom in the vorticity distribution is eliminated. Zabusky has suggested the name *contour dynamics* for this topic. In this section, we review some of the results obtained so far in the mathematical theory (see also Zabusky 1981b).

The equations of motion for a vortex patch with vorticity ω_0 can be obtained by considering the continuum limit of Equations (6) for identical vortices. The integral over area, which arises from the right-hand side, can then be transformed using Green's theorem, and an integral equation for the motion of points on the boundary, $\mathcal{C}$, is obtained. This equation is

$$\dot{z} = -\frac{\omega_0}{2\pi} \oint_{\mathcal{C}} d\zeta \log|z - \zeta| \tag{14}$$

(Deem & Zabusky 1978, Su 1979). If the boundary is such that the radial distance from the centroid, $R(\theta)$, is a single-valued function of the polar angle θ, Equation (14) can be rewritten as

$$\frac{\partial}{\partial t} R(\theta) = \frac{\omega_0}{2\pi} \int_0^{2\pi} d\phi \, (\log \rho) R(\phi) [(Q(\theta) - Q(\phi)) \cos(\phi - \theta)$$

$$+ (1 + Q(\theta)Q(\phi)) \sin(\phi - \theta)], \tag{15}$$

where $\rho^2 = R^2(\theta) + R^2(\phi) - 2R(\theta)R(\phi)\cos(\phi - \theta)$ and $Q(\theta) = \partial \log R(\theta)/\partial\theta$. For a uniformly rotating boundary, $\partial R/\partial t = \Omega \, dR/d\theta$. Deem & Zabusky (1978) discretized the contour integral in (15) and numerically searched for steady states. They found the ellipse of Kirchhoff but also shapes with three, four, and five bulges that look like regular polygons with rounded corners and curved sides (see Figure 1a). Their investigation suggests that there exist steady states, which they call *V-states*, with m bulges for any integer m ($m = 2$ corresponds to the ellipse). Essentially, Equation (14) defines the law of propagation for a wave motion on the boundary of the vortex. The steadily rotating vortices correspond to standing waves. Burbea & Landau (1982), whose work is described later, suggest the name *Kelvin waves*.

In order to find analytical approximations to the bounding contours, Su (1979) proposed a scheme in which the function $R(\theta)$ is expanded in a Fourier series and successive expansion coefficients are calculated perturbatively. This method has recently been extended by Malafronte (1981), and certain infinite-order approximants to the contour shape have been found by summing the dominant terms in such series exactly. The results are expressed in terms of generalized hypergeometric functions. For a vortex with three bulges ($m = 3$), for example,

$$R(\theta) \cong (6\beta\cos\theta)^{-1}\left\{1 + 2\sin\left(\frac{2}{3}\sin^{-1}(\beta\sqrt{27}\cos3\theta) - \frac{\pi}{6}\right)\right\}, \qquad (16a)$$

where β is related to the frequency of rotation, Ω, through another series

$$\Omega = \omega_0\left(\frac{1}{3} - \frac{1}{2}\beta^2 - \frac{11}{2}\beta^4 - \frac{13477}{96}\beta^6 + \dots\right). \qquad (16b)$$

Using these formulae, and similar ones for $m = 4$, to first calculate β given Ω and to then calculate $R(\theta)$, Malafronte (1981) obtains agreement (to within 3% in the worst case) with the values of area, circumference, and minor and major axes obtained numerically by Deem & Zabusky (1978).

The development of other steady solutions from the circular vortex as the rotation rate is increased suggests casting the problem in the language of *bifurcation theory*. This has been done by McKee (1981), working from Equation (14) (see also Burbea 1980). McKee also considers the problem of bifurcations from the elliptical vortex, a problem addressed many years ago by Love (1893), who found the values of the eccentricity at which such bifurcations might occur.

The most elegant and possibly deepest view of the problem, however, is due to Burbea (1980, 1982a,b; also Burbea & Landau 1982) who employs tools from the theory of complex functions. In particular, he considers the conformal mapping that maps the exterior of the unit disk onto the exterior of the steady-state vortex. This mapping is now expanded in terms of *Faber polynomials*, and the entire theory can ultimately be written as a system of nonlinear equations between the *Grunsky coefficients* of the appropriately normalized mapping function. The reader is referred to the papers cited for technical details. With this analytical apparatus in place, the problem of stability can be formulated in terms of well-defined linear operators (Burbea & Landau 1982), and perturbation schemes for calculating the boundary shapes can be developed (Burbea & Hebert 1982).

One of the important results to emerge from these researches is that the range of angular frequency over which a steady state with $m \geqslant 3$ can exist is finite. As shown by Kelvin (see Lamb 1932, p. 231), the infinitesimal wave perturbation with m bulges has angular frequency $\Omega^{(m)} = (\omega_0/2)(m-1)/m$. The finite-amplitude Kelvin waves with m bulges have frequencies Ω in the interval $\Omega^{(m-1)} \leqslant \Omega \leqslant \Omega^{(m)}$ for $m \geqslant 2$. It was originally reported in the literature that the waves occupy this entire band of frequencies (see Landau 1981, Zabusky 1981b, Burbea 1982b, Burbea & Landau 1982) as is in fact true for $m = 2$ (Kirchhoff 1876). It was also

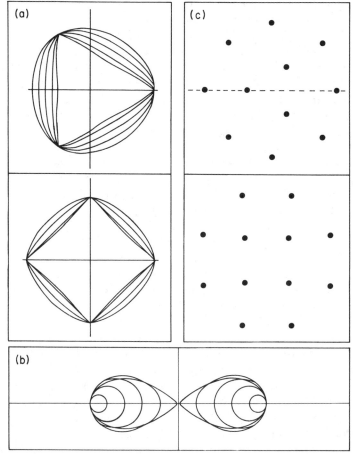

Figure 1 Vortex configurations in steady rotation (*a*) Kelvin waves, five with $m = 3$, four with $m = 4$ [Note that according to recent results of Wu et al. (in preparation) the innermost contours are spurious.]; (*b*) five different doubly connected, rotating *V*-states (after Zabusky 1981b); (*c*) point-vortex patterns 11_1 and 12_3 in Campbell & Ziff's (1978) catalog. "Ring-numbers" are 3-8 and 4-8, respectively, but 11_1 only has an axis of symmetry (dashed line).

reported that one has linear instability over exactly three quarters of this frequency range: if $\Omega_c^{(m)} \equiv (3\Omega^{(m)} + \Omega^{(m-1)})/4$, we have instability for $\Omega^{(m-1)} < \Omega < \Omega_c^{(m)}$. For $m = 2$, this result is due to Love (1893). However, a very recent manuscript by Wu, Overman & Zabusky (in preparation) raises serious doubts about these conclusions (for $m \geqslant 3$) and claims that the frequency domain of existence is much smaller than originally stated. In fact, the detailed study by Wu et. al. (in preparation) suggests that the innermost contours shown in Figure 1a are spurious, due largely to numerical effects.

A significant phenomenon observed in numerical computations of the initial-value problem for a perturbed steady-state vortex patch (Deem & Zabusky 1978, Landau 1981, Zabusky 1981b), one that is probably related to the finite domain of existence for the steady states, is the feature of *filamentation*. This is essentially the ability of the surface waves on the boundary contour of the vortex to amplify and ultimately break, ejecting a thin arm of vortex fluid into the surrounding potential flow. Presently, it is not clear whether this process leads to the appearance of a singularity on the vortex boundary and thus to a kind of ill-posedness of the initial-value problem. Some of the numerical calculations show the thin ejected filament beginning to roll-up like a finite vortex sheet. In the *Kelvin-Helmholtz instability* problem for an infinite, periodic sheet, singularities probably do appear after a finite time (Moore 1979, Meiron et al. 1982). The techniques used in these two papers would seem applicable to the problem of filamentation in contour dynamics.

Two Vortices

The motion of two point vortices in unbounded flow is discussed in textbooks (Lamb 1932, Sommerfeld 1964, Batchelor 1967). Exact solutions with two vortices of opposite circulations placed symmetrically with respect to an axis of symmetry of the boundaries, modeling the interaction of a vortex pair with an orifice for example, have appeared in the literature (Greenhill 1878, Karweit 1975, Sheffield 1977). Such solutions are atypical in the sense that integrability (and hence "solvability") only results by exploiting a *discrete symmetry* of the boundaries and/or initial configuration. For a general boundary (not a circle or straight line), the two-vortex problem is believed to be nonintegrable. In our earlier notation, $N_* = 1$. However, we postpone a discussion of nonintegrable behavior until it appears for unbounded flow ($N_* = 3$), and in this section we concentrate again on phenomena associated with finite cores.

The case of uniform vortices may be treated using the method of contour dynamics, and steady states can be calculated numerically. These represent *desingularized* versions of two point vortices. In particular, they rotate (without change of shape) about the center of vorticity when the total circulation is nonzero, and they translate uniformly when the two patches have opposite circulations. Such vortex pairs have been found by Deem & Zabusky (1978), Pierrehumbert (1980), and by Wu et al. (in preparation). Particularly interesting are a family of solutions, found by Saffman & Szeto (1980) (see also Zabusky 1980, Landau 1981), with two vortex patches of the same circulation rotating steadily about the centroid halfway between them. They range from a pair of small, distant, nearly circular disks to a pair of teardrop-shaped vortices forming a "figure eight" (see Figure 1b).

There are several numerical experiments in the literature (e.g. Roberts & Christiansen 1972, Christiansen 1973, Rossow 1977, Zabusky et al. 1979) in which two uniform vortices are followed forward in time. The initial conditions usually chosen contain two circular disks of vorticity of radius R with a center-to-center separation $D(>2R)$. If I denotes the continuum limit of the invariant in Equation (12) for a single disk, calculated with the origin at the disk center, the outcome of this initial-value problem depends critically on the value of the dimensionless parameter $\lambda \equiv (\Gamma D^2 / 2I)^{1/2} = D/R$, where Γ is the total circulation of one of the disks. It is found that for large D, the vortices rotate about one another for several periods. For smaller D, on the other hand, the vortices "quickly" (on a time scale D^2/Γ) wrap around one another and a single region of vorticity develops. The crossover occurs at $\lambda = 3.5$, approximately, so the two initial disks need not touch at all. Rossow (1977) introduced the apt term *convective merging* to describe this process. Note that it is apparently a mechanism available within the framework of the 2DEE; it does not require any viscous effects. For nonuniform vortices, a similar phenomenology is observed but the value of λ at the crossover depends on the vorticity profile in the initial disks.

Convective merging survives as an important mechanism in vortex rows, where it is usually called *pairing* (Winant & Browand 1974), and it plays an essential role in our understanding of the dynamics of vortex structures in a shear layer as is discussed later. It should be emphasized that the resulting single vortex due to a merging or pairing event is in principle much more complicated than the vortex patches from which it came. Due to the conservation laws of the 2DEE, the merged vortex must have a structure like a "jelly-roll," with entrained irrotational fluid wound in tightly with the rotational vortex fluid from the two constituents. Viscosity will smooth out this structure.

In spite of its importance, our understanding of why this merging process takes place at all is still imperfect. An appeal to statistical mechanics is helpful: The work of Lundgren & Pointin (1977), for example, shows only single-region vorticity distributions as statistical equilibra. Hence one can argue that a two-region initial state will try to equilibrate to a single-region final state. The phenomenon of merging also illustrates very clearly two competing but interconnected mechanisms in two-dimensional flow: on one hand, the generation of larger scales from smaller scales (two small vortices fuse to one large vortex), which in two-dimensional turbulence is responsible for the *inverse energy cascade* (see Kraichnan & Montgomery 1980 or Rhines 1979 for a discussion); on the other hand, the drawing out of long spiral arms ("induced" filamentation!) as the vortices wrap around each other is the real space agent of the *enstrophy cascade*, discussed in wave-number space by Batchelor (1969) and Kraichnan (1967) many years ago. See also the seminal paper by Fjørtoft (1953) and its refinement by Merilees & Warn (1975).

It would be most helpful to have an exact solution of the dynamical equations describing vortex merger or pairing. And here the "figure eight" equilibria of Saffman & Szeto (1980) and others may have a significant role to play, since it seems possible to elucidate from them the merging mechanism by performing a perturbation analysis. Careful numerical experiments in this vein have recently been performed by Overman & Zabusky (1982a). It is unfortunately also possible that the inviscid problem of vortex merging for vortex patches is ill-posed and that a small viscosity must be introduced to avoid singularities after a finite time.

Three Vortices

The motion of three point vortices in unbounded flow is integrable, as we have seen, and the problem of three vortex patches has hardly been investigated. In this section, we therefore concentrate on point vortices. We describe the results available for the three-vortex problem and discuss their significance. Recent work on this problem originates from a paper by Novikov (1975) in which the motion of three identical vortices is considered. The essential step is to shift attention from the individual, "absolute," vortex positions and focus on their relative configuration by considering the *separations* $\ell_{\alpha\beta} = |z_\alpha - z_\beta|$, where $1 \leqslant \alpha < \beta \leqslant 3$. It is possible from Equations (6) to derive equations of motion for the $\ell_{\alpha\beta}$. For an arbitrary number of vortices N, these are

$$\frac{d}{dt}\ell_{\alpha\beta}^2 = \frac{2}{\pi}\sum_\gamma{}''\kappa_\gamma\sigma_{\alpha\beta\gamma}A_{\alpha\beta\gamma}\left(\frac{1}{\ell_{\beta\gamma}^2} - \frac{1}{\ell_{\gamma\alpha}^2}\right),\tag{17}$$

where the sum extends over all vortices $\gamma \neq \alpha, \beta$; $A_{\alpha\beta\gamma}$ is the area of the triangle with vertices at z_α, z_β, and z_γ; and $\sigma_{\alpha\beta\gamma}$ is its orientation ($+1$ for z_α, z_β, and z_γ appearing counterclockwise in the plane). Clearly the orientations will only change sign when a vortex triangle collapses to a line segment. Except for such occurrences, Equations (17) form a closed dynamical system giving the evolution of the separations. For general N, there are $\frac{1}{2}N(N-1)$ different separations $\ell_{\alpha\beta}$ but only $2N-3$ of these are independent. There are two integrals involving only the separations, viz. H [Equation (8)] and

$$\frac{1}{2} \sum_{\alpha,\beta=1}^{N} \kappa_\alpha \kappa_\beta \ell_{\alpha\beta}^2 = \left(\sum_{\alpha=1}^{N} \kappa_\alpha \right) I - Q^2 - P^2 \qquad (18)$$

[cf. Equations (11), (12)].

It is clear from the form of Equations (17) that the three-vortex problem is of particular significance. If we think of the vortex separations in a many-vortex system as defining the *scales of motion* in the flow, Equations (17) show that the three-vortex interaction is the lowest-order dynamical process capable of exciting new scales. Furthermore, the rate of change of a separation is decomposed through Equations (17) into a sum of contributions, each arising from a *triplet interaction*. It follows, essentially from this observation, that both the quantity in Equation (18) and H are conserved individually by every interacting vortex triple so that Equations (17) have what may be called *detailed conservation laws*. Similar properties of the Euler equation when written in terms of interacting wavelet triads have been stressed by Kraichnan (1973). It is interesting that both the basic wavelet triad and the vortex triple define integrable dynamical systems. Chaotic behavior and, ultimately, turbulence arise from the couplings between these integrable constituents.

IDENTICAL VORTICES   There are two kinds of steadily rotating configurations. In the stable one, discussed by Kelvin long ago (see Kelvin 1910), the vortices are at the corners of an equilateral triangle. The second type, which is unstable, has the vortices arranged on a line, with one halfway between the other two. These equilibria essentially determine the structure of the system phase space. States that start as not too large perturbations of the triangle pulsate and rotate but never collapse to a collinear configuration. The orientation σ_{123} is an integral of the motion for such initial conditions. For larger values of the energy (Hamiltonian), the vortices do become collinear. The relative motion is periodic in time except right at the crossover from one regime to the other, where the vortices asymptotically relax to one of the collinear equilibria.

The above results were found by Novikov (1975), who also gave explicit integral formulae for oscillation periods, etc., using a geometrical

approach. The essential ingredient of Novikov's analysis is the interpretation of the triple $(\ell_{12}, \ell_{23}, \ell_{31})$ as a point in three-dimensional space,
constrained by (8) and (18) to lie on the intersection of two surfaces.
There is a further "non-holonomic" constraint from the fact that ℓ_{12},
ℓ_{23}, and ℓ_{31} are sides of a triangle and therefore must obey triangle
inequalities. The end result is that the point $(\ell_{12}, \ell_{23}, \ell_{31})$ in *separation
space* must remain inside some physically meaningful region and there
move along one of a family of trajectories that arise as the intersections
of level surfaces for the integrals of motion. Novikov's (1975) analysis has
been extended to vortices on a nonrotating sphere by Bogomolov (1979).

A recent development, due to Aref & Pomphrey (1980, 1982), is to
apply the methods of classical mechanics, in particular the theory of
canonical transformations, to the N degrees of freedom Hamiltonian
system (7)–(8). For arbitrary N ($\geqslant 3$), a sequence of canonical transformations can be found that reduces this system to a Hamiltonian with
$N - 2$ degrees of freedom. The procedure makes use of the involution
property (13). An analogous method has been found by Khanin (1982)
for $N = 4$. For $N = 3$, a system with one degree of freedom emerges. The
effective Hamiltonian can be interpreted as describing a certain nonlinear
oscillator with an amplitude (the action variable in the formalism)
proportional to the product of orientation and area of the vortex triangle.
A compact solution of this oscillator problem is found in terms of Jacobi
elliptic functions (Aref & Pomphrey 1982).

ARBITRARY CIRCULATIONS   The geometrical approach of Novikov (1975)
was generalized by Aref (1979) to cover three-vortex systems with arbitrary circulations. It is possible to perform all the geometrical constructions in a plane by introducing a system of *trilinear coordinates* as is
suggested by (18) for $N = 3$, viz.

$$\kappa_1 \kappa_2 \ell_{12}^2 + \kappa_2 \kappa_3 \ell_{23}^2 + \kappa_3 \kappa_1 \ell_{31}^2 = \text{constant}. \tag{19}$$

The three trilinear coordinates are each proportional to one of the
summands on the left-hand side. The triangle inequalities restrict the
physically realizable portion of the phase plane to a certain subregion
dependent on κ_1, κ_2, and κ_3. Curves corresponding to constant H
[Equation (8)] can now be constructed. The reader is referred to the
original paper (Aref 1979) for examples of such phase-plane plots.

The overall topology of the resulting diagram depends again essentially
on the positioning and type of singular points corresponding to steady
states. Analysis shows that the qualitative appearance of the entire phase
plane is governed solely by the different combinations of signs of the

three symmetric functions:

$$s_1 \equiv \kappa_1 + \kappa_2 + \kappa_3, \tag{20a}$$

$$s_2 \equiv \kappa_1\kappa_2 + \kappa_2\kappa_3 + \kappa_3\kappa_1, \tag{20b}$$

$$s_3 \equiv \kappa_1\kappa_2\kappa_3, \tag{20c}$$

with the sign of s_2 being of major importance.

VORTEX COLLAPSE   The special case when $s_2 = 0$ merits particular attention. The phase-plane diagram then reveals that it is possible for the three vortices to move in such a way that the triangle spanned by them does not change its shape. In particular, the three vortices may collide at the center of vorticity after a finite time (Aref 1979). Since they are traced out by a sequence of similar triangles, the vortex trajectories in this case are logarithmic spirals. The constant in (19), a measure of the angular momentum of the system, must equal zero for this phenomenon of *vortex collapse* to occur (infinite self-similar expansion is also possible). Novikov & Sedov (1979b) have considered generalizations of such motions to systems of four and five vortices, with the circulations satisfying the analogous condition $\sum_{i \neq j} \kappa_i \kappa_j = 0$. However, the significance of this type of motion, in particular its relation to convective merging, is still unclear.

PAIR-POINT COLLISIONS   The rich phenomenology that is observed in the interactions of vortex rings (Kambe & Takao 1971, Fohl & Turner 1975, Oshima 1978, Saffman 1981) has the dynamics of vortex pairs as its more modest counterpart in two dimensions. The mutually induced velocities of the two constituents in a pair of opposite point vortices give it qualities reminiscent of a classical particle. It is thus natural to inquire into the fate of a vortex pair when it encounters other vortices, and the simplest problem is that of the interaction with a single vortex of the same circulation as one of those in the pair. This is simply the three-vortex problem with $\kappa_1 = \kappa_2 = -\kappa_3$, and it is amenable to the type of analysis described above. Complete quantitative details are given in Aref (1979). The qualitative description is worth recalling here. We start with a vortex pair, vortices 1 and 3 say, far removed from the "target" vortex 2. Assume that 13 is headed in the general direction of 2, that initially (really as time tends to $-\infty$) $\ell_{13} \equiv d$, and that the distance of 2 from the line of free propagation of 13 is ρ. In analogy with the collision dynamics of massive particles, we may think of ρ as an impact parameter. It is then intuitively clear that if $\rho \gg d$, the pair will translate essentially as if vortex 2 were absent and will only experience a slight deflection of its path. On the other hand, when ρ and d are comparable, all three vortices come into close contact, and the analysis shows that the target (vortex 2) may pair up with the incoming vortex of opposite circulation (vortex 3) and escape with it to infinity, leaving its initial partner (vortex 1) behind. Using an easily understood symbolic notation, $13 + 2 \rightarrow 1 + 32$. For obvious reasons, this second mode is referred to as an *exchange collision* and the first mode as a *direct collision*. At the crossovers from direct to exchange collision, the system cannot "decide" how to match up the vortices to form a freely moving pair, and the result is that all three vortices asymptotically become trapped in a steadily rotating configuration, which is clearly unstable. A completely analogous phenomenology occurs in the collision dynamics of two (and in fact an arbitrary number of) vortex pairs, as we shall see.

It should be emphasized that the techniques developed so far analyze the *relative* motion of three vortices. The *absolute* motion can be obtained by quadratures, but only in special cases do we have detailed information on individual particle trajectories. The article by Charney (1963) provides a few numerically generated examples and also a discussion of applications to atmospheric hydrodynamics.

HISTORICAL NOTE: WORK OF GRÖBLI AND SYNGE   While preparing this article I became aware that much of the recent work on the three-vortex problem was predated by two important but almost entirely forgotten papers. The first is the Göttingen dissertation by Gröbli (1877), which is referred to in later editions of Kirchhoff's lectures, by Love (1901) and also by Lamb (1932), though always with a qualifying remark, e.g. that the paper just contains "other interesting examples of rectilinear vortex systems." Gröbli's 86-page work deserves much more credit than that! He first shows explicitly how the three-vortex problem can be reduced to quadratures for arbitrary circulations. It follows, of course, that the system is integrable, a fact pointed out independently by Poincaré (1893) in his treatise on vortex motion. [The formal proof through Equation (13) is a "modern" statement.] Gröbli (1877) then goes on to consider a series of special cases. He discusses identical vortices, where he describes precisely the construction that Novikov (1975) was to rediscover a century later! He gives integral formulae similar to Novikov's. [The recent solution in terms of elliptic functions (Aref & Pomphrey 1982) is apparently new.] Gröbli (1877) also discusses the pair-point collision problem and describes the phenomenology of direct and exchange collisions. He shows vortex trajectories corresponding to one of the asymptotic trappings in an illustration. He also considers the case $\kappa_1 = 2\kappa_2 = -2\kappa_3$, calculating an example of the collision between a neutral vortex pair and a single vortex of larger circulation.

Gröbli (1877) goes on to determine all possible steadily rotating configurations. He then seeks motions in which the vortex triangle contracts (or expands) in size without change of shape. This quickly leads to the phenomenon of vortex collapse, which he illustrates.

Neither Gröbli nor his contemporaries could know how much of the problem had actually been uncovered by these particular case studies. The general classification scheme described above (Aref 1979) seemed to provide an answer to this more difficult question for the first time. However, writing for the first volume of the *Canadian Journal of Mathematics*, Synge (1949) had already addressed precisely the issues left open by Gröbli's (1877) work, of which he was well aware.[2] His analysis again makes use of a geometrical phase-diagram approach but instead of surfaces intersecting in three dimensions, Synge proposes "another and better representation by trilinear coordinates in a plane." (His actual definitions differ from those mentioned above.) The important role of the sign of s_2 [Equation (20b)] is clearly described by Synge (1949), and the general topology of his phase-plane diagrams is discussed in terms of this parameter. In summary, there is little in the recent papers by Novikov (1975) and Aref (1979) to surprise the conscientious student of the works by Gröbli (1877) and Synge (1949)! I hope that these two overlooked contributions to vortex motion theory may finally begin to receive the recognition they deserve.

Vortex Polygons and Related Equilibria

It is of considerable interest to determine which of the motions described for two or three vortices can be generalized to systems with an arbitrary number of vortices, since such results give us insight into the structure of the accessible phase-space region for a many-vortex system. A relatively simple class of motions for N vortices are those in which the configuration as a whole rotates or translates without change of shape. We have already discussed such motions for one or two vortex patches and for three point vortices. Here, we review work on rigidly moving configurations with several point vortices. Kelvin (1910) referred to this general topic as *vortex statics*. We call configurations of this type *vortex equilibria*.

The case of equilibria with identical vortices was promoted by Kelvin in connection with his (erroneous) theory of "vortex atoms." Thomson (1883) in his Adams Prize essay considered *vortex polygons*, configurations with N identical point vortices at the corners of a regular N-gon. He proved the first version of a theorem, since named for him, that if $N > 7$,

---

[2] I am indebted to R. G. Littlejohn for pointing out Synge's work.

such an equilibrium is unstable to infinitesimal perturbations. A natural extension is to place a vortex (of arbitrary circulation) at the center. Several studies of the stability of this type of equilibrium have appeared. The reader is referred to Havelock (1931), Khazin (1976), Mertz (1978), and the comprehensive study by Morikawa & Swenson (1971). (See also Bauer & Morikawa 1976.)

For large N (and identical vortices), polygons, whether face-centered or not, become unstable. On the other hand, a simple variational argument shows that there must exist at least one stable equilibrium for every N. Campbell & Ziff (1978, 1979) have performed an exhaustive numerical search for these equilibria. They claim to have found all stable (and some unstable) equilibria for $N \leqslant 30$ (and for $N = 37$ and $N = 50$). Their results are collected in a fascinating catalog that contains pictures as well as characterizations in terms of *ring numbers*. It turns out, as one might expect, that the vortices are arranged approximately on concentric circles, "atomic shells" in Kelvin's metaphor, and the numbers of vortices on each circle or ring give apt characterizations of the visual impressions created by the equilibria. However, closer inspection reveals that many of them do not have rotational symmetry at all. In fact, all of the configurations have an *axis of symmetry* but many apparently do not have any higher symmetry (see Figure 1c). At the present time, our analytical understanding of these interesting states is incomplete. Campbell (1981) has extended the investigation to stability properties of the numerically determined equilibria. This work was primarily motivated by experiments on vortices in superfluid He, performed by Yarmchuk et al. (1979), in which uniformly rotating equilibria with $N = 2,\ldots,11$ were visualized. Palmore (1982) considers the application of *Morse theory* to estimate the number of all equilibria as a function of N.

Unstable equilibria with the vortices on a line exist for every N. For identical vortices, the positions along the line are given by the zeros of the Nth *Hermite polynomial*, a result discovered long ago by Stieltjes (for point charges on a line) and rediscovered by Calogero (1978) in connection with extensive researches on the zeros of classical polynomials and integrable many-body systems. These equilibria provide finite N versions of the infinite row of equally spaced vortices ubiquitous in discussions of shear-layer dynamics. Their stability can be determined analytically using matrix and eigenvalue relations given by Bruschi (1979).

Zabusky (1981b) has conjectured that to every point-vortex equilibrium there corresponds a family of *desingularized* steady states made up of vortex patches. To obtain these, one replaces the point vortices by tiny disks with the appropriate vorticities so that corresponding circulations in the two systems agree. The disks will remain approximately circular so long as they are very small. As they increase in size their shapes become distorted. So far, this conjecture "works" for vortex pairs and also for certain periodic configurations such as vortex rows and vortex "streets." Finite-area versions of these latter cases were obtained by Saffman & Szeto (1981) and Saffman & Schatzman (1981).

Conversely, as the number of constituents, N, tends to infinity, point-vortex equilibria converge to finite-area vortices, or vortex sheets if the points are on a line, in steady motion. However, all the configurations found by Campbell & Ziff (1978, 1979), for example, must tend to finite-area vortices with circular symmetry, since the trajectories of the elemental point vortices are circles. Hence, there are steady states of uniform vorticity, e.g. Kirchhoff's ellipse, that do not emerge as a limit of point-vortex equilibria.

CHAOTIC FEW-VORTEX SYSTEMS

The realization that deterministic dynamical systems with very few degrees of freedom can display chaotic behavior has had a profound influence on our theoretical understanding of transition to turbulent flow in various contexts. The seminal paper by Lorenz (1963) has led to a large body of theoretical and experimental work centered around the concept of a *strange attractor*. A recent review of this field of investigation is due to Lanford (1982). Since essentially all the models are dissipative, as are most systems studied in fluid mechanics, and since only a few modes are retained, which are thus all substantially affected by the dissipation, these model systems pertain to *transitional flows*.

There is a parallel and contemporaneous development for nondissipative, primarily Hamiltonian, systems with close ties to celestial mechanics. A seminal paper here is the study by Hénon & Heiles (1964) in which nonintegrability is described for a certain Hamiltonian system with two degrees of freedom, but many of the ideas go back to Poincaré. General discussions of nonintegrable or chaotic behavior in conservative systems may be found in the monographs by Arnold & Avez (1968) and Moser (1973). The proceedings edited by Jorna (1978) and Helleman (1980) contain much additional information on chaos in both dissipative and nondissipative systems. In this section, we review the concept of *chaotic motion* as it relates to vortex dynamics. In the next section, we argue that some of these results are relevant to *turbulent flows*.

Four Vortices

It is well known that a promising way to produce a chaotic dynamical system is to perturb (i.e. couple to new degrees of freedom) an integrable system with two unstable states connected by phase-space trajectories, a so-called *saddle connection* (see the article by M. V. Berry in Jorna 1978; see also Moser 1973, Holmes 1980).

As described earlier, the problem of three identical point vortices has three collinear equilibria, which may be designated 123, 312, and 231 according to the positions of the vortices on the line. (Half a period of rotation later, 123 becomes 321.) The phase-plane analyses of this problem show that 123 can evolve into 231, and conversely 231 can evolve into 123 for certain infinitesimal changes in the initial state that preserve the integrals of motion. Hence these collinear equilibria form saddle connections of the aforementioned type. Synge (1949) commented that "these oscillations between configurations which differ only through interchange of vortices of equal strength appear rather interesting."(!) Thus we expect that adding a time-dependent velocity field to the three-vortex system will in general produce chaotic dynamics.[3] The question arises whether coupling to a fourth vortex is sufficiently "general." Examples such as Equations (3) show that the answer is not obvious a priori.

The arguments just given motivate consideration of a system of three identical vortices perturbed by a fourth vortex with very small circulation. However, the first work to appear (Novikov & Sedov 1978) dealt immediately with the case of four identical vortices.[4] In that paper, chaotic behavior was suggested on the basis of numerical experiments that check the rate of separation of phase-space trajectories for very close initial conditions. Exponential separation is a hallmark of nonintegrable behavior. A more revealing test is the construction of a *Poincaré section*, which in this problem can be done in the following intuitively appealing way: We return to the notion of a space of separations, which now consists of sextuples $(\ell_{12}, \ell_{13}, \ell_{23}, \ell_{14}, \ell_{24}, \ell_{34})$. We then note that there are three constraints on such a sextuple. First, by Euclidean geometry there are only five independent distances between four points in a plane. Second, there is the dynamical conservation law (18), and third, the system conserves H [Equation (8)] as it evolves. Thus, the available region in *separation space* is at most a three-dimensional manifold. Sections through it obtained by recording the values of the pair (ℓ_{13}, ℓ_{23}) whenever ℓ_{12} increases through some preselected value will then tell us, in the usual way, whether the system is integrable or not (see Jorna 1978 for background).

A section of this type was calculated numerically by Inogamov & Manakov (1979, unpublished), who did not draw a definite conclusion,

---

[3] P. J. Holmes immediately made this suggestion after a seminar I gave in the fall of 1977.
[4] But see the appendix by Ziglin to the recent article by Khanin (1982).

and a slightly different section was computed by Novikov & Sedov (1979a) that again suggested chaotic behavior. A comprehensive investigation by Aref & Pomphrey (1980) shows both a section with points arranged on several apparently regular *tori* and another at higher energy with completely *chaotic splatter* of section points.

IDENTICAL VORTICES AND NONLINEAR OSCILLATORS   The reduction of degrees of freedom by successive canonical transformations, which was mentioned briefly in connection with the problem of three identical vortices, was developed primarily with the four-vortex problem in mind (Aref & Pomphrey 1980, 1982; also Khanin 1982). In terms of the action-angle variables of this theory, the effective Hamiltonian governing the motion is remarkably similar in form to Hamiltonians that have been studied in connection with model systems of *coupled nonlinear oscillators* (see the article by J. Ford in Jorna 1978). There is a reasonably complete body of theory describing the appearance of chaotic behavior in such systems in terms of a phenomenon known as *resonance overlap* (Chirikov 1979).

It turns out in the four-vortex problem that a certain subset of solutions can be obtained exactly. These solutions again show saddle connections in which a collinear state, e.g. 1234, is connected by two phase-space trajectories to either 2143 or 1324. In the three-vortex problem, the available degrees of freedom were exhausted in producing such a topology. For four vortices, on the other hand, the saddle connections reside in a subspace defined by the *discrete symmetry* $z_1 = -z_2$, $z_3 = -z_4$ (inversion symmetry about the origin), which is preserved by the equations of motion. Breaking this symmetry introduces the full-fledged four-vortex problem, which in turn is chaotic. The reader is referred to the original paper (Aref & Pomphrey 1982) for a complete discussion.

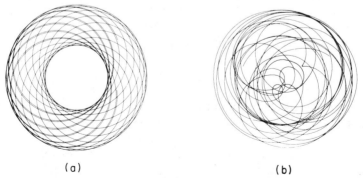

(a)   (b)

Figure 2   Trajectories of one of four identical vortices in (a) regular (integrable) and (b) irregular (chaotic) motion (cf. Aref & Pomphrey 1982).

For point vortices, the actual trajectory in the (x, y)-plane directly represents the phase-space motion, as we have seen. Thus all the complexity of chaotic motion is immediately related to actual, physical trajectories. It should therefore be possible to discern at once clear differences in qualitative features of the motion, depending on initial conditions, by simply looking at the trajectories followed by one of four identical vortices. Figure 2 shows two examples. In Figure 2a, the initial state has the inversion symmetry mentioned previously and the motion is quasiperiodic. In Figure 2b, on the other hand, the initial state leads to chaotic motion and the trajectory is highly irregular.

It would appear that the simple qualitative effect exemplified in Figure 2 is potentially important for a variety of applications. For example, Novikov & Sedov (1978) propose that cyclones in the atmosphere can participate in chaotic motion and that this possibility captures the essence of the *predictability* problem usually discussed in completely different terms (Leith & Kraichnan 1972). Other examples that come to mind include the work on *vortex sound* (Powell 1964), wherein the point-vortex-model studies performed so far (Stüber 1970, Sayeed-Ur-Rahman 1971, Crighton 1972) have been concerned with integrable motion. What happens to the characteristics of the emitted sound for vortex motion in the chaotic regime? Another example is the pressure force exerted on a solid boundary by a vortex-dominated flow, recently modeled using point vortices by Conlisk & Rockwell (1981). (See also the related model study by Rogler 1978). The extent to which chaotic motion is damped by finite core effects and viscosity is an important issue that should be addressed in the future.

In the next section, we assume that the important role played by saddle connections and the notion that systems of (identical) vortices resemble coupled nonlinear oscillators survive as the number of vortices is increased, and thus that these features pertain to the dynamics of an infinite row of vortices. It is suggested that the ideas presented here (appropriately extended) are relevant to recent experiments on shear layers, in particular the controlled vortex mergings of Ho & Huang (1982).

THE RESTRICTED PROBLEM: ADVECTION OF A PASSIVE MARKER   It was argued earlier that a system of three identical vortices with a weak fourth vortex should be nonintegrable, and the following question arises: what happens as this fourth vortex is reduced to strength zero and thus becomes just another fluid particle in the unsteady potential flow, "a speck of dust" advected by the three vortices but not influencing them in any way? This problem, which Gröbli (1877) specifically writes that he will not consider (!), was called the *restricted four-vortex* problem by Aref & Pomphrey (1980) in analogy with the terminology used in celestial mechanics. The three vortices will of course move in exactly the same integrable way as they did before. The question is what the motion of the advected particle will be like. If celestial mechanics is a reliable guide, we expect nonintegrable behavior as in the restricted two-body problem described, for example, in Moser's (1973) text. This expectation was indeed borne out in the numerical experiments reported by Aref & Pomphrey (1980) and in the simultaneous appearance of an analytical proof of nonintegrability by Ziglin (1980).[5]

The analysis of the restricted four-vortex problem is related to the investigations reviewed recently by Cantwell (1981) under the heading *entrainment diagrams* and, in general, to the subject of particle trajectories in flows. Such studies are of considerable importance to topics like flow visualization, pollutant dispersal in the atmosphere or oceans, and all kinds of stirring and mixing problems. They seem to represent a class of problems in which dynamical-systems theory has a significant role to play. The point of view that emerges with particular clarity in the present discrete vortex model is that particle advection by steady flows corresponds to a kind of unforced, nonlinear oscillation and thus to integrable dynamics, whereas unsteady flows lead to forced oscillations, parametric resonance, and ultimately chaos.

VORTEX PAIR COLLISIONS   As we have seen, several families of vortex pairs that propagate without change of shape are available, and the question to be addressed now, inspired by the concept of *solitons* (Scott et al. 1973, Zabusky 1981a), is what happens to such pairs when they collide with one another? This is again a four-vortex problem with the important difference from the case of identical vortices that the available phase (or configuration) space is no longer bounded.

Collisions between pairs of point vortices have been studied by Gröbli (1877), Love (1894), Hicks (1922), Friedrichs (1966) and Acton (1976). All these studies make use of the fact that if the two pairs have a common axis of symmetry, the problem can be reduced to an integrable two-particle system. Indeed it is then just the problem of two vortices moving in a half-plane. Love (1894) analyzed the case when the two coaxial pairs move in the same direction and all vortices have the same absolute circulation. He was interested in elucidating the "leap-frogging" motion sometimes observed with two vortex rings wherein alternately one

---

[5]According to J. E. Marsden (private communication) this proof is incomplete for reasons that are briefly described in Holmes & Marsden (1982). Similar criticism applies to the extension of the proof mentioned in footnote 4.

ring passes through the other (Oshima 1978, Yamada & Matsui 1978, Maxworthy 1979, Saffman 1981). Love (1894) obtained the result that all four point vortices must always be collinear at some stage during the motion with one pair situated inside the other. He showed that the ratio of the sizes of the two pairs at this instant, λ/Λ say, is a decisive parameter. If $3-2\sqrt{2} < \lambda/\Lambda < 3+2\sqrt{2}$, the leap-frogging mode will arise. For λ/Λ outside this range, which corresponds to pairs very disparate in size at the instant of slip-through, the pairs simply separate never to interact again. This latter case then corresponds closely to the interaction of two solitons, whereas the leap-frogging case is reminiscent of a *breather mode*.

The case when coaxial pairs collide head-on is not covered by Love's (1894) analysis but may be found in Gröbli's (1877) study, and the necessary formulae can also be extracted from Acton's (1976) paper. Leap-frogging is no longer possible but there is now a crossover from a *direct*, slip-through collision to an *exchange* scattering, which qualitatively can be described by the symbolic formula $14+23 \rightarrow 13+24$. The resulting pairs (13 and 24) head away from the common axis of the initial pairs (14 and 23). The experiment by Tatsuno & Honji (1977) seems to capture half of this sequence. The time-reversed version gives the motion of a pair *impinging* on an infinite flat plate with an arbitrary angle of incidence. The case of a vortex pair impinging normally on a flat plate is shown on p. 223 of Lamb's (1932) treatise. This case was extended to vortex patches by Saffman (1979).

Consider a system with $N = 2n$ vortices making up n vortex pairs. For the case when all vortices have the same absolute value of the circulation (i.e. for $N = 4$, $\kappa_1 = \kappa_2 = -\kappa_3 = -\kappa_4$), it is again possible to reduce the number of degrees of freedom from N to $N-2$ by a sequence of canonical transformations (Aref & Pomphrey, in preparation). The technique is similar to that mentioned for identical vortices but makes use of the fact that $[Q, P] = 0$ [see Equations (11)] instead of (13). The reduced Hamiltonian contains Q and P in it as parameters. The existence of the invariant I [Equation (12)] now shows rigorously that the two-pair problem ($n = 2$) is integrable if $Q = P = 0$. This condition is equivalent to the existence of a *center of symmetry*. With the above indexing, we have $z_1 = -z_2$ and $z_3 = -z_4$, an inversion symmetry that is preserved by the equations of motion. This particular case of two pair collisions in which the total *impulse* (or momentum) vanishes was apparently overlooked in the classical literature. It is analyzed by Aref & Pomphrey (in preparation) who show that the collision dynamics again displays regimes of *direct* and *exchange* scattering. At the crossover, *asymptotic trapping* into a steadily rotating, rhombic configuration occurs. This is the smallest of a family of equilibria found long ago by Havelock (1931), who called them *double alternate rings*. These motions may be generalized to an arbitrary number of vortex pairs (Aref, submitted for publication).

The quest for solitonlike vortex pairs in geophysical flows led Larichev & Reznik (1978) (see also Berestov 1979) to a family of solutions of the quasi-geostrophic potential vorticity equation, which they called *two-dimensional Rossby solitons*, although they did not consider the properties of these distributed vorticity, dipole-eddies during collisions. The Rossby solitons need not translate at all, and in this limit they become identical to a set of vortex pairs found earlier by Stern (1975), which he named *modons*. In another limit, when the Coriolis parameter goes to zero, we recover the classical vortex pairs mentioned earlier with Bessel-function vorticity distributions in the two lobes. These are discussed by Lamb (1932) and by Batchelor (1967), who presents them as a continuum analog of the point-vortex pair. It turns out that solutions of this type are of interest in plasma physics as well, where the barotropic potential vorticity equation is sometimes known as the *Hasegawa-Mima equation*.

Collisions between Rossby "solitons" have been studied numerically by Makino et al. (1981) in the context of plasma physics, by McWilliams & Zabusky (1982) in the context of atmospheric flows, and by Aref (1980) for the 2DEE. Overman & Zabusky (1982b; see also Zabusky 1981a) have studied collisions between translating "V-states"—coaxial pairs with uniform vorticity lobes. The results of all these investigations are that such eddies are not exactly solitons but that there does exist a "slip-through" regime in which they survive collisions without too severe changes in shape and internal vorticity profile. Pairs with a more concentrated vorticity distribution survive better than the piecewise-uniform pairs. The "ringing" left over from the collision is analogous to the appearance of *dispersive tails* in certain one-dimensional field equations that support solitary waves that are not quite solitons (Bona et al. 1980). In the "leap-frogging" regime, on the other hand, convective merging between like-signed lobes takes place and, as first shown by Aref (1980), the pair collisions become *inelastic*. (In experiments, similar processes augmented by viscous effects apparently make leap-frogging of vortex rings difficult to observe; see also Maxworthy 1972, 1979.) Such behavior is completely at variance with the concept of solitons.

Returning once again to point-vortex pairs, one would expect chaotic behavior to ensue when the discrete symmetries mentioned above ($P = Q = 0$ or coaxiality) are broken. Experience with other scattering problems then indicates that one can have either *nonintegrable scattering* as in the restricted three-body problem of celestial mechanics (see Moser 1973 or M. V. Berry's article in Jorna 1978) or *integrable scattering*, although the system is not completely integrable, as in the so-called Störmer problem [Moser (unpublished) 1963, Dragt & Finn 1976]. Finally, it is also possible that this particular choice of vortex strengths leads to a completely integrable system. In our earlier notation, N_* may depend on the circulations. Analytical and numerical studies are currently underway to decide this issue (Aref & Pomphrey, in preparation).

Non - Integrability of Euler's Equation

We come now to the important question of what the discrete-vortex results imply for the 2DEE, in particular whether this equation is integrable or not [Novikov & Sedov 1978, Inogamov & Manakov (unpublished) 1979, Aref & Pomphrey 1982]. Integrability would have profound consequences for our perception of two-dimensional flows and would presumably provide considerable embarrassment to the theory of two-dimensional turbulence.

Let us first note that if the point vortices gave rise to fields ψ, ζ that were smooth, there would be no problem deciding the issue. For assuming the 2DEE to be integrable, through application of the inverse scattering method (cf. Scott et al. 1973) or some generalization thereof for example, the restriction to fields ψ, ζ that are vortex-decomposed would a fortiori be integrable. But for such fields, the 2DEE reduces to the vortex equations of motion, which we know to be nonintegrable. We used a variant of this argument for Burgers' equation to argue that Equations (3) are integrable, although this can also be shown independently.

However, as commented upon earlier, the point-vortex decomposition is singular and one may worry that the nonintegrability observed in the four-vortex problem is a peculiar feature of this singular system. As a warning, one can point to the vortex-collapse solutions that would seem to suggest that for certain three-vortex initial conditions, the 2DEE can have a singularity after a finite time. But such a conclusion is definitely incorrect, for, according to a classical theorem by Wolibner (1933), smooth initial data of this form give rise to solutions of the 2DEE that are regular for all time!

It is possible, however, to base a rather strong argument against integrability of the 2DEE on the idea of *desingularization* mentioned earlier: Suppose we take initial states of four identical point vortices but replace the δ-function singularities by very peaked, yet arbitrarily smooth, approximating functions. Such functions exist that are only nonzero on a disk of arbitrarily small radius but still have derivatives of all orders. The initial state just constructed is obviously as smooth as needs be and, on the other hand, is as close to the four point-vortex system as we desire. It seems to follow (but this requires a formal proof) that by choosing the radii of the disks as small as necessary, we can make this smooth distribution of vorticity follow the point-vortex system for as long as we wish. We conclude that for such initial states, the 2DEE has transients of arbitrarily long duration that are indistinguishable from chaotic solutions. This argument guarantees nonintegrability at a pragmatic level, but I suspect more is true. Other comments may be found in Aref & Pomphrey (1982).

A couple of recent papers pointing in different directions are due to Olver (1982) and Ebin (1983). These are both highly mathematical studies that work directly with the partial differential equation and make little reference to vortices. For a general discussion of integrability, chaos, turbulence, and finite-time singularities in fluid mechanics, see the recent review by Frisch (1982).

THE MANY-VORTEX PROBLEM

As the number of elemental point vortices is increased, individual trajectories become of less concern and attention is shifted to collective modes exhibited by groups of many vortices. For simple initial conditions, such as a band of vorticity of one sign or two bands of opposite circulation, these collective modes take the form of clusters or regions of vorticity that, like galaxies emerging from a system of point masses, take on an identity of their own and become the basic building blocks of the flow. We now have a chaotic dynamics with many degrees of freedom, a flow situation that we previously agreed to call *turbulent*. But the surprise is that this regime does not correspond to a completely disorganized distribution of vorticity. Rather we find flows with considerable structural features. As the reader most likely knows, the stimulus for studying the two-dimensional case has come from the experimental discovery of so-called *coherent vortex structures* in certain laboratory and geophysical shear flows. There is a large and growing literature on this subject (see the review by Cantwell 1981; also see Roshko 1976, Kovasznay 1978, and Siggia & Aref 1980 for varied points of view on the interplay between "statistics" and "structures").

It is important to stress that the purely two-dimensional problem, which is our main concern here, differs in essential ways from its "real life," geophysical or laboratory counterpart. The point is that the "real" flows usually have scales of motion that are fully three-dimensional and the action of these back on the two-dimensional large scales is not captured correctly in two-dimensional hydrodynamics. It is unfortunately still necessary to state this distinction so explicitly, since the literature contains recent claims concerning the applicability of two-dimensional flow solutions to turbulent shear flows with three-dimensional small scales. In my view, most such claims are exaggerated and misleading.

Nevertheless, the subject of two-dimensional, turbulent, shear flows would be devoid of physical interest were there not certain points of contact with experiment. Indeed, when the "real" flow is essentially two-dimensional on all scales, theoretical ideas and numerical results from two-dimensional hydrodynamics do apply, as we shall see. These instances ultimately provide the motivation for pursuing the subject. As Rhines (1979) put it, writing about geostrophic turbulence, "Without this kinship, two-dimensional turbulence would be of less interest, a kind of stochastic complement to 19th century hydrodynamics."

Our restriction to shear-flow situations in this section reflects the historical fact that homogeneous, isotropic, two-dimensional turbulence was formulated and is usually discussed wholly in terms of statistical theory (Batchelor 1969, Kraichnan 1967, Leith 1968). Later attempts (e.g. Saffman 1971) to produce analogous results using arguments based on vortex dynamics have so far led to different answers. The computation by Sedov (1976) to investigate an earlier vortex dynamic interpretation due to Novikov (1975) of Fjørtoft's theorem (Fjørtoft 1953, Merilees & Warn 1975) and the calculation by Siggia & Aref (1981) of forced, homogeneous, isotropic, two-dimensional turbulence using vortex methods should also be mentioned as recent examples of a more mechanistic point of view.

Recalling that the many-vortex system is Hamiltonian and nonintegrable, we may be tempted to dispense with dynamics altogether and invoke *statistical mechanics* as originally suggested by Onsager (1949). This leads to a very interesting theory, in particular for a bounded domain. However, equivalent results may often be deduced with greater ease by using Fourier modes instead of vortices. See the recent review by Kraichnan & Montgomery (1980) for a full discussion of this topic. Statistical mechanics provides an indication of the type of state toward which a given initial distribution of vorticity will tend. Frequently, however, we are more interested in the mechanisms, modes, and states through which the system attempts to reach this ultimate equilibrium (which may in itself be relatively "uninteresting"). This argument suggests the *kinetics of vortices* as a logical next step, but there is relatively little work along those lines.

Novikov (1975) considered a system of identical vortices and wrote down the *BBGKY hierarchy* for the multiparticle distribution functions. Introducing a generating functional for these distributions, Novikov (1975) showed that it satisfies an equation of motion that is essentially the *Hopf equation* (cf. Monin & Yaglom 1975) for velocity fields induced by discrete vortices. This suggests an interesting, if not altogether surprising, connection between vortex systems and the description of turbulent flows using functional methods.

In the more specialized context of a free shear layer, Takaki & Kovasznay (1978) considered a kinetic model in which vortex pairings occur probabilistically. Aref & Siggia (1981) use a kinetic equation in the relaxation time approximation to elucidate properties of the disintegration of a vortex street.

In the following subsections, work is reviewed relating to the shear layer and the vortex street. However, before taking up that discussion, we briefly consider the use of vortices as a computational device.

Vortex Methods for Flow Simulation

The possibility of transcribing two-dimensional, incompressible hydrodynamics [Equation (4) with $v = 0$] into a system of ordinary differential equations [Equations (6)] has been made the basis of algorithms for numerical flow computations for a very long time, starting with the calculations of Rosenhead (1931) and Westwater (cf. Batchelor 1967) prior to the availability of digital computers. The basic equations in the form (6) have the disadvantage that the number of operations per time step necessary to compute the vortex velocities from their positions scales as N^2, where N is the number of vortices. Hence in practice $N \sim 100$ has defined an upper limit for computations that employ Equations (6) directly [see e.g. Abernathy & Kronauer 1962 ($N = 42$), Michalke 1964 ($N = 72$), Acton 1976 ($N = 96$)]. The comprehensive calculation by Ashurst (1979) employed an increasing number of point vortices, from 1 to 800 as the calculation progressed, and required 250 hours on a CDC 6600 computer! Furthermore, the steeply increasing velocity that a point vortex produces in its immediate vicinity makes Equations (6) a rather *stiff* system of differential equations. And although integration schemes now exist that can handle such systems satisfactorily, too much time is spent computing details of close vortex encounters that are usually of little relevance to the smoothed-out, average vorticity that is the objective of the calculation. Algorithms have thus been developed that in a systematic way eliminate the singularity at the vortices by replacing each point vortex with one of finite area. Such schemes usually ignore the deformations of the finite-area vortex that in reality should take place, and simply move it along without change of shape. Thus the decompositions into vortices provided by these schemes involve approximations; in particular, they do have truncation errors associated with them. Physically, such errors act like an artificial "eddy-damping" on the smallest scales. The ideas behind the use of such approximate decompositions were outlined by Chorin (1973).

A particularly elegant resolution of the problems associated with the operation count of Equations (6) and (to a degree) with the singularity of point vortices is achieved in the *cloud-in-cell* technique that has arisen in several fields. It was described for vortices by Christiansen (1973). In this method, the calculation of the stream-function from the vorticity, i.e. inversion of Poisson's equation, is performed on a grid superimposed over the vortices by using a fast transform method. This eliminates the singularity at the vortex positions and reduces the operation count from $O(N^2)$ to $O(N\log_2 N)$. However, considerable resolution at subgrid scales survives and advantages of retaining the Lagrangian particles are realized. Examples of calculations employing this technique appear in Roberts & Christiansen (1972), Christiansen & Zabusky (1973), Baker (1979), Aref & Siggia (1980, 1981), and Aref (1980), where further methodological details can be found. A comprehensive review of numerical methods that use vortex decompositions, with the same title as the heading of this subsection, is due to Leonard (1980).

Vortex sheets have consistently presented special problems for vortex methods, as is shown by a long list of contributions to the literature (Birkhoff & Fisher 1959, Kuwahara & Takami 1973, Chorin & Bernard 1973, Moore 1974, Fink & Soh 1978, Baker 1979, 1980, van de Vooren 1980). The inability of a row of point vortices to roll-up as expected of a vortex sheet suggested serious inadequacies with vortex methods. However, it now appears that many of the problems stem ultimately from the nature of the vortex sheet itself, which, at least for the case of an infinitely long sheet, seems to develop a singularity after a finite time (Moore 1979, Meiron et al. 1982). This is not in conflict with regularity results, such as Wolibner's (1933) theorem, since the vorticity distribution in a vortex sheet is a singular initial condition. Ironically, discretizing the sheet into point vortices all of one sign removes the singularity! A lucid discussion of the problems with numerical computations of vortex sheet roll-up has recently been given by Moore (1981). (Note that the term "chaotic" is used in Moore's 1981 paper to designate motions of the elemental vortices in which they behave very differently from the elements of a vortex sheet. Chaos in the sense of nonintegrable behavior, as discussed in this article for point-vortex systems, is neither explicitly suggested nor checked, although the two phenomena are undoubtedly related.)

To conclude this subsection, the topic of which could easily be made the basis for a separate review, we briefly mention the recent literature on vortex calculations of interfacial motions in certain *stratified flow* problems. The main observation here is that a sharp interface, separating contiguous regions of homogeneous fluid with different physical properties, within which we have potential flow, is a tangential discontinuity and hence a vortex sheet. The difference from the calculation of a vortex sheet in ordinary two-dimensional flow is that the circulation now changes as the sheet evolves. This time development of circulation is given by an auxiliary integral or integro-differential equation depending on the problem. Meng & Thomson (1978) considered a variety of such problems using schemes adapted to the limit of very weak stratification (in which case the auxiliary equation reduces to an algebraic or ordinary differential equation). Similar ideas were used by Hill (1975) to calculate the effects of buoyancy on the fluid entrained by a vortex pair. Baker et al. (1980) have extended the method to an arbitrary, finite-density jump in the *Rayleigh-Taylor* problem. They show that the iteration scheme necessary to calculate the vortex-sheet strength at every time step is convergent, and they produce several computational profiles extending well into the nonlinear regime with good precision. Coupling the iteration scheme with a cloud-in-cell calculation of vortex velocities yields a code capable of following collective interactions between several interfacial structures. Aref & Tryggvason (in preparation) consider the formation and interaction of *fingers* in the finite-amplitude *Taylor-Saffman* instability using this approach. Although these implementations are all relatively recent, the basic ideas behind using vortex methods to calculate sharp interfaces in stratified flows go back several decades (Birkhoff 1954, 1962, Josselin de Jong 1960).

Results for Free Shear Flows

SHEAR LAYERS   The flow configuration that has received the most attention, both experimentally and theoretically, is the free shear layer. It comes in essentially two varieties: the spatially evolving flow downstream of a splitter plate and the temporally evolving flow that starts from a tangential discontinuity in the velocity. We shall consistently refer to the spatial evolution as a *mixing layer* and reserve *shear layer* for the temporal problem. We also agree to avoid the singularities of the mathematical idealization to completely inviscid flow by considering the limit of small but nonzero viscosity or equivalently high Reynolds number. The dynamics of large-scale motions, which is the topic of concern here, is then still predominantly inertial. This is the type of system to which most of the recent experiments, calculations, and theories relate.

There is a general subdivision of tasks in that the mixing layer is the primary object of experimental study, whereas most theories and numerical computations refer to the unsteady shear layer. This is largely a matter of convenience, and Ashurst (1979) has in fact computed a mixing-layer flow, while Thorpe (1968) has realized a time-evolving shear layer experimentally. It is important to stress however, as has been pointed out by many authors, that the two problems are not equivalent in the sense that there is no Galilean transformation that will take us from one to the other. The two flows are obviously closely related and the standard transcription of results between mixing layer and shear layer is frequently used.

Considering first the initial-value problem posed by a thin band of vorticity, the well known *Kelvin-Helmholtz* instability mechanism suggests a roll-up into a line of periodically spaced, concentrated vortices. In the present case, a predominant wave number will prevail and it seems sufficient in a numerical calculation to follow just one (or at most a few) period(s) by considering a computational domain with periodic boundaries in one direction. Computations of roll-up to a finite-amplitude vortex using this geometry and various algorithms appear in Amsden & Harlow (1964), Christiansen (1973), Patnaik et al. (1976) (see also Corcos & Sherman 1976), and Aref & Siggia (1980). According to these calculations, this initial stage of the motion apparently proceeds very much *in phase* along the entire shear layer. When completed, we have a row of periodically spaced, identical vortices. If we replace each of them by a point vortex, we obtain once again a steady state and linearized stability analysis can be applied. If we rely on the numerical experiments and look more closely at the finite-area vortices that were produced by the roll-up, we see that they will not in general be in a steady state but rather will have a complex system of waves continually propagating around within each individual vortex region. Initially these time-periodic motions are completely in phase between one vortex and the next, but eventually time-stepping errors destroy the exact periodicity. I tend to believe that this *loss of phase coherence* along a shear layer is physically correct, but more experimental and theoretical work is called for. In any case, even a slight lack of periodicity within the row of vortices is amplified tremendously by what happens next.

As was already mentioned in our discussion of the dynamics of two vortices, a row of identical vortices is subject to an instability mechanism wherein neighboring vortices wrap around each other and ultimately fuse, the process of vortex pairing (Winant & Browand 1974). In a representation by point vortices, a simple linearized stability analysis (cf. Lamb 1932) shows that the most unstable mode has the shortest wavelength that can be supported by a line of periodically spaced nodes. Since this most unstable wave involves only pairs of neighboring vortices and is periodic along the row, it can be reduced to a two-body problem and thus calculated at finite amplitude. The vortices are then seen to exchange places in pairs, so that if the vortices initially are labeled *...abcdef...*, they become transposed to *...badcfe...*. There is of course no merging for point vortices, but the "cat's eye" trajectories do bring a pair of neighbors much closer together than they were initially.

It is important to emphasize that in this most unstable mode, the motion is again completely in phase along the row: the pairs *ab*, *cd*, *ef*, etc., move in exactly the same way. This periodicity is a discrete symmetry within the infinite system that is broken by the full set of interactions. Furthermore, the mode itself represents a motion connecting (as $t \to \pm \infty$) two unstable, steady states, i.e. we have once again a saddle connection. Thus we must expect that when the exact periodicity is destroyed, chaotic motions will result. In the present case, these show up as an amplification of the loss of periodicity or, extrapolating to finite-area vortices, a substantial *phase incoherence* of pairing events along the row. This is consistent with the results of numerical experiments (compare Figures 18 and 19 of Aref & Siggia 1981). Both in real and numerical experiments on a row of two-dimensional vortices, it is very difficult to produce pairing events that are exactly in phase along the row. The ideas reviewed earlier about instability to chaotic motion when saddle connections are disrupted provide mathematical mechanisms to explain this qualitative result. Indeed, in the turbulent regime it is essential that the vortex pairings get out of phase, since simple calculations (Aref & Siggia 1980) show that the contribution to the Reynolds stress from a single pair of vortices can have substantial excursions from its average value. If vortex pairings along a shear layer were somehow locked into phase with one another, the average Reynolds-stress profile would not have a simple "one-bump" appearance.

In light of these remarks, most of the proposed models and theories of two-dimensional shear-layer dynamics appear rather unconvincing. Such models are usually based on the energetics of merging (or some other dynamical process, e.g. the "tearing" suggested by Moore & Saffman 1975) for a single pair of vortices and are made to apply to the full shear layer by assuming that the results can be periodically continued to the entire row. Examples of discussions in this format include the studies by Moore & Saffman (1975), Jimenez (1980), Ferziger (1980), and Pierrehumbert & Widnall (1981). The computation by Acton (1976) of the merging of just two vortex regions belongs in this category as well. These papers contain much information on vortex motion of independent interest; however, I contend that because of an instability to phase decorrelation between vortex pairings, the shear layer must be seen as a many-body system and cannot be properly understood on the basis of two-body dynamics periodically continued. (This point is amplified substantially when we look at the results of large, numerical simulations.)

Similar conclusions have been reached in experimental studies of mixing
layers (Dimotakis & Brown 1976), but the above criticism relates to the
purely two-dimensional problem and does not depend on the presence of
three-dimensional small scales. Models not open to these criticisms
include the kinetic approach of Takaki & Kovasznay (1978) and the

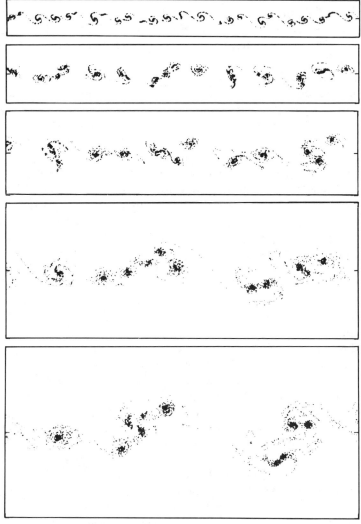

Figure 3 Turbulent vortex motion. Instantaneous plots of vortex positions in a numerical
simulation of the two-dimensional, unsteady shear layer (from Aref & Siggia 1980). Frames
do not represent the physical boundaries of the flow region, which are essentially "at
infinity."

successive equilibration model in Aref & Siggia (1980, Section 4), which has recently been improved upon by Kuzmin et al. (1982).

The literature contains three reasonably large numerical vortex simulations. The direct-summation, mixing-layer calculation by Ashurst (1979) has already been mentioned. A grid-based algorithm, the "center-to-center" method with $N = 750$ particles, was used by Delcourt & Brown (1979). Finally, the cloud-in-cell technique, with 4096 vortices on a 256^2 grid, was used by Aref & Siggia (1980). For various technical reasons, discussed in the second and third of these papers, the calculation by Aref & Siggia (1980) probably has the least amount of dissipation. Although a conclusive study is still lacking, this seems to have important consequences for the size and distribution of vortex structures that ensue.

In the pictures produced by Ashurst (1979), the vortex regions are all roughly aligned, with their centers on a ray emanating from the tip of the splitter plate. Vortex pairings of course induce momentary departures from this line, but the merged vortex evolves to take up its place in the row. This aligned geometry of vortices is consistent with the famous shadowgraphs by Brown & Roshko (1974) and with many other visual impressions from experiment. Koochesfahani et al. (1979) actually measured the distribution of center positions about the above-mentioned "line" and, excluding events that were manifestly pairing excursions from the data, obtained a very narrow distribution. However, the numerical simulations by Aref & Siggia (1980) and Delcourt & Brown (1979) show a qualitatively different picture. Here, instead of a row of large vorticity regions, one sees clusters of smaller concentrated vortices (each in turn made up of a large number of elemental vortices). And the pairing is not between two neighboring vortex regions but consists of a mingling of neighboring clusters, with merging of vortices taking place subsequently within a cluster (see Figure 3). Some recent experimental studies of mixing layers in the region close to the splitter plate, where the flow is still highly two-dimensional, have indeed shown features that are consistent with such a more *intermittent vorticity distribution* (Dimotakis et al. 1981, Martinez-Val 1981) but are difficult to reconcile with the canonical, single-row picture.

The proposed rationalization of these relationships between numerical simulations and experiments is the following: As long as the experimental flow is largely two-dimensional, results from quasi-inviscid calculations like those of Delcourt & Brown (1979) and Aref & Siggia (1980) apply. However, once the flow becomes fully turbulent (in the traditional sense[6]) and the vortices contain within them fully three-dimensional

---

[6]Some will claim that vortex structures are entirely obliterated by then. I do not want to enter that discussion here.

small scales (clearly visible in Brown & Roshko's 1974 pictures), the "eddy viscosity" due to these motions is no longer captured correctly by quasi-inviscid dynamics. A certain morphological resemblance with experiment can, however, be obtained by introducing diffusive and dissipative mechanisms into the vortex simulation as is done in Ashurst's (1979) work. (See also comments below on statistical self-similarity.)

A few additional remarks should be made: First, the vortices found numerically and experimentally (Browand & Weidman 1976) are not uniform but have rather concentrated vorticity profiles. Estimates that specifically assume uniform vorticity are thus not necessarily reliable. Most of the previously quoted theories based on two vortices also assume uniform vorticity. Second, flow visualization gives a passive marker trace, not a vorticity distribution. Hama (1962), for example, reported on numerical experiments in which a marker seemed to indicate roll-up into discrete vortices in a shear flow although no such process was going on! Williams & Hama (1980) have extended the investigation and suggest that streaklines may give an erroneous impression of vortex pairing. Aref & Siggia (1980) show the evolution of a line of marker during vortex pairing along with the vorticity distribution itself. In the final picture of the sequence, one large blob of marker is seen enveloping two much smaller vortices. In our earlier discussion of the restricted four-vortex problem we mentioned mechanisms that allow organized vortex motion to produce a diffuse marker cloud (see also the model calculation of Jimenez 1980).[7] Hence we must reiterate the familiar warnings of not relying blindly on flow visualization to mark vortices. Improvements in experimental technique (e.g. Breidenthal 1979, Dimotakis et al. 1981, Frish & Webb 1981) are highly desirable and eagerly awaited. Third, the flows produced numerically by Aref & Siggia (1980) were shown to be *statistically self-similar* at the level of low-order velocity correlations. However, the universal functions that describe this self-preserving, two-dimensional, turbulent shear flow differ from those that have been established experimentally for high-Reynolds-number mixing layers with three-dimensional small scales. Thus, there can be no facile analogies between the coherent vortex structure in a so-called plane mixing layer (with three-dimensional small scales) and its purely two-dimensional counterpart. As a corollary, models of the fully turbulent mixing layer based on two-dimensional hydrodynamics cannot be very realistic. Using steady-state solutions of the 2DEE, commonly with piecewise-uniform vorticity, to interpret experimental observations is altogether too simplistic.

[7]Amsden & Harlow (1964) refer to "the relative orderliness of Eulerian representation over Lagrangian."

The appearance of vortex clusters, and thus vortex triplings, quadruplings, etc., instead of just vortex pairings, can be rationalized by recalling that although the pairing mode is the most unstable for a row of point vortices, higher "multi-vortex" modes are not much less unstable and should become excited on slightly longer time scales. Conversely, and this has technological importance, such multi-vortex modes may be excited artificially as in the work of Ho & Huang (1982). When this is done, novel phenomena, such as frequency-locking between excitation and response, present themselves. Some of these are reminiscent of the behavior seen in systems of coupled, nonlinear oscillators, and the formal connections established earlier between the motion of four identical vortices and a system of two coupled nonlinear oscillators are worth recalling.

THE VORTEX STREET   As a final topic, consider the dynamics of two parallel vorticity bands of opposite circulation. Appropriately perturbed, they roll up into the characteristic, staggered *vortex street* configuration similar to that observed experimentally in the wake of a bluff body. This particular flow was one of the first to be modeled by replacing each finite-area vortex with a point vortex. A well-known result of the ensuing theory, due to von Kármán (see Lamb 1932), is that the point-vortex configuration is unstable to infinitesimal perturbations unless the ratio of row separation b to intrarow vortex separation a has a particular value $b/a = .28055....$ Subsequent work by Kochin (see Kochin et al. 1964) in which the stability analysis was carried to higher order showed that the point-vortex street is in fact always unstable (see also Domm 1956). This unfortunate result leaves us with little understanding of why vortex streets maintain themselves at all (cf. Wille 1960).

Leaving on one side the important problems connected with the formation of a street by vortex-shedding (see Bearman & Graham 1980 for a report on a recent conference), we only consider here the double, alternate bands of vorticity as an initial-value problem. The early vortex computations by Abernathy & Kronauer (1962) suggested that with the proper initial perturbation, a vortex-street configuration would always form. A recent numerical study at considerably higher resolution by Aref & Siggia (1981) suggests, however, that modes other than street formation can arise (compare also the discussion by Weihs 1973). In these calculations and also in the computations by Zabusky & Deem (1971), the *Kármán ratio*, b/a, observed in cases where a street does form is very different from its "classical," point-vortex value (see also Saffman & Schatzman 1981). Furthermore, as for the shear layer, the vortices that evolve in simulations of the initial-value problem are not steady but support internal wavelike motions. Zabusky & Deem (1971) (see also Zabusky 1977) suggest that such time-dependent undulations, which they call *vortex nutation*, have experimental support.

Once the vortices have formed, one might expect pairing within the rows individually to occur, and indeed such modes can be found. They were first seen in numerical simulations by Christiansen & Zabusky (1973) and later, somewhat more convincingly, by Aref & Siggia (1981). In these numerical pairing modes the streamwise scale of the vortex street exactly doubles. The calculations were inspired by earlier experiments of Taneda (1959) (see also Durgin & Karlsson 1971) in which structural transitions in a vortex street are reported. However, in Taneda's experiments the ratio of longitudinal scales is Reynolds-number-dependent and not necessarily 2:1. The modes seen numerically and in experiment are thus presumably related but certainly not identical. Additional work on this question would be helpful.

When the vortex street breaks down, experiments show that the flow becomes substantially three-dimensional. In two-dimensional hydrodynamics, on the other hand, the vortices survive and dominate the flow in later stages of evolution. Some convective merging between likesigned vortices takes place, but a new mechanism wherein pairs of opposite circulation join up and propagate away from the vortex-street region is found to play an important role. This *pair dissolution* process is discussed at length by Aref & Siggia (1981). In its early stages, it appears not inconsistent with the experimental observations of Papailiou & Lykoudis (1974). As mentioned previously, there are analytical solutions describing breakdown into vortex pairs of Havelock's (1931) double alternate rings, which are finite, circular counterparts of the vortex street (Aref, submitted for publication). The possibility of forming freely propagating vortex pairs when a vortex street breaks down in two-dimensional flow seems to offer an interesting mechanism for several applications in geophysical fluid dynamics.

The unsatisfactory state of a theory of vortex streets has led to the conjecture (Christiansen & Zabusky 1973, Zabusky 1977) that the wavelike motions that are set up on the finite-area vortices of the street produce a damping of the instability found for point vortices. There is some support for this idea from numerical integrations of the initial-value problem (Christiansen & Zabusky 1973) and from a recent stability calculation by Saffman & Schatzman (1982). These authors suggest that the mechanism of *Arnold diffusion* (see Chirikov 1979 or the article by M. A. Lieberman in Helleman 1980) may provide a relevant explanation for the slow instability phenomenon associated with breakdown of the Kármán point-vortex street.

CONCLUDING REMARKS

The topic of two-dimensional vortex motion will undoubtedly continue to flourish and the possibility of detailed numerical calculations will continue to provide significant material for theoretical contemplation. In this article, I have also repeatedly stressed the budding impact of ideas from the theory of dynamical systems. I submit that this new alliance can offer a potent source of inspiration for the future.

Two-dimensional hydrodynamics presents a considerable richness of worthy problems but also a wide variety of important potential applications. And while we must guard against two-dimensional models and theories becoming too simplistic, we definitely need not succumb to the disillusioned views of Abbott's[8] hapless hero, who in his despair concludes that "all the substantial realities of Flatland itself appear no better than the offspring of a diseased imagination, or the baseless fabric of a dream."

ACKNOWLEDGMENTS

It is a pleasure to acknowledge the profound impact that E. D. Siggia and N. Pomphrey have had on my understanding of the topics treated here. I am furthermore indebted to G. R. Baker, F. K. Browand, J. Burbea, F. Calogero, L. J. Campbell, T. G. McKee, J. D. Meiss, M. V. Morkovin, E. A. Novikov, G. Tryggvason, and N. J. Zabusky for varied productive interactions.

This work was supported by National Science Foundation grant MEA 81-16910 to Brown University.

Literature Cited

Abernathy, F. H., Kronauer, R. E. 1962. The formation of vortex streets. *J. Fluid Mech.* 13:1–20

Acton, E. 1976. The modelling of large eddies in a two-dimensional shear layer. *J. Fluid Mech.* 76:561–92

Amsden, A. A., Harlow, F. H. 1964. Slip instability. *Phys. Fluids* 7:327–34

Aref, H. 1979. Motion of three vortices. *Phys. Fluids* 22:393–400

Aref, H. 1980. Coherent features by the method of point vortices. *Woods Hole Oceanogr. Inst. Tech. Rep. WHOI-80-53*, pp. 233–49

Aref, H., Pomphrey, N. 1980. Integrable and chaotic motions of four vortices. *Phys. Lett. A* 78:297–300

Aref, H., Pomphrey, N. 1982. Integrable and chaotic motions of four vortices. I. The case of identical vortices. *Proc. R. Soc. London Ser. A* 380:359–87

Aref, H., Siggia, E. D. 1980. Vortex dynamics of the two-dimensional turbulent shear layer. *J. Fluid Mech.* 100:705–37

Aref, H., Siggia, E. D. 1981. Evolution and breakdown of a vortex street in two dimensions. *J. Fluid Mech.* 109:435–63

[8] E. A. Abbott (1838–1926), author of *Flatland, A Romance of Many Dimensions* (republished by Dover, New York, 1952).

Arnold, V. I., Avez, A. 1968. *Ergodic Problems of Classical Mechanics.* New York/Amsterdam: Benjamin. 286 pp.

Ashurst, W. T. 1979. Numerical simulation of turbulent mixing layers via vortex dynamics. In *Turbulent Shear Flows I*, ed. F. Durst et al., pp. 402–13. Berlin/Heidelberg/New York: Springer

Baker, G. R. 1979. The "cloud-in-cell" technique applied to the roll up of vortex sheets. *J. Comput. Phys.* 31:76–95

Baker, G. R. 1980. A test of the method of Fink and Soh for following vortex sheet motion. *J. Fluid Mech.* 100:209–20

Baker, G. R., Meiron, D. I., Orszag, S. A. 1980. Vortex simulations of the Rayleigh-Taylor instability. *Phys. Fluids* 23:1485–90

Basset, A. B. 1888. *A Treatise on Hydrodynamics.* Cambridge: Deighton, Bell & Co. 264 pp.

Batchelor, G. K. 1956. On steady laminar flow with closed streamlines at large Reynolds number. *J. Fluid Mech.* 1:177–90

Batchelor, G. K. 1967. *An Introduction to Fluid Dynamics*, Chapt. 7. Cambridge Univ. Press. 615 pp.

Batchelor, G. K. 1969. Computation of the energy spectrum in homogeneous two-dimensional turbulence. *Phys. Fluids* 12:233–39 (Suppl. 2)

Bauer, L., Morikawa, G. K. 1976. Stability of rectilinear geostrophic vortices in stationary equilibrium. *Phys. Fluids* 19:929–42

Bearman, P. W., Graham, J. M. R. 1980. Vortex shedding from bluff bodies in oscillatory flow: A report on Euromech 119. *J. Fluid Mech.* 99:225–45

Berestov, A. L. 1979. Solitary Rossby waves. *Izv. Acad. Sci. USSR Atmos. Oceanic Phys.* 15:443–47

Birkhoff, G. 1954. Taylor instability and laminar mixing. *Los Alamos Sci. Rep. LA-1862.* 75 pp.; Appendices in *Rep. LA-1927.* 91 pp.

Birkhoff, G. 1962. Helmholtz and Taylor instability. In *Hydrodynamic Instability, Proc. Symp. Appl. Math.*, 13:55–76. Providence: Am. Math. Soc. 319 pp.

Birkhoff, G., Fisher, J. 1959. Do vortex sheets roll up? *Rend. Circ. Mat. Palermo* 8:77–90

Bogomolov, V. A. 1979. Two-dimensional fluid dynamics on a sphere. *Izv. Acad. Sci. USSR Atmos. Oceanic Phys.* 15:18–22

Bona, J. L., Pritchard, W. G., Scott, L. R. 1980. Solitary wave interaction. *Phys. Fluids* 23:438–41

Breidenthal, R. 1979. Chemically reacting, turbulent shear layer. *AIAA J.* 17:310–11

Browand, F. K., Weidman, P. D. 1976. Large scales in the developing mixing layer. *J. Fluid Mech.* 76:127–44

Brown, G. L., Roshko, A. 1974. On density effects and large structure in turbulent mixing layers. *J. Fluid Mech.* 64:775–816

Bruschi, M. 1979. On the algebra of certain matrices related to the zeros of Hermite polynomials. *Lett. Nuovo Cimento* 24:509–12

Burbea, J. 1980. Vortex motions and conformal mappings. In *Nonlinear Evolution Equations and Dynamical Systems, Lecture Notes in Physics*, 120:276–98. Berlin/Heidelberg/New York: Springer

Burbea, J. 1982a. On patches of uniform vorticity in a plane of irrotational flow. *Arch. Ration. Mech. Anal.* 77:349–58

Burbea, J. 1982b. Motions of vortex patches. *Lett. Math. Phys.* 6:1–16

Burbea, J., Hebert, D. J. 1982. A perturbation calculation for the shapes of rotational regions of vorticity. *J. Math Anal. Appl.* In press

Burbea, J., Landau, M. 1982. The Kelvin waves in vortex dynamics and their stability. *J. Comput. Phys.* 45:127–56

Calogero, F. 1978. Motion of poles and zeros of special solutions of nonlinear and linear partial differential equations and related "solvable" many-body problems. *Nuovo Cimento B* 43:177–241

Campbell, L. J. 1981. Transverse normal modes of finite vortex arrays. *Phys. Rev. A* 24:514–34

Campbell, L. J., Ziff, R. M. 1978. A catalog of two-dimensional vortex patterns. *Los Alamos Sci. Lab. Rep. No. LA-7384-MS.* 40 pp.

Campbell, L. J., Ziff, R. M. 1979. Vortex patterns and energies in a rotating superfluid. *Phys. Rev. B* 20:1886–1902

Cantwell, B. J. 1981. Organized motion in turbulent flow. *Ann. Rev. Fluid Mech.* 13:457–515

Charney, J. G. 1963. Numerical experiments in atmospheric hydrodynamics. In *Experimental Arithmetic, High Speed Computing and Mathematics, Proc. Symp. Appl. Math.*, 15:289–310. Providence: Am. Math. Soc. 396 pp.

Chirikov, B. V. 1979. A universal instability of many-dimensional oscillator systems. *Phys. Rep.* 52:263–379

Choodnovsky, D. V., Choodnovsky, G. V. 1977. Pole expansions of nonlinear partial differential equations. *Nuovo Cimento B* 40:339–53

Chorin, A. J. 1973. Numerical study of slightly viscous flow. *J. Fluid Mech.* 57:785–96

Chorin, A. J. 1980. Vortex models and boundary layer instability. *SIAM J. Sci. Stat. Comput.* 1:1–21

Chorin, A. J., Bernard, P. S. 1973. Discretization of a vortex sheet with an example of roll-up. *J. Comput. Phys.* 13:423–29

Christiansen, J. P. 1973. Numerical simulation of hydrodynamics by the method of point vortices. *J. Comput. Phys.* 13:363–79

Christiansen, J. P., Zabusky, N. J. 1973. Instability, coalescence and fission of finite-area vortex structures. *J. Fluid Mech.* 61:219–43

Clements, R. R., Maull, D. J. 1975. The representation of sheets of vorticity by discrete vortices. *Prog. Aeronaut. Sci.* 16:129–46

Conlisk, T., Rockwell, D. 1981. Modelling of vortex-corner interaction using point vortices. *Phys. Fluids* 24:2133–42

Corcos, G. M., Sherman, F. S. 1976. Vorticity concentration and the dynamics of unstable free shear layers. *J. Fluid Mech.* 73:241–64

Crighton, D. G. 1972. Radiation from vortex filament motion near a half plane. *J. Fluid Mech.* 51:357–62

Deem, G. S., Zabusky, N. J. 1978. Vortex waves: Stationary "V states," interactions, recurrence, and breaking. *Phys. Rev. Lett.* 40:859–62

Delcourt, B. A. G., Brown, G. L. 1979. The evolution and emerging structure of a vortex sheet in an inviscid and viscous fluid modelled by a point vortex method. In *2nd Symp. Turbul. Shear Flows, Imperial Coll., London*, pp. 14.35–40

Dimotakis, P. E., Brown, G. L. 1976. The mixing layer at high Reynolds number: Large structure dynamics and entrainment. *J. Fluid Mech.* 78:535–60

Dimotakis, P. E., Debussy, F. D., Koochesfahani, M. M. 1981. Particle streak velocity field measurements in a two-dimensional mixing layer. *Phys. Fluids* 24:995–99

Domm, U. 1956. Über die Wirbelstrassen von geringster Instabilität. *Z. Angew. Math. Mech.* 36:367–71

Dragt, A. J., Finn, J. M. 1976. Insolubility of trapped particle motion in a magnetic dipole field. *J. Geophys. Res.* 81:2327–40

Durgin, W. W., Karlsson, S. K. F. 1971. On the phenomenon of vortex street breakdown. *J. Fluid Mech.* 48:507–27

Ebin, D. G., 1983. Integrability of perfect fluid motion. *Comm. Pure Appl. Math.* 36: In press

Ferziger, J. H. 1980. Energetics of vortex rollup and pairing. *Phys. Fluids.* 23:1–4

Fink, P. T., Soh, W. K. 1978. A new approach to roll-up calculations of vortex sheets. *Proc. R. Soc. London Ser. A* 362:195–209

Fjørtoft, R. 1953. On changes in the spectral distribution of kinetic energy in two-dimensional non-divergent flow. *Tellus* 5:225–30

Fohl, T., Turner, J. S. 1975. Colliding vortex rings. *Phys. Fluids* 18:433–36

Friedrichs, K. O. 1966. *Special Topics in Fluid Dynamics*, Chapt. 19. New York/London/Paris: Gordon & Breach. 177 pp.

Frisch, U. 1982. Fully developed turbulence and singularities. In *Chaotic Behavior in Deterministic Systems*, Les Houches Summer School. Amsterdam/New York/Oxford: North-Holland. In press

Frish, M. B., Webb, W. W. 1981. Direct measurement of vorticity by optical probe. *J. Fluid Mech.* 107:173–200

Greenhill, A. G. 1878. Plane vortex motion. *Q. J. Math.* 15:10–29

Gröbli, W. 1877. *Specielle Probleme über die Bewegung geradliniger paralleler Wirbelfäden*. Zürich: Zürcher und Furrer. 86 pp.

Hall, M. G. 1972. Vortex breakdown. *Ann. Rev. Fluid Mech.* 4:195–218

Hally, D. 1980. Stability of streets of vortices on surfaces of revolution with a reflection symmetry. *J. Math. Phys.* 21:211–17

Hama, F. R. 1962. Streaklines in a perturbed shear flow. *Phys. Fluids* 5:644–50

Havelock, T. H. 1931. The stability of motion of rectilinear vortices in ring formation. *Philos. Mag.* (7) 11:617–33

Helleman, R. H. G., ed. 1980. *Nonlinear Dynamics*, Ann. NY Acad. Sci., Vol. 357

Helmholtz, H. 1858. On integrals of the hydrodynamical equations which express vortex-motion. Transl., P. G. Tait, 1867, in *Phil. Mag.* (4) 33:485–512

Hénon, M., Heiles, C. 1964. The applicability of the third integral of motion: Some numerical experiments. *Astron. J.* 69:73–79

Hicks, W. M. 1922. On the mutual threading of vortex rings. *Proc. R. Soc. London Ser. A* 102:111–31

Hill, F. M. 1975. A numerical study of the descent of a vortex pair in a stably stratified atmosphere. *J. Fluid Mech.* 71:1–13

Hill, M. J. M. 1884. On the motion of fluid, part of which is moving rotationally and part irrotationally. *Philos. Trans. R. Soc. London* 175:363–411

Ho, C-M., Huang, L-S. 1982. Subharmonics and vortex merging in mixing layers. *J. Fluid Mech.* 119:443–73

Holmes, P. J. 1980. Averaging and chaotic motions in forced oscillations. *SIAM J. Appl. Math.* 38:65–80

Holmes, P. J., Marsden, J. E. 1982. Horseshoes in perturbations of Hamiltonian systems with two degrees of freedom. *Comm. Math. Phys.* 82:523–44

Jimenez, J. 1980. On the visual growth of a turbulent mixing layer. *J. Fluid Mech.* 96:447–60

Jorna, S., ed. 1978. *Topics in Nonlinear Dynamics. A Tribute to Sir Edward Bullard.* New York: Am. Inst. Phys. 404 pp.

Josselin de Jong, G. de. 1960. Singularity distributions for the analysis of multiple-fluid flow through porous media. *J. Geophys. Res.* 65:3739–58

Kambe, T., Takao, J. 1971. Motion of distorted vortex rings. *J. Phys. Soc. Jpn.* 31:591–99

Karweit, M. 1975. Motion of a vortex pair approaching an opening in a boundary. *Phys. Fluids* 18:1604–6

Kelvin, Lord. 1910. *Mathematical and Physical Papers*, Vol. 9 (No. 10, 12). Cambridge Univ. Press. 563 pp.

Khanin, K. M. 1982. Quasi-periodic motions of vortex systems. *Physica D* 4:261–69

Khazin, L. G. 1976. Regular polygons of point vortices and resonance instability of steady states. *Sov. Phys. Dokl.* 21:567–69

Kida, S. 1975. Statistics of a system of line vortices. *J. Phys. Soc. Jpn.* 39:1395–1404

Kirchhoff, G. R. 1876. *Vorlesungen über Matematische Physik*, Vol. 1. Leipzig: Teubner. 466 pp.

Kochin, N. E., Kibel, I. A., Roze, N. V. 1964. *Theoretical Hydromechanics*, Chapt. 5. New York: Interscience. 577 pp.

Koochesfahani, M. M., Catherasoo, C. J.,

Dimotakis, P. E., Gharib, M., Lang, D. B. 1979. Two-point LDV measurements in a plane mixing layer. *AIAA J.* 17:1347–51

Kovasznay, L. S. G. 1978. Large scale structure in turbulence: A question or an answer. In *Structure and Mechanisms of Turbulence I, Lecture Notes in Physics*, 75:1–18. Berlin/Heidelberg/New York: Springer

Kraichnan, R. H. 1967. Inertial ranges in two-dimensional turbulence. *Phys. Fluids* 10:1417–23

Kraichnan, R. H. 1973. Helical turbulence and absolute equilibrium. *J. Fluid Mech.* 59:745–52

Kraichnan, R. H., Montgomery, D. 1980. Two-dimensional turbulence. *Rep. Prog. Phys.* 43:547–619

Kuwahara, K., Takami, H. 1973. Numerical studies of two-dimensional vortex motion by a system of point vortices. *J. Phys. Soc. Jpn.* 34:247–53

Kuzmin, G. A., Likhachev, O. A., Patashinsky, A. Z. 1982. Structural turbulence in a free shear layer. In *Structural Turbulence*, ed. M. A. Goldschtik (In Russian). Engl. transl. in Inst. Thermophys. Novosibirsk. Preprint 81–97

Lamb, H. 1932. *Hydrodynamics*, Chapt. 7. New York: Dover. 738 pp. 6th ed.

Landau, M. 1981. *The structure and stability of finite area vortex regions of the two-dimensional Euler equations.* PhD thesis. Univ. Pittsburgh. 126 pp.

Lanford, O. III. 1982. The strange attractor theory of turbulence. *Ann. Rev. Fluid Mech.* 14:347–64

Larichev, V. D., Reznik, G. M. 1978. Two-dimensional Rossby soliton: An exact solution. *POLYMODE News* 19:3

Leibovich, S. 1978. The structure of vortex breakdown. *Ann. Rev. Fluid Mech.* 10:221–46

Leith, C. E. 1968. Diffusion approximation for two-dimensional turbulence. *Phys. Fluids* 11:671–73

Leith, C. E., Kraichnan, R. H. 1972. Predictability of turbulent flows. *J. Atmos. Sci.* 29:1041–58

Leonard, A. 1980. Vortex methods for flow simulation. *J. Comput. Phys.* 37:289–335

Lin, C. C. 1941. On the motion of vortices in two dimensions. I. Existence of the Kirchhoff-Routh function. *Proc. Natl. Acad. Sci. USA* 27:570–75

Lin, C. C. 1943. *On the Motion of Vortices in Two Dimensions.* Toronto Univ. Press. 39 pp.

Lorenz, E. N. 1963. Deterministic nonperiodic flow. *J. Atmos. Sci.* 20:130–41

Love, A. E. H. 1893. On the stability of certain vortex motions. *Proc. London Math. Soc.* 25:18–42

Love, A. E. H. 1894. On the motion of paired vortices with a common axis. *Proc. London Math. Soc.* 25:185–94

Love, A. E. H. 1901. Hydrodynamik: Theoretische Ausführungen. *Encykl. Math. Wiss.*, 4(3):86–147. Leipzig: Teubner

Lundgren, T. S., Pointin, Y. B. 1977. Statistical mechanics of two-dimensional vortices. *J. Stat. Phys.* 17:323–55

Makino, M., Kamimura, T., Taniuti, T. 1981. Dynamics of two-dimensional solitary vortices in a low-β plasma with convective motion. *J. Phys. Soc. Jpn.* 50:980–89

Malafronte, T. A. 1981. *Steady motion of a non-circular vortex.* PhD thesis. Brown Univ., Providence. 100 pp.

Martinez-Val, R. 1981. Vorticity distribution in the mixing layer. *Phys. Fluids* 24:2117–18

Maxworthy, T. 1972. The structure and stability of vortex rings. *J. Fluid Mech.* 51:15–32

Maxworthy, T. 1979. Comments on "Preliminary study of mutual slip-through of a pair of vortices." *Phys. Fluids* 22:200–1

McKee, T. G. 1981. *Some contributions to the self-interaction of a vortex cylinder.* PhD thesis. Brown Univ., Providence. 267 pp.

McWilliams, J. C., Zabusky, N. J. 1982. Interactions of isolated vortices. I. Modons colliding with modons. *Geophys. Astrophys. Fluid Dyn.* 19:207–27

Meiron, D. I., Baker, G. R., Orszag, S. A. 1982. Analytic structure of vortex sheet dynamics. I. Kelvin-Helmholtz instability. *J. Fluid Mech.* 114:283–98

Meng, J. C. S., Thomson, J. A. L. 1978. Numerical studies of some nonlinear hydrodynamic problems by discrete vortex element methods. *J. Fluid Mech.* 84:433–53

Merilees, P. E., Warn, H. 1975. On energy and enstrophy exchanges in two-dimensional non-divergent flow. *J. Fluid Mech.* 69:625–30

Mertz, G. J. 1978. Stability of body-centered polygonal configurations of ideal vortices. *Phys. Fluids* 21:1092–95

Michalke, A. 1964. Zur Instabilität und nichtlinearen Entwicklung einer gestörten Scherschicht. *Ing. Arch.* 33:264–76

Monin, A. S., Yaglom, A. M. 1975. *Statistical Fluid Mechanics: Mechanics of Turbulence,* ed. J. L. Lumley, Vol. 2, Chapt. 10. Cambridge: MIT Press. 874 pp.

Moore, D. W. 1974. A numerical study of the roll-up of a finite vortex sheet. *J. Fluid Mech.* 63:225–35

Moore, D. W. 1979. The spontaneous appearance of a singularity in the shape of an evolving vortex sheet. *Proc. R. Soc. London Ser. A* 365:105–19

Moore, D. W. 1981. On the point vortex method. *SIAM J. Sci. Stat. Comput.* 2:65–84

Moore, D. W., Saffman, P. G. 1975. The density of organized vortices in a turbulent mixing layer. *J. Fluid Mech.* 69:465–73

Morikawa, G. K. 1960. Geostrophic vortex motion. *J. Meterol.* 17:148–58

Morikawa, G. K., Swenson, E. V. 1971. Interacting motion of rectilinear geostrophic vortices. *Phys. Fluids* 14:1058–73

Moser, J. 1973. *Stable and Random Motions in Dynamical Systems.* Princeton Univ. Press. 198 pp.

Murty, G. S., Rao, K. S. 1970. Numerical study of the behavior of a system of parallel line vortices. *J. Fluid Mech.* 40:595–602

Novikov, E. A. 1975. Dynamics and statistics of a system of vortices. *Sov. Phys. JETP* 41:937–43

Novikov, E. A. 1980. Stochastization and collapse of vortex systems. *Ann. NY Acad. Sci.* 357:47–54

Novikov, E. A., Sedov, Yu. B. 1978. Stochastic properties of a four-vortex system. *Sov. Phys. JETP* 48:440–44

Novikov, E. A., Sedov, Yu. B. 1979a. Stochastization of vortices. *JETP Lett.* 29:677–79

Novikov, E. A., Sedov, Yu. B. 1979b. Vortex collapse. *Sov. Phys. JETP.* 50:297–301

Oberhettinger, F., Magnus, W. 1949. *Anwendung der elliptischen Funktionen in Physik und Technik,* Chapt. 4. Berlin: Springer. 126 pp.

Olver, P. J. 1982. A nonlinear Hamiltonian structure for the Euler equations. *J. Math. Anal. Appl.* In press

Onsager, L. 1949. Statistical hydrodynamics. *Nuovo Cimento* 6:279–87 (Suppl.)

Oshima, Y. 1978. The game of passing through of a pair of vortex rings. *J. Phys. Soc. Jpn.* 45:660–64

Overman, E. A. II, Zabusky, N. J. 1982a. Evolution and merger of isolated vortex structures. *Phys. Fluids.* 25:1297–1305

Overman, E. A. II, Zabusky, N. J. 1982b. Translation, interaction and scattering of Euler equation V-states via contour dynamics. *J. Fluid Mech.* In press

Palmore, J. I. 1982. Relative equilibria of vortices in two dimensions. *Proc. Natl. Acad. Sci. USA* 79:716–18

Papailiou, D. D., Lykoudis, P. S. 1974. Turbulent vortex streets and the entrainment mechanism of the turbulent wake. *J. Fluid. Mech.* 62:11–31

Patnaik, P. C., Sherman, F. S., Corcos, G. M. 1976. A numerical simulation of Kelvin-Helmholtz waves of finite amplitude. *J. Fluid Mech.* 73:215–40

Peskin, C. S., Wolfe, A. W. 1978. The aortic sinus vortex. *Fed. Proc.* 37:2784–92

Pierrehumbert, R. T. 1980. A family of steady translating vortex pairs with distributed vorticity. *J. Fluid Mech.* 99:129–44

Pierrehumbert, R. T., Widnall, S. E. 1981. The structure of organized vortices in a free shear layer. *J. Fluid Mech.* 102:301–13

Poincaré, H. 1893. *Théorie des Tourbillons*, ed. G. Carré, Chapt. 4. Paris: Deslis Frères. 211 pp.

Powell, A. 1964. Theory of vortex sound. *J. Acoust. Soc. Am.* 36:177–95

Rhines, P. B. 1979. Geostrophic turbulence. *Ann. Rev. Fluid Mech.* 11:401–41

Roberts, K. V., Christiansen, J. P. 1972. Topics in computational fluid mechanics. *Comput. Phys. Commun.* 3:14–32 (Suppl.)

Roberts, P. H., Donnelly, R. J. 1974. Superfluid mechanics. *Ann. Rev. Fluid Mech.* 6:179–225

Rogler, H. 1978. The interaction between vortex-array representations of free stream turbulence and semi-infinite flat plates. *J. Fluid Mech.* 87:583–606

Rosenhead, L. 1931. The formation of vortices from a surface of discontinuity. *Proc. R. Soc. London Ser. A* 134:170–92

Roshko, A. 1976. Structure of turbulent shear flows: A new look. *AIAA J.* 14:1349–57

Rossow, V. J. 1977. Convective merging of vortex cores in lift-generated wakes. *J. Aircr.* 14:283–90

Saffman, P. G. 1971. On the spectrum and decay of random two-dimensional vorticity distributions at large Reynolds number. *Stud. Appl. Math.* 50:377–83

Saffman, P. G. 1979. The approach of a vortex pair to a plane surface in inviscid fluid. *J. Fluid Mech.* 92:497–503

Saffman, P. G. 1981. Dynamics of vorticity.

J. Fluid Mech. 106:49–58

Saffman, P. G., Baker, G. R. 1979. Vortex interactions. *Ann. Rev. Fluid Mech.* 11:95–122

Saffman, P. G., Schatzman, J. C. 1981. Properties of a vortex street of finite vortices. *SIAM J. Sci. Stat. Comput.* 2:285–95

Saffman, P. G., Schatzman, J. C. 1982. Stability of a vortex street of finite vortices. *J. Fluid Mech.* 117:171–85

Saffman, P. G., Szeto, R. 1980. Equilibrium shapes of a pair of equal uniform vortices. *Phys. Fluids* 23:2339–42

Saffman, P. G., Szeto, R. 1981. Structure of a linear array of uniform vortices. *Stud. Appl. Math.* 65:223–48

Sayeed-Ur-Rahman. 1971. Berechnung der Schallerzeugung beim frontalen Zusammenstoss zweier Wirbelpaare. *Acustica* 24:50–54

Scott, A. C., Chu, F. Y. F., McLaughlin, D. W. 1973. The soliton: A new concept in applied science. *Proc. IEEE* 61:1443–83

Sedov, Yu. B. 1976. Evolution of the energy spectrum of an eddy system. *Izv. Akad. Nauk SSSR Mekh. Zhidk. Gaza* 11:43–48 (In Russian)

Sheffield, J. S. 1977. Trajectories of an ideal vortex pair near an orifice. *Phys. Fluids* 20:543–45

Siggia, E. D., Aref, H. 1980. Scaling and structures in fully turbulent flows. *Ann. NY Acad. Sci.* 357:368–76

Siggia, E. D., Aref, H. 1981. Point-vortex simulation of the inverse energy cascade in two-dimensional turbulence. *Phys. Fluids* 24:171–73

Singh, K. R. 1954. Path of a vortex round the rectangular bend of a channel with uniform flow. *Z. Angew. Math. Mech.* 34:432–35

Sommerfeld, A. 1964. *Mechanics of Deformable Bodies*, Chapt. 4. New York/London: Academic. 396 pp.

Stern, M. E. 1975. Minimal properties of planetary eddies. *J. Mar. Res.* 33:1–13

Stuart, J. T. 1967. On finite amplitude oscillations in laminar mixing layers. *J. Fluid Mech.* 29:417–40

Stüber, B. 1970. Schallabstrahlung und Körperschallanregung durch Wirbel. *Acustica* 23:82–92

Su, C. H. 1979. Motion of fluid with constant vorticity in a singly-connected region. *Phys. Fluids* 22:2032–33

Synge, J. L. 1949. On the motion of three vortices. *Can. J. Math.* 1:257–70

Takaki, R., Kovasznay, L. S. G. 1978. Statistical theory of vortex merger in the two-dimensional mixing layer. *Phys. Fluids* 21:153–56

Taneda, S. 1959. Downstream development of the wakes behind cylinders. *J. Phys. Soc. Jpn.* 14:843–48

Tatsuno, M., Honji, H. 1977. Two pairs of rectilinear vortices. *J. Phys. Soc. Jpn.* 42:361–62

Thomson, J. J. 1883. *A Treatise on the Motion of Vortex Rings.* London: Macmillan. 124 pp.

Thorpe, S. A. 1968. A method of producing a shear flow in a stratified fluid. *J. Fluid Mech.* 32:693–704

van de Vooren, A. I. 1980. A numerical investigation of the rolling-up of vortex sheets. *Proc. R. Soc. London Ser. A* 373:67–91

Weihs, D. 1973. On the existence of multiple Kármán vortex-street modes. *J. Fluid Mech.* 61:199–205

Whitham, G. B. 1974. *Linear and Nonlinear Waves*, Chapt. 4. New York: Wiley. 636 pp.

Widnall, S. E. 1975. The structure and dynamics of vortex filaments. *Ann. Rev. Fluid Mech.* 7:141–65

Wille, R. 1960. Kármán vortex streets. *Adv. Appl. Mech.* 6:273–87

Williams, D. R., Hama, F. R. 1980. Streaklines in a shear layer perturbed by two waves. *Phys. Fluids* 23:442–47

Winant, C. D., Browand, F. K. 1974. Vortex pairing: A mechanism of turbulent mixing-layer growth at moderate Reynolds number. *J. Fluid Mech.* 63:237–55

Wolibner, W. 1933. Un théorème sur l'existence du mouvement plan d'un fluide parfait, homogène, incompressible, pendant un temps infiniment long. *Math. Z.* 37:698–726

Yamada, H., Matsui, T. 1978. Preliminary study of mutual slip-through of a pair of vortices. *Phys. Fluids* 21:292–94

Yarmchuk, E. J., Gordon, M. J. V., Packard, R. E. 1979. Observation of stationary arrays in rotating superfluid helium. *Phys. Rev. Lett.* 43:214–17. Also in *Phys. Today* 32(12):21

Zabusky, N. J. 1977. Coherent structures in fluid dynamics. In *The Significance of Nonlinearity in the Natural Sciences*, ed. A. Perlmutter, L. F. Scott, pp. 145–205. New York: Plenum. 423 pp.

Zabusky, N. J. 1980. Contour-Dynamics: A boundary-integral evolutionary method for incompressible dissipationless flows. In *Numerical Methods for Engineering, GAMNI 2nd Int. Congr.*, ed. E. Absi et al., pp. 503–13. Paris: Dunod

Zabusky, N. J. 1981a. Computational synergetics and mathematical innovation. *J. Comput. Phys.* 43:195–249

Zabusky, N. J. 1981b. Recent developments in contour dynamics for the Euler equations. *Ann. NY Acad. Sci.* 373:160–70

Zabusky, N. J., Deem, G. S. 1971. Dynamical evolution of two-dimensional unstable shear flows. *J. Fluid Mech.* 47:353–79

Zabusky, N. J., Hughes, M. H., Roberts, K. V. 1979. Contour dynamics for the Euler equations in two dimensions. *J. Comput. Phys.* 30:96–106

Ziglin, S. L. 1980. Nonintegrability of a problem on the motion of four point vortices. *Sov. Math. Dokl.* 21:296–99

Ann. Rev. Fluid Mech. 1983. 15:391–427

THE FORM AND DYNAMICS OF LANGMUIR CIRCULATIONS

Sidney Leibovich

Sibley School of Mechanical and Aerospace Engineering,
Cornell University, Ithaca, New York 14853

> "...*the water itself is rippled by the wind. I see where the breeze dashes across it by the streaks...*"

> Henry Thoreau, *Walden* (1854)

INTRODUCTION

When the wind blows at modest speeds over natural bodies of water, numerous streaks or slicks nearly parallel to the wind direction may appear on the surface. This form of surface streakiness is commonplace, and under favorable conditions it is readily apparent to the casual observer. The streaks result from the collection of floating substances—seaweed, foam from breaking waves, marine organisms, or organic films—into long narrow bands. Flotsam makes the bands visible directly, and compressed films make them visible by the damping of capillary waves, thereby giving the bands a smoother appearance. Naturalists and seafarers often note color variations of the sea due to minute marine organisms. Bainbridge (1957) cited many old descriptions of long narrow "bands," "streaks," or "lanes" including several by Darwin in 1839 during the voyage of the *Beagle*. James Thomson (1862) described observations of streaks made jointly with his brother, Lord Kelvin, in a paper that also indicated increased abundances of marine life below the streaks. The first connection between the wind and streak directions, among the authors cited by Bainbridge, was made by Collingwood (1868): "if a moderate breeze were blowing and the sea

391

0066-4189/83/0115-0391$02.000

raised, instead of a uniform pellicle, the dust would be arranged in long parallel lines, bands or streaks, extending unbroken as far as the eye could reach, and always taking the direction of the wind."

In a remarkable series of experiments, Langmuir (1938) showed that these streaks, or "windrows," are the visible manifestations of a parallel series of counterrotating vortices in the surface layers of the water with axes nearly parallel to the wind. The motions strongly resemble thermal convection rolls. In fact, with weak winds and thermally unstable conditions, buoyancy may be the motive force, the resulting thermal convection being organized into rolls aligned with the wind by shear in the water according to experimentally and theoretically established characteristics of thermally unstable Couette flow. Langmuir, and numerous investigators after him, showed that the rolls also form with nearly equal ease under conditions of stable density stratification, and he concluded that the motions were mechanically driven, ultimately by the wind. The motions are now commonly known as Langmuir circulations, regardless of the mechanism responsible for their formation, but those of mechanical origin are peculiarly Langmuir's.

His observations led Langmuir to believe that the vortices he discovered are largely responsible for the formation of thermoclines and the maintenance of mixed layers in lakes (and presumably also in the ocean). If this is true, they assume a special kind of significance. This is partly because of the importance of the mixed layer and the heat, mass, and momentum transport processes within it. Even more intriguing, however, is the attribution of a dominant mixing role to a flow structure whose apparent orderliness and coherence suggests the possibility of a deterministic explanation and treatment. The expectations and hopes raised by this suggestion are clearly shared by those interested in the more recently discovered coherent structures in turbulent flows; it is possible that Langmuir circulations are related to coherent streamwise vortices in turbulent wall layers. At present, while the existence and many features of Langmuir circulations are generally accepted, experimental evidence that they accomplish a large part of the stirring of the mixed layer is ambiguous.

Langmuir circulations have been the subject of previous reviews by Faller (1971) and Pollard (1977); these reviews will be helpful to the reader. Pollard's review provides a valuable summary of observations and an assessment of the state of theories as they existed in 1976. Development has been rapid since that time: new field observations, laboratory experiments, and theoretical work are now available. These are described in what follows, along with the earlier work.

THEORIES

Many mechanisms for Langmuir circulations have been proposed since Langmuir first reported his observations. Most of these easily can be shown to be nonessential to the phenomenon, and have therefore been dismissed. These early ideas are briefly mentioned, followed by theories not so easily discounted, although serious objections can also be raised about some members of this latter group. Before reviewing the theories, however, I first give a sketch of the phenomenon these theories must explain.

Qualitative Features of Observed Langmuir Circulations

A composite of observed characteristics of Langmuir circulations is given here without documentation; the sources from which the picture is drawn are reviewed in a subsequent section.

Langmuir circulations form, aligned within a few degrees of the wind direction, within a few minutes of onset of a wind of 3 m s^{-1} or faster. Circulations form regardless of surface heating or cooling; with surface cooling (thermally unstable conditions), the circulations may form at a lower wind speed and the subsurface motion may be more vigorous, but for wind speeds of 3 m s^{-1} or faster, surface heating or cooling appears to only slightly modify the strength and form of the resulting circulations. The spacing of windrows ranges from a meter or two up to hundreds of meters; the factors determining this spacing have yet to be conclusively determined. A hierarchy of spacings is often observed, with smaller, more irregular, and less well-defined streaks occurring between stronger and more widely spaced streaks. When there is a hierarchy, the evidence suggests that the small scales continually form and are slowly swept up into the more permanent larger scales. The circulations are in the form of parallel vortices with axes aligned with the wind direction, as shown in the sketch in Figure 1a. The depth of penetration of the cells appears to be limited to the first significant density gradient in the water body, and there is evidence suggesting that the aspect ratio ($L/2D$) of the largest cells is not significantly different from unity, where L is the spacing between windrows and D is the depth of penetration.

Surface streaks are lines of surface convergence, and downwelling (vertical) motion takes place below them. Figure 1b is a photograph of bands of oil collected in convergences; such collections often occur in oil spills at sea. The downwelling takes the form of powerful jets, with maximum speeds $|w_d|$ of roughly one percent of the wind speed, or one third to one quarter of the maximum wind-induced surface currents. The widths of the jets seem to be quite small compared to streak separations. Surface-sweeping speeds are comparable to the downwelling speeds and probably decay rapidly with depth. The surface-current component (u) in the wind direction is noticeably larger in the streaks above the downwelling jet; the u anomaly is thus also comparable to the downwelling speed $|w_d|$.

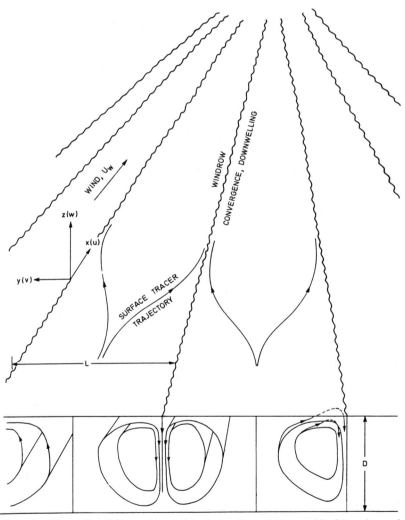

Figure 1a   Illustration of Langmuir circulations showing notation used in this review and surface and subsurface motions.

Craik & Leibovich (1976) presented a list that abstracts from this description a minimum of qualitative features that any viable theory of Langmuir circulations must be able to reproduce.

LC (1)   A parallel system of vortices aligned with the wind must be predicted.

LC (2)   A means must be given by which these vortices are driven by the wind.

LC (3)   The resulting cells must have the possibility of an asymmetric structure with downwelling speeds larger than upwelling speeds.

LC (4)   Downwelling zones must be under lines where the wind-directed surface current is greatest.

LC (5)   The Langmuir circulations must have maximum downwelling speeds comparable to the mean wind-directed surface drift.

These still constitute a minimal checklist against which to compare a theory.

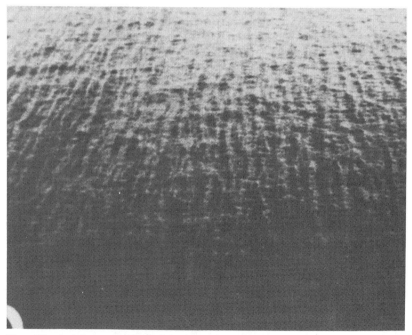

Figure 1b   Aerial photograph of oil bands in the Gulf of Mexico collected by Langmuir-circulation sweeping of the surface. The oil originated from the IXTOC I well blowout. The photograph is taken from Atwood et al. (1980).

Early Ideas Concerning Mechanism

Thermal convection has been frequently suggested as the mechanism driving Langmuir circulations, and this idea was discussed at some length by Csanady (1965). Langmuir (1938) himself discounted the idea, clearly stating his belief that the wind was the responsible agent. Thermal convection cannot explain circulations arising under thermally stable or neutral conditions, as Csanady (1965) notes, and has been rejected as a primary mechanism.

Effects produced by surface films have been proposed by Welander (1963) and Kraus (1967) as candidates for a mechanism; neither author reduced his thoughts to mathematical terms. Welander suggested that surface slicks would modify the wind field near the water surface. Damping of capillary waves in slicks produces a smoother surface and presumably a higher wind speed. This in turn is supposed to accelerate the water, producing a secondary inflow into the slick region. This hypothetical sequence, in itself, is hardly clear, since enhancement of wind speed presumably results from a reduction in drag, and therefore the water, if anything, should decelerate at the slicks. Furthermore, no wind-speed anomalies of the sort proposed are known to be correlated with windrows. Consequently, this mechanism has been discounted (see also Pollard 1977). Kraus (1967) suggested that radiation pressure due to the damping of capillary waves in slicks could lead to acceleration of slicks in the wind direction. Several arguments have been advanced to show that this is unlikely (Pollard 1977), but perhaps the most telling is the energy budget given by Myer (1971). It is also worth noting that circulation patterns form in laboratory experiments, such as Faller's (1969), in the absence of films.

The possibility that streaks result from roll vortices in the atmosphere, presumably originating in a paper by Woodcock & Wyman (1946), has been discounted by Stommel (1951), Langmuir (in work cited by Myer 1971), and Myer (1971). While such atmospheric vortices do arise, their positions seem to bear no relation to the streaks, and their effect is too weak to account for the water motion.

Concepts involving purely irrotational wave effects were put forward by Stewart & Schmitt (1968) and Faller (1969), but Langmuir circulations are fundamentally rotational, as Faller (1971) noted in retracting his proposal, and these suggestions make no provision for the necessary streamwise vorticity. It should be noted, however, that Faller's (1969) paper also contains experimental results and a clear statement of the essential physical ingredients required to produce Langmuir circulations, and it therefore remains a key paper in the development of the subject.

Instabilities of the Ekman layer are known to take the form of convective rolls at a small angle to the wind, and were proposed by Faller (1964) as a mechanism for Langmuir circulations. The growth rates of such instabilities are too small to account for Langmuir circulations. More recently, Gammelsrod (1975) again advanced this idea. Although more than one physical situation may correspond to the basic flow he considered, only one seems germane to Langmuir circulations in the ocean. This contemplates a balance between viscous and Coriolis forces in a surface layer bounded below by a strong stable density gradient at a depth H small compared to the Ekman scale depth. The density gradient is assumed to act as an impenetrable boundary to vertical motions, but it does not affect the basic Ekman-layer structure. In this case, Ekman's classic solution (1905) yields a current in the mixed layer that is essentially unidirectional and uniformly sheared. Gammelsrod considered the linearized stability of this current to roll motions parallel to the current direction. The inclusion of the Coriolis acceleration is crucial, since in its absence no instability occurs. When the dimensionless shear rate $f^{-1}\Delta U/H$ is sufficiently large (where ΔU is the basic current variation across the layer with thickness H and f is the Coriolis parameter) Gammelsrod found that small inviscid disturbances will amplify on a time scale consistent with the growth time of Langmuir cells.

This conclusion is at odds with the results of more complete treatments of Ekman layer instability by Lilly (1966) and by Faller & Kaylor (1966). In particular, their analyses show that their "parallel" modes (corresponding to Gammelsrod's solutions) have growth rates of the order of $f^{3/4}$. Gammelsrod's model produces growth rates that have finite limits as $f \to 0$, or as the Rossby number of the basic flow tends to infinity. A significant Coriolis effect in such a situation is most peculiar and contradicts previous experience. It indicates a discontinuity of the growth rate as f increases from zero, since at $f = 0$ the basic state, plane Couette flow, is known to be stable. There is therefore a puzzle concerning Gammelsrod's stability analysis which bears further examination. Leibovich & Radhakrishnan (1977) and Pollard (1977) have questioned the applicability of Gammelsrod's theory to Langmuir circulation on other grounds.

Wave-Current Interaction Theories

Theories involving the distortion of vortex lines in the current by the action of surface waves have been extensively explored recently, and appear to be capable of predicting the observed features of Langmuir circulations. Two independent but related theories, that of Garrett (1976) and the Craik-Leibovich theory, have been advanced. In both approaches, the motion in the upper layers is assumed to consist of a nearly irrotational wave field and a weaker rotational current, and circulations are driven through nonlinear interactions between the waves and the currents. The theories differ in the mechanisms by which the interactions engender circulations. Common ground shared by these theories is discussed by Leibovich (1980). This paper also discusses a similar model due to Moen (1978) that combines certain features of the Garrett and Craik-Leibovich theories.

Garrett's model depends upon diffraction of surface waves by the u-current anamoly: according to a WKBJ analysis, the surface waves will amplify in the current-anomaly region. Such a picture is consistent with an increase in wave height in streaks reported by Myer (1971) in observations in Lake George. The indicated amplification of waves suggests preferential wave breaking and consequent momentum transfer from the wave field to the currents in streaks. This process is assumed to occur and is modeled in an ad hoc fashion. A wave-current interaction also leads to a force on the water toward the streaks. The combination of this surface convergence and the momentum transfer due to wave breaking leads to an instability below the zone of significant wave activity, provided frictional effects are accounted for there. Thus an infinitesimal anomaly would tend to amplify and produce motions of roll type, with axes presumably aligned with the general direction of wave propagation; this, in turn, will be in the wind direction if the waves are generated by the wind.

Garrett's model thus depends upon the modification of the wave field by weak horizontal variations of current speed, and upon preferential wave breaking. The existence of each effect has recently been cast in doubt, the first on theoretical grounds and the second on experimental grounds. The amplification of waves by a sheared current, predicted by the WKBJ method, may be misleading. Smith (1980) examined the question without appeal to the WKBJ approximation, and was unable to find the amplification predicted by the approximate method. Although his findings cannot be regarded as conclusive, the basis for this principal ingredient of the mechanism is at present without a firm theoretical foundation. The assumption of preferential wave breaking within streaks appears to be contradicted by experiments conducted on Loch Ness by Thorpe & Hall (1980) and on Lake-of-the-Woods by Kenney (1977); these authors find wave breaking to occur with equal frequency between streaks and in them.

The Craik-Leibovich, or "CL," theories have been shown to produce circulatory motions by two distinct theoretical mechanisms, both of which depend upon wave-current interactions. The theories are "rational" in the sense that they are based upon a set of nonlinear equations derived from the Navier-Stokes equations by perturbation procedures that formally indicate the level of error anticipated. The equations defining the theories were first set out by Craik & Leibovich (1976), and evolved from an earlier attempt by Craik (1970) and its critique by Leibovich & Ulrich (1972).

Following Leibovich & Ulrich (1972), the motion in the surface layer is assumed to be dominated by the orbital motion of irrotational surface gravity waves. The rotational currents that are the objects of the investigation are assumed to be smaller, typically comparable to the Stokes mass drift

$$\mathbf{u}_s = \left\langle \int^t \mathbf{u}_w \, dt \cdot \nabla \mathbf{u}_w \right\rangle, \tag{1}$$

where $\mathbf{u}_w$ is the velocity vector of the orbital motion induced by the water waves, the angle brackets correspond to an appropriate averaging operation, and $\mathbf{u}_s$ is the resulting Stokes drift (Phillips 1977) associated with the surface waves. Vorticity in the water body may arise from currents whose origins are unspecified, or, as explored by subsequent workers, by diffusion whose source is an applied wind stress at the water surface. Surface waves perturb this configuration, producing a vorticity fluctuation correlated with the waves, and the nonlinear interaction of the fluctuating vorticity with the surface waves produces a stretching and rotation of the vortex lines. By averaging, assuming the waves are either periodic or random but statistically stationary, it is found that the rectified effects of the waves arise from additional advection and stretching of mean vorticity by the wave Stokes drift, as anticipated in the elementary inviscid treatment of Leibovich & Ulrich (1972).

The CL model is extended to allow time development of the currents and is made more systematic in Leibovich (1977a), and a further extension allowing for density stratification (under the Boussinesq approximation) is given by Leibovich (1977b). In these papers, the effects of turbulent fluctuations are crudely represented by constant eddy diffusivities, but more sophisticated turbulence models could be invoked.

The governing equations for the rectified water motion under the Boussinesq approximation and the assumption of constant eddy diffusivities of momentum (ν_T) and heat (α_T) are (Leibovich 1977b)

$$\partial \mathbf{v}/\partial t + \mathbf{v} \cdot \nabla \mathbf{v} = -\nabla \pi + \mathbf{u}_s \times \mathrm{curl}\, \mathbf{v} + \beta g \theta \mathbf{k} + \nu_T \nabla^2 \mathbf{v}$$

$$\partial \theta/\partial t + \mathbf{v} \cdot \nabla \theta = -w \overline{T}' + \alpha_T \nabla^2 \theta \tag{2}$$

$$\nabla \cdot \mathbf{v} = 0.$$

Here the temperature distribution in the absence of circulations is $\overline{T}(z)$, θ is the temperature perturbation, β is the coefficient of thermal expansion, g is the acceleration of gravity, $\mathbf{v}$ is the mean velocity vector in the currents, and π is a modified pressure term that includes the mean pressure as well as terms involving wave kinetic energy. These equations reduce to the usual Navier-Stokes equations of a Boussinesq fluid as $\mathbf{u}_s \rightarrow 0$, that is at sufficiently great depth. Near the surface, gravity waves must normally be included. It is possible to modify these equations to include Coriolis forces; for constant-density water, this has been done by Huang (1979), who showed that the Ekman layer is altered by the vortex force, and by Leibovich (1980).

The equations (2) governing the mean flow are the full equations for the instantaneous flow, altered only by the appearance of an apparent "vortex force" (Leibovich 1977b, 1980)

$$\mathbf{f} = \mathbf{u}_s \times \mathbf{\omega}, \tag{3}$$

where $\mathbf{\omega}$ is the mean vorticity. The waves are essentially unaffected by the currents if the conditions contemplated by the theory are met (as they are generally expected to be in the ocean and in lakes) so that they may be specified and $\mathbf{u}_s$ computed a priori, and the equations are therefore closed.

Heuristic Discussion of Two "Vortex Force" Mechanisms

When the wind blows in a fixed direction over water of unlimited horizontal extent and depth that is otherwise undisturbed, symmetry dictates the development of a surface wave field with unidirectional Stokes drift aligned with the wind or $\mathbf{u}_s = U_s\mathbf{i}$ (see Figure 1a for coordinate system). Neglecting Coriolis accelerations, the total momentum imparted to the water will be similarly aligned, and the horizontally averaged velocity will be $\mathbf{u} = U(z, t)\mathbf{i}$, with positive y-vorticity $\mathbf{\omega} = (\partial U/\partial z)\mathbf{j}$. The vortex force/mass

$$\mathbf{f}_v = \mathbf{k} U_s(\partial U/\partial z) \tag{4}$$

is oriented vertically upward and is formally analogous to a buoyancy force. If U_s depends only on depth z, then the vortex force can be balanced by the analog of a hydrostatic pressure gradient; this configuration represents a developing unidirectional current. If the Stokes drift varies across the wind (y-direction), however, the vortex force cannot be balanced by pressure, and an overturning is induced. Surface waves in a "short-crested" sea have this character. The directional spectrum of a wind-generated sea is symmetric with respect to the wind (Longuet-Higgins 1962). If it were bimodal, with peaks of wave energy at angles

$\pm\,\theta$ with respect to the wind, and if the wave spatial structure were to remain coherent for times sufficient to carry out the averaging operation inherent in the Stokes drift (many times a typical wave period), then the Stokes drift would be spatially periodic with a crosswind wave number of $2k\,\sin\theta$, where k is a characteristic wave number of the surface waves. This horizontally periodic wave drift produces a torque due to horizontal variations of vortex force that directly drives roll motions; this is the mechanism proposed by Craik & Leibovich (1976) and called the CL 1 mechanism by Faller & Caponi (1978). An alternate kinematic interpretation of the mechanism is that described by Leibovich & Ulrich (1972); vortex lines associated with the current are deformed by the Stokes drift, producing streamwise vorticity periodically (in y) alternating in sign. Both the dynamic and kinematic explanations of the CL 1 mechanism are represented pictorially in Figure 2, which also includes a schematic

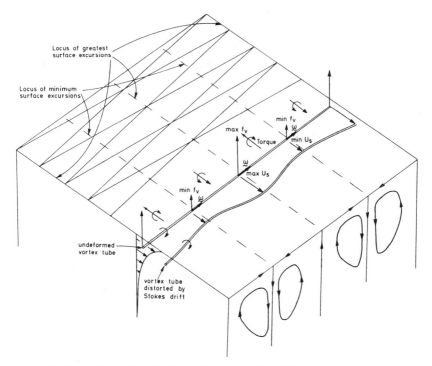

Figure 2 Sketch illustrating the direct-drive mechanism of Langmuir-circulation generation. Wave crests of the assumed crossed-wave pattern are shown; the Stokes drift is higher along lines joining crest intersections and the Stokes-drift variations distort vortex lines of the primary current. The resulting variations of vortex force create a torque leading to overturning.

diagram of idealized "crossed-wavetrains" assumed in calculations with the CL 1 mechanism.

A second mechanism, called CL 2 by Faller & Caponi (1978), was originally suggested by Craik (1977) and further explored by Leibovich (1977b). It requires no coherent surface-wave structure. Circulations are produced via the vortex force as an inviscid instability of the unidirectional current. If the waves lack a coherent spatial structure, $\mathbf{u}_s = U_s(z)\mathbf{i}$; the time development of the waves under the action of the wind is ignored, as it has been in all of the investigations carried out with wave-interaction models. The vortex force (4) can now be balanced by a vertical pressure gradient. Since U_s and U typically decrease monotonically with depth below the surface, the vortex force does as well, and so the joint effects of typical distributions of U_s and U are directly analogous to a statically unstable density distribution. One may therefore anticipate that the rectilinear current is unstable if dissipation is sufficiently weak. (By contrast, if the waves and current are *opposed*, then the vortex force is *stabilizing*.)

Both the dynamics and kinematics of onset of CL 2 mechanisms are traced schematically in Figure 3. Suppose an infinitesimal spanwise

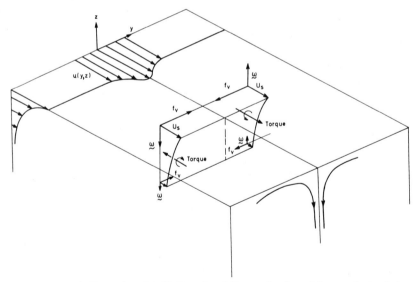

Figure 3 Sketch illustrating the CL 2, or instability mechanism of Langmuir-circulation generation. The Stokes drift is horizontal, but decays in depth. Streamwise vorticity is induced by the Stokes-drift rotation of vertical vorticity associated with spanwise perturbations of the current. Variations of vortex force caused by the current perturbation create torques leading to overturning.

irregularity $u(y, z, t)$ is present in an otherwise horizontally uniform current $U(z)$. This produces vertical vorticity $[\omega_z = -(\partial u / \partial y)]$ and a horizontal vortex-force component $-U_s \omega_z \mathbf{j}$ that is directed toward the planes of maximum u. This causes an acceleration toward these planes, where, by continuity, the fluid must sink. Assuming that $\partial U / \partial z > 0$ and ignoring shear stresses, conservation of x-momentum for a vanishingly thin slab of fluid centered on the convergence plane shows that as the fluid sinks, u must increase. Thus, in the absence of frictional effects, a current anomaly will lead to a convergence and thereby be amplified, which in turn further amplifies the convergence. Kinematically, the vertical vorticity is rotated and stretched by the Stokes drift, leading to convergence and amplification of the anomaly.

DIRECT-DRIVE MECHANISM   Craik & Leibovich (1976) set out the theoretical basis for the CL 1 mechanism, explored aspects of the linearized CL equations, and presented numerical results for weakly nonlinear motions in infinitely deep water assuming "crossed-waves" and invariance of the rectified motions along the wind direction (x). The theory was restricted by an assumption of time-independent motion, which led to a need to specify the horizontally averaged drift in an ad hoc manner. (A steady horizontally averaged current corresponding to an applied surface stress is not possible in infinitely deep water.) This restriction precluded the calculation of a complete current system arising from a fully defined physical problem.

These restrictions were removed by Leibovich (1977a), who extended the theory to include time evolution of the coupled (wind-directed) currents and circulations. The formulation of the problem as an initial-value problem, with currents and circulations initially zero and initiated by a step function in surface stress, results in a well-posed mathematical problem. Assuming invariance of the wave field in the x- (wind) direction and its symmetry with respect to the x-axis, the problem is independent of x and any emerging circulations must be in the form of rolls. Assuming a constant wind stress for $t \geqslant 0$ corresponding to a friction velocity u_*, surface waves with characteristic frequency σ, wave number k and characteristic wave amplitude a, and an eddy viscosity v_T, the initial-value problem was shown to depend upon a single dimensionless parameter

$$\text{La} = \left(v_T^3 k^2 / \sigma a^2 u_*^2 \right)^{1/2} \tag{5}$$

This parameter, which Leibovich called the "Langmuir number," expresses a balance between the rate of diffusion of streamwise vorticity and the rate of production of streamwise vorticity by the vortex stretching accomplished by the Stokes drift: it also has the usual interpretation of an inverse Reynolds number. The reduction of the initial-value problem to dependence on a single dimensionless parameter (plus an angle representing the directional properties of the waves if this feature is invoked) is a consequence of x-invariance and implies adoption of the scalings shown in Table 1. Other scalings can be adopted, but these lead to the greatest simplifications. They emphasize a balance between vortex force and the applied shear stress when x-variations are negligible; they are therefore appropriate also to problems involving the CL 2 instability mechanism when the same set of assumptions is applicable. Leibovich (1977a) assumed the water to be infinitely deep and of constant density, and the Stokes drift to result from an idealized narrow spectrum with energy concentrated in a pair of uniform wave trains propagating at equal and opposite angles to the wind. The fully nonlinear problem was explored numerically by Leibovich & Radhakrishnan (1977). When the Coriolis force is neglected the problem has no steady limit, because in infinitely deep water, there is no bottom friction to balance the applied surface stress: thus the momentum of the water body increases linearly with time. Nevertheless, a definite current structure forms near the surface within about 10 T_d, where T_d is the time scale from Table 1, and changes little thereafter. The continuing momentum increase is accommodated by cellular mixing of momentum to ever-deeper waters. Horizontal spacing of convergence lines is fixed by the Stokes drift and shows its (assumed) periodicity. Downwelling occurs below convergence lines coinciding with lines of *minimum* Stokes drift, and upwelling under lines of *maximum* Stokes drift, as first shown by Leibovich & Ulrich (1972) in their simple inviscid model. All qualitative features [LC (1) to LC (5)] are reproduced. Maximum surface-sweeping speeds are close to the maximum downwelling speeds, and maximum downwelling speeds range from slightly less than twice to more than seven times the maximum upwelling speed, depending upon the wave angle θ and elapsed time. Sharply peaked surface-current anomalies occur over downwelling zones, and exceed the minimum surface speeds midway between streaks by 55–75%. In addition, Leibovich (1977a) shows that the horizontally averaged

Table 1   Scalings for the CL equations

Length	Time (T_d)	Wind-directed component of rectified current	Crosswind velocity component of rectified currents
k^{-1}	$(v_T/\sigma)^{1/2}/aku_*$	$u_*^2/v_T k$	$(au_*/v_T)(\sigma v_T)^{1/2}$

currents very near the surface predicted by this deterministic model have the logarithmic profile characteristic of wall-bounded turbulent flows. Logarithmic behavior of wind-drift current profiles is known experimentally to occur (Bye 1965, Wu 1975, Shemdin 1972).

The assumption of a spatial pattern of the Stokes drift coherent with any given line on the water surface is basic to the CL 1 theory and is regarded as its weakest point (see the discussions of Pollard 1977, Leibovich & Radhakrishnan 1977, and Leibovich 1977b). Patterns of the sort contemplated do arise in wind-generated seas (Kinsman 1965, p. 543), and a possible theoretical basis for bimodal directional spectra has been advanced by Longuet-Higgins (1976) and Fox (1976), based on the resonant interaction theory of Davey & Stewartson (1974). The formation of a Stokes-drift pattern requires a given wave pattern to be phase-locked for several wave periods. While this is plausible, the formation of well-developed circulations by the CL 1 mechanism seems to require the Stokes drift pattern to remain fixed for times of the order of $10\ T_d$, which typically is hundreds of wave periods. Phase-locking for times as long as this apparently is not expected in a wind-generated sea, although it can be achieved in the laboratory (Faller 1978).

Mobley (1977) has attempted to check the CL 1 theory by a direct, three-dimensional numerical simulation, including "crossed" surface waves, using a finite difference form of the full Navier-Stokes equations. He was unable to find evidence of the CL 1 mechanism. The numerical task is clearly formidable, and uncertainities have been expressed (Faller & Caponi 1978) about the reliability of the computations. Putting other numerical questions aside, it appears from the data listed in Mobley (1977) that the finest spatial resolution used leads to grid Reynolds numbers larger than 55 (Mobley cites smaller values, but the formula he invokes for grid Reynolds number is in error and grossly underpredicts it), and is much too coarse to resolve the motion. The numerical scheme is inaccurate as a consequence; it also produces artificial viscosity, and it appears that the computed results may be contaminated by this effect.

INSTABILITY MECHANISMS (CL 2)   The linear stability of a unidirectional sheared current in the presence of a parallel, spanwise uniform Stokes drift has been explored for water of constant density by Craik (1977) and for density-stratified water by Leibovich (1977b) using special cases of Equations (2). Both papers deal only with rolls periodic in the crosswise directions and estimate (but do not compute) conditions for marginal stability for constant current shear and constant Stokes-drift gradient based upon an analogy with thermal convection. In addition, Craik

(1977) computed growth rates assuming inviscid fluid for several sample current and Stokes-drift profiles. Leibovich (1977b) also computed growth rates ignoring viscosity and heat conductivity, and discussed stability criteria and characteristics for general temperature, Stokes-drift, and current profiles.

The principal result found by Leibovich (1977b) for an inviscid, nonconducting fluid of infinite depth is this: the system is stable if

$$M(z) = U_s'(z)U'(z) - N^2(z) \tag{6}$$

is everywhere negative, and it is unstable otherwise. Here $N = \sqrt{\beta g \overline{T}'(z)}$ is the Brunt-Väisälä frequency of the basic state, and U_s' and U' are the vertical gradients of the Stokes drift and of the shear currents. If unstable, and if the maximum value of M is obtained at the surface, then the maximum growth rate σ_{max} is given by

$$\sigma_{max} = \sqrt{M(0)} \tag{7}$$

and is obtained for waves of infinitesimal length. In an unstable system with a stable density stratification, $N^2 \geqslant 0$, there will be a characteristic depth below which no disturbances penetrate. The layer above is unstable to waves of all lengths in the crosswind direction, with very short wavelength disturbances felt only in a correspondingly thin layer near the surface and most subjected to the neglected dissipative effects. Disturbances of greater length grow more slowly, but penetrate to greater depths.

According to (6), stability occurs for

$$\text{Ri}^* = \min\left[N^2(z) / U'(z) U_s'(z) \right] > 1 \tag{8}$$

when the minimum is taken over depth. This is in the form of a gradient Richardson number, with the geometric mean of U' and U_s' replacing the usual shear.

Leibovich & Paolucci (1980a,b, 1981) have explored the CL 2 instability with nonzero eddy diffusivities; the basic state considered was constant positive N^2 and the exact time-dependent, unidirectional current solution of (2) in infinitely deep water with a surface stress applied at $t = 0$ and held constant thereafter. The dimensionless problem is characterized by the Langmuir number La of Equation (5), and a characteristic Richardson number

$$\text{Ri} = \beta g \overline{T}' / \left[(au_* k)^2 \sigma / v_T \right]. \tag{9}$$

This is similar to Equation (8), with U' based upon the current shear at the surface and the characteristic Stokes drift gradient based upon $a^2 k^2 \sigma$. In these papers, the Stokes drift was taken to be $U_s = 2a^2 k \sigma \exp(2kz)$, twice the Stokes drift of a single wave train with wave parameters (a, k, σ); the factor of 2 was introduced to facilitate comparison with the crossed-wave solution of Leibovich & Radhakrishnan (1977). Thus (9), as used by Leibovich & Paolucci, is 4 times (8), and inviscid stability would be expected for Ri $\geqslant 4$ in this case. Numerical results for linear-stability limits given in Table 2 were found by Leibovich & Paolucci (1981) for two-dimensional rolls. They found the energy-stability limit $La_G^{-1} = 1.46$, occurring at $k_G^{-1} = 0.32$; this result is independent of Ri (not, as they indicated, a weak function of Ri). The energy- and linear-stability limits are close; subcritical instability in the narrow gap between them is not ruled out. Consideration of general three-dimensional disturbances (Leibovich & Paolucci 1980b) leads to the conclusion that they are more stable on an energy-stability basis, and that two-dimensional rolls are likely to be preferred, in agreement with observation. One would expect the inverse critical Langmuir number (La_c^{-1}) to be infinite for Ri $\geqslant 4$, based upon the inviscid theory, but recent unpublished computations by Paolucci indicate otherwise.

Leibovich & Paolucci (1980a) traced the fully nonlinear evolution of the instability mechanism for constant N, with Ri $= 0.1$ and La $= 0.01$, using a computational domain with width comparable to the wavelength of the most unstable linear mode. S. K. Lele (private communication) has discovered a programming error in their computer code. The solutions presented are strictly valid only for Ri $= 0$, and the temperature profiles and mixing computations must be interpreted as those of a passive scalar. The solutions obtained are qualitatively like the direct-drive solutions of Leibovich & Radhakrishnan (1977), but with one interesting difference. Small-scale, very weak perturbations deliberately introduced grew not only in strength, but also cascaded to larger scales until they reached the maximum permitted by the finite size of the computational domain.

Cases with constant N, together with other examples with nonconstant N simulating preexisting mixed layers bounded by "thermoclines," have been computed by Leibovich & Lele (1982) using a corrected version of the Leibovich & Paolucci (1980a) computer code. For constant N (twice the nominal value used by Leibovich & Paolucci 1980a) and Ri* $= 0.05$,

Table 2   Critical inverse Langmuir number La_c^{-1} and critical wave number k_c of linear theory at various Richardson numbers

Ri	La_c^{-1}	k_c
0.00	1.52	0.32
0.10	1.58	0.31
0.25	1.66	0.30

the results are qualitatively similar to those reported by Leibovich & Paolucci (1980a), the principal difference being a reduction in the penetration rate as the convective motions mix the upper layers and form a layer of strong temperature gradient. An example of an instantaneous streamline pattern associated with the cellular convective motion is shown in Figure 4, while Figure 5 shows the horizontally averaged temperature that evolves from the initially linear profile. The development of a "thermocline" is of particular interest. The incorrect solution predicted a similar thermocline development, but the error precluded any dynamic effect. For $Ri^* \geq 0.125$, new features emerge; growth rates are smaller and the unstable motions are oscillatory. The motions appear qualitatively to be a mixture of monotonically growing circulations in homogeneous water and internal waves. For larger values of Ri^* (≥ 0.25), but well below the inviscid criterion (8), the system appears to be stable at the Langmuir number of 0.01 used in the calculations.

Preexisting thermoclines are found by Leibovich & Lele to act as an impenetrable "bottom" for convective motion, provided the temperature gradient is sufficiently strong and the thermocline is sufficiently thick. This is in accord with Langmuir's (1938) expectations.

FURTHER REMARKS ON THEORY   Waves and currents can exist in the absence of wind, of course, and their directions need not then be related. No work has yet been done using the CL theories when $\mathbf{u}_s$ is not parallel to the horizontally averaged current $\mathbf{U}$. Heuristic considerations of the action of the vortex force suggest that instability could occur whenever $\mathbf{u}_s \cdot \mathbf{U} \geq 0$, although there is no longer any reason to believe that rolls would be favored.

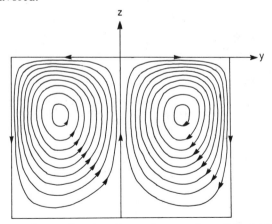

Figure 4   Computed streamlines in a Langmuir convection cell in fluid with a statically stable density stratification.

Other instabilities arising from the density-stratified form of the CL equations are possible, but have yet to be systematically explored. The linearized equations governing stability of a steady current $U(z)$ parallel to a Stokes-drift distribution U_s in a water body with temperature gradient $\bar{T}'$ are given in Leibovich (1977b); disturbances in the form of rolls aligned with U satisfy Equations (18) of that paper. Assuming a strong thermocline exists at a depth d and acts as a stress-free "bottom" to convective motion, and that the vertical gradients $U', U_s', \bar{T}'$ can be regarded as constants above the thermocline, Equations (18) of Leibovich (1977b) can be rescaled and rearranged in the following form

$$\left(\frac{\partial}{\partial t} - \nabla^2\right)\nabla^2 w = -R_U u_{yy} + R_T \theta_{yy} \quad \text{(a)}$$

$$\left(\frac{\partial}{\partial t} - \nabla^2\right)u = -w, \quad \text{(b)}$$

$$\left(\frac{\partial}{\partial t} - \frac{1}{\Pr}\nabla^2\right)\theta = -w, \quad \text{(c)}$$

$$\nabla^2 = \left(\partial^2/\partial y^2 + \partial^2/\partial z^2\right) \quad \text{(d)}$$

$$R_U = U'U_s'd^4/v_T^2, \quad R_T = \beta g\bar{T}'d^4/v_T^2, \quad \Pr = v_T/\alpha_T,$$

(10)

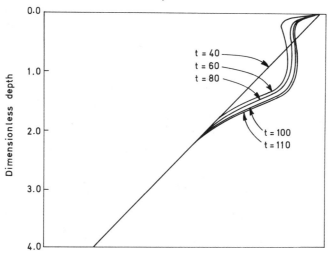

Horizontally averaged temperature
(arbitrary units)

$t = 40$
$t = 60$
$t = 80$

$t = 100$
$t = 110$

Dimensionless depth

Figure 5 Horizontally averaged temperature at various times after onset of Langmuir circulations by the instability mechanism. The initial state is one of constant temperature gradient. A strong temperature gradient, or thermocline, eventually evolves.

where w is the scaled vertical velocity component, u the scaled perturbation to the basic current U, and θ a scaled perturbation temperature. The scales used to make the equations dimensionless are the layer depth d for length, the momentum diffusion time d^2/v_T for time, v_T/d for vertical velocity, $U'd$ for the x-component (u), and $\overline{T}'d$ for temperature. The parameter R_U is in the form of a Rayleigh number based upon a "temperature" gradient ($U'U_s'/\beta g$), and R_T is essentially the negative of a conventional Rayleigh number. These equations are precisely those describing double diffusion of u and θ. The diffusivity of u is typically expected to be larger than that for θ, i.e. one expects Pr > 1. By reference to more familiar thermohaline double-diffusion problems (Turner 1973, Chap. 8), u is seen to act as temperature and θ as salt. The "Richardson number" of Equation (8) is Ri* = R_T/R_U, and the CL instabilities discussed so far treat only positive R_T and R_U. The literature on double diffusion indicates a number of other possibilities, shown schematically in the diagram of Figure 6 adapted from Baines & Gill (1969). Features in the right-hand quadrant $R_T, R_U > 0$ found in double-diffusion prob-

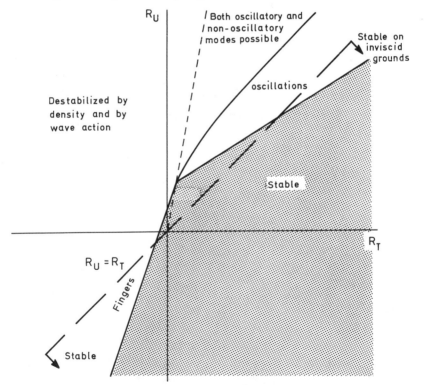

Figure 6 Diagram illustrating parametric regions in which wave-induced instabilities may be expected by analogy with double-diffusive instabilities.

lems have been also found for the Langmuir-circulation problem, including regions of monotonic growth as well as oscillatory growth (Leibovich & Lele 1982). The second quadrant, in which both the vortex force and temperature gradient are destabilizing, is intuitive. The "fingering" regions in the third quadrant correspond to a stabilizing vortex force (U' and U_s' opposed) and destabilizing thermal conditions leading to overall conditions $U'U_s' - \beta g \bar{T}' < 0$, which are stable if diffusion is neglected but are unstable when $\nu_T > \alpha_T$. This represents a new possibility for instability that presumably can exist in the ocean, but has yet to be explored.

Another possibility that has yet to be treated by the CL theories arises when the energy density of the surface waves has a nonzero horizontal divergence. This possibility is anticipated in the depth-averaged surface-convergence force derived by Garrett (1976). In the CL theories, it is represented by the adoption of a Stokes-drift velocity vector that has a nonzero vertical component.

Craik (1982a) has shown that the weak viscous Eulerian current induced by water waves without wind is unstable by the CL 2 mechanism, and that this may be also the case for the similar currents induced near a density interface by internal waves. Craik notes that this instability may be the cause of the scatter found in experiments on the drift current of surface water waves. These possibilities arise in the absence of wind, and are interesting to contemplate.

An alternate derivation of the CL equations was given by Leibovich (1980), using the exact "generalized Lagrangian mean" theory of Andrews & McIntyre (1978). The "GLM" theory presents a general set of equations incorporating the effects of wavy oscillations on the mean flow without the need for assumptions concerning the relative size of irrotational and rotational contributions. Leibovich (1980) suggested that this approach might be the route to a generalization of the CL theories to situations involving strongly sheared mean flows (when the waves can no longer be regarded as irrotational) or when finite wave amplitudes must be accounted for. Craik (1982b) has developed this line of thought farther, and has shown that such theoretical extensions are indeed possible. While the question of coherent structures in turbulent wall layers has not been addressed yet from this point of view, one cannot help but wonder if their physics can be most efficiently expressed by similar formalisms.

FIELD OBSERVATIONS

Measurements at sea are difficult to obtain. These difficulties and the lack of comprehensive theoretical hypotheses have resulted in data that are invariably incomplete in some way. To determine the structural form

of a circulation cell, measurements of temperatures and three components of velocity are required at a number of points sufficient to resolve
the motion. To educe the structure of a "typical" circulation cell, the
conditional sampling of a large number of cells formed under similar
environmental conditions is required. To search for a mechanism for the
formation of the cells, environmental conditions coexisting with the
circulations and conceivably related to them must be determined simultaneously. The experimental task is clearly formidable, and it is not
surprising that the available information is fragmentary.

The most significant environmental parameters seem to be wind speed
and the distribution of density with depth. Most, but not all, observations of windrows include measurements of wind speed; some also
include temperature profiles. Some information (Katz et al. 1965, Myer
1971) suggests the sea state is also important, but as this is strongly
coupled to the wind, it is difficult to isolate its independent significance;
only a few observers have described surface-wave activity. Observed
Langmuir-circulation characteristics include windrow spacing, speeds of
surface tracers, time required for realignment of windrows after a shift in
wind direction, time required for windrow formation, vertical velocities
below convergences and divergences, and the angle between the local
wind direction and surface streaks. A directory of observations is given in
Table 3. Some of those listed are of an incidental or anecdotal nature, but
they substantiate more systematic studies.

Table 3 A chronology of field observations of Langmuir circulations (see Figure 1a for
notation)

Investigator	Location(s)	Main observables reported
Langmuir (1938)	North Atlantic	L, row width time response
Langmuir (1938)	Lake George, N.Y.	U_w, (u, v, w) currents, L, water temperature profiles
Woodcock (1944)	Gulf of Mexico	v, u anomaly, air and water temperature
Woodcock (1950)	Atlantic	Drawdown of surface sargassum as function of U_w
Van Straaten (1950)	Shallow tidal flats in North Sea	L, v
Stommel (1951)	Ponds on Cape Cod	Reorientation time
Sutcliffe et al. (1963)	Atlantic off Cape Cod; Atlantic off Bermuda	U_w, $\lvert w_d \rvert$

Table 3   (*continued*)

Investigator	Location(s)	Main observables reported
Welander (1963)	Baltic Sea	U_w, u, v, reorientation
Faller & Woodcock (1964)	Atlantic 39°N, 60°W	L, surface heat flux, U_w
Faller (1964)	Atlantic off Provincetown, Mass.	Angle between rows and wind
Williams (1965)	Ocean	L, u, v
Ichiye (1965)	Atlantic off Long Island	L
Katz et al. (1965)	N.Y. Bight, Pacific off Argentina	L, U_w, u anomaly Sea state, reorientation
Ichiye & Plutchak (1966)	Long Island Sound	L, U_w
Owen (1966)	Pacific	L, air and water temperature
Ichiye (1967)	Ocean	Time required for row formation (measure of v), u anomaly
McLeish (1968)	Pacific off California	Row orientations
Scott et al. (1969)	Lake George	U_w, $|w_d|$, temperature perturbations below streaks
Myer (1969, 1971)	Lake George	U_w, air and water temperatures, temperature perturbations below streaks, $|w_d|$
Gordon (1970)	Plantagenet Bank off Bermuda	U_w, u anomaly
Assaf et al. (1971)	Atlantic off Bermuda	L, mixed-layer depth
Sutcliffe et al. (1971)	Sargasso Sea	Increased particulates below convergence
Maratos (1971)	Monterey Bay	L, U_w, angle between rows and wind, mixed-layer depth, u, $|w_d|$, air temperatures
Harris & Lott (1973)	Lake Ontario	U_w, $|w_d|$, air and water temperatures
Kenney (1977)	Lake-of-the-Woods	u, v, w, U_w, L, air and water temperatures
Filatov et al. (1981)	Lake Ladoga, Soviet Union	$|w_d|$, U_w
Thorpe & Hall (1982)	Loch Ness	Temperature perturbations below streaks, L, acoustic scattering from bubbles

Threshold Wind Speed

Surface streaking is generally reported to occur only when the wind speed exceeds some threshold value, most frequently placed at 3 m s^{-1} (Walther 1967). This value is not critical however; Myer (1971) observed streaking at lower wind speeds, with thermally unstable conditions requiring a somewhat lower threshold speed than necessary in thermally stable conditions. Ichiye (1967) found streaks "even in the calm sea when there are pronounced swells," a situation that presumably indicates no more than a slight breeze. Since visual detection of streaks is necessarily subjective, a variability in minimum required wind speeds is not surprising. Presumably the strength of surface-sweeping currents in Langmuir circulation scales with wind speed, and there will be some minimum sweeping speed required to overcome the random surface-speed fluctuations that resist organization of surface tracers into rows.

Surface Patterns: Spacing, Orientation, Formation Times, and Persistence

SPACING   Streaks have been identified either visually by the collection of surface tracers such as foam, seaweed, or organic films in convergence lines, or by infrared photography, which renders surface-temperature anomalies visible.

Langmuir (1938) reports that streak orientations are parallel to the wind, and spacings in Lake George are between 5 and 25 m with a variability associated with seasonal changes. He also writes of the difficulty in determining spacing: "Quantitative measurements of streak spacings are difficult because between the well-defined streaks there are numerous smaller and less well-defined streaks. Just as large waves have smaller waves upon them, it appears that the surfaces of the larger vortices contain smaller and shallower vortices. The patterns of streaks on the lake surface are slowly changing; some growing, some dying out."

Spacing data in lakes are reported by Scott et al. (1969), Myer (1969, 1971), Csanady & Pade (1969), Kenney (1977), and Thorpe & Hall (1982). The range of cell spacings quoted by Langmuir is generally confirmed in lakes and ponds, although spacings of less than a meter have been reported (Myer 1969).

Reported spacings in the ocean range from 2 to hundreds of meters. Very regular rows of zooplankton 1.5 m apart have been reported by Owen (1966), but he attributed these to thermohaline convection since the pattern was apparently not formed by wind action. The largest scales reported are the massive collections of seaweed in "bands 2 to 6 m wide with spacing from 1100-200 m" seen by Langmuir (1938) in the North

Atlantic in 1927 (it would appear that the 1100 figure is a misprint, and should have read 100), and the 280-m spacing observed by Assaf et al. (1971) near Bermuda.

Spacing hierarchies are reported in the ocean by Assaf et al. (1971), Williams (1965), McLeish (1968), and Katz et al. (1965), and in lakes by Langmuir (1938), Scott et al, (1969), Myer (1969, 1971), and Harris & Lott (1973). The time evolution of a set of marked windrows suggests a dynamical cascade from smaller scales towards some dominant large scale. Harris & Lott (1973), for example, report that the distance between streaks increased with time in Lake Ontario, and that new streaks appeared between streaks already marked. This is supported by oceanic measurements of Williams (1965) and Ichiye (1967). Figure 2 from Williams's report summarizes change in spacing with time. "Young" stripes form with spacings that peak at 2 m and 5 m; as they age, the fraction of small streaks decreases, the number at 5 m grows, and large scales seem to make their appearance. Ichiye's (1967) Figure 4 seems to illustrate the same phenomenon. This kind of consolidation of surface markers suggests a gradual sweeping of the surface (including smaller-scale cells) by the largest scale, and Ichiye's (1967) experiment confirms Harris & Lott's (1973) assumption of the reformation of smaller scales, which can be made visible by periodically reintroducing surface tracers.

The spacing of windrows was presumed by Langmuir (1938) and many others to be proportional to the depth of penetration of the cellular pattern that they mark, in analogy with other types of convective motion. For example, the linearly most unstable Benard roll cells would correspond to a spacing L/D of roughly 2. Langmuir believed, with some support from his measurements, that the largest cells penetrated essentially to the thermocline, which suggests a correlation of windrow spacing and depth to the thermocline.

The hypothetical relationship between spacing of the largest windrow scales and the mixed-layer depth has yet to be conclusively established, but there are no data to dispute it either. Langmuir noted that spacings are generally greater in the fall when the thermocline is deepest than in the spring and early summer when it is shallow. His own "velocity indicator" measurements, however, showed a penetration depth that fell short of the thermocline. Scott et al. (1969) found that a significant maximum in the temperature gradient typically exists between the free surface and the thermocline in Lake George. They found that streak spacing correlates with the depth to this first stable layer, and concluded that heat is readily transported to this depth by Langmuir circulations, but not deeper. Assaf et al. (1971) found spacings of the largest streak scales to be 280 m, with a mixed-layer depth of 200 m or $L/D = 1.4$.

The results of Maratos (1971) on the maximum ratio of spacing to thermocline depth are consistent with this; his (single) extreme realization yields a ratio $L/D = 1.66$, with all other realizations ranging between .66 and 1.42.

The thesis that the thermocline limits the maximum penetration of cells cannot be ruled out by a lack of correlation (found to be the case by Faller & Woodcock and by Maratos) between L and D. Clearly other dynamical processes are at work during the growth phase of cells, and the thermocline has no effect until their penetration depth approaches it. Thus, a significant correlation between spacing and thermocline depth would be expected only if the thermocline depths were comparable to or smaller than the other dynamically important length scales in the problem (which are not obvious a priori). A test of Langmuir's hypothesis is to be had by dealing with an ensemble consisting of pairs of *maximum* spacings and thermocline depths. There is, to date, no evidence to contradict Langmuir's hypothesis; rather, the evidence available must be regarded as supporting it.

A statistically significant correlation between mean spacing and windspeed is reported by Faller & Woodcock (1964). They take this to be $L = (4.8 \text{ s}) \times U_w$, where U_w is the windspeed at some unspecified height near the sea surface; the formula is not appropriate for $U_w < 3 \text{ m s}^{-1}$, for which no rows were observed. Katz et al. (1965) present data suggesting the possibility of a positive correlation with smaller slope L/U_w, and Maratos (1971) found the correlation $L = 0.1 \text{ m} + (2.8 \text{ s}) \times U_w$. In all of these cases, the number of samples and the range of wind speeds were small, and the data show considerable scatter. Scott et al. (1969) considered the question of spacing-windspeed correlation but did not find it to be statistically significant.

ORIENTATIONS   Windrows are always nearly aligned with the wind direction, with maximum angular differences between pairs of streaks being no more than 20°. Whether there is a systematic deviation from the wind direction has been discussed by Faller (1964), who analyzed oceanic data from the Northern Hemisphere and found a systematic deflection of $13° \pm 2°$ to the right of the wind. This conclusion is partially supported by Katz et al. (1965), who found similar angular deviations in 4 out of 6 observations; however, the orientation data of Welander (1963), Maratos (1971), and Williams (1965) in the ocean, of Langmuir (1938) and Walther (1967) in Lake George, and of Kenney (1977) in Lake-of-the-Woods show that windrows are aligned within a few degrees of the wind, with no significant systematic bias to one side or the other. If Langmuir circulations develop in a well-defined Ekman layer, then they will presumably show the marked influence of the skewed velocities of the underlying Ekman spiral. This may explain the occasional angular deviations that have been reported.

If long stretches of streaks are well marked and viewed from a sufficiently distant perspective, they frequently form a network of intersecting lines. This point was emphasized by McLeish (1968) and was also made by Williams (1965) and Harris & Lott (1973). The length that an individual streak line can be traced between intersections is seldom mentioned. Williams (1965) estimates (subjectively) that the ratio of length to a characteristic spacing was of order 20 in his experiments. Kenney (1977) classifies his windrows by this ratio, and defines " very regular" ones as those with length-to-spacing ratios in excess of 100, viewed from the top of a 20-m tower; he also comments that the regularity seldom is perceptible when the windrows are viewed from close to the water surface.

FORMATION TIMES   The time scales for formation of Langmuir circulations can be inferred from the times required for surface rows to reorient themselves after a shift in wind direction. All investigators indicate that this is a rapid process, the original pattern being destroyed and a new one reforming, aligned with the new wind direction, within a few minutes of the wind shift. Stommel (1951) found that small-scale windrows on ponds were reoriented within 1 or 2 minutes after a shift in wind direction. Maratos (1971) reports a single instance of a rapid shift in wind direction, followed by a reorientation of streaks within 2 to 4 minutes. Welander (1963) observed that 8-m-wide streaks in the Baltic were reoriented within 10 minutes of a shift in wind direction; the newly formed streaks also had a spacing of 8 m. Langmuir (1938) noted that the reorientation of a large-scale windrow pattern in the Atlantic occurred within 20 minutes of a shift in wind direction. Katz et al. (1965) observed reorientation in one experiment off the Argentine coast: a change in wind direction without change of speed occurring within a half-hour period resulted in a realignment of rows, the new rows being of smaller scale. They found similar results in other experiments.

The extremely rapid reorientation response times seen by Stommel (1951) led him to speculate that the rows marked a process confined to a layer very near the surface ("top few inches"). In the wind-shift observation described by Welander (1963), neutrally buoyant floats placed at depths of 1 and 2 m continued moving in the old wind direction, although surface floats were carried into the reoriented windrows. Faller (1981) has pointed out that the rapid response is presumably due to the smaller scales, which may be expected to be shallower and capable of rapid response. Response times of the order seen in the field are not inconsistent with Faller's (1978) laboratory measurements, or with the CL models described above.

PERSISTENCE   Accounts of observations give the impression that individual streaks maintain their identity for substantial periods of time, but this is seldom explicitly discussed. Kenney (1977) used 20-minute reels of film in his study, and found that under steady wind conditions individual streaks remained identifiable for periods in excess of the time required to expose the film. In fact, he observed identifiable streaks for periods in excess of one hour (private communication) during his investigations in Lake-of-the-Woods.

Surface and Subsurface Current Structure

All investigators who report surface-tracer speeds have indicated that the windward (u in Figure 1a) component is larger in streaks than between them. This current anomaly, Δu, has been measured or inferred in a few cases, but no systematic studies of the complete current system have been undertaken. Ichiye (1967) estimates that speeds in streaks were twice those outside; Harris & Lott (1973) measured values of Δu up to 6 cm s^{-1}. Anomalies were estimated to have been 1–3 cm s^{-1} in observations discussed by Gordon (1970), 3 cm s^{-1} by Williams (1965), and 10 cm s^{-1} by Assaf et al. (1971). Unfortunately, these values were reported without reference to other components of the current system or to the prevailing wind speeds.

The order of magnitude of the horizontal surface "sweeping" component v in a system of Langmuir vortices can be inferred from the time required to collect tracers introduced at the surface into rows. This has been estimated in a few cases. Langmuir (1938) found that leaves placed midway between streaks spaced 12–20 m apart took about 5 minutes to reach the streaks, which gives an average sweeping speed of about 3 cm s^{-1}. Since the sweeping both at the streak and at the line midway between is zero, or nearly so, the peak speeds must have been somewhat greater than the average. Ichiye (1967) also noticed that computer cards placed across the wind were swept into rows in a few minutes, and one can infer from his paper that the cards had to travel laterally about 10 m to reach convergence lines. Woodcock (1944) estimated that all of the surface tracers introduced at the surface were swept into convergence zones in 3–5 minutes: one can also infer from his paper that the row spacing was on the order of 20 m, as in Langmuir's case.

Subsurface sweeping motions toward vertical planes marked by surface convergence lines have been observed by Langmuir (1938) and Harris &

Lott (1973) but speeds have not been estimated. Langmuir found that a neutrally buoyant drogue suspended at depths up to 5 m gradually drifted under a streak, but one suspended at 10 m had no such tendency. Since the sweeping velocity presumably must change sign away from the streak plane at some depth, this is consistent with a cellular motion extending to depths of 10 m or more. The picture is seconded by theory (Leibovich & Radhakrishnan 1977, Leibovich & Paolucci 1980a), which shows that maximum sweeping speeds toward convergences very near the surface greatly exceed lateral currents away from streak planes at greater depth. Harris & Lott (1973) also found that drogues suspended at 5 to 6 m drifted toward planes below streaks.

Kenney (1977) observed the sweeping motions of drogues suspended at several depths in the shallow water of Lake-of-the-Woods. All drogues were swept toward planes beneath streaks, including those near the bottom. The return motion anticipated at depth was not seen; this curious observation has yet to be explained and clearly merits further investigation.

The best-documented cellular current component is the vertical velocity below streaks. This has been inferred by Myer (1969, 1971) by timing isotherm displacements in the upper 7 m of Lake George; he found descending currents in narrow "jets" below streaks of about 3 cm s^{-1} under conditions of thermal stability, and 5 cm s^{-1} under thermally unstable conditions. These data are consistent with direct velocity measurements taken by Sutcliffe et al. (1963), Harris & Lott (1973), and Rjanzhin (1980) and Filatov et al. (1981), all of which probably should be considered more reliable. Harris & Lott (1973) traced aluminum plates ballasted for neutral buoyancy; this device was first used by Langmuir (1938), who measured descending currents of 2 to 3 cm s^{-1} below streaks and rising currents of 1 to 1.5 cm s^{-1} midway between adjacent streaks. Rjanzhin and Filatov et al. (1981) used "Sutcliffe" floats, first used by Sutcliffe et al. (1963). This device measures the drag on a circular disk by the buoyancy force on a partially submerged buoyant element; it therefore measures the maximum vertical speed within a distance from the free surface determined by the size of the device. The device used by Sutcliffe et al. (1963) is capable of measuring vertical currents up to 6 cm s^{-1}, after which the device is fully submerged. Filatov et al. (1981) give a nonlinear correlation between wind speed and downwelling speed with coefficients depending upon the gradient Richardson number, a measure of the surface heat flux, in the air near the surface. It appears that they may have forced the correlation to pass through zero at zero wind speed, a refinement that might be relaxed since circulations were observed only for wind speeds larger than a threshold of about 3 m s^{-1}. None of these authors report depths at which the maximum downwelling current is attained.

Data taken by Sutcliffe et al. (1963), Harris & Lott (1973), and Filatov et al. (1981) are shown in Figure 7. Kenney (1977) has pointed out a possible error in the data reduction in Sutcliffe et al. (1963); the drag expression printed in their paper is inconsistent with the aerodynamic form of the drag coefficient used for the disk. If this is not merely a misprint, then downwelling speeds in Figure 7 as taken from their paper should be increased by a factor of $\sqrt{2}$. There is a great deal of scatter in the data shown in Figure 7; clearly the downwelling currents increase with wind speed, but other factors are obviously important. The more vigorous downwelling that occurs with surface heating in the Harris & Lott data is contrary to expectations, as they point out, but they also indicate uncertainties in determining the heat flux. If each data set in Figure 7 is individually considered, with those attributed to thermally stable and unstable conditions treated as distinct, downwelling current data are well fitted by a linear relationship with wind speed. The two straight lines in Figure 7 represent a compromise; the dashed line is a least-squares linear fit to all of the Sutcliffe et al. (1963) and Harris & Lott (1973) data taken together, and the solid line is the corresponding fit to all of the Filatov et al. (1981) data. Both composite data sets are well fitted in this way. The goodness of fit to each data set shows that

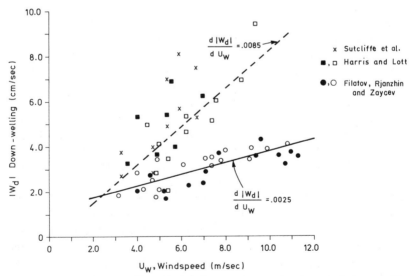

Figure 7 Measured downwelling speeds below streaks as a function of wind speed. The open squares and circles correspond to surface heating, closed symbols to surface cooling.

downwelling increases in an essentially linear fashion with wind speed in the range shown, provided other factors, such as the density structure of the water column, are held more or less fixed.

We also note that Woodcock (1950) inferred downwelling speeds of 3 to 7 cm s^{-1} by observing abundances of surface sargassum, which are submerged by vertical speeds of this order. Maratos (1971) has inferred average upwelling speeds of 0.8 cm s^{-1} by monitoring the sedimentation rate of fine sand in cells driven by a wind speed of 6 m s^{-1}. Additional data taken using unspecified methods are presented in Scott et al. (1969).

Vertical Penetration of Cells

Observations of drogue motions by Langmuir (1938) and Harris & Lott (1973), referred to in the previous section, indicate sweeping motions at depths of 5 to 6 m toward planes below surface streaks. Theoretical models then suggest that weaker cellular motions extend to depths that are a small multiple of these figures, and therefore comparable to the depth of the thermocline and of the streak spacings they observed.

More systematic inferences of the penetration depth have been made by Myer (1969, 1971) and Thorpe & Hall (1982) by measurements of temperature anomalies.

Myer (1969, 1971) found isotherm displacements of up to 3.5 m under stable thermal conditions and significantly larger displacements under unstable thermal conditions. These displacements occurred in narrow "jets" below surface streaks. The depth of penetration was inferred by the deepest isotherm undergoing a displacement beneath streaks, and was estimated to be between 2 to over 7 m under stable conditions. Since Myer's thermistor string was limited to a 7-m length, deeper effects could not be determined.

Harris & Lott (1973) found difficulties in making similar measurements on Lake Ontario; their thermistors did not always show temperature anomalies below streaks in thermally stable conditions. Thorpe & Hall (1982) found similar difficulties, and found that attempts to directly correlate temperature anomalies with surface streaks failed: the expected anomaly signal was weaker than the background "noise" of thermal fluctuations due to other effects. By careful data analysis, however, they were able to extract a statistically significant correlation between surface streaks and temperature anomalies extending throughout the depth of the mixed layer. Isotherm displacements are small, however, amounting to no more than 60 cm, and the temperature anomalies are "very much less than the amplitude of the r.m.s. variation" of the measured temperature. This isotherm displacement is much less than that computed by

Leibovich & Paolucci (1980a) using the CL 2 instability model. They note, however, that the thermal stratification was much larger than that used in the computations. It is also important in this regard to note again the error in Leibovich & Paolucci (1980a) that led to an overprediction of isotherm displacements.

Thorpe & Hall (1982) also measured acoustic scattering from submerged bubbles produced by breaking surface waves, and found that the average acoustic cross section of the bubbles is higher below streaks (at depths between 0.5 and 2.3 m) than between them. Since Thorpe & Hall (1980) found that waves break with equal frequency in windrows and between them, wave breaking provides a uniformly distributed bubble source. The variations in acoustic cross sections are therefore regarded as substantial evidence of downwelling motion below streaks.

LABORATORY EXPERIMENTS

Faller (1969) was the first to demonstrate convincingly that organized rolls could be established mechanically in the laboratory by wind and wave action; McLeish (1968) alludes to possibly similar experiments, but provides little detail. Faller's demonstrations, conducted in a small wind-wave tank, are of particular interest, since they showed that circulations required only current shear and waves. Waves and currents simultaneously produced by wind action led to longitudinal rolls.

Mechanically produced waves by themselves did not lead to circulations: a slight current shear produced by slow draining of the tank also did not, by itself, produce circulations. If mechanically driven waves were then introduced into a slightly sheared flow, however, vigorous convection was observed as soon as the waves reached the test section. Thus, Faller demonstrated that wind stress is not necessary, and that the role of the wind may be simply to introduce the two principal ingredients, current shear and waves, into the water body.

By various experiments Faller (1969) also showed that thermal convection was not necessary, and that the circulations produced by wave and current shear were substantially more vigorous than thermal convection organized by shear.

The generation of Langmuir-like circulations in the laboratory by wind and waves was studied in more detail by Faller & Caponi (1978). Roll motions with axes parallel to the wind readily formed in their wind-wave tank. The crosswind spacings, λ_c, of convergence zones in the cellular motion were determined, and a clear dependence upon the wavelength of the surface waves, λ_w, was demonstrated. Their data, and several oceanic measurements, are fitted reasonably well by the relation

$$\lambda_c/H = 4.8\left[1 - \exp(-0.5\lambda_w/H)\right], \tag{11}$$

where H is the mean depth of the water in the wind-wave tank, or the estimated depth of the mixed layer for the oceanic cases.

In transient studies, they noted that circulations began within seconds after arrival of the waves. The bands that first formed gradually increased in width by a factor of about 3 before reaching the asymptotic state reported in Equation (11) above. Whether this change in spacing was associated with an increase in λ_w was not made clear. The authors concluded that the mechanism underlying Langmuir circulations is wave-related, and that the scale of the surface waves plays a role in setting the scale of the circulations.

Experiments designed to test the CL 1 mechanism have been reported by Faller (1978), and similar experiments have been performed by S. Mizuno (unpublished) with similar conclusions. Faller generated a crossed-wave pattern using a specially designed mechanical wavemaker, and a current was created by blowing a light wind over the water surface. Circulations did not form with either wind alone or waves alone, but regular, steady, and reproducible Langmuir circulations formed when both were present. The circulations were marked by floats on the surface, which were swept into convergence lines, and by dye crystals on the bottom. In accordance with the CL 1 mechanism, upwelling occurred beneath lines of maximum surface excursions (lines traversed by points of wave intersections) and downwelling occurred midway between. Reversals of the wind direction reversed the direction of circulation in the cells, another feature inherent in the CL 1 model. Direct measurements of most of the parameters occurring in the CL equations could be made, and u_* and v_T were inferred by indirect means. Using measured and inferred parameters, the time scale T_d (see Table 1) was estimated and found to be 29 s, while La was estimated to be 0.59. The experimental response time was estimated by the increase of the surface-sweeping speed with time after switching on either wind or waves. The exponential time constant found this way was $T = 12$ s. This was compared with the value of 10 T_d quoted by Leibovich & Radhakrishnan (1977) for the time required for essentially steady conditions to be obtained at the surface in numerical experiments at La = 0.01. Clearly 10 T_d is considerably larger (24 times) than the observed exponential rise time, but the two values refer to different quantities, and ought not to be directly compared. Furthermore, as the small time solution of Leibovich (1977a) makes clear, the rate at which the surface-sweeping component develops initially is proportional to $\sqrt{\text{La}}$. Thus the surface-sweeping speed at La = 0.59

builds up at a rate 7.7 times faster than the corresponding quantity at La = 0.01. It therefore appears that the theory and experiment are probably in rough agreement as far as time scales are concerned.

Further details of Faller's (1978) experiments are given in Faller & Cartwright (1982). This report also mentions preliminary results of experiments to test the CL 2 instability mechanism. A completely stable shear flow is found to be destabilized by the addition of a very small amplitude [$O(1\text{ mm})$] monochromatic plane wave. The disturbance takes the form of longitudinal rolls, and breakdown is rapid [$O(1\text{ min})$], with time scale decreasing with increasing wave amplitude. The results are said to be in reasonable agreement with the theoretical stability diagrams of Leibovich & Paolucci (1981), despite some differences between the experimental conditions and those assumed in the theoretical analysis.

Conclusions

The present status of understanding of Langmuir circulations can be summarized as follows:

1. Their existence in lakes and the ocean is well established, although their detailed structure remains to be determined.
2. They are convective in nature and mechanical in origin, and are driven by the wind.
3. Evidence exists to suggest that these convective motions may be an important mixing mechanism in the upper ocean, but the existing experimental case is not strong. In view of the potential significance of identifying this phenomenon as a major contributor to upper-ocean mixing, further work is strongly recommended.
4. Laboratory experiments and recent theoretical concepts show promise of explaining the origin and mechanism of naturally occurring Langmuir circulations.

This natural phenomenon is clearly extremely complex, and cannot be isolated for study in the field from its equally complex environment. As a result, it would appear that further advances in understanding will come most rapidly by a combination of controlled laboratory experiments and theory.

ACKNOWLEDGMENTS

Professor A. J. Faller provided information on unpublished work; I thank him for permission to quote from it, and for his valuable comments on a draft of this paper. Mr. Sanjiva Lele willingly provided assistance when it was needed and I found his help valuable. My research on Langmuir circulations, and the writing of this review, have been sponsored by the Physical Oceanography Program of the US National Science Foundation under Grant OCE 79-15232; I am grateful to the Foundation for their continuing support. Supplementary support for the writing of this paper was provided by the US Office of Naval Research, Physical Oceanography Program (ONR 422PO), under Contract N000 14-80-C-0079.

Literature Cited

Andrews, D. G., McIntyre, M. E. 1978. An exact theory of nonlinear waves on a Lagrangian-mean flow. *J. Fluid Mech.* 89:609–46

Assaf, G., Gerard, R., Gordon, A. L. 1971. Some mechanisms of oceanic mixing revealed in aerial photographs. *J. Geophys. Res.* 76:6550–72

Atwood, D. K., Benjamin, J. A., Farrington, J. W. 1980. The mission of September 1979 RESEARCHER/PIERCE IXTOC-I Cruise and the physical situation encountered. In *Preliminary Results from the September 1979 RESEARCH-ER/PIERCE IXTOC-I Cruise,* ed. D. K. Atwood. *US Dept. Commerce NOAA/RD/MP3,* Boulder, Colo.

Bainbridge, R. 1957. The size, shape, and density of marine phytoplankton concentrations. *Cambridge Philos. Soc. Biol. Rev.* 32:91–115

Baines, P. G., Gill, A. E. 1969. On thermohaline convection with linear gradients. *J. Fluid Mech.* 37:289–306

Bye, J. A. T. 1965. Wind driven circulation in unstratified lakes. *Limnol. Oceanogr.* 10:451–58

Collingwood, E. 1868. Observations on the microscopic algae which cause discoloration of the sea in various parts of the world. *Trans. R. Micr. Soc.* 16:85–92

Craik, A. D. D. 1970. A wave-interaction model for the generation of windrows. *J. Fluid Mech.* 41:801–21

Craik, A. D. D. 1977. The generation of Langmuir circulations by an instability mechanism. *J. Fluid Mech.* 81:209–23

Craik, A. D. D. 1982a. The drift velocity of water waves. *J. Fluid Mech.* 116:187–205

Craik, A. D. D. 1982b. Wave-induced longitudinal-vortex instability in shear flows. *J. Fluid Mech.* In press

Craik, A. D. D., Leibovich, S. 1976. A rational model for Langmuir circulations. *J. Fluid Mech.* 73:401–26

Csanady, G. T. 1965. Windrow studies. *Rept. No. PR26,* Great Lakes Inst., Univ.

Toronto, Ont. 82 pp.

Csanady, G. T., Pade, B. 1969. Windrow observations. In *Dynamics and Diffusion in the Great Lakes,* ed. G. T. Csanady. Univ. Waterloo. 142 pp.

Davey, A., Stewartson, K. 1974. On three-dimensional packets of surface waves. *Proc. R. Soc. London Ser. A* 338:101–10

Ekman, V. W. 1905. On the influence of the earth's rotation on ocean currents. *Ark. Mat. Astron. Fys.* 2:1–54

Faller, A. J. 1964. The angle of windrows in the ocean. *Tellus* 16:363–70

Faller, A. J. 1969. The generation of Langmuir circulations by the eddy pressure of surface waves. *Limnol. Oceanogr.* 14:504–13

Faller, A. J. 1971. Oceanic turbulence and the Langmuir circulations. *Ann. Rev. Ecol. Syst.* 2:201–36

Faller, A. J. 1978. Experiments with controlled Langmuir circulations. *Science* 201:618–20

Faller, A. J. 1981. The origin and development of laboratory models and analogues of the ocean circulation. In *Evolution of Physical Oceanography,* ed. B. A. Warren, C. Wunsch, 16:462–79. Cambridge, Mass.: MIT Press. 623 pp.

Faller, A. J., Caponi, E. A. 1978. Laboratory studies of wind-driven Langmuir circulations. *J. Geophys. Res.* 83:3617–33

Faller, A. J., Cartwright, R. W. 1982. Laboratory studies of Langmuir circulations. *Tech. Rept. BN-985,* Inst. Phys. Sci. Technol., Univ. Md.

Faller, A. J., Kaylor, R. E. 1966. A numerical study of the instability of the laminar Ekman boundary layer. *J. Atmos. Sci.* 23:466–80

Faller, A. J., Woodcock, A. H. 1964. The spacing of windrows of Sargassum in the ocean. *J. Mar. Res.* 22:22–29

Filatov, N. N., Rjanzhin, S. V., Zaycev, L. V. 1981. Investigation of turbulence and Langmuir circulation in Lake Ladoga. *J. Great Lakes Res.* 7:1–6

Fox, M. J. H. 1976. On the nonlinear transfer of energy in the peak of a gravity-wave spectrum. II. *Proc. R. Soc. London Ser. A* 348:467–83

Gammelsrod, T. 1975. Instability of Couette flow in a rotating fluid and origin of Langmuir circulations. *J. Geophys. Res.* 80:5069–75

Garrett, C. J. R. 1976. Generation of Langmuir circulations by surface waves—a feedback mechanism. *J. Mar. Res.* 34:117–30

Gordon, A. L. 1970. Vertical momentum flux accomplished by Langmuir circulations. *J. Geophys. Res.* 75:4177–79

Harris, G. P., Lott, J. N. A. 1973. Observations of Langmuir circulations in Lake Ontario. *Limnol. Oceanogr.* 18:584–89

Huang, N. E. 1979. On the surface drift currents in the ocean. *J. Fluid Mech.* 91:191–208

Ichiye, T. 1965. Diffusion experiments in coastal waters using dye techniques. *Symp. Diffusion Oceans Fresh Waters, Lamont Geophys. Lab., Palisades, N.Y.*, pp. 54–67

Ichiye, T. 1967. Upper ocean boundary-layer flow determined by dye diffusion. *Phys. Fluids* 10:S270–77 (Suppl.)

Ichiye, T., Plutchak, N. B. 1966. Photodensimetric measurements of dye concentration in the ocean. *Limnol. Oceanogr.* 11:364–70

Katz, B., Gerard, R., Costin, M. 1965. Responses of dye tracers to sea surface conditions. *J. Geophys. Res.* 70:5505–13

Kenney, B. C. 1977. *An experimental investigation of the fluctuating currents responsible for the generation of windrows.* PhD thesis. Univ. Waterloo, Ont. 163 pp.

Kinsman, B. 1965. *Wind Waves: Their Generation and Propagation on the Ocean Surface.* Englewood Cliffs, N.J.: Prentice-Hall. 676 pp.

Kraus, E. B. 1967. Organized convection in the ocean surface layer resulting from slicks and wave radiation stress. *Phys. Fluids* 10:S294–97 (Suppl.)

Langmuir, I. 1938. Surface motion of water induced by wind. *Science* 87:119–23

Leibovich, S. 1977a. On the evolution of the system of wind drift currents and Langmuir circulations in the ocean. Part 1. Theory and the averaged current. *J. Fluid Mech.* 79:715–43

Leibovich, S. 1977b. Convective instability of stably stratified water in the ocean. *J. Fluid Mech.* 82:561–85

Leibovich, S. 1980. On wave-current interaction theories of Langmuir circulations.

J. Fluid Mech. 99:715–24

Leibovich, S., Lele, S. K. 1982. Thermocline erosion and surface temperature variability due to Langmuir circulations. *FDA Rep. #82-07*, Sibley School Mech. & Aerosp. Eng., Cornell Univ., Ithaca, N.Y.

Leibovich, S., Paolucci, S. 1980a. The Langmuir circulation instability as a mixing mechanism in the upper ocean. *J. Phys. Oceanogr.* 10:186–207

Leibovich, S., Paolucci, S. 1980b. Energy stability of the Eulerian-mean motion in the upper ocean to three-dimensional perturbations. *Phys. Fluids* 23:1286–90

Leibovich, S., Paolucci, S. 1981. The instability of the ocean to Langmuir circulations. *J. Fluid Mech.* 102:141–67

Leibovich, S., Radhakrishnan, K. 1977. On the evolution of the system of wind drift currents and Langmuir circulations in the ocean. Part 2. Structure of the Langmuir vortices. *J. Fluid Mech.* 80:481–507

Leibovich, S., Ulrich, D. 1972. A note on the growth of small scale Langmuir circulations. *J. Geophys. Res.* 77:1683–88

Lilly, D. K. 1966. On the stability of Ekman boundary flow. *J. Atmos. Sci.* 23:481–94

Longuet-Higgins, M. S. 1962. The directional spectrum of ocean waves, and processes of wave generation. *Proc. R. Soc. London Ser. A* 265:286–315

Longuet-Higgins, M. S. 1976. On the nonlinear transfer of energy in the peak of a gravity-wave spectrum: a simplified model. *Proc. R. Soc. London Ser. A* 347:311–28

Maratos, A. 1971. *Study of the near shore surface characteristics of windrows and Langmuir circulation in Monterey Bay.* MS thesis. US Nav. Postgrad. Sch., Monterey, Calif.

McLeish, W. 1968. On the mechanisms of wind-slick generation. *Deep-Sea Res.* 15:461–69

Mobley, C. D. 1977. *A numerical model of wind and wave-generated Langmuir circulations.* PhD thesis. Univ. Md., College Park. 170 pp.

Moen, J. 1978. *A theoretical model for Langmuir circulations.* PhD thesis. Univ. Southampton. 235 pp.

Myer, G. E. 1969. A field investigation of Langmuir circulations. *Proc. 12th Conf. Great Lakes Res., Ann Arbor, Mich.*, pp. 625–63

Myer, G. E. 1971. *Structure and mechanics of Langmuir circulations on a small inland lake.* PhD dissertation. State Univ. N.Y., Albany

Owen, R. W. Jr. 1966. Small-scale, horizontal vortices in the surface layer of the sea. *J. Mar. Res.* 24:56–65

Phillips, O. M. 1977. *Dynamics of the Upper Ocean.* Cambridge Univ. Press. 336 pp. 2nd. ed.

Pollard, R. T. 1977. Observations and theories of Langmuir circulations and their role in near surface mixing. In *A Voyage of Discovery: George Deacon 70th Anniversary Volume,* ed. M. Angel, pp. 235–51. Oxford: Pergamon

Rjanzhin, S. V. 1980. On the energy and depth of penetration of Langmuir circulation. *Izv. All Union Geogr. Soc.* 112:46–54 (In Russian)

Scott, J. T., Myer, G. E., Stewart, R., Walther, E. G. 1969. On the mechanism of Langmuir circulations and their role in epilimnion mixing. *Limnol. Oceanogr.* 14:493–503

Shemdin, O. H. 1972. Wind-generated current and phase speed of wind waves. *J. Phys. Oceanogr.* 2:411–19

Smith, J. 1980. *Waves, currents, and Langmuir circulations.* PhD thesis. Dalhousie Univ., Halifax, N. S. 242 pp.

Stewart, R., Schmitt, R. K. 1968. Wave interaction and Langmuir circulations. *Proc. 11th Conf. Great Lakes Res., Cent. Great Lakes Studies, Univ. Wis.,* pp. 31–32 (Abstr.)

Stommel, H. 1951. Streaks on natural water surfaces. *Weather* 6:72–74

Sutcliffe, W. H., Baylor, E. R., Menzel, D. W. 1963. Sea surface chemistry and Langmuir circulations. *Deep-Sea Res.* 10:233–43

Sutcliffe, W. H. Jr., Sheldon, R. W., Prakash, A., Gordon, D. C. Jr. 1971. Relations between wind speed. Langmuir circulation and particle concentration in the ocean. *Deep-Sea Res.* 8:639–43

Thomson, J. 1862. On the calm lines often seen on a rippled sea. *Philos. Mag. Ser. 4.* 24:247–48

Thoreau, H. D. 1854. *Walden; Or, Life in the Woods,* p. 170. In *Walden and Other Writings of Henry David Thoreau,* ed. B. Atkinson, 1950 Modern Library Edition. New York: Random House. 732 pp.

Thorpe, S. A., Hall, A. J. 1980. The mixing layer of Loch Ness. *J. Fluid Mech.* 101:687–703

Thorpe, S. A., Hall, A. J. 1982. Observations of the thermal structure of Langmuir circulation. *J. Fluid Mech.* 114:237–50

Turner, J. S. 1973. *Buoyancy Effects in Fluids.* Cambridge Univ. Press. 367 pp.

Van Straaten, L. M. J. J. 1950. Periodic patterns of rippled and smooth areas on water surfaces, induced by wind action. *K. Ned. Akad. Wet. Ser. B* 53:2–12

Walther, E. G. 1967. *Wind streaks.* MSc thesis. State Univ. N.Y., Albany. 31 pp.

Welander, P. 1963. On the generation of wind streaks on the sea surface by action of surface film. *Tellus* 15:67–71

Williams, K. G. 1965. Turbulent water flow patterns resulting from wind stress on the ocean. *NRL Memo. Rep. 1653.* US Nav. Res. Lab.

Woodcock, A. H. 1944. A theory of surface water motion deduced from the wind-induced motion of the Physalia. *J. Mar. Res.* 5:196–205

Woodcock, A. H. 1950. Subsurface pelagic Sargassum. *J. Mar. Res.* 9:77–92

Woodcock, A. H., Wyman, J. 1946. Convective motion in air over the sea. *Ann. NY Acad. Sci.* 48:749–76

Wu, J. 1975. Wind-induced drift currents. *J. Fluid Mech.* 68:49–70

Ann. Rev. Fluid Mech. 1983. 15:429–59

THE TURBULENT WALL JET
—MEASUREMENTS AND
MODELING

B. E. Launder

Department of Mechanical Engineering, University of Manchester Institute of Science and Technology, Manchester, England

W. Rodi

Institut für Hydromechanik, Universität Karlsruhe, Karlsruhe, Federal Republic of Germany

INTRODUCTION

The turbulent wall jet, even limiting attention to the topographically simple cases beloved of academics, arguably provides more puzzles for those seeking an ordered set of rules to describe turbulence than any other class of turbulent shear layer. Formally, we can regard a wall jet as a boundary layer in which, by virtue of the initially supplied momentum, the velocity over some region in the shear layer exceeds that in the free stream.

Wall jets are of great and diverse engineering importance. While some of the best known and most challenging applications lie in the field of advanced airfoil design, these aerodynamic roles are far outweighed in number and, probably, in economic importance by the use of wall jets in problems of heating, cooling or ventilating—areas where traditionally design has proceeded unfettered by any deep concern about the turbulence structure of the flows in question.

Two major industrial applications are the film-cooling of the lining walls of gas-turbine combustion chambers and of the leading stages of

429

0066-4189/83/0115-0429$02.00

the turbine itself. In both cases the aim is to introduce a cool layer of fluid adjacent to a solid surface in order to provide protection from a hot external stream. In practice, for reasons of constructional strength, it is not possible to provide a continuous cooling slot extending over the full lateral extent of the region to be cooled. A series of short slots or holes is thus employed. One wants, therefore, a rapid lateral spreading of coolant to fill in the "gaps" but a low mixing rate in the direction normal to the surface in order that the coolant be diluted only slowly. The most widespread use of the wall jet for heat and (more importantly) mass transfer modification must be in the automobile demister. On a larger scale, one may mention the deflectors used in conditioned air-circulation systems in the home or factory. The design and positioning of such deflectors becomes especially crucial in large-scale, one-of-a-kind applications such as a concert auditorium. Clearly here, as in the other examples cited, a thorough knowledge of the fluid mechanics of the wall jet—including the turbulence field it generates—can materially improve the design of such devices.

The authors have recently undertaken a review of experiments on turbulent wall jets, the aim being to identify suitable test cases for the Stanford Conference on Complex Turbulent Flows (Launder & Rodi 1981, which, for brevity, we hereafter cite as LR1). The present article extends that review in three ways: it gives more attention to the physical processes, it considers an important type of flow—the radial wall jet—which time limitations forced us to exclude from our earlier review, and it examines how well current calculation schemes succeed in mimicking the measured behavior of wall jets. Although the two articles are best read in conjunction with one another, a certain amount of material is carried over from the earlier paper to allow the present treatment to be self-contained.

THE PLANE TWO-DIMENSIONAL WALL JET

Figure 1 shows a typical distribution of mean velocity across a wall jet and serves to define nomenclature. The wall jet is usefully thought of as a two-layer shear flow comprising an inner region (from the wall to y_m) in which the flow exhibits similarities in structure with that of a conventional turbulent boundary layer, and an outer layer (extending from y_m to the outer edge of the flow) where the shear-layer character is more like that of a free shear flow than one bounded by a wall. In making this point some further qualification is perhaps needed, for it is often helpful to regard the outer 60% or so of a regular boundary layer as akin to a free shear layer. The essential differences in the case of a wall jet are that the outer-region shear stress is of the opposite sign from that near the wall, with a maximum shear-stress level that is usually several times larger than the wall stress itself.

This two-layer model of the wall jet provides a basis for detecting unexpected features in the shear flow. There is no shortage of these, for the two regions strongly interact. We limit attention for the moment to the wall jet developing in stagnant surroundings, the case that has been most extensively studied. In LR1, after the elimination of data showing signs of lack of two-dimensionality, the rate of growth of the outer layer, characterized by $dy_{1/2}/dx$, was reported as

$$\frac{dy_{1/2}}{dx} = 0.073 \pm 0.002. \tag{1}$$

This rate of growth is more than 30% below the consensus value for the plane free jet (Rodi 1975). The difference arises, we believe, from the damping of turbulent velocity fluctuations in the direction normal to the wall[1] due to turbulent pressure reflections from the wall itself; this topic

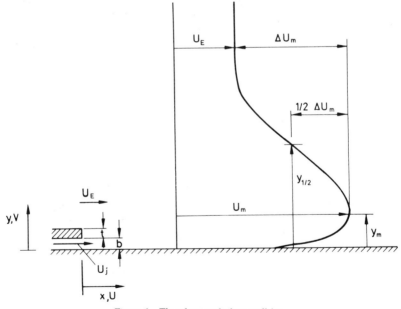

Figure 1   The plane turbulent wall jet.

[1] LR1 reports a wide variation among the experimental distributions of $\overline{v^2}$, the normal stress perpendicular to the wall. The most credible data, however, all show a level at $y = y_{1/2}$ approximately one half that found in the free jet in stagnant surroundings (Patel 1970).

is taken up further in the section on mathematical modeling. There is some suggestion of a very weak Reynolds number dependence (Tailland & Mathieu 1967), a twofold variation in Re_m (the Reynolds number based on U_m and y_m) producing a 4% decrease in spreading rate. This apparent influence of viscosity on the outer-layer structure is really a "displacement" effect due to the weak though incontrovertible influence of Re_m on dy_m/dx. If $(y_{1/2} - y_m)$ were adopted instead as the outer-layer length scale, the variation in its rate of growth with Reynolds number would become negligible[2].

The most frequently reported influence of the outer region on the inner is the displacement of the position of zero shear stress from the position of maximum velocity (where it would occur in a laminar wall jet) to about two-thirds y_m. This feature appears to have been noticed first by Mathieu (1959), who concluded that the growth of the inner region was incompatible with a two-dimensional momentum balance unless the shear stress at y_m was of about the same magnitude as and of opposite sign from that at the wall itself. Simple mixing-length arguments (Launder 1969) suggest that in an asymmetric shear flow with a velocity extremum the shear stress passes through zero on the side of the position of zero velocity gradient, where the length scales are smaller. The observed behavior in the wall jet accords with this proposition.

Following well-established practice in boundary layers, workers have usually tried to represent the measured inner-layer velocity profiles in the form

$$\frac{U}{U_\tau} = A \log \frac{yU_\tau}{\nu} + B, \tag{2}$$

though a wide variation in the values of the constants A and B have been proposed. This variation arises from two main causes: errors in measuring the wall shear stress τ_w and thus U_τ (defined as $\sqrt{\tau_w/\rho}$), and attempting to fit (2) over a too large portion of the inner region. LR1 concluded that the force-balance measurements of Alcaraz (1977) (see also Alcaraz et al. 1968) provided the most reliable values of wall stress and these closely supported the Preston-tube data of Bradshaw & Gee (1960) and Kamemoto (1974).

A Preston tube can only give reliable measurements of wall stress if the flow past it follows the same scaled velocity distribution as in the calibrating experiments. The fact that the friction factor obtained by a Preston tube agrees with the force-balance data provides an indication

[2] An earlier review by Narasimha et al. (1973), which reports a dependency of $dy_{1/2}/dx$ on slot Reynolds number raised to the power -0.18, retained data sets discarded by LR1 on the grounds of poor two-dimensionality.

that close enough to the wall the mean velocity profile is the same as in a boundary layer on a flat plate or in pipe flow (the latter being the flow habitually used for calibration). That is to say, in a wall jet, A and B take the "universal" values of about 5.5 and 5.45 proposed by Patel (1965). The fully developed velocity profiles reported by Bradshaw & Gee (1960) and Guitton (1970) are in fact consistent with the universal profiles only for values of yU_τ/ν up to about 50. In marked contrast with a flat-plate boundary layer, the region of the profile over which the usual semilogarithmic profile applies does not increase as the wall jet develops downstream. The explanation of this paradox appears to be that the wall stress decreases faster than the stress in the outer region, so that the ratio of maximum to wall shear stress increases indefinitely in the streamwise direction. Now, the larger this ratio, the greater the encroachment of the outer region on the inner and it is this that prevents the growth in the universal region, even though $y_m U_\tau/\nu$ does increase (slowly) with downstream distance.

The strong influence of the outer region on the inner is equally evident in the distributions of turbulent kinetic energy (Guitton 1970, Wilson & Goldstein 1976, Alcaraz[3] 1977): the energy decreases smoothly from a maximum level at $y \simeq 0.7 y_{1/2}$ as the wall is approached, with no hint of the near-wall peaking found in a boundary layer.

The presence of an external stream reduces the relative strength of the outer region; to first order, the ratio of the maximum stress to the wall stress is reduced by the factor $(1 - U_e/U_m)^2$. As a result, the inner region is thicker relative to the case of stagnant surroundings, and the impact of the outer region on the inner is less. If the external-stream velocity is uniform, the flow is not self-preserving; indeed, if the development continues far enough, the velocity maximum will be consumed and the flow will revert to a conventional boundary layer. By suitably tailoring the external-stream velocity, however, a family of self-preserving wall jets can be established in which U/U_m remains invariant as the flow develops downstream. This family has been extensively studied by Newman and his colleagues at McGill University (Patel & Newman 1961, Gartshore & Newman 1969, Irwin 1973). The last of these papers reports an exceptionally thorough and self-consistent study that was selected as a test case for the 1981 Stanford Conference. The mean flow in this case exhibits a substantial near-wall region where the mean velocity convincingly follows the logarithmic law (with Patel's values of A and B). The study is perhaps most useful for the detailed measurements it provides of the turbulence

---

[3]Alcaraz's measurements were made on a curved surface, though at $x/h = 30$ the influence of curvature appears to have been barely significant.

field, an example of which is provided in the turbulent kinetic-energy balance shown in Figure 2. The two regions of intense turbulence energy generation are clearly evident in the inner and outer regions. The measured distribution of velocity-induced turbulent diffusion broadly suggests a transfer from regions of high energy to low. The "total" diffusion curve, which represents the combined effects of both velocity-induced and pressure-induced diffusion and has been obtained as the closing term from measured values of all the others, shows a similar variation. Irwin's wall-jet studies also included a series of non-equilibrium flows where a high-speed jet of fluid was injected beneath a thick external boundary layer developing in a strongly adverse pressure gradient. One of these flows is discussed later in the section on modeling.

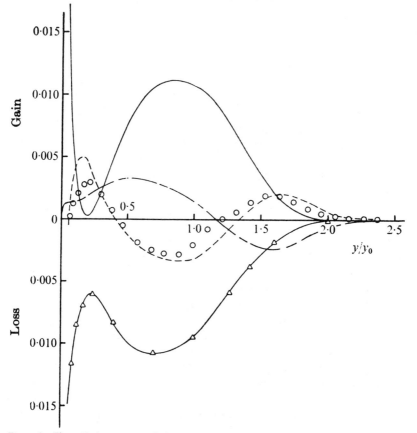

Figure 2 The turbulence energy balance in an equilibrium wall jet (Irwin 1973): —·—convection; —production; ○○○ velocity diffusion; ----total diffusion (by difference); Δ——Δ dissipation. (Reproduced by permission of Cambridge University Press.)

THE WALL JET ON CURVED SURFACES

Although the striking deflection of a plane jet toward an adjacent convex wall is generally associated with the name of Coanda, Young (1800) appears to have been the first to write about it. For an account of the early studies on this topic, including notable contributions by Reynolds (1870) and Lafay (1929), the reader is referred to the reviews by Newman (1961) and Wille & Fernholz (1965). In terms of turbulent transport, the convex streamline curvature intensifies turbulent mixing in the outer part of the flow, where the fluid's angular momentum decreases with radius, and diminishes it in the inner region. This preferential amplification further increases the importance of turbulent diffusion from the outer to the inner region; indeed, in one of Guitton & Newman's (1977) experiments the position of zero shear stress is displaced to about $0.1 y_{\mathrm{m}}$.

It is only in the last twenty years that experimenters have had flow conditions sufficiently well under control for the data to be usable for testing hypotheses about turbulent transport. The cases considered have been the nominally two-dimensional flows around a cylinder and a logarithmic spiral. A major problem has been to achieve a flow that actually is two-dimensional. The difficulties arise from two sources: (a) flow irregularities—"instabilities"—which tend to develop a short distance after discharge; and (b) secondary flow caused by slow-moving fluid in the side-wall boundary layer moving under centripetal force down the side walls, thus causing flow convergence in the center of the test plate. Fekete (1963) found that to eliminate the first a blunt (i.e. square-ended) nozzle lip was needed, a feature uniformly adopted in subsequent studies. The secondary-flow problem can be reduced if a small slot-height to flow-width ratio is adopted (a value less than .005 is desirable). Guitton & Newman (1977) also bled off the side-wall boundary layers at intervals, thus directly attacking the cause of the secondary motion. The measurements of Alcaraz (1977) made on the first 20 degrees of arc of a cylinder 5.8 m in radius show clearly that the outer-region stresses are markedly amplified even when the radius of curvature is two orders of magnitude larger than the local shear-layer thickness. At 90 slot heights from the origin, peak levels of outer-region shear stress and kinetic energy are fully 60% larger than for the case of a plane wall. The experiments of Fekete (1963) and Wilson & Goldstein (1976) on a cylinder and Guitton & Newman (1977) on a logarithmic-spiral surface provide information on flows where the ratio of shear-layer width to radius of curvature was up to two orders of magnitude greater. For a logarithmic spiral, the local radius of curvature is proportional to distance from the flow origin, an arrangement that leads to the development of a self-preserving wall jet growing linearly with downstream distance. The rate of growth of these equilibrium layers is highly sensitive to the ratio of shear-layer thickness to radius of curvature. The target growth law adopted for the 1981 Stanford conference was

$$y_{1/2}/x = 0.073 + 0.8(y_{1/2}/R) - 0.3(y_{1/2}/R)^2. \tag{3}$$

Figure 3 compares three sets of measurements for the root mean square turbulent velocity fluctuations in the stream direction and normal to the wall. It appears that the intensity of streamwise fluctuations "saturates" at a level only a little above that measured by Alcaraz (1977) at his most downstream station. In contrast, the intensity perpendicular to the wall shows a sustained increase and exhibits a peak level nearly as large as for the streamwise fluctuations in the case of the logarithmic spiral with $x/R = 1$ (indeed, larger according to the Wilson-Goldstein cylinder data at 130 degrees). In LR1, we suggest that a contributory factor to the different growth patterns of these intensities is that pressure reflections from the wall, which on a plane surface so effectively impede the transfer of energy into the component of turbulence perpendicular to the wall, become less effective on a convex surface, at any rate when the shear-layer width is of the same order as the radius of curvature.

Alcaraz's experiment at a slot Reynolds number of 4×10^4 also includes profiles of all the important triple moments except $\overline{w^2 v}$. There is very little variation in these profiles from one station to another when they are nondimensionalized by the corresponding second moments $(\overline{v^2 u}/\overline{v^2}\sqrt{\overline{u^2}}$, etc.). This interesting result seems to indicate that the triple moments, even in as complex a flow as the curved wall jet, may be fairly well approximated in terms of the Reynolds stresses and their gradients.

On a concave surface the extra strains associated with curvature produce the inverse effects to those found for the convex case. Two careful studies have been undertaken by Guitton (1964) and Spettel et al. (1972), though in neither case is the flow structure reported in significant detail.

THE RADIAL WALL JET

A radial wall jet can be generated by fluid leaving a radial nozzle placed on a wall or by a circular jet impinging on a surface. The flow is axisymmetric and is directed radially outward along the wall. Radial wall jets are used in many applications to effect cooling, heating, drying, leaching of solids, and toughening of glass, and the flow is also of practical importance in connection with the vertical take-off of aircraft.

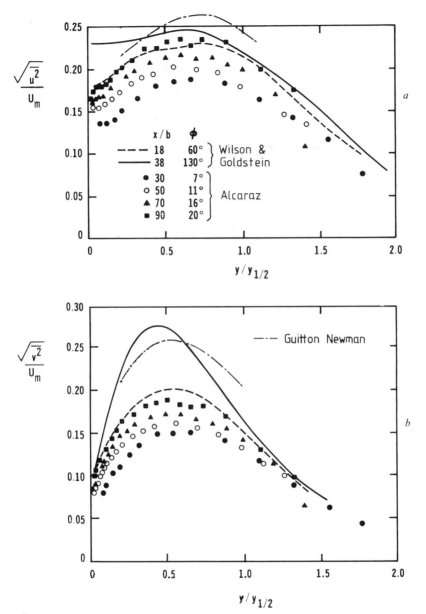

Figure 3 Development of turbulence intensities in a curved wall jet. (*a*) Streamwise intensity; (*b*) normal intensity

A self-preserving free jet exhibits a linear rate of growth with radius and a maximum velocity U_m that varies as r^{-1}. For the wall jet, Glauert (1956) and Poreh et al. (1967) have shown that the presence of wall friction changes the analysis somewhat, leading to $y \propto r^m$ and $U_m \propto r^{-n}$, where m and n differ slightly from unity by amounts that depend on the exact assumptions made in the analysis. Experiments on radial wall jets have been carried out by Bakke (1957), Bradshaw & Love (1959), Hodgson (1965), Poreh et al. (1967), Jayatelleke (1969), Baker (1969), Ng (1971), Witze (1974), Tanaka & Tanaka (1977), and Codazzi et al. (1981). In most of these experiments, the radial increase in jet width was found to be linear within measurement uncertainty. When the low values of Bakke (1957), Jayatelleke (1969), and Codazzi et al. (1981) are excluded, the rate of spread in the similarity region at sufficiently large radial distances from the symmetry axis was measured to lie in the range

$$dy_{1/2}/dx = 0.09 \pm .005,$$

with no particular dependence on the way in which the flow was generated. It does appear that the wall jet requires longer to reach a fully developed state when a radial-nozzle discharge, rather than an impinging jet, is used. The recent detailed hot-wire study by Codazzi et al. extended only 26 slot heights downstream of discharge and shows both a slower growth rate (0.073) and lower Reynolds stresses.

The rate of spread of 0.09 is some 20% higher than that observed for the plane wall jet (though about 15% lower than the free "constrained" radial jet; see Witze & Dwyer 1976). If the rate of spread were the same, it is readily demonstrated, assuming self-preservation, that the dimensionless shear stress $\overline{uv}/U_m^2$ would have to be twice as high in the radial wall jet as in the corresponding plane case. The turbulence measurements do indeed indicate that the shear-stress level is more than twice as high as in the plane jet, and the same is true for other turbulent stresses also. The following maximum values have been measured, with the values for the plane case given in parentheses:

$$\overline{u^2}/U_m^2 = 0.09\text{--}0.12(0.04), \qquad \overline{v^2}/U_m^2 = 0.048\text{--}0.084(0.015\text{--}0.025)$$

and

$$k/U_m^2 = 0.1\text{--}0.14(0.04\text{--}0.05).$$

As in the case of the plane wall jet, there is considerable variation in the measured normal fluctuations $\overline{v^2}$. Since the dimensionless velocity profile is roughly the same as in the plane jet, the "eddy" viscosity $\nu_t \equiv -\overline{uv}/\partial U/\partial y$ must also be more than twice as high, and because $\nu_t \sim k^{1/2}L$, the length scale L of the turbulent motion is larger by a factor of at least $\sqrt{2}$. This has, of course, considerable consequences for the calculation of radial wall jets, as is discussed below.

The shapes of the velocity and turbulent stress profiles are very similar to those measured in plane wall jets, and the few measurements of the friction coefficient indicate that this also behaves much the same as in the plane case. Overall, the influence of the radial geometry is seen mainly in the outer region.

THREE-DIMENSIONAL WALL JETS

The flow considered is that developed when fluid issues from a nozzle of finite width along a plane wall, as indicated in Figure 4. Newman et al. (1972) showed from similarity considerations (neglecting wall friction) that the two half-widths $y_{1/2}$ and $z_{1/2}$ and the reciprocal of U_m should grow linearly with distance from the orifice. Beyond the initial-development region, experiments confirm this behavior. The wall jet's most noticeable feature, well brought out in the photograph of Figure 5c, is that it exhibits a far greater lateral rate of spread than it does normal to the wall. From a review of experiments available in 1979, LR1 recommended $dy_{1/2}/dx = 0.048$ and $dz_{1/2}/dx = 0.26$, a ratio of spreading rates of about about 5.5:1. Indeed, a more recent study by Davis & Winarto (1980) reports a ratio that approaches 8.5:1 at large distances from the nozzle. These workers measured higher levels of momentum transport by turbulence in the lateral direction $(\overline{uw})$ than normal to the plate $(\overline{uv})$, and it was to this augmented lateral mixing that they attributed the aniso-

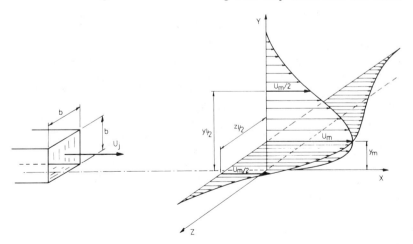

Figure 4   The three-dimensional wall jet.

a

b

Figure 5 Normal and lateral spreading of the three-dimensional wall jet. (*a*) Bursts in unstable laminar wall jet, Re = 600; (*b*) turbulent wall jet, Re = 1100; (*c*) turbulent wall jet, Re = 4000; (*d*) angle of lateral spread, Karlsruhe and UMIST experiments.

c

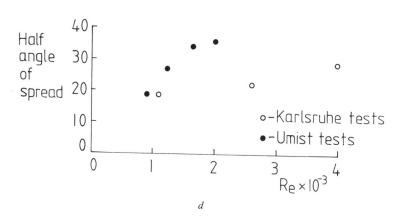

d

tropic growth pattern. However, if this behavior were due principally to turbulent mixing, fluid at the edges of the wall jet would move by entrainment toward the jet center. In fact, the surface "streamline" photograph of Newman et al. (1972) reveals a quite different picture: there is a strong flow divergence away from the symmetry plane, with the angle of the streamlines near the jet edge being nearly as large as the jet spreading angle. We might suspect that, with such a strong lateral motion, continuity would require that the y-component of velocity should be directed toward the wall. That this is indeed the case may be inferred from the fact that $dy_{1/2}/dx$ is some 35% less than for the plane two-dimensional wall jet, even though the level of streamwise-turbulence intensity is 20% higher than in the two-dimensional case (Newman et al. 1972). The three-dimensional wall jet is thus an example of a straight flow where a substantial and sustained streamwise vorticity is created. In fact, if one reexamines Davis & Winarto's (1980) data from this point of view, one is led to conclude that the "shear-stress" data represent correlations between the instantaneous rather than turbulent velocities; as such, the results would describe the momentum transport due to both the secondary mean motion and to turbulence, the former effect being the predominant one.

Let us now consider the possible physical causes of this strong streamwise vorticity in the fully developed flow. The starting point will be the equation describing the rate of increase of streamwise vorticity, which is obtained by taking the curl of the Reynolds equation. For a steady uniform density flow, this may be written

$$\frac{D\Omega_x}{Dt} = \underset{(A)}{\Omega_x \frac{\partial U}{\partial x}} + \underset{(B)}{\Omega_y \frac{\partial U}{\partial y}} + \underset{(C)}{\Omega_z \frac{\partial U}{\partial z}}$$

$$\underbrace{+ \frac{\partial^2}{\partial y \partial z}\left(\overline{w^2} - \overline{v^2}\right) + \frac{\partial^2 \overline{vw}}{\partial y^2} - \frac{\partial^2 \overline{vw}}{\partial z^2}}_{(D)} \qquad (4)$$

$$\underset{(E)}{+ \nu \left(\frac{\partial^2 \Omega_x}{\partial y^2} + \frac{\partial^2 \Omega_x}{\partial z^2}\right)},$$

where $\Omega_x \equiv (\partial V/\partial z - \partial W/\partial y)$, $\Omega_y \equiv (\partial W/\partial x - \partial U/\partial z)$ and $\Omega_z \equiv (\partial U/\partial y - \partial V/\partial x)$ are the component mean vorticities defined so that positive vorticity Ω_j represents anticlockwise rotation of a fluid element viewed by an observer at a position where x_j is smaller than for the element.

Term A in Equation (4) represents vorticity amplification by streamwise stretching and in this flow, since $\partial U / \partial x$ is predominantly negative, it will generally oppose the amplification of streamwise vorticity of whatever sign. Vortex-line-bending effects are provided by terms B and C. The effect of this process is more clearly seen by canceling $(\partial U / \partial y)(\partial U / \partial z)$, which appears as the primary contribution in both terms but with opposite sign. Thus

$$\Omega_y \frac{\partial U}{\partial y} + \Omega_z \frac{\partial U}{\partial z} = \frac{\partial W}{\partial x} \frac{\partial U}{\partial y} - \frac{\partial V}{\partial x} \frac{\partial U}{\partial z}. \tag{5}$$

For the axisymmetric free jet (where the plane $y = 0$ is a surface of symmetry) the two terms exactly balance one another at all points in the flow, with the result that no streamwise vorticity is created. The application of the no-slip boundary at $y = 0$, however, destroys this symmetry. Even if one initially postulates similar growth rates of $y_{1/2}$ and $z_{1/2}$, it is readily appreciated that over the inner region ($y < y_m$) the first term on the right of Equation (5) is easily the dominant one, for V is small compared with W ($\partial V / \partial y = 0$ at the wall) and (assuming $y_{1/2} \simeq z_{1/2}$) $\partial U / \partial y$ is still an order of magnitude larger than $\partial U / \partial z$. The component W will be positive over the inner region in an "equal growth rate" scenario; $\partial W / \partial x$ is thus negative (velocity levels decrease with distance from discharge), $\partial U / \partial y$ positive, and thus the net streamwise vorticity source negative. The action of this source is to strengthen the z-component of velocity. A corollary is that W will be larger than required for isotropic growth rates and thus V will be smaller. This has a knock-on effect on the outer region, where we may now conclude that the first term on the right of (5) will again be the larger one over most of the flow; thus the vorticity source will be positive ($\partial U / \partial y$ and $\partial W / \partial x$ both negative), producing a lateral velocity profile roughly as suggested in Figure 6a.

It must be emphasized that the assessment of terms (B) and (C) in Equation (4) related to the developed flow only. In the vicinity of discharge, $\partial V / \partial x$ and $\partial W / \partial x$ will both be positive as the jet begins to spread, producing vorticity sources in the opposite sense from that found in fully developed flow, which would generate a secondary velocity distribution roughly as shown as Figure 6b. (The inner-region source is the predominant one since in the initial development, over the outer region, the two terms on the right of (5) will nearly cancel.) The process is a self-amplifying one whose strength may be judged by the rapid increase in the lateral growth rate following transition shown in Figure 5b.

It is appropriate here to mention a first cousin of the three-dimensional wall jet, namely a jet issuing from a rectangular nozzle that is bounded by an upper and lower wall (Foss & Jones 1968, Foss & Holdeman 1975).

Here, too, an outward secondary motion is observed near the walls with a return flow in toward the midplane. The authors attribute this phenomenon, by arguments analogous to those presented above, to the reorienting of vortex lines.

It is not yet clear to what extent—if at all—the self-preserving far-field behavior of the three-dimensional wall jet depends, through streamwise convection [the left-hand side of (4)], on the initially generated streamwise vorticity. In certain free shear flows (trailing vortices from a wing, for example) the initial development leaves a permanent imprint on the downstream flow. Flows along walls that develop to a self-preserving state normally forget about their origins, however. Experiments give no

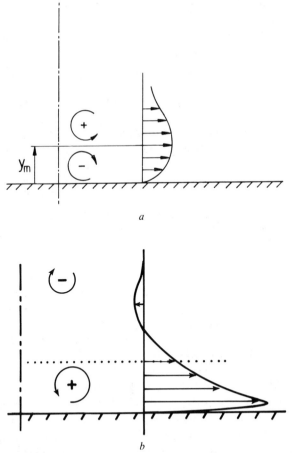

Figure 6 Streamwise vorticity generation in the three-dimensional wall jet. (*a*) Developed flow region; (*b*) developing region.

conclusive answer on this matter: there are small differences among the rates of development in the different experiments (LR1), but apart from the clear cut effect of Reynolds number discussed later, any possible systematic dependence on initial conditions is obscured by experimental uncertainty.

There is of course another potential source of streamwise vorticity provided by the inhomogeneities in the Reynolds stress field [term D in Equation (4)]. In fully developed flow in straight ducts, these terms are solely responsible for the weak but influential secondary motions that are created. There is unfortunately nothing approaching a complete mapping available of the Reynolds stress field in a three-dimensional wall jet. In the far field, it seems that the term $\partial^2(\overline{w^2} - \overline{v^2})/\partial y\,\partial z$ is of the same order of magnitude as the vortex-line-bending process discussed above. Integration over the outer and inner regions of the wall jet shows that the term respectively contributes a net vorticity source to these regions of plus and minus $(\overline{w^2} - \overline{v^2})$ evaluated at $y = y_m$, $z = 0$. Since $\overline{w^2}$ is greater than $\overline{v^2}$ (by about $4 \times 10^{-3}U_m^2$ at the point in question), the process appears to reinforce that due to vortex line bending indicated in Figure 6a. In the inner region, however, this net effect is the result of a positive source over most of the region being outweighed by a concentrated negative region in the viscosity-affected sublayer where viscous diffusion (E) will allow leakage to the wall. There is, moreover, the contribution of the shear stress $\overline{vw}$, for which no data are available but which cannot be presumed of lesser importance than $(\overline{w^2} - \overline{v^2})$. These complications prevent any firm conclusions about the importance of Reynolds stresses in the y-z plane on the flow's development. There is, however, some evidence that can be gathered from computation. Very recently Kebede (1982) solved the self-similar form of the equations for the three-dimensional wall jet by adapting a standard two-dimensional elliptic solver. Both laminar and turbulent flows were considered, but the turbulence model assumed an isotropic viscosity; consequently, there was sensibly no contribution from the modeled Reynolds stress field to streamwise-vorticity generation. The cross-stream velocity pattern shown in Figure 7 is qualitatively in accord with the earlier discussion: the flow in the outer region is directed toward the wall and the peak z-direction outflow occurs near $y = y_m$. Despite the qualitative agreement, however, there are serious quantitative differences, for the computations give a ratio of $z_{1/2}$:$y_{1/2}$ of less than 2:1. It thus appears that the Reynolds stress field may well make a significant contribution to the generation of streamwise vorticity.

Although the theme of this article is the turbulent wall jet, Kebede's (1982) laminar results deserve mention. At a Reynolds number of 120 (based on U_m and $y_{1/2}$), the secondary flows, while in generally the same sense as for turbulent flows, are an order of magnitude smaller. The explanation appears to be that $\partial W/\partial x$ and $\partial V/\partial x$ are much smaller and, moreover, the peak velocity occurs much further from the wall than in turbulent flow (where the effective viscosity increases strongly with distance from the wall). The vorticity-generating terms in Equation (5) are thus smaller and more nearly in balance over the inner region than for turbulent flow.

A further aspect of the three-dimensional wall jet that still awaits a complete explanation is the influence of Reynolds number on the rate of spread, particularly in the lateral direction. Newman et al. (1972) observed no detectable change in spreading rate as the discharge Reynolds number was raised from 2800 to 16,400. Hofer (1979), however, found that as the Reynolds number was raised from 800 to 4400 the spreading half-angle increased from 19° to 34°. In an attempt to clarify the situation, flow visualization experiments have been undertaken at our institutions. At Karlsruhe, the study was carried out in a water tank 1 m wide and several meters long, and with a water depth of 35 cm. The bottom and the side walls of the tank were made of glass. A square outlet was produced by machining a square groove into a perspex block, which was then glued flush on the glass bottom of the tank. The outlet width

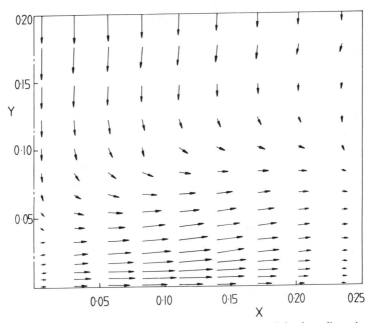

Figure 7 Cross-stream velocity directions in computations of the three-dimensional turbulent wall jet (Kebede 1982).

was 2.6 mm and the groove was 22 cm long, connected at its end to a constant-head tank filled with water to which dye had been added. With this arrangement, discharge Reynolds numbers in the range 500–4000 could be obtained. A mirror was placed under the tank so that simultaneous pictures could be taken of the top and side view of the wall jet. The UMIST apparatus, about two thirds the scale of that at Karlsruhe, was built to a very similar design except that the grooved block was incorporated within the constant-head tank rather than the discharge tank. In this latter case, discharge of the wall jet was essentially that from a small hole at the base of a semi-infinite vertical wall, with boundary conditions $V = W = 0$ at $x = 0$.

Experiments in both tanks exhibited a significant dependence of the rate of spread on Reynolds number; this is clearly evident in the photographs of Figure 5 taken in the Karlsruhe tank. At Re = 600, shown in Figure 5a, the flow is in marginally unstable laminar motion and the bursts suggest a roughly equal rate of spread in each direction. At Re = 1100 the wall jet becomes turbulent with a lateral rate of spread approximately twice that normal to the plate, while at Re = 4000 the lateral:normal spreading rate exceeds 3. Figure 5d shows, however, that the dependence on Reynolds number is by no means the same in the two apparatuses. In the UMIST tank, with a plane back wall at $x = 0$, the asymptotic spreading rate appears to have been reached by a Reynolds number of 2000, while in the Karlsruhe apparatus the spreading still seems somewhat short of its final value at Re = 4000. We believe that the differences in the observed behavior are attributable to the different conditions external to the jets at the discharge plane. Indeed, in support of this view, we note that Hofer's (1979) wall jets exited from a tube fixed to the wall of the test plate, while a large back plate is visible in the photographs from the study by Newman et al. (1972).

THE NUMERICAL CALCULATION OF TURBULENT WALL JETS

The various fundamental flows discussed in earlier sections establish the wall jet as a significantly more complex flow than, say, a simply strained boundary layer. We consider here how well its features can be imitated by current viscous-flow calculation methods. Our survey includes, besides the basic flows already discussed, some more complicated wall jets that are more representative of those found in engineering practice. We limit our enquiry, however, to calculation schemes that solve discretized forms of the mean momentum and continuity equations. In the former,

Reynolds stresses appear as unknowns that the computor must approximate as best he can—or chooses to—with a turbulence model. For two-dimensional parabolic flows, at any rate, one should be able to attribute blame for discrepancies in computation to weaknesses in the turbulence model. In more complex flows, particularly in three dimensions, numerical errors are often of an uncertain magnitude; it will then usually not be possible to distinguish failures in the turbulence model from inaccuracies arising from discretization and truncation.

Boussinesq Viscosity Models (BVMs)

All the applications described in this section assume a Boussinesq effective viscosity, v_t. Two main approaches to calculating v_t have been followed and these are discussed in what follows.

MIXING-LENGTH HYPOTHESIS   The first attempts to calculate wall jets with a differential method made use of Prandtl's mixing-length hypothesis (Patankar & Spalding 1967), a Boussinesq viscosity model (BVM) in which the turbulent kinematic viscosity v_t is obtained from the local level of the "mixing length" l_m and the mean velocity gradient:

$$v_t = l_m^2 \left| \frac{\partial U}{\partial y} \right|.$$

The mixing length is a user-prescribed parameter and the calculations of Patankar & Spalding (1967) were carried out with the ramp function:

$$l_m = \kappa y \qquad y < \lambda y_e / \kappa,$$
$$l_m = \lambda y_e \qquad \lambda y_e / \kappa \leqslant y < y_e,$$

where y_e is the location of the "edge" of the shear layer (where the velocity deviates from the free stream velocity by 1% of the maximum velocity difference across the layer) and the two constants are given the values $\kappa = 0.435$ and $\lambda = 0.09$. This ramp distribution was developed for and used successfully in calculating conventional turbulent boundary layers. (The ramp distribution was perhaps suggested by the similarity between the outer part of the boundary layer and a free shear flow, together with the knowledge that the mixing-length prescription $l_m = \lambda y_e$, with y_e the distance from the axis of symmetry to the edge and λ again equal to 0.09, satisfactorily predicts the growth of the plane jet in stagnant surroundings.) That the indicated mixing-length distribution should predict the correct spreading rate for the plane wall jet in still air (Patankar & Spalding 1967) is a quite remarkable achievement. Apparently, the linear drop-off of mixing length near the wall effectively imitates the damping of the normal velocity fluctuations, thereby reducing the shear stress by the right amount. The physical result, while impressive, must be considered fortuitous. The frailty of the success is illustrated by the fact that for the radial wall jet, Patankar & Spalding (1967) had to use substantially different constants in their ramp function. A more extensive set of computations by Sharma (1972) covered conical wall jets with half-angles ranging from 10° to 90° (the latter value representing a radial wall jet). He found uniformly poor agreement with experiment if a mixing-length distribution giving good agreement for the flat-plate boundary layer was adopted.

The mixing-length model has been applied by Pai & Whitelaw (1971) to film-cooling problems. Straightforward application of the mixing-length hypothesis would have led to zero turbulent diffusivity (and hence zero heat flux) at locations of velocity maximum or minimum (e.g. in the wake behind a lip). Accordingly, in the region of the velocity extremum a "bridged" diffusivity was used having regard for neighboring diffusivity levels. Rastogi (1972) applied the Patankar & Spalding ramp function to calculate curved wall jets. As other workers computing curved flows have found, however, the effect of the secondary strain associated with curvature has to be empirically augmented by an order of magnitude to bring reasonable agreement with data. The precise value required for the empirical coefficient varies from flow to flow.

TWO-EQUATION MODELS   In an effort to avoid the more obvious limitations of the mixing-length hypothesis it is nowadays common to obtain the turbulent viscosity from the Kolmogorov proposal:

$$v_t = c_\mu k^{1/2} L,$$

where c_μ is a constant, k is the local turbulent kinetic energy, and L is a length scale representative of the energy-containing motion. The quantities k and L are obtained from approximate transport equations that are solved simultaneously with the mean flow. Several such models have been developed, but only two have been applied extensively to the calculation of wall jets, the $k \sim \varepsilon$ model (Jones & Launder 1972, Launder & Spalding 1974) and the $k \sim kL$ model (Rodi & Spalding 1970, Ng 1971). Both schemes solve a transport equation for turbulence energy; they differ only in their choice of the second turbulence variable, kL or ε (which may be interpreted as $k^{3/2}/L$). Although the semiempirical transport equations for these two variables are similar in form, with source and sink processes as well as transport terms, that proposed for ε is the simpler and, perhaps for that reason, is the most widely adopted. Overall, with a single set of empirical constants, two-equation models do achieve a significantly greater width of "predictability" than does the mixing-length hypothesis. It must be stated, however, that when applied to the plane wall jet in stagnant surroundings the rate of spread obtained with the $k \sim \varepsilon$ model is calculated to be 30% greater than experiment. The main cause of this discrepancy is the wall's damping of the normal velocity fluctuations, not just in the immediate vicinity of the surface but in the outer "free-shear-layer" region beyond the velocity maximum. This physical process is not accounted for with the Kolmogorov formula for v_t if c_μ is a constant. This weakness is clearly evident in Figure 8, which considers spreading rates for the family of equilibrium wall jets in which, by suitably tailoring the streamwise pressure gradient, the ratio of maximum to external-stream velocity is held constant. The figure also suggests that the error is much less important when a substantial external velocity is present. The $k - kL$ model manages to predict correctly the plane wall jet in stagnant surroundings largely, we may suppose, because Ng (1971) tuned his wall-correction function (a term included only for flows along walls) by reference to this flow. It does less well for the radial wall jet, where a too small level of viscosity is calculated. In contrast, Sharma (1972) found that the $k \sim \varepsilon$ model satisfactorily reproduced the spreading of a family of wall jets developed in stagnant surroundings on cones ranging from 10° to 90° half-angle. As we noted earlier, the mixing-length computations by the same worker exhibited a far too slow rate of spread.

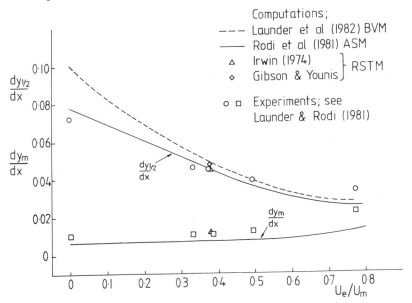

Figure 8 Rate of spread of equilibrium wall jets in a moving stream.

When the describing set of transport equations is simply transformed to a curvilinear coordinate system it is seen from Figure 9 that the $k \sim \varepsilon$ model seriously underpredicts the sensitivity of the wall jet to streamline curvature. (The experimental line relates to the equilibrium wall jet on a logarithmic spiral surface.) This failure is endemic to all turbulence models using a Boussinesq stress-strain relation unless the coefficient c_μ is itself made sensitive to curvature. Although the problem can be reduced by the introduction of a curvature correction to the ε equation (e.g. Launder et al. 1977), the proper way of removing the difficulty is by moving to a closure scheme based on the Reynolds stress transport equations.

A further basic weakness of the Boussinesq stress-strain law has emerged in relation to the three-dimensional wall jet. Kebede's (1982) numerical computation has scrupulously included such usually neglected terms in the mean momentum equation as the variation of mean pressure across the shear layer. His results, however, do not reproduce the observed large asymmetry in the rates of growth of $y_{1/2}$ and $z_{1/2}$. As our earlier discussion indicated, the physical causes of this asymmetry are not fully understood. If, however, the Reynolds stress field is making a major contribution to the generation of streamwise vorticity, the phenomenon cannot be mimicked with a Boussinesq viscosity model.

It was mentioned earlier that the shear stress does not vanish at the velocity maximum but closer to the wall. This is yet another feature of a

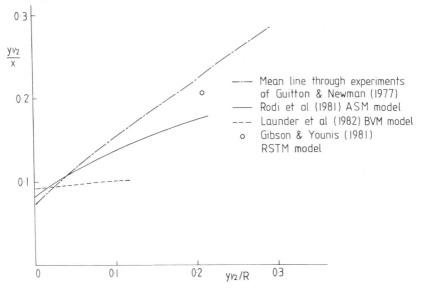

Figure 9   Rate of spread of equilibrium wall jets on a logarithmic spiral surface.

wall jet that cannot be predicted with any BVM treatment—though the consequences of this failure are often not of practical importance. A BVM computation usually seems to predict the position of zero shear stress quite well (but necessarily gets the position of maximum velocity wrong)—so the shear-stress profile, which dictates the growth, will often be satisfactorily calculated.

Having considered some of the underlying weaknesses of a BVM treatment, we now examine computations obtained with the more elaborate and physically realistic Reynolds-stress closures. Before doing so, however, we should mention that the two-equation BVM treatment represents, for the computor, an attractive level at which to tackle industrial-type wall-jet problems where resort must be made to iterative solution schemes. Representing the turbulent stresses through a diffusion-type transport law promotes convergence and removes the need to store the Reynolds stresses over the field. Among applications of this type are (a) the explorations of Matthews & Whitelaw (1971) of film-cooling in geometries where, due to the presence of steps and thick discharge lips, flow recirculations are present, and (b) the various ventilating-design studies in both two and three dimensions by Nielsen and his coworkers (Nielsen et al. 1978, Nielsen 1979). There are many such problems where all that is needed is to obtain an answer accurate within 30%. In these cases, use of a BVM, perhaps judiciously optimized in the light of certain experimental data, will probably remain the preferred approach for several years to come.

Applications of Reynolds-Stress Closures and Derivatives

An exact equation for the rate of change of Reynolds stress is readily obtained from the Navier-Stokes equation, e.g. Hinze (1959). Although most of the stress-modifying processes on the right of that equation are not directly knowable—and thus require "modeling"—a very important one is: the rate of generation of $\overline{u_i u_j}$ by the interaction of Reynolds stresses and velocity gradients, i.e.

$$\text{Generation} = -\left(\overline{u_i u_k} \frac{\partial U_j}{\partial x_k} + \overline{u_j u_k} \frac{\partial U_i}{\partial x_k} \right).$$

This term alone implies far more subtle responses of the stress field to mean strain than is allowed by a BVM. Moreover, the fact that such a major element in the equation can be treated exactly offers encouragement that, at this level, closures can be devised with a moderate width of generality.

Of the other terms in the stress transport equation, the most important for imitating the behavior of wall jets is the correlation between the fluctuating pressure field and the fluctuating strain, $\overline{p/\rho\left(\partial u_i/\partial x_j + \partial u_j/\partial x_i\right)}$. The term makes no direct contribution to the turbulence energy level since, by continuity, $\partial u_k/\partial x_k = 0$. It thus acts as a distributor of kinetic energy from one direction to another while, in the shear-stress equations, it serves to diminish the correlation between the velocity fluctuations. The process therefore works to randomize the turbulent field. It is quite clearly a powerful agency, for in most *free* shear flows, less than half the turbulence energy resides in streamwise fluctuations, even though essentially all the energy extracted from the mean flow enters turbulence through this component. In flow along a rigid boundary we find a somewhat different behavior. Pressure reflections from the wall—the "echo effect"—severely impede the transfer of energy from the streamwise direction to that normal to the surface; consequently, the streamwise turbulent normal stress contains some 60% of the total energy and is about 5 times as large as that normal to the wall. It turns out to be of great importance to include a model of this wall-reflection (or wall-damping) process in calculating wall jets.

Space limitations preclude a review of closure methodology in the present paper. The closures of Launder et al. (1975) and Gibson & Launder (1978) have formed the basis of most wall-jet computations to date. We discuss in parallel the applications of Reynolds stress transport models (RSTMs), which arise from a formal term-by-term approximation of the unknown processes in the exact equation, and algebraic stress models (ASMs), in which, in addition, the stress-transport processes (convection and diffusion) are approximated in terms of the corresponding transport of turbulent kinetic energy. The latter simplification means that the scalar k is the only Reynolds-stress element that needs to be found from a transport equation; the stress components are then obtained from a set of algebraic equations. In fairly simple strains, ASMs can be organized to a form analogous to the $k \sim \varepsilon$ BVM except that the coefficient c_μ becomes a function of various dimensionless parameters of the turbulence field; this rearrangement significantly simplifies numerical solution. A specific ASM development is given by Ljuboja & Rodi (1980).

The first application of an RSTM to a wall jet was by Hanjalić & Launder (1972) with a model that accurately reproduced the development of the flat-plate boundary layer and the plane jet in stagnant surroundings. However, for the wall jet, although the calculation correctly predicted the separation between the surfaces of zero shear stress and maximum velocity, the level of shear stress in the outer region was excessive, causing a rate of spread that was 20% too high. The physical origin of the discrepancy was the absence of any model of turbulent pressure reflections from the wall; through their influence on the pressure-containing correlations in the stress-transport equations, these pressure reflections strongly damp the intensity of turbulent fluctuations normal to the wall. With a model of wall-damping included (Launder et al. 1975), Reece (1977) and Irwin (1974) achieved the correct rate of development for both the wall jet and the turbulent boundary layer with the same model. More recently, Gibson & Younis (1981) have computed Irwin's (1973) equilibrium wall jet using a somewhat different pressure-reflection model (Gibson & Launder 1978). They report satisfactory agreement on growth rates (see Figure 8), and the mean velocity distribution shown in Figure 10 is in excellent agreement with Irwin's measurements. The algebraic stress treatment of Ljuboja & Rodi (1980) and Rodi et al. (1981) gives reasonably satisfactory agreement with measured spreading rates for the whole family of equilibrium boundary layers shown in Figure 8, although because of the crude approximation of stress diffusion in ASM schemes, the velocity maximum is somewhat closer to the wall than in the measurements.

When the Reynolds-stress equations are transformed into a curvilinear coordinate system following a curved surface, curvature terms appear naturally. Irwin (1974) (see also Irwin & Arnot Smith 1975) and Gibson & Younis (1981) have calculated the development of wall jets on curved surfaces using their transformed Reynolds-stress equations with no extra

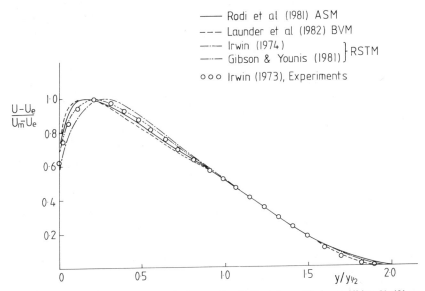

Figure 10 Comparison of mean velocity profiles for Irwin's equilibrium wall jet, $U_e/U_m = 0.38$.

empirical input to account for curvature effects. (Irwin neglected curvature terms in the mean flow equations, however, and hence his calculations are appropriate only for wall jets with mild curvature.) The model used by Gibson & Younis (1981) has been used to calculate successfully curved wall boundary layers (Gibson et al. 1981) and the curved mixing layer (Gibson & Rodi 1981). Its application to Alcaraz's wall jet on a large radius cylinder has shown that curvature effects are indeed well accounted for, at least over the first 20° of arc considered in the experiment. Irwin's (1974) calculations were for Fekete's (1963) experiments, which extended to more than 200 degrees. In this case the computations underestimated the growth rate in the later stages of development, possibly because of the neglect of curvature terms in the mean equations. For the case of the equilibrium wall jet developed on a logarithmic spiral, Figure 9 shows that the Gibson-Younis computation does predict a strong increase in spreading rate, but still not quite as strong as in the experiments. The distributions of the individual stresses in this self-preserving flow are also in good agreement with the measurements.

Rodi et al. (1981), with an ASM treatment, have also computed the development on a logarithmic spiral; their calculations, too, correctly reflect the sensitivity of the growth rate to streamline curvature. Agreement deteriorates when the curvature becomes very severe ($x/R \sim 1$), though not much more so than for the stress-transport model. It must be acknowledged that an algebraic stress model does not reflect accurately the very high diffusive transport of stress from the destabilized outer region to the damped interior, and this may be contributing to the poorer agreement in highly curved flows.

As a final application, Figure 11 reproduces one of Irwin's (1974) RSTM computations of a blown boundary layer developing in a strongly adverse pressure gradient. Starting with the measured initial conditions (shown at the top left of the figure), the calculations succeed very well in reproducing the development of this complex flow. Further applications of the model to flows of this type have been made by Saripalli & Simpson (1980).

It is apparent from this short survey that models derived from the Reynolds stress transport equation (whether of RSTM or ASM form) have been more successful than those based on the Boussinesq stress-strain relation in imitating the wall jet's behavior. A crucially important element in these closures is the modeling of turbulent pressure reflections from the wall: it is this process that is responsible for the slower growth rate of the plane wall jet than the free jet and for the very strong sensitivity of the flow to streamline curvature. Generally, ASM treat-

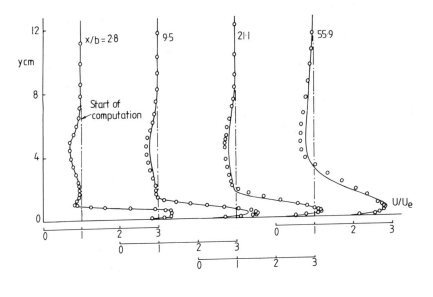

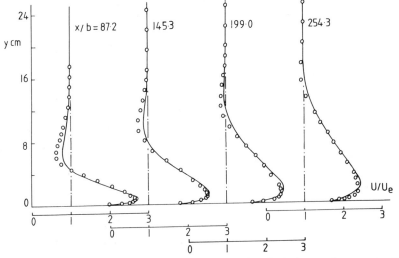

Figure 11 Prediction of development of blown boundary layer in an adverse pressure gradient (Irwin 1974).

ments have acquitted themselves practically as well as the RST models; where computing times are a significant factor, there is clearly an advantage to their use. Whether existing models of this type will be equally successful in reproducing the measured characteristics of radial and three-dimensional wall jets is a question whose answer we await with interest.

ACKNOWLEDGMENTS

Our thanks to Mr. D. Bierwirth at the University of Karlsruhe and Mr. D. C. Jackson at UMIST for assistance with the flow visualization studies of the three-dimensional wall jet. We also acknowledge several helpful discussions with Professor V. C. Patel on the spreading mechanisms driving this flow.
Authors' names appear alphabetically.

Literature Cited

Alcaraz, E. 1977. *Contribution a l'étude d'un jet plan turbulent évoluant le long d'une paroi convexe a faible courbure*. Thèse d'etat. Univ. Claude Bernard, Lyon, Fr.

Alcaraz, E., Guillermet, G., Mathieu, J. 1968. Mesures de frottement à la paroi à l'aide de tubes de Preston et d'une balance. *C. R. Acad. Sci. Paris Ser. B* 266:432–34

Baker, E. 1969. Influence of mass injection on turbulent flow near walls. *Prog. Heat Mass Transfer* 1:99–192

Bakke, P. 1957. Experimental investigation of wall jet. *J. Fluid Mech.* 2:467–72

Bradshaw, P., Gee, M. T. 1960. Turbulent wall jets with and without an external stream. *Aeronaut. Res. Counc. R & M 3252*

Bradshaw, P., Love, E. M. 1959. The normal impingement of a circular air jet on a flat surface. *Aeronaut. Res. Counc. R & M 3205*

Codazzi, D., Teitgen, R., Burnage, H. 1981. Jet turbulent pariétal axisymetrique. Similitude de l'écoulement hors de la couche limite. *C. R. Acad. Sci. Ser. II* 293: 103–6

Davis, M. R., Winarto, H. 1980. Jet diffusion from a circular nozzle above a solid plane. *J. Fluid Mech.* 101:201–21

Fekete, G. I. 1963. Coanda flow in a two-dimensional wall jet on the outside of a circular cylinder. *Mech. Eng. Dept. Rep. No. 63-11*, McGill Univ., Montreal

Foss, J. F., Holdeman, J. D. 1975. Initiation, development, and decay of the sec-

ondary flow in a bounded jet. *J. Fluids Eng.* 97:342–52

Foss, J. F., Jones, J. B. 1968. Secondary flow effects in a bounded rectangular jet. *J. Basic Eng.* 90:214–48

Gartshore, I., Newman, B. G. 1969. The turbulent wall jet in an arbitrary pressure gradient. *Aeronaut. Q.* 20:25

Gibson, M. M., Launder, B. E. 1978. Ground effects on pressure fluctuations in the atmospheric boundary layer. *J. Fluid Mech.* 86:491–511

Gibson, M. M., Rodi, W. 1981. A Reynolds-stress closure model of turbulence applied to the calculation of a highly curved mixing layer. *J. Fluid Mech.* 103:161–82

Gibson, M. M., Younis, P. A. 1981. Calculation of a turbulent wall jet on a curved wall with a Reynolds stress model of turbulence. *Proc. 3rd Symp. Turbul. Shear Flows, Davis, Calif.*

Gibson, M. M., Jones, W. P., Younis, P. A. 1981. Calculation of turbulent boundary layers on curved surfaces. *Phys. Fluids* 24:386–95

Glauert, M. B. 1956. The wall jet. *J. Fluid Mech.* 1:625

Guitton, D. E. 1964. Two-dimensional turbulent wall jets over curved surfaces. *Mech. Eng. Dept. Rep. No. 64-7*, McGill Univ., Montreal

Guitton, D. E. 1970. *Some contributions to the study of equilibrium and non-equilibrium wall jets over curved surfaces*. PhD thesis. McGill Univ., Montreal

Guitton, D. E., Newman, B. G. 1977. Self-preserving turbulent wall jets over convex surfaces. *J. Fluid Mech.* 81:155

Hanjalić, K., Launder, B. E. 1972. A Reynolds stress model of turbulence and its application to thin shear flows. *J. Fluid Mech.* 52:609–38

Hinze, J. O. 1959. *Turbulence.* New York: McGraw-Hill. 1st ed.

Hodgson, T. H. 1965. *Pressure fluctuations in flow turbulence.* PhD thesis. Univ. London

Hofer, K. 1979. Turbulente Wandstrahlen mit Auftrieb. *Mitt. Versuchsanst. Wasserbau, Hydrol., Glaziol., Eidg. Tech. Hochsch., Zürich 42*

Irwin, H. P. A. H. 1973. Measurements in a self-preserving wall jet in a positive pressure gradient. *J. Fluid Mech.* 61:33

Irwin, H. P. A. H. 1974. *Measurements in blown boundary layers and their prediction by Reynolds stress modelling.* PhD thesis. McGill Univ., Montreal

Irwin, H. P. A. H., Arnot Smith, P. 1975. Predictions of the effect of streamline curvature on turbulence. *Phys. Fluids* 18:624–30

Jayatelleke, C. L. V. 1969. The influence of Prandtl number and surface roughness on the resistance of the laminar sub-layer to momentum and heat transfer. *Prog. Heat Mass Transfer* 1:193

Jones, W. P., Launder, B. E. 1972. Prediction of laminarization with a 2-equation model of turbulence. *Int. J. Heat Mass Transfer* 15:301

Kamemoto, K. 1974. Investigation of turbulent wall jets over logarithmic spiral surfaces. (1) Development of jets and similarity of velocity profile (2) Properties of flow near the wall. *Bull. JSME* 17:333

Kebede, W. 1982. *Numerical computations of the 3-dimensional wall jet.* MSc dissertation. Univ. Manchester

Lafay, M. A. 1929. Contribution a l'étude de l'effet Chilowsky. *Mem. de l'Artillerie VIII*, 3ᵉ fasceau 385

Launder, B. E. 1969. The Prandtl-Kolmogorov Model of turbulence with the inclusion of second-order terms. *J. Basic Eng.* 91D:855

Launder, B. E., Rodi, W. 1981. The turbulent wall jet. *Prog. Aeorosp. Sci.* 19: 81–128

Launder, B. E., Spalding, D. B., 1974. The numerical computation of turbulent flows. *Comput. Meth. Appl. Mech. Eng.* 3:269

Launder, B. E., Leschziner, M. A., Sindir, M. 1982. The UMIST/UCD computa-tions for the AFOSR-HTTM-Stanford Conference on Complex Turbulent Flows. *Univ. Manchester Inst. of Sci. Technol. Rep. TFD/82/1.* To be published in *Proc. Stanford Conf. Complex Turbul. Flows*

Launder, B. E., Priddin, C. H., Sharma, B. I. 1977. The calculation of turbulent boundary layers on curved and spinning surfaces. *J. Fluids. Eng.* 99:231–39

Launder, B. E., Reece, G. J., Rodi, W. 1975. Progress in the development of a Reynolds stress turbulence closure. *J. Fluid Mech.* 68:537–66

Ljuboja, M., Rodi, W. 1980. Calculations of turbulent wall jets with an algebraic stress model. *J. Fluids Eng.* 102:350

Mathieu J. 1959. *Contribution a l'étude aerothermique d'un jet plan évoluant en presence d'une paroi.* These de Docteur ès Sciences. Univ. Grenoble

Matthews, L., Whitelaw, J. H. 1971. The prediction of film cooling in the presence of recirculating flows with a 2-equation model of turbulence. *Mech. Eng. Dept. Rep. EHT/TN/A/35.* Imperial Coll., London

Narasimha, R., Narayan, K. Y., Parthasarathy, S. P. 1973. Parametric analysis of turbulent wall jets in still air. *Aeronaut. J.* 77:335

Newman, B. G. 1961. The deflexion of plane jets by adjacent boundaries—Coanda effect. In *Boundary Layer and Flow Control.* London: Pergamon

Newman, B. G., Patel, R. P., Savage, S. B., Tjio, H. K. 1972. Three-dimensional wall jet originating from a circular orifice, *Aeronaut. Q.* 23:187

Ng, K. H. 1971. *Predictions of turbulent boundary-layer development using a two-equation model of turbulence.* PhD thesis. Univ. London

Nielsen, P. F. 1979. The distribution of air velocity in large rooms with small side-wall mounted supply openings. *15th Int. Congr. Refrig. Pap. E1/8,* Venice

Nielsen, P. V., Restivo, A., Whitelaw, J. H. 1978. The velocity characteristics of ventilated rooms. *J. Fluids Eng.* 100:291

Pai, B. R., Whitelaw, J. H. 1971. The prediction of wall temperature in the presence of film cooling. *Int. J. Heat Mass Transfer.* 14:409–22

Patankar, S. V., Spalding, D. B. 1967. *Heat and Mass Transfer in Boundary Layers.* London: Morgan-Grampian

Patel, R. P. 1970. *A study of two-dimensional symmetric and asymmetric turbulent shear flows.* PhD thesis. McGill Univ., Montreal

Patel, R. P., Newman, B. G. 1961. Self-preserving two-dimensional turbulent jets and wall jets in a moving stream. *Mech. Eng. Dept. Rep. No. Ae5*, McGill Univ., Montreal

Patel, V. C. 1965. Calibration of the Preston tube and limitations on its use in pressure gradients. *J. Fluid Mech.* 23:185

Poreh, M., Tsuei, Y. G., Cermak, J. E. 1967. Investigation of a turbulent radial wall jet. *J. Appl. Mech.* 34:457–63

Rastogi, A. K. 1972. *Effectiveness and heat-transfer downstream of three-dimensional film cooling slots.* PhD thesis. Univ. London

Reece, G. J. 1977. *A generalized Reynolds-stress model of turbulence.* PhD thesis. Univ. London

Reynolds, O. 1870. Suspension of ball by jet of fluid. *Proc. Manchester Lit. Phil. Soc.* 9:114

Rodi, W. 1975. A review of experimental data of uniform density free turbulent boundary layers. In *Studies in Convection*, ed. B. E. Launder, 1:70–165. London: Academic

Rodi, W., Spalding, D. B. 1970. A two-parameter model of turbulence and its application to free jets. *Waerme-Stoffuebertrag.* 3:85

Rodi, W., Celik, I., Demuren, A. O., Scheuerer, G., Shirani, E., Leschziner, M. A., Rastogi, A. K. 1981. Calculations for the 1980/81 AFOSR-HTTM-Stanford Conference on Complex Turbulent Flows. *Univ. Karlsruhe Rep. SFB80/T/*

199. To be published in *Proc. Stanford Conf. Complex Turbul. Flows*

Saripalli, K. R., Simpson, R. L. 1980. Investigation of blown boundary layers with an improved wall jet system. *NASA Contract. Rep. 3340*

Sharma, R. N. 1972. *Momentum and mass transfer in turbulent conical wall jets.* PhD thesis, ITT, Kapur

Spettel, F., Mathieu, J., Brison, J. 1972. Tensions de Reynolds et production d'energie cinetique turbulent dans les jets parietaux sur parois planes et concaves. *J. Méc.* 11:403

Tailland, A., Mathieu, J. 1967. Jet parietal. *J. Méc.* 6:103

Tanaka, T., Tanaka, E. 1977. Experimental studies for a radial turbulent jet (2nd report, Wall jet on a flat smooth plate). *Bull. JSME* 20:209–15

Wille, R., Fernholz, H. 1965. Report on the first European Mechanics Colloquium on the Coanda Effect. *J. Fluid Mech.* 23:801

Wilson, D. J., Goldstein, R. J. 1976. Turbulent wall jets with cylindrical streamwise surface curvature. *J. Fluids Eng.* 96:550

Witze, P. O. 1974. *A study of impinging axisymmetric turbulent flows: the wall jet, the radial jet, and opposing free jets.* PhD thesis. Univ. Calif., Davis

Witze, P. O., Dwyer, H. A. 1976. The turbulent radial jet. *J. Fluid Mech.* 75:401

Young, T. 1800. *Outlines of experiment and inquiries respecting sound and light.* Presented at R. Soc., London, Jan.

Ann. Rev. Fluid Mech. 1983. 15:461–512

FLOW IN CURVED PIPES

S. A. Berger and L. Talbot

Department of Mechanical Engineering, University of California, Berkeley, California 94720

L.-S. Yao

Department of Mechanical and Aerospace Engineering, Arizona State University, Tempe, Arizona 85281

1. INTRODUCTION

Although the systematic theoretical and experimental exploration of flow in curved conduits is of fairly recent origin, it has long been appreciated that the flow is considerably more complex than that in straight conduits. The earliest observation of this was in fact for open channel flow, where the effects of curvature are most evident and striking (Thomson 1876, 1877). In this century, Grindley & Gibson (1908) noticed the curvature effect on the flow through a coiled pipe in experiments on the viscosity of air. Williams et al. (1902) observed that the location of the maximum axial velocity is shifted toward the outer wall of a curved pipe. Later, Eustice (1910, 1911) demonstrated the existence of a secondary flow by injecting ink into water flowing through a coiled pipe.

There are a whole host of areas where such flows are of practical importance and where questions are raised regarding them. For example, since curved sections invariably arise in all piping systems, it is important to know the pressure drop in the developing and fully developed parts of the flow if one is to predict the pumping power needed to overcome curvature-induced pressure losses. Because the secondary motions can be expected to enhance heat exchange between the fluid and its surroundings, knowledge of the magnitude of this effect is important in designing

461

0066-4189/83/0115-0461$02.00

heat exchangers. Blood flow in the human arterial system has been of particular interest to fluid mechanicians in recent years (Pedley 1980). The largest vessel in this system, the aorta, is highly curved. Of particular interest concerning the flow in this vessel is the entry length—the distance required for the flow to become fully developed—and the location of sites of maximum and minimum wall shear stress. The entry length is relevant to the flow in other curved blood vessels, since there is some question whether there is sufficient distance between bifurcations for the flow in arteries to become fully developed; the secondary motions in curved flows would be expected to be important in this context. The sites of extrema in the wall shear are important in the aorta and other large curved arteries because these may be the sites of cholesterol buildup on vessel walls, and thus may play a major role in atherogenesis.

In this article we review as comprehensively as space permits what is known about the flow in curved tubes and pipes. In particular, we discuss developing and fully developed flows, both steady and unsteady, in rigid infinitely coiled pipes; the flow in finite bends; thermal effects; and more briefly multiphase flow, the flow of non-Newtonian fluids, and the effect of flexible walls. Laminar incompressible flows will be our primary concern. We do not consider curved open-channel flows, since they differ enough in analysis and application to receive separate attention. Our main emphasis will be on tubes or pipes with circular cross sections. Finally, while we make reference to experimental results throughout, we have thought it wise to devote a separate section to them, with emphasis on the most recent studies and experimental techniques.

Secondary flows appear whenever fluid flows in curved pipes or channels. Although such secondary motions can arise in a perfect inviscid fluid as a result of a nonuniform distribution of velocity at the entrance to the bend (Squire & Winter 1949, Hawthorne 1951), throughout this review we shall consider real viscous fluids for which the secondary flow can be attributed principally to the effect of the centrifugal pressure gradient in the main flow acting on the relatively stagnant fluid in the wall boundary layer.

We begin with the steady, laminar, incompressible flow in a rigid circular pipe of uniform curvature.

2.   STEADY, LAMINAR FLOW IN A RIGID PIPE

Consider a circular pipe with cross-sectional radius a coiled in a circle of radius R. We introduce the toroidal coordinate system (r', α, θ) shown in Figure 1, where r' denotes the distance from the center of the (circular) cross section of the pipe, α the angle between the radius vector and the plane of symmetry, and θ the angular distance of the cross section from the entry of the pipe. The corresponding velocity components are (u', v', w'). The following nondimensional variables are defined

$$r = \frac{r'}{a}, \qquad s = \frac{R\theta}{a}, \qquad t = \frac{\overline{W}_0 t'}{a}$$

$$\mathbf{q} = \frac{\mathbf{q}'}{\overline{W}_0}, \qquad p = \frac{p'}{\rho \overline{W}_0^2}, \tag{1}$$

(primes denote dimensional quantities, unprimed quantities are dimensionless) where t is the time, $\mathbf{q} = (u, v, w)$, p is the pressure, ρ the density, and $\overline{W}_0$ the mean axial velocity in the pipe. For convenience we begin by writing down the full governing equations for viscous flow in such a curved pipe, without yet imposing the restriction that the flow is steady; these are

$$u_r + \frac{u}{r} \frac{1 + 2\delta r \cos \alpha}{1 + \delta r \cos \alpha} + \frac{v_\alpha}{r} - \frac{\delta v \sin \alpha}{1 + \delta r \cos \alpha} + \frac{w_s}{1 + \delta r \cos \alpha} = 0, \tag{2a}$$

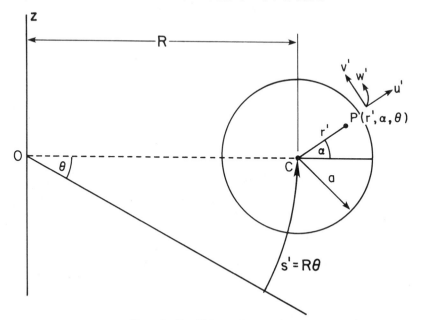

Figure 1   Toroidal coordinate system.

$$u_t + uu_r + \frac{vu_\alpha}{r} + \frac{wu_s}{1 + \delta r \cos \alpha} - \frac{v^2}{r} - \underline{\frac{\delta w^2 \cos \alpha}{1 + \delta r \cos \alpha}}$$

$$= -p_r - \frac{1}{\text{Re}} \left[\left(\frac{1}{r} \frac{\partial}{\partial \alpha} - \frac{\delta \sin \alpha}{1 + \delta r \cos \alpha} \right) \left(v_r + \frac{v}{r} - \frac{u_\alpha}{r} \right) \right.$$

$$\left. - \frac{u_{ss}}{(1 + \delta r \cos \alpha)^2} + \frac{1}{1 + \delta r \cos \alpha} \left(w_{rs} + \frac{\delta w_s \cos \alpha}{1 + \delta r \cos \alpha} \right) \right], \tag{2b}$$

$$v_t + uv_r + \frac{vv_\alpha}{r} + \frac{wv_s}{1 + \delta r \cos \alpha} + \frac{uv}{r} + \underline{\frac{\delta w^2 \sin \alpha}{1 + \delta r \cos \alpha}}$$

$$= -\frac{p_\alpha}{r} + \frac{1}{\text{Re}} \left[\frac{v_{ss}}{(1 + \delta r \cos \alpha)^2} - \frac{w_{s\alpha}}{r(1 + \delta r \cos \alpha)} + \frac{\delta w_s \sin \alpha}{(1 + \delta r \cos \alpha)^2} \right.$$

$$\left. + \left(\frac{\partial}{\partial r} + \frac{\delta \cos \alpha}{1 + \delta r \cos \alpha} \right) \left(v_r + \frac{v}{r} - \frac{u_\alpha}{r} \right) \right], \tag{2c}$$

$$w_t + uw_r + \frac{\delta uw \cos \alpha}{1 + \delta r \cos \alpha} + \frac{vw_\alpha}{r} - \frac{\delta vw \sin \alpha}{1 + \delta r \cos \alpha} + \frac{ww_s}{1 + \delta r \cos \alpha}$$

$$= -\frac{p_s}{1 + \delta r \cos \alpha} + \frac{1}{\text{Re}} \left[\left(\frac{\partial}{\partial r} + \frac{1}{r} \right) \left(w_r + \frac{\delta w \cos \alpha}{1 + \delta r \cos \alpha} \right) + \frac{w_{\alpha\alpha}}{r^2} \right.$$

$$\left. - \frac{1}{r} \frac{\partial}{\partial \alpha} \left(\frac{\delta w \sin \alpha}{1 + \delta r \cos \alpha} \right) - \left(\frac{\partial}{\partial r} + \frac{1}{r} \right) \frac{u_s}{1 + \delta r \cos \alpha} - \frac{1}{r} \frac{\partial}{\partial \alpha} \left(\frac{v_s}{1 + \delta r \cos \alpha} \right) \right], \tag{2d}$$

(subscripts denote derivatives) where

$$\delta = \frac{a}{R} \quad \text{and} \quad \text{Re} = \frac{a\overline{W}_0}{v}. \tag{3}$$

Here (2a) is the continuity equation and (2b–d) the momentum equations in the r, α, and s directions, respectively. [The reason for underlining the terms in (2b, c) is explained below.]

Nearly all analyses of this problem use as entry conditions either (a) constant total pressure across the cross section at the entrance:

$$\left. \begin{array}{l} u = v = 0, \\ w = \dfrac{w'}{\overline{W}_0} = \dfrac{1}{1 + \delta r \cos \alpha} \end{array} \right\} \text{ at } \theta = 0, \tag{4a}$$

or (b) constant injection velocity:

$$u = v = 0,$$
$$w = \frac{w'}{W_0} = 1 \quad \left.\right\} \text{ at } \theta = 0. \tag{4b}$$

(Occasionally, particularly if a straight section precedes the curved pipe, a Poiseuille profile is used as the entry condition.)

The boundary conditions of no slip at the pipe wall are

$$u = v = w = 0 \quad \text{at} \quad r = \frac{r'}{a} = 1. \tag{5}$$

Equations (2), (4a) or (4b), and (5) constitute the full formulation of the problem.

For the remainder of Section 2 we consider only steady flow, so the time derivative terms on the left-hand sides of (2b–d) are set equal to zero.

2.1   Fully Developed Flow

2.1.1   LOOSELY COILED PIPE   At this point nearly all analyses assume that the pipe has only slight curvature, i.e. $\delta = a/R \ll 1$. Before dropping the $O(\delta)$ terms in (2a–d) compared with $O(1)$ terms, it is necessary to rescale the velocities to make the centrifugal-force terms in (2b,c), the underlined terms, of the same order of magnitude as the viscous and inertial terms, since it is the centrifugal force that drives the secondary motion. The transformation that accomplishes this is

$$(u, v, w) \rightarrow (\delta^{1/2}\hat{u}, \delta^{1/2}\hat{v}, \hat{w}). \tag{6}$$

In what follows we shall omit the caret sign, it being understood that the velocities that appear are those on the right-hand side of (6). It is also convenient to rescale the axial distance as follows:

$$s = \frac{R\theta}{a} = \delta^{-1/2}z. \tag{7}$$

The rescaled forms of (2) are

$$\frac{\partial u}{\partial r} + \frac{u}{r}\left(\frac{1+2\,\delta r\cos\alpha}{1+\delta r\cos\alpha}\right) + \frac{1}{r}\frac{\partial v}{\partial\alpha} - \frac{\delta v\sin\alpha}{1+\delta r\cos\alpha} + \frac{1}{1+\delta r\cos\alpha}\frac{\partial w}{\partial z} = 0,$$

$$\tag{8a}$$

$$u\frac{\partial u}{\partial r} + \frac{v}{r}\frac{\partial u}{\partial \alpha} + \frac{w}{1+\delta r\cos\alpha}\frac{\partial u}{\partial z} - \frac{v^2}{r} - \frac{w^2\cos\alpha}{1+\delta r\cos\alpha}$$

$$= -\frac{1}{\delta}\frac{\partial p}{\partial r} - \frac{2}{\kappa}\left\{\left(\frac{1}{r}\frac{\partial}{\partial\alpha} - \frac{\delta\sin\alpha}{1+\delta r\cos\alpha}\right)\left(\frac{\partial v}{\partial r} + \frac{v}{r} - \frac{1}{r}\frac{\partial u}{\partial\alpha}\right)\right.$$

$$-\frac{\delta}{(1+\delta r\cos\alpha)^2}\frac{\partial^2 u}{\partial z^2}$$

$$+\frac{1}{1+\delta r\cos\alpha}\left(\frac{\partial^2 w}{\partial r\partial z} + \frac{\delta\cos\alpha}{1+\delta r\cos\alpha}\frac{\partial w}{\partial z}\right)\bigg\},$$

$$(8b)$$

$$u\frac{\partial v}{\partial r} + \frac{v}{r}\frac{\partial v}{\partial\alpha} + \frac{w}{1+\delta r\cos\alpha}\frac{\partial v}{\partial z} + \frac{uv}{r} + \frac{w^2\sin\alpha}{1+\delta r\cos\alpha}$$

$$= -\frac{1}{r\delta}\frac{\partial p}{\partial\alpha} + \frac{2}{\kappa}\left\{\frac{\delta}{(1+\delta r\cos\alpha)^2}\frac{\partial^2 v}{\partial z^2} - \frac{1}{1+\delta r\cos\alpha}\frac{1}{r}\frac{\partial^2 w}{\partial\alpha\partial z}\right.$$

$$+\frac{\delta\sin\alpha}{(1+\delta r\cos\alpha)^2}\frac{\partial w}{\partial z}$$

$$+\left(\frac{\partial}{\partial r} + \frac{\delta\cos\alpha}{1+\delta r\cos\alpha}\right)\left(\frac{\partial v}{\partial r} + \frac{v}{r} - \frac{1}{r}\frac{\partial u}{\partial\alpha}\right)\bigg\},$$

$$(8c)$$

$$u\frac{\partial w}{\partial r} + \frac{\delta uw\cos\alpha}{1+\delta r\cos\alpha} + \frac{v}{r}\frac{\partial w}{\partial\alpha} - \frac{\delta\sin\alpha}{1+\delta r\cos\alpha}uw + \frac{w}{1+\delta r\cos\alpha}\frac{\partial w}{\partial z}$$

$$= -\frac{1}{1+\delta r\cos\alpha}\frac{\partial p}{\partial z} + \frac{2}{\kappa}\left\{\left(\frac{\partial}{\partial r} + \frac{1}{r}\right)\left(\frac{\partial w}{\partial r} + \frac{\delta w\cos\alpha}{1+\delta r\cos\alpha}\right) + \frac{1}{r^2}\frac{\partial^2 w}{\partial\alpha^2}\right.$$

$$-\frac{1}{r}\frac{\partial}{\partial\alpha}\left(\frac{\delta w\sin\alpha}{1+\delta r\cos\alpha}\right) - \left(\frac{\partial}{\partial r} + \frac{1}{r}\right)$$

$$\times\frac{\delta}{1+\delta r\cos\alpha}\frac{\partial u}{\partial z} - \frac{1}{r}\frac{\partial}{\partial\alpha}\left(\frac{\delta}{1+\delta r\cos\alpha}\frac{\partial v}{\partial z}\right)\bigg\},$$

$$(8d)$$

where κ, the Dean number, is defined as

$$\kappa = 2\delta^{1/2}\text{Re} = \left(\frac{a}{R}\right)^{1/2}\frac{2a\overline{W}_0}{\nu}.$$

$$(9)$$

From these governing equations we see that two parameters, κ and δ, characterize the flow in curved pipes. The Dean number κ is equal to the ratio of the square root of the product of the inertia and centrifugal forces to the viscous force. Since the secondary flow is induced by centrifugal forces and their interaction primarily with viscous forces, κ is a measure of the magnitude of the secondary flow. We note that no secondary flow will be induced for an inviscid fluid. The parameter δ is a more detailed measure of the effect of geometry and the extent to which the centrifugal force varies on the cross section. Thus δ affects the balance of inertia, viscous, and centrifugal forces, and while its influence has been studied much less than that of κ, it can play a major role in curved-pipe flows. This is discussed in Section 2.1.2.

We note that (8b) and (8c) are of the form

$$O(1) = -\frac{1}{\delta}\frac{\partial p}{\partial r}, \qquad O(1) = -\frac{1}{r\delta}\frac{\partial p}{\partial \alpha}. \tag{10}$$

Thus when $\delta \ll 1$, we may write

$$p = p_0(z) + \delta p_1(r, \alpha, z) + \ldots, \tag{11}$$

i.e. the major contribution to the axial pressure gradient may be separated from the cross-sectional component.

We can now consider the loosely coiled pipe limit by setting $\delta = 0$ in (8). Using (11), this yields the following reduced equations:

$$\frac{\partial u}{\partial r} + \frac{u}{r} + \frac{1}{r}\frac{\partial v}{\partial \alpha} + \frac{\partial w}{\partial z} = 0, \tag{12a}$$

$$u\frac{\partial u}{\partial r} + \frac{v}{r}\frac{\partial u}{\partial \alpha} + w\frac{\partial u}{\partial z} - \frac{v^2}{r} - w^2\cos\alpha$$
$$= \frac{-\partial p_1}{\partial r} - \frac{2}{\kappa}\left\{\frac{1}{r}\frac{\partial}{\partial \alpha}\left(\frac{\partial v}{\partial r} + \frac{v}{r} - \frac{1}{r}\frac{\partial u}{\partial \alpha}\right) + \frac{\partial^2 w}{\partial r\,\partial z}\right\}, \tag{12b}$$

$$u\frac{\partial v}{\partial r} + \frac{v}{r}\frac{\partial v}{\partial \alpha} + w\frac{\partial v}{\partial z} + \frac{uv}{r} + w^2\sin\alpha$$
$$= -\frac{1}{r}\frac{\partial p_1}{\partial \alpha} + \frac{2}{\kappa}\left\{\frac{-\partial^2 w}{r\,\partial\alpha\,\partial z} + \frac{\partial}{\partial r}\left(\frac{\partial v}{\partial r} + \frac{v}{r} - \frac{1}{r}\frac{\partial u}{\partial \alpha}\right)\right\}, \tag{12c}$$

$$u\frac{\partial w}{\partial r} + \frac{v}{r}\frac{\partial w}{\partial \alpha} + w\frac{\partial w}{\partial z} = \frac{-\partial p_0}{\partial z} + \frac{2}{\kappa}\left\{\left(\frac{\partial}{\partial r} + \frac{1}{r}\right)\frac{\partial w}{\partial r} + \frac{1}{r^2}\frac{\partial^2 w}{\partial \alpha^2}\right\}. \tag{12d}$$

The only nondimensional parameter appearing in these equations is the Dean number κ, which plays the role of a "Reynolds number" of the flow and leads to the so-called Dean-number similarity, first noted by Dean (1928). Note that this is not the case for curved-pipe flows in general, but only when $\delta \ll 1$.

The flow is fully developed when it no longer varies with axial position. The governing equations, obtained by setting all z-derivatives, except that of the pressure, equal to zero in (12a–d), are

$$\frac{\partial u}{\partial r} + \frac{u}{r} + \frac{1}{r}\frac{\partial v}{\partial \alpha} = 0, \tag{13a}$$

$$u\frac{\partial u}{\partial r} + \frac{v}{r}\frac{\partial u}{\partial \alpha} - \frac{v^2}{r} - w^2\cos\alpha = -\frac{\partial p_1}{\partial r} - \frac{2}{\kappa}\frac{1}{r}\frac{\partial}{\partial \alpha}\left(\frac{\partial v}{\partial r} + \frac{v}{r} - \frac{1}{r}\frac{\partial u}{\partial \alpha}\right), \tag{13b}$$

$$u\frac{\partial v}{\partial r} + \frac{v}{r}\frac{\partial v}{\partial \alpha} + \frac{uv}{r} + w^2\sin\alpha = -\frac{1}{r}\frac{\partial p_1}{\partial \alpha} + \frac{2}{\kappa}\frac{\partial}{\partial r}\left(\frac{\partial v}{\partial r} + \frac{v}{r} - \frac{1}{r}\frac{\partial u}{\partial \alpha}\right), \tag{13c}$$

$$u\frac{\partial w}{\partial r} + \frac{v}{r}\frac{\partial w}{\partial \alpha} = -\frac{\partial p_0}{\partial z} + \frac{2}{\kappa}\left\{\left(\frac{\partial}{\partial r} + \frac{1}{r}\right)\frac{\partial w}{\partial r} + \frac{1}{r^2}\frac{\partial^2 w}{\partial \alpha^2}\right\}. \tag{13d}$$

The velocity components are now functions of r and α only; it then follows from (13d) that $\partial p_0/\partial z$ is independent of z and hence we can write $p_0(z) = -Gz$, where G is a constant.

Equation (13a) can be satisfied automatically by introducing a stream function for the secondary flow, defined by

$$u = \frac{1}{r}\frac{\partial \psi}{\partial \alpha}, \qquad v = -\frac{\partial \psi}{\partial r}. \tag{14}$$

Substitution into (13d) yields

$$\nabla_1^2 w - \frac{\kappa}{2}\frac{\partial p_0}{\partial z} = \frac{\kappa}{2r}(\psi_\alpha w_r - \psi_r w_\alpha), \tag{15}$$

while cross-differentiation of (13b,c) and elimination of the pressure yields

$$\frac{2}{\kappa}\nabla_1^4\psi + \frac{1}{r}\left(\psi_r\frac{\partial}{\partial \alpha} - \psi_\alpha\frac{\partial}{\partial r}\right)\nabla_1^2\psi = -2w\left(\sin\alpha\, w_r + \frac{\cos\alpha}{r}w_\alpha\right), \tag{16}$$

where

$$\nabla_1^2 \equiv \frac{\partial^2}{\partial r^2} + \frac{1}{r}\frac{\partial}{\partial r} + \frac{1}{r^2}\frac{\partial^2}{\partial \alpha^2}. \tag{17}$$

The boundary conditions are

$$\psi = \psi_r = w = 0 \quad \text{at} \quad r = 1. \tag{18}$$

Equations (15) and (16) for w and ψ, together with boundary conditions (18), a formulation originally given by Dean (1928), completely determine the problem for fully developed flow for a loosely coiled pipe.

2.1.1.1 *Small Dean number* For small values of the Dean number, Dean (1927, 1928) solved the above problem by expanding the solution in a series in powers of the Dean number, i.e.

$$w = \sum_{n=0}^{\infty} \kappa^{2n} w_n(r, \alpha), \qquad \psi = \kappa \sum_{n=0}^{\infty} \kappa^{2n} \psi_n(r, \alpha). \tag{19}$$

The leading term for w, w_0, is Poiseuille flow in a straight tube; the leading term for ψ is $O(\kappa)$. This series expansion in κ is equivalent to perturbing the equivalent Poiseuille flow and calculating the influence of the inertia terms by successive approximation, and is thus only appropriate for slightly, or loosely, coiled pipes or tubes. Figure 2 shows, for a typical case, secondary streamlines in the pipe cross section and

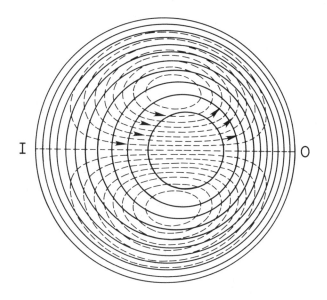

————— CONTOURS OF CONST. AXIAL VELOCITY

—————— SECONDARY STREAMLINES

Figure 2 Secondary streamlines and axial-velocity contours at low Dean number. I denotes inner bend, O outer bend.

contours of constant axial velocity. The flow consists of a pair of counterrotating helical vortices, placed symmetrically with respect to the plane of symmetry. This secondary flow pattern arises because the centrifugally-induced pressure gradient, approximately uniform over the cross section, drives the slower-moving fluid near the wall inward, while the faster-moving fluid in the core is swept outward. The vortices are almost symmetrically placed with respect to $\alpha = \pm\,\pi/2$, with the stagnation points (or "centers") of the secondary vortices located on these lines at $r \approx 0.43$, while the position of maximum axial velocity moves outward to $r \approx 4.2 \times 10^{-4}K$, $\alpha = 0$.

The nondimensional volume flux in the pipe depends only on Dean number and is given by

$$\frac{Q}{\pi a^2 \overline{W}} = 1 - 0.0306\left(\frac{K}{576}\right)^2 + 0.0120\left(\frac{K}{576}\right)^4 + O\left[\left(\frac{K}{576}\right)^6\right], \qquad (20)$$

where $K = (2a/R)(W_{\max}a/v)^2$, $W_{\max}$ is the maximum velocity in a straight pipe of the same radius under the same axial pressure gradient, and $\overline{W}$ is the mean velocity in the same pipe (so $\overline{W} = W_{\max}/2$). K is a variant of the Dean number, in the form first used by Dean (see Section 2.1.1.2 below). The first term on the right-hand side represents the Poiseuille straight-pipe value. Not surprisingly, we see that the effect of curvature is to reduce the flux. For example, when $K = 576$ the flux is reduced in consequence of the curvature by about 2%.

One outstanding feature of Dean's series solution is that the flow resistance is not affected by the first-order terms and is increased only by the second-order terms. This flow resistance is generally represented nondimensionally in terms of the friction factor, or ratio, λ (White 1929), the ratio of the resistance in the curved pipe (λ_c) to the resistance in a straight pipe (λ_s) carrying the same flux. Since this is equal to $(Q_c/Q_s)^{-1}$, where Q_c and Q_s are the fluxes in the curved and straight pipes, respectively, under the same pressure gradient, it follows from (20) that

$$\lambda = \frac{\lambda_c}{\lambda_s} = \left(\frac{Q_c}{Q_s}\right)^{-1} = 1 + 0.0306\left(\frac{K}{576}\right)^2 - 0.0110\left(\frac{K}{576}\right)^4 + \cdots \qquad (21)$$

Dean conjectured, from a consideration of the coefficients in (20), that his small-Dean-number series solution was valid for $K \lesssim 576$, a statement subsequently confirmed experimentally by White (1929) based on consideration of (21). Experiments (White 1929, Taylor 1929, Adler 1934) have also confirmed that the flow in a curved pipe or tube is much more stable than that in a straight pipe or tube, so whereas the critical Reynolds number for the latter is typically about 2000, in a curved tube, of even small curvature, it may be larger by a factor of two or more. For example, Taylor found that for $a/R = 1/31.9$, the critical Reynolds number was about 5000.[1] This corresponds to $K \approx 1.6 \times 10^6$ and therefore approximately delineates the laminar regime over which calculated results would be of interest. Since the upper limit of validity of Dean's series solution ($K = 576$) obviously falls far short of covering the full laminar-flow range, we next consider solutions for larger values of Dean number. Before doing so, however, we digress for a moment to consider the question of the appropriate definition of Dean number.

2.1.1.2 *Definition of Dean number* In Equation (9), the Dean number κ was defined as

$$\kappa = 2\delta^{1/2} \text{Re} = \left(\frac{a}{R}\right)^{1/2}\left(\frac{2a\overline{W}_0}{v}\right),$$

where $\overline{W}_0$ is the mean axial velocity in the pipe, while the original form of the Dean number was defined by Dean (1928) as

$$K = 2\left(\frac{a}{R}\right)\left(\frac{aW_0}{v}\right)^2 = \frac{2W_0^2 a^3}{v^2 R}, \tag{22}$$

where W_0 is defined only as a constant having the dimensions of a velocity. If we take $W_0 = \overline{W}_0$, then K and κ are related by $\kappa = (2K)^{1/2}$. For fully developed flow it follows from (2b–d) that the axial pressure gradient is constant, say $\partial p'/R\,\partial\theta = -G$ (G constant), in dimensional terms. We can then define a nondimensional constant, $C = Ga^2/\mu W_0$, and rewrite (22) as

$$K = \frac{2a^3}{v^2 R}\left(\frac{Ga^2}{\mu C}\right)^2. \tag{23}$$

If, following Dean (1928), we specify W_0 as the maximum velocity (W_{max}) in a straight pipe of the same radius and with the same pressure gradient, then $C = 4$ and (23) becomes

$$K = \frac{2a^3}{v^2 R}\left(\frac{Ga^2}{4\mu}\right)^2 = 2\left(\frac{a}{R}\right)\left(\frac{Ga^3}{4\mu v}\right)^2 = \frac{G^2 a^7}{8\mu^2 v^2 R} \tag{24}$$

whereas if we simply set $C = 1$, (23) becomes

$$K = \frac{2a^3}{v^2 R}\left(\frac{Ga^2}{\mu}\right)^2. \tag{25}$$

---

[1] This effect of curvature seems never to have been explained. For a phenomenological conjecture based on recent ideas about coherent structures in turbulent flow, see Coles (1981). This can be contrasted with an earlier alternative explanation given by Lighthill (1970).

Since Dean's original work, most investigators have used not only all of the above variants of the Dean number, but to make matters even worse they have used $2a$ rather than a, or vice versa, in $\delta = a/R$ and the definition of Reynolds number, or they have used κ^2 or $K^{1/2}$ rather than κ or K. This obviously makes for considerable confusion in reading and interpreting the literature. [Van Dyke (1978) gives a useful compilation of the relationships between the various versions used in some of the most often cited papers.]

Versions of Dean number based on $\overline{W}_0$, the mean axial velocity, are natural for the experimentalist because this quantity, being readily measured, provides a more convenient characterization of the flow than the more difficult to measure pressure gradient. For fully developed flow it makes little difference whether one uses a form of Dean number based on $\overline{W}_0$ or any of the variants based on G, i.e. (23), since $C = $ constant for such a flow. For these flows most theoretical and numerical investigators, beginning with McConalogue & Srivastava (1968), have used the square root of (25) and denoted this by D, i.e.

$$D = \left(\frac{2a^3}{\nu^2 R} \right)^{1/2} \frac{Ga^2}{\mu} = 4 \left(\frac{2a}{R} \right)^{1/2} \frac{Ga^3}{4\mu\nu}, \tag{26}$$

which is related to the original Dean number, (24), by $D = 4K_{\text{Dean}}^{1/2}$. [Equations (20) and (21) can be rewritten in terms of D by replacing each of the factors $(K/576)^2$ by $(D/96)^4$.] The relationship between K or D and κ is more complicated, and depends on the flux ratio itself, according as

$$\frac{Q_c}{Q_s} = \frac{\kappa}{\left(\frac{1}{2} K \right)^{1/2}} = 2^{5/2} \frac{\kappa}{D} \tag{27}$$

(McConalogue 1970, Van Dyke 1978). Thus, while the relationship between K or D and κ is known a priori for values of Dean number small enough that $\overline{W}_0$ is related to G approximately as in Poiseuille flow, for larger values the relationship between these Dean numbers can only be determined a posteriori after the solution, and hence the flux ratio, is calculated.

For flows that are not fully developed the situation is even more complicated, for while $\overline{W}_0$ is a true constant for a pipe of given cross section and fixed flow conditions, the pressure gradient is, in general, a function both of axial location and position in the cross section. (For a loosely coiled pipe, to lowest order, see (11), it only depends on axial location.) Consequently C, the ratio of G to W_0, whatever the choice of W_0, is not a constant; moreover, this ratio is not known until the solution is obtained—the relationship between G and W_0 being one of the principal objects of the investigation. Therefore, the forms of Dean number that follow from (23), e.g. (24) and (25), can be used for other than fully developed flows only if one specifies G in such a way that it is a constant. Since there is no a priori way of knowing how to do this so as to make the resulting Dean number properly characterize the flow, and in any case measurement of any such pressure gradient is likely to be difficult, it would appear that basing Dean number on the mean axial velocity, such as in (9), is preferable. To avoid the confusion described above we strongly suggest that future studies use the form (9) whether the flow is fully developed or not. This suggestion notwithstanding, the concerned reader should be wary in interpreting the literature—past, present, and future!

Finally, we note that in terms of the Dean number D, defined by (26), the upper limit of validity of Dean's series solution, corresponding to $K = 576$, is $D = 96$, while the upper limit of the laminar-flow regime is approximately $D = 5000$. We now discuss the work that has attempted to bridge this gap.

2.1.1.3 *Intermediate Dean number* Since, as we shall see in the next section, boundary-layer concepts are applicable to the flow at large Dean number, whereas for intermediate values one has to solve the full nonlinear equations (15) and (16) numerically, solutions in this range historically followed much of the well-known large-Dean-number work.

McConalogue & Srivastava (1968) obtained numerical solutions for the fully developed flow for the range $96 < D \leq 600$ by expanding in Fourier series (in α), then integrating the resulting ordinary differential equations (in r) numerically. Since then, many papers have reported finite-difference solutions, e.g. Truesdell & Adler (1970), Akiyama & Cheng (1971), Greenspan (1973), Austin & Seader (1973), Patankar et al. (1974), Collins & Dennis (1975), Zapryanov & Christov (1977), Dennis & Ng (1980, 1982), and Dennis (1980). Among these the most extensive, and certainly among the most reliable, are those of Collins & Dennis (1975), who solved (15) and (16) for the full laminar range $D = 96$–5000. Figure 3 shows the secondary streamlines and contours of constant axial velocity for $D = 96$ and 606 from the solutions of McConalogue & Srivastava (1968). The outward movement of the location of maximum axial velocity as D increases is evident, and while the secondary streamlines continue to manifest a vortex structure for $D = 606$, the beginnings of secondary and axial boundary layers are clearly discernable at the larger D value. This is more dramatically exhibited in Figure 4 for $D = 5000$, taken from Collins & Dennis, which shows also a much greater distortion of the secondary streamlines.

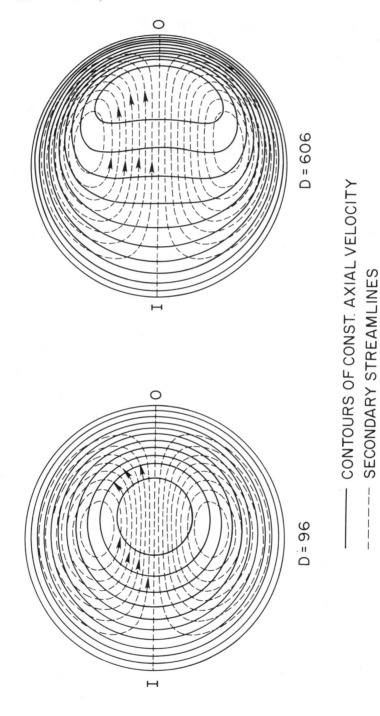

CONTOURS OF CONST. AXIAL VELOCITY
SECONDARY STREAMLINES

D = 606

D = 96

Figure 3 Secondary streamlines and axial-velocity contours at low and intermediate Dean numbers (McConalogue & Srivastava 1968). I denotes inner bend, O outer bend.

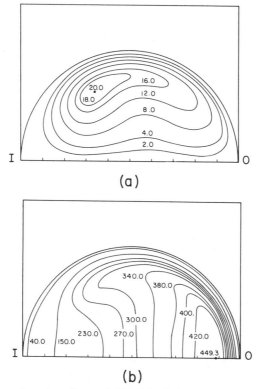

Figure 4 (*a*) Secondary streamlines and (*b*) axial-velocity contours at large Dean number, $D = 5000$ (Collins & Dennis 1975). I denotes inner bend, O outer bend.

A different approach was taken by Van Dyke (1978), who extended Dean's series by computer up to 24 terms and demonstrated that the series converges for $D < 96.8$. A Domb-Sykes plot and an Euler transformation was then employed to extend the series for λ_c/λ_s to $D = \infty$; the series agrees well with all the numerical solutions as well as the experimental data up to $\kappa \simeq 100$ (see Figure 7). This analysis predicts that

Table 1 Asymptotic friction ratio for large Dean number

λ_c/λ_s	Reference
$0.1064 \, \kappa^{1/2}$	Adler (1934)
$0.09185 \, \kappa^{1/2}$	Barua (1963)
$0.1080 \, \kappa^{1/2}$	Mori & Nakayama (1965)
$0.1033 \, \kappa^{1/2}$	Itō (1969)
$0.47136 \, \kappa^{1/4}$	Van Dyke (1978)

λ_c/λ_s grows asymptotically as $\lambda_c/\lambda_s \sim 0.47136\kappa^{1/4}$, which, as is illustrated later, is at variance with all other existing solutions for large Dean numbers (see Table 1).

2.1.1.4 *Large Dean number*   As the Dean number increases, the centers of the two vortices move to the outer bend, $\alpha = 0°$. There is a considerable reduction in flow in the curved pipe compared with a straight pipe. Adler (1934) observed that boundary layers exist on the two halves of the torus. His interpretation was that the two boundary layers, on the upper and lower halves, collide at the innermost point of the cross section, separate there, and form a reentrant jet that moves outward through the core. The boundary-layer model proposed indicates that λ_c/λ_s should be proportional to $\kappa^{1/2}$ (see Table 1), where $\kappa = (2a\overline{W}_0/\nu)(a/R)^{1/2}$. Subsequently, Barua (1963), Mori & Nakayama (1965), and Itō (1969) improved the boundary-layer model; they all found that $\lambda_c/\lambda_s \propto \kappa^{1/2}$. The coefficient of proportionality in this relationship differs slightly for different boundary-layer models (see Table 1). In Barua's model, the flow is assumed to consist of an inviscid core plus a thin boundary layer, with the additional assumption that the flow in the core lies in planes parallel to the plane of symmetry; this picture of the flow is not inconsistent with the numerical results of Collins & Dennis (see, for example, Figure 3, $D = 606$) discussed in the previous section. The Pohlhausen momentum-integral method is used to solve the boundary layer, which is coupled to the inviscid-core flow by assuming that the maximum circumferential velocity occurs at $\alpha = \pm 90°$. For a large D flow this assumption is not rigorously correct. Barua's integral solution indicates that the two circumferential boundary layers separate from the pipe wall about 27° away from the inner bend. This separation location according to Barua is in reasonable agreement with data of H. Squire (1954, unpublished) and Adler (1934). Itō also used the Pohlhausen integral method but with less restrictive conditions on the inviscid core. His model indicates that the fluid enters the secondary boundary layers within about 35° from the outer bend, i.e. $-35° \leqslant \alpha \leqslant 35°$; it also predicts that the boundary-layer thickness remains almost constant until it increases sharply in the neighborhood of the inner bend, $\alpha = 180°$. Itō's solution, however, unlike Barua's, does not predict separation.

It will be useful at this point to describe briefly how the flow pattern evolves as the Dean number increases. For small-Dean-number flows, the circumferential velocity attains its maximum value at $\alpha = \pm 90°$ and the two vortices are symmetric with respect to these values of α. As the Dean number increases, (secondary) boundary layers develop on the wall, with fluid entering these boundary layers near the outer bend and leaving near the inner bend. Further increase of the Dean number, and hence the centrifugal force, leads to an increase in circumferential velocity and to more fluid being sucked into the secondary boundary layers near the outer bend. The secondary boundary layers adjust by thinning near the outer bend and thickening near the inner bend. Simultaneously, the locations of the maximum circumferential velocity move toward the outer bend. The two vortices are skewed by the adjustment of the secondary boundary layers. With further increase of the Dean number the secondary flow boundary layers near the inner bend thicken further, and according to some accounts (Barua 1963, Yao & Berger 1975), eventually separate, the width of the separation region gradually increasing for even larger D. No analytical model has yet been developed that includes the effect of this boundary-layer separation on the inviscid-core flow; more importantly, however, as has been mentioned above and will be further discussed below, the existence of this separation is very much in dispute. Interestingly, predictions of flow resistance by the various integral models, whether or not they predict separation, agree reasonably well among themselves and with the experimental data (see Table 1). This is probably because the viscous shear is small near the inner bend whatever the nature of the flow in this region and does not contribute much to the total flow resistance.

The interest in large-Dean-number curved-pipe flows has continued; however, in the past decade, far more attention has been given to finding the appropriate, consistent scales for each of the subregions of the flow field and to obtaining solutions, e.g. for the boundary layer, with greater precision than is possible using momentum-integral methods. We can only briefly sketch here these more recent developments; for a fuller account, see Pedley (1980).

We begin with the observation that the numerical results for intermediate D discussed in Section 2.1.1.3 suggest that the structure of the flow at large D is an inviscid rotational core surrounded by a thin boundary layer. If one seeks scales such that in the core inertia forces balance pressure gradients and in the boundary layer there is a balance between inertia, viscous, and centrifugal forces, then one finds that $w_{core} = O(D^{2/3})$, $\psi = O(D^{1/3})$, and the boundary-layer thickness is $O(D^{-1/3})$. (These scales are consistent with the numerical results of Collins & Dennis.) Using these scales, one can then obtain from (15) and (16) the governing equations for the core and the boundary layer. The core equation can be solved up to an arbitrary constant that can be determined by matching with the boundary layer. The equations for the boundary layer have been solved by Itō (1969) using a Pohlhausen momentum-integral technique, by Riley & Dennis (1976) using finite differences, and by other investigators. All such solutions break down either at $\alpha = \pi$ or α considerably less than π. This is not necessarily due to separation of the boundary layer, because, for example, Itō does not predict separation, and local scaling arguments (Pedley 1980) suggest that the solution should be good up to $\pi - \alpha = O(D^{-1/8})$. Thus we are left with a number of open questions: (*a*) does the boundary layer extend right up to $\alpha = \pi$, and if so is there a collision of boundary layers at this point?; (*b*) is there an interaction of the boundary layer with the core near $\alpha = \pi$?; (*c*) does a reentrant jet form at $\alpha = \pi$ (see Section 2.2), and if so what is its effect upon the core? Some of these issues are addressed in the recent work of Stewartson et al. (1980) and Stewartson & Simpson (1982). Clearly, the flow structure at large D, and in particular the local structure and local shear variation near the inner bend, is not yet fully delineated and requires further study. This difficulty is not uncommon in other laminar flows at large Reynolds numbers. On the other hand, and on a more optimistic note, for correlating the flow resistance the available numerical solutions and boundary-layer models are probably sufficient, and have been verified by the experimental data.

2.1.2 FINITE δ It has already been pointed out that in general the flow in curved pipes depends on two parameters: the Dean number and the curvature ratio δ. While the influence of the latter has been studied much less than that of the former, it can play a major role in these flows.

Topakoglu (1967) solved (15) and (16) by expanding w and ψ in a double series in K [the Dean number defined by (24)] and δ, and obtained analytically the first few terms in the expansion. Larrain & Bonilla (1970) extended the double series to fourteenth order by computer. They studied the effect of curvature for $0 \leqslant \delta \leqslant 0.2$. The influence of δ is more profound for small K. For example, the percentage increment of the flow resistance for $\delta = 0.1$ can be seven times larger than that for $\delta \rightarrow 0$ at $K \simeq 18$. The absolute increment of the flow resistance increases with K for larger δ. The relative increment, however, diminishes rapidly when $K > 100$. Another interesting feature pointed out by Larrain & Bonilla is that the flow resistance is actually less than that in a straight pipe for a nonzero δ when K is about unity.

It is noteworthy that the series solution can be obtained for $\delta < 0.2$ and may not be valid for larger δ. The limit of small-Dean-number flow can be achieved by taking $\delta \rightarrow 0$ or Re $\rightarrow 0$ if $\delta < 0.2$. That this is not the case for $\delta > 0.2$ suggests that the flow distribution for $\delta > 0.2$ may be quite different from that of Dean flow.

The boundary-layer models discussed in Sec. 2.1.1.4 were formulated in the limit $\delta \rightarrow 0$. Nunge & Lin (1973) reported Lin's (1972) extension of

Itō's model to study the effect of nonzero δ. Lin indicated that a significant δ-effect can appear when δ is as small as 0.07. A surprising finding is that the flow resistance for a strongly curved pipe, say $\delta = 0.5$, is less than that of a loosely coiled pipe, $\delta = 0.01$ when $K > 85$. Truesdell & Adler (1970) and Austin & Seader (1973) have solved (15) and (16) numerically, using relaxation methods, for small but finite values of δ ($\delta \leqslant 0.2$) and moderate values of Dean number. In contrast to the Lin result, the finite-difference solution of Austin & Seader shows that the flow resistance for $\delta = 0.2$ is slightly higher than that for $\delta = 0.01$. Despite the discrepancy of these two solutions for larger δ, they agree well for $\delta = 0.01$ over the range $100 \leqslant D \leqslant 400$. A comparison of Austin & Seader's solution with White's data for $0.0049 \leqslant \delta \leqslant 0.66$ shows that the variation of the flow resistance for $0.01 \leqslant \delta \leqslant 0.2$ falls within the scatter of the experimental data. The Truesdell & Adler and Austin & Seader solutions both demonstrate that the dependence of the flow resistance, and the flow in general, on δ is much smaller than that on K. The variation of λ_c/λ_s with δ seems to be no more than 10%, based on the limited available information for $\delta < 0.1$. On the other hand, it seems unlikely that a small-Dean-number flow can exist for large δ. We discuss this problem further in Section 4. Suffice it to say that a careful numerical study of (15) and (16) over a wider variation of δ for intermediate K would be useful in more fully establishing the importance of δ.

In his monograph on flow in the large blood vessels, Pedley extends existing theory, wherever possible, to nonvanishingly small values of δ.

2.1.3 HELICAL PIPES Although this fact is not always explicitly stated, almost universally the analyses described above are, strictly speaking, valid only for toroidally curved pipes. The question then arises as to whether the results of such analyses can be applied to helical pipes. Helical pipes have both a curvature and a torsion or twist, and therefore there is in addition to a (modified) Dean number a second independent parameter, a measure of the torsion. An early attempt to address this question was that of Truesdell & Adler (1970), who presented ad hoc quantitative arguments suggesting that their numerical solutions for a toroidally coiled tube were applicable to a helically coiled tube of moderate pitch if the radius of curvature of the toroidal tube was replaced by that of the helical tube. Wang (1981) and Murata et al. (1981) addressed this question more rigorously by obtaining, in a nonorthogonal helical coordinate system, solutions for steady, fully developed flow in helical pipes of small curvature. They both obtain analytic solutions for low Dean number and small torsion by successive approximation, and Murata et al. for larger Dean number and torsion by numerical methods. Although Murata et al. are able to conclude, as suggested by Truesdell & Adler, that the resistance formula for a toroidally curved pipe is also applicable to a helically coiled pipe if the curvature of the helically coiled pipe is used in place of that of the toroidally curved one, their results and those of Wang show that there are significant differences, particularly in the secondary flow patterns, at very small and very large values of the (modified) Dean number.

Numerical solutions for arbitrary curvature ratios and finite pitch over the whole laminar-flow regime have been obtained by Manlapaz & Churchill (1980). They find the effect of pitch is significant only for coils for which the increase in elevation per revolution is greater than the radius of the coil. They also develop a correlating equation for the friction factor, or ratio, for all Reynolds numbers, all ratios of coil radius to tube radius, and all ratios of pitch to coil radius for which the flow remains laminar.

2.1.4   FINAL REMARKS   In their continuing study of fully developed curved-pipe flow by numerical methods, Dennis and co-workers (Dennis & Ng 1982) again solved (15) and (16) for $96 < D < 5000$, but this time by Fourier-series expansion in α and series truncation of the resulting ordinary differential equations. For $D < 956$, Dennis & Ng find the usual symmetric pair of counterrotating vortices and their results fully agree with the earlier Collins & Dennis (1975) solution. The same type of solution is also found for $D > 956$, but in addition a second family of solutions sometimes emerges in which the secondary flow has a four-vortex pattern consisting of two symmetrical vortex pairs. This non-uniqueness, the existence of both a two-vortex and a four-vortex mode for a large enough value of D, had previously been reported (Masliyah 1980) for a curved pipe with semicircular cross section (the flat surface forming the outer bend). This dual four-vortex solution for a curved pipe with a circular cross section was discovered independently by Nandakumar & Masliyah (1982). Nandakumar & Masliyah show that bifurcation of the two-vortex solution into two- and four-vortex solutions at higher Dean numbers occurs for other cross sections as well. All of the above authors are in agreement that their results fall within the framework of the bifurcation theory of Benjamin (1978a, b). In each case they find the point at which the solution bifurcates into dual solutions in the manner suggested by Benjamin, from above, i.e. by *decreasing* the flow parameter, which in this case is the Dean number.

Interestingly, Dennis & Ng (1982) find that the values of the friction ratio λ corresponding to the four-vortex solutions are very close to those of the two-vortex solutions.

2.2   Entry Flow

As has already been pointed out, apart from a purely academic interest, the entry flow into a curved pipe is of great practical importance in that it informs us as to (*a*) the distance required for the flow to become fully developed, (*b*) the contribution of the entry region to the overall pressure drop, and (*c*) the sites of extrema in the wall shear. In addition, investigation of the flow development in the entry region might help to resolve some of the unsolved questions of the fully developed flow at large Dean number (see Section 2.1.1.4).

Whereas the entry flow in a straight tube or pipe has been extensively studied (see e.g. Schlichting 1979), the same cannot be said for a curved tube, no doubt in part because the fully developed flow is so much more complicated and even now is not fully investigated and understood.

We begin with a qualitative description of the development of the flow in a curved tube (Singh 1974, Yao & Berger 1975). When the fluid first enters the pipe from, say, a large reservoir the central core is not influenced by viscosity, the effects of which are confined to a thin layer near the walls. This boundary layer develops initially like that in a straight pipe. Thus, immediately downstream of the entrance the flow field consists of two regions: (*a*) an inviscid core in which the centrifugal force, due to the curved motion of the main body of fluid, is balanced by a pressure-gradient force directed toward the center of curvature; and (*b*) a thin boundary layer in which there is a balance between viscous and inertia forces. The inwardly directed pressure force acting on the core is impressed on the boundary layer, inducing a transverse, or azimuthal, flow in the boundary layer from the outside of the bend toward the inside. The displacement effect of the growing boundary layer accelerates the flow in the core, while the (second-order) azimuthal flow in the boundary layer induces a cross-flow from the inside to the outside of the bend. Thus the effect of curvature is to induce a secondary flow, with slower-moving fluid in the boundary layer on the wall moving inward, and faster-moving fluid in the core moving outward. The secondary boundary layer thus acts as a sink receiving fluid near the outer wall and a source of fluid near the inner wall.

The picture of the developing flow presented to this point is valid for all values of the Dean number up to a distance of $O(a)$ from the entrance. Beyond this point, the nature of the flow development depends on the Dean number. We defer consideration of this subsequent development of the flow, and first discuss the $O(a)$ region immediately downstream of the entrance, for which an analytical solution was given by Singh (1974). This solution is based on the observation that since curvature effects are of second order in this $O(a)$ region, the solution can be obtained as a perturbation of the developing flow in a straight tube.

Singh begins with entry conditions (4a), corresponding to uniform total pressure, which requires that the static pressure behave as

$$p = -\tfrac{1}{2}(1 + \delta r \cos \alpha)^{-2}. \tag{28}$$

These entry conditions are exact solutions of the governing differential equations, (2); they therefore also represent the first-order solution in the inviscid core.

The first- and higher-order boundary-layer solutions and second- and higher-order core solutions are found by writing the solutions for these regions as double perturbation series in δ and β ($= \mathrm{Re}^{-1/2}$) about the straight-pipe solution and matching them using the techniques of matched asymptotic expansions. The analytic solution is given up to $O(\beta\delta)$, the lowest interaction term between tube curvature and the boundary-layer displacement effect.

Typical secondary-flow streamlines at different axial locations calculated from Singh's solution are shown in Figure 5. Initially there is an inward flow from the entire pipe circumference, due to the displacement effect of the boundary layer and resultant acceleration of the core flow, and, analytically, two singularities arise in the central region, a nodelike sink at the origin (point O) and a saddle-point-like stagnation point (point P); the latter singularity moves outward as the fluid moves downstream, eventually vanishing as the cross-flow becomes established.

One of the principal results of the Singh analysis is the prediction of a crossover in the location of the maximum axial wall shear, or skin

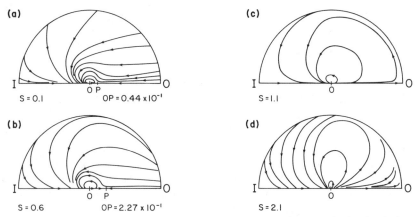

Figure 5 Streamlines in various cross-sectional planes near the entry in the developing flow in a curved tube (Singh 1974). I denotes inner bend, O outer bend.

friction, from the inner to the outer bend. The wall shear is larger initially at the inner wall because (*a*) the axial velocity is larger at the inner wall (since the initial profile is that of an inviscid vortex), and (*b*) the boundary layer is thinner there because of the shorter wall length, at the same *s*. A switch to the outer wall occurs at some location downstream of the entrance because (*a*) the point of maximum axial velocity moves outward, (*b*) the boundary layer is thinned at the outer wall because of the axial acceleration, and (*c*) the boundary layer is thickened at the inner wall and at the same time the external velocity decreases there. From the boundary-layer solution, the axial skin friction is found to be

$$\tau_{ws} = -\left(\frac{\mu\rho\overline{W}_0^3}{2as}\right)^{1/2}[0.4696 + \delta\cos\alpha(0.2562s^2 - 0.9392)], \qquad (29)$$

from which Singh calculates the crossover point to be located at $s \approx 1.9$ or $R\theta \approx 1.9a$, independent both of δ and β. This was verified experimentally by means of an electrochemical technique (Choi et al. 1979, Talbot & Wong 1982). The measurements, however, do suggest possibly some Reynolds-number effect, the crossover point moving slightly toward the inlet as Re increases, probably due to the boundary-layer displacement effect.

Singh shows that the solution for uniform-entry conditions, (4b), can be obtained by minor modification of the above solution for inviscid vortex entry. In this case the crossover occurs at $s \approx 0.95$, about half the value for the first case.

Beyond the distance $O(a)$ from the inlet, as mentioned above, the flow development depends on the Dean number. For small Dean number the centrifugal force, which is of second order in this initial $O(a)$ region, remains of second order as the fluid flows further downstream. The boundary layer continues to grow downstream, uneventfully, until it fills the tube and the flow is then fully developed, in a flow regime characterized by the axial length scale a Re. Thus for a curved pipe at low Dean number, the entry length is of the same order as that in a straight pipe. The analytical solution for this flow region should be obtainable by perturbing the entry flow in a straight pipe and treating the secondary flow as a higher-order effect, in a manner similar to that of Dean (1928) for the fully developed flow.

Singh's solution breaks down when $s = O(\delta^{-1/2})$ [when a term of $O(s\delta^2)$ in the boundary-layer solution becomes of $O(1)$], or, dimensionally, $R\theta = O(a/\sqrt{a/R}) = O(\sqrt{aR})$. This is the point beyond which the curvature-induced centrifugal forces, initially small, become as important as inertia and viscous forces in the boundary layer, and because of the displacement effect upon the core flow, they become equally important there as well. (Note that this means if $\delta \approx 0.2$, e.g. the maximum value in the aortic arch, Singh's solution is valid only for $s \lesssim 2$, about one diameter from the entrance.) Since $\sqrt{aR} = (a\,\mathrm{Re})/\kappa$, for $\kappa = O(1)$, $\sqrt{aR} = O(a\,\mathrm{Re})$ and the domain of validity of Singh's solution extends over most of the developing region, i.e. the flow essentially becomes fully developed before the centrifugal forces become important. For this case, then, the additional mixing due to the secondary flow appears to have little effect on the entry length, which is of the same order as that in a straight tube $[O(a\,\mathrm{Re})]$. For larger Dean number, one can develop a matched asymptotic expansion for the region $s = O(\delta^{-1/2})$ based on centrifugal forces being as important as viscous and inertia forces in the boundary layer. Such an expansion apparently breaks down (Singh, private communication) at a distance $R\theta = O(a\,\mathrm{Re})$, beyond which presumably the final transition to fully developed flow occurs.

For large Dean number the flow development is different. Centrifugal effects are as important as viscosity and inertia effects almost from the beginning. Much of the flow development occurs within a distance $O(\sqrt{aR})$, where these three forces balance, and since now $\sqrt{aR} \ll a\,\mathrm{Re}$, the entry length for this case is much shorter than that for a straight pipe. The magnitude of the secondary flow increases to the point that its displacement effect becomes larger than that of the axial flow. The cross-flow becomes stagnation-point-like near the outer bend. The convective effect of this locally stagnant flow prevents the secondary boundary layer from growing. Thus, the secondary boundary layer will remain thin, except possibly near the inner bend, as the flow asymptotically approaches the fully developed state (Barua-like flow; Barua 1963). The boundary layer near the outer generator acts as a reservoir, receiving the fluid moving toward the outer wall. The secondary boundary layers on the upper and lower halves of the tube either separate somewhat ahead of $\alpha = \pm 180°$ or collide at that point, forming an outwardly directed, or reentrant, jet.

Based on the above description of the flow, Yao & Berger (1975) have presented a model for the developing flow in a curved pipe for large Dean number. They derive two sets of equations; one for the inviscid-core flow and one for the three-dimensional boundary layer. They assume that the axial-velocity profile of the inviscid-core flow varies approximately linearly with distance from the vertical axis $\alpha = \pm \pi/2$, the coefficients of this linear expression being functions of axial distance down the pipe and determined from mass and momentum considerations and matching with the solution in the boundary layer. They also assume that the central cross-flow is essentially parallel to the plane of symmetry, an assumption employed by Barua (1963) for the fully developed flow and subsequently confirmed for the developing flow by the numerical solution of Yeung (1980). The Kármán-Pohlhausen integral method is used to solve the three-dimensional boundary-layer equations with a uniform-inlet axial velocity. Yao & Berger find downstream of the region of distance $O(\sqrt{aR})$ from the inlet a third region of $O(\kappa^{1/2})$ where the developing flow approaches fully developed form[2]. Also, they show that the thickness of the fully developed boundary layer is $O(\kappa^{-1/2})$. This indicates that the axial flow is accelerated by a factor $1 + O(\kappa^{-1/2})$ due to the axial displacement effect. Due to the acceleration of the axial flow inside the boundary layer, the axial velocity of the inviscid-core flow slightly decreases in this region. Yao & Berger's integral solution of the three-dimensional boundary layer also predicts that separation of the secondary boundary layer occurs at an axial distance about $0.01\, a \cdot (\kappa/\delta)^{1/2}$ from the inlet. The width of the separation zone is found to grow downstream, asymptotically becoming approximately 54° wide (measured from the inner bend). Finally, their results show that the pressure drop in the entry region is less than the loss of pressure head necessary to maintain the fully developed flow, a consequence, perhaps, of the fact that the cross-flow in the entry region is smaller than that for the fully developed flow.

A careful comparison of Singh's boundary-layer equations, which are valid in $R\theta \sim O(a)$, with the three-dimensional boundary-layer equations of Yao & Berger valid in $R\theta \sim O(\sqrt{aR})$ reveals that Singh's equations are a special case of the latter for small secondary flow and are included in the latter equations in $R\theta \sim O(\sqrt{aR})$. Therefore Singh's solution is valid for $\theta \leq C_1\sqrt{\delta}$. The constant C_1 can in principle be determined by comparing Singh's solution with the solution of these boundary-layer equations in $R\theta \sim O(\sqrt{aR})$. Unfortunately, Yao & Berger's solution is too approximate to accurately determine C_1. Stewartson et al. (1980) numerically integrated the three-dimensional boundary-layer equations starting at the pipe entrance with the assumption that interaction between the boundary layer and the core flow is negligible in $R\theta \sim O(\sqrt{aR})$. They did not compare their solution with Singh's. However, Singh's solution indicates that the axial-velocity profiles at $\alpha = \pm 90°$ are not affected by the secondary flow in the region $R\theta \sim O(a)$. The numerical solution shows that this is true for $\theta \leq 0.3\sqrt{\delta}$, indicating that Singh's solution is valid for $C_1 = 0.3$.

The boundary-layer solution of Stewartson et al. (1980) for the developing flow was undertaken to clarify the nature of the fully developed

---

[2]Pedley (1980) demonstrates that this is the distance required for the transport of secondary vorticity from the boundary layer into the core.

flow for large Dean number (see Section 2.1.1.4). An interesting feature of their solution is the vanishing of the axial skin friction at
$R\theta = 0.943a/\sqrt{\delta}$ at the inner generator, $\alpha = 180°$. Since both the circumferential velocity and the circumferential skin friction are no longer
zero beyond this point, the authors conclude that the two secondary
boundary layers collide here, and they conjecture that a radial jet is set
up conveying fluid from the inner bend into the core region, the flow
structure being much like that near the equator of a spinning sphere. The
structure of the local singularity that occurs at this separation point of
the axial boundary layer is explored further in Stewartson & Simpson
(1982); considerable progress is made in elaborating this structure but
only for points upstream of the separation point. Also, contrary to
Stewartson et al., where the boundary-layer equations were integrated
beyond the point of zero axial shear, Stewartson & Simpson conclude
that the forward integration must stop at this point. To examine experimentally the behavior of the flow near this predicted collision point,
Talbot & Wong (1982) measured the wall shear along $\alpha = 180°$ in a
curved pipe of $\delta = 1/7$. They found that the wall shear dips to a
minimum along this inner generator at a distance $0.96a/\sqrt{\delta}$ from the
inlet, a location in good agreement with the numerical solution. Upstream of this point the predicted and measured wall shear stresses are
also in good agreement; downstream of it, however, although both
increase, there is substantial discrepancy between the measured values
and those calculated (but now discounted).

Clearly, the nature of the developing flow in curved pipes for large
Dean numbers is far from understood. The difficulty is primarily associated with the behavior of the boundary layers near the inner bend and
with the interaction of these boundary layers with the core flow.

Before closing this section, we should mention the few fully numerical
solutions for developing flow. Patankar et al. (1974) have applied their
well-known finite-difference method to solve the "parabolized" Navier-
Stokes equations for developing and fully developed flows in curved
pipes. These equations were also solved numerically for this problem by
Rushmore (1975). Using a finite-difference method of his own design,
Liu (1976, 1977) has solved the full elliptic Navier-Stokes equations for
one particular developing flow, flow into a 90° elbow from a reservoir of
uniform total pressure, but for the fairly small Dean number $\kappa \approx 179$.
Yeung (1980) has solved the entry-flow problem numerically for large
Dean number by dividing the flow field into a boundary-layer region and
an inviscid-core region. Notwithstanding the value of these solutions,
they are still too crude to throw much light on the problem of the
analytic structure of the developing, and fully developed, flow at large

Dean number, in particular near the inner generator. Ideally, for developing flow this would come from a fully three-dimensional solution of the full Navier-Stokes equations, with enough computational points to adequately resolve the thin wall boundary layers and the local structure near $\alpha = 180°$. The problem, akin to that for high-Reynolds-number finite-difference calculations, is how to do this in such a way that numerical viscosity does not "wash out" the details of this inner-bend flow structure.

Entry conditions   Thus far we have mentioned at least three different entry or inlet conditions: (a) uniform total pressure, i.e. an inviscid vortex, (b) uniform axial velocity, and (c) straight-pipe Poiseuille flow. (In all three cases the radial and circumferential velocity components are taken to be identically zero.) Each of these has been employed as an inlet condition in experimental, analytical, and numerical investigations.[3] Strictly speaking, the inlet conditions can only be determined by solving the developing flow in the pipe simultaneously with the flow development before the fluid enters the pipe. The assumption of a profile at the entrance is then purely a pragmatic device, but necessarily only an approximate one, to avoid the necessity to deal with this obviously much more complicated flow situation. (For some indication of situations in which this simplified approach is inadequate, see Section 4).

Immediate to the inlet one would expect the flow development to depend on the choice of inlet conditions, so comparisons of different analyses or experiments for the developing flow in curved tubes is not always possible. This should not be a problem for the fully developed flow, which should be independent of (i.e. have "forgotten") the inlet conditions. The recent experimental results of Agrawal et al. (1978; see Section 8) indicate that whatever the configuration upstream of the entrance to the curved section, the flow very rapidly takes on a potential-vortex structure, before any flow development due to curvature occurs, so in future studies there is some reason to favor this as the inlet condition. It is interesting to note in this connection that in the numerical solution of Yeung (1980) the assumed uniform-entry profile developed into a potential vortex almost immediately downstream of the entrance.

Entry Length   Of particular interest relative to curved pipe flows is the entry length, the length required for the flow to become fully developed, and its relationship to that for a straight pipe. For a straight pipe, the

---

[3] We note again that although this situation is not considered separately here, nonuniform velocity distributions at the entrance to a curved pipe can cause secondary flows to arise even in the absence of viscosity (Squire & Winter 1949, Hawthorne 1951).

entry length is commonly quoted as $l_s = 0.25a\,\mathrm{Re}$ (Fargie & Martin 1971), where $\mathrm{Re} = Ua/v$, and U is the uniform axial velocity at the entry to the pipe. Yao & Berger (1975) predict that at large Dean number the entry length in a curved pipe is $l_c = e_1(\kappa/\delta)^{1/2}a = e_1a(2\,\mathrm{Re})^{1/2}\delta^{-1/4}$, where e_1 is weakly dependent on δ and Re lies between 2 and 4. Thus

$$\frac{l_c}{l_s} = 8e_1\kappa^{-1/2}, \tag{30}$$

so in the limit of very large Dean numbers, $l_c/l_s \ll 1$. For practical values of the parameters, however, this may not be the case; for example, if $\delta = 0.05$, $\mathrm{Re} = 2000$, then $\kappa = 894$, and $l_c/l_s = 0.584$, so the entry length is approximately half that for a straight pipe. Pedley (1980) has pointed out that there is at least partial experimental support for this functional dependence of l_c on Re and δ in the work of Olson (1971) and Agrawal et al. (1978). These investigators find the flow to be virtually fully developed at the last station measured: $s = 50$ for Olson, $s = 57$ for Agrawal et al; evaluation of $\mathrm{Re}^{1/2}\delta^{-1/4}$ for each of these yields 46 for Olson, 50 for Agrawal et al. Pedley also gives a physical argument supporting this scaling.

Finally, we note that the value of l_c predicted by Yao & Berger satisfies for large Dean number the inequality

$$\delta^{-1/2} \ll \frac{l_c}{a} \propto \mathrm{Re}^{1/2}\delta^{-1/4} \ll \mathrm{Re}, \tag{31}$$

where the rightmost expression represents the entry length in a straight pipe and the leftmost expression the range of validity of Singh's solution (1974).

3.   UNSTEADY, LAMINAR FLOW IN A RIGID PIPE

Since flow in the aorta is pulsatile, and a description of flow in this blood vessel has been one of the prime motivations for much of the recent work on curved tubes, it is not surprising that a good part of this work has focused on the analysis of unsteady flow, although, as we shall see, with considerably less success than in the steady case. We are able here to do no more than give the briefest outline of this work, referring the reader to the relevant literature. As before we consider separately fully developed and developing flows, beginning with the former. Such studies may also have nonphysiological applications, e.g. in the analysis of fluidic devices, hydraulic control lines, etc.

3.1   Fully Developed Flow

3.1.1   OSCILLATORY FLOW   We assume that the dimensionless pressure gradient p_s in (2d) varies as

$$- p_s = G + \bar{\alpha}^2 \hat{W} \cos \bar{\alpha}^2 t, \tag{32}$$

where G is the (constant) dimensionless steady component of the pressure gradient, ω the frequency of the oscillation, and $\hat{W}$ a characteristic velocity of the fluid that would result from the oscillation. We are also now replacing the nondimensionalizing velocity $\overline{W}_0$ in Equations (1) by v/a. This means that time is now nondimensionalized with respect to the diffusion time a^2/v. The parameter $\bar{\alpha}$, equal to $a\sqrt{\omega/v}$, is called the Womersley parameter in the blood-flow literature.[4]

If one assumes that the tube is loosely coiled, $\delta \ll 1$, Equations (2) again reduce so that a stream function can be defined as in (14), and one obtains as the governing equations, corresponding to (15) and (16),

$$\nabla_1^2 w' + D + \bar{\alpha}^3 (2R_s)^{1/2} \cos \tau = \frac{1}{r} (\psi_\alpha w'_r - \psi_r w'_\alpha) + \bar{\alpha}^2 w'_\tau, \tag{33}$$

and

$$\nabla_1^4 \psi + \frac{1}{r} \left(\psi_r \frac{\partial}{\partial \alpha} - \psi_\alpha \frac{\partial}{\partial r} \right) \nabla_1^2 \psi = - w' w'_y + \bar{\alpha}^2 \nabla_1^2 \psi_\tau, \tag{34}$$

where $y = r \sin \alpha$; $\tau = \bar{\alpha}^2 t$; D is the Dean number, defined as $D = (2\delta)^{1/2} G$; and R_s is a new dimensionless parameter, defined as $R_s = \delta \hat{W}^2/\bar{\alpha}^2$ or $\delta \hat{W}_a^2/\omega v$, where $\hat{W}_a = v\hat{W}/a$. [For compatibility with other studies, we have changed the definitions of nondimensional scaled velocities; instead of (6), the scaling now is $(u, v, w) \to (u, v, (2\delta)^{-1/2} w')$. We have, however, followed Pedley (1980), rather than many other investigators, in the notation and definitions, in particular that of $\bar{\alpha}$, because the connection with physiological flows of interest is more direct.]

We see from (33) and (34) that three independent parameters, D, $\bar{\alpha}$, and R_s, characterize these flows, and so there are a large number of possible flow situations. Since even a direct numerical treatment of these nonlinear equations for arbitrary values of the parameters would be formidable indeed, asymptotic analysis for different limiting cases is an attractive alternative; this has been done for a large number of cases by Smith (1975), only some of which are discussed below.

3.1.1.1   *Steady limit*   This case corresponds to small perturbations about the steady flow; for small Dean number, it can be treated by a double

---

[4]We have followed standard notation in denoting the Womersley parameter by α, but with an overbar to avoid confusion with the azimuthal polar angle α.

series expansion in $\bar{\alpha}^2$ and D. For large Dean number, one must use boundary-layer ideas patterned after those described earlier for steady flow. For these analyses, see Smith (1975).

This steady limiting case, in which $\bar{\alpha} \ll 1$, is inappropriate for the flow in large arteries because there the values of $\bar{\alpha}$ are large and the amplitude of the oscillating component of the motion is at least as large as the mean component. Below we briefly consider flows in these situations, beginning with the case of purely oscillatory motion. For a much more complete account of this work, see Pedley (1980), from which the summary below borrows heavily.

3.1.1.2 Zero mean pressure gradient: purely oscillatory motion

This case, for which G and hence D are zero, was first treated by Lyne (1971); it allows one to expose the effect of a purely oscillatory pressure gradient. For large $\bar{\alpha}$ the centrifugal forces cause a mean secondary flow to develop in the Stokes layer on the wall ($r = 1$), much like the steady streaming that develops in other high-frequency oscillatory viscous flows, which in turn drives steady secondary motions in the core. The flow in the latter region is governed by the "secondary-streaming Reynolds number" R_s, and can be analyzed analytically only in the limits of small and large R_s. For small R_s the core flow is a simple Stokes-flow problem, whereas for large R_s (the more physiologically relevant limit) the effects of viscosity are confined to thin boundary layers [of thickness $O(R_s^{-1/2})$] near the walls and along the plane of symmetry, the remainder of the core being inviscid and of uniform vorticity (of opposite sign in the upper and lower halves). The pattern of steady secondary streaming in the core, from the outside to the inside of the bend, predicted by Lyne in this large R_s limit, was also confirmed experimentally by him and Munson (1975).

A different approach to this purely oscillatory case was that of Zalosh & Nelson (1973), who linearized the Navier-Stokes equations by expansion in powers of an amplitude parameter, equal to $\delta \hat{W}^2$ in our notation and assumed to be small, and solved them by (numerical inversion of) finite Hankel transforms. The analysis is valid only for $R_s \ll 1$ but for all values of $\bar{\alpha}$. Closed-form solutions are obtained in the limits $\bar{\alpha} \to 0$ and $\bar{\alpha} \to \infty$. Claiming there to be a discrepancy between the equations and results quoted in Zalosh & Nelson, Mullin & Greated (1980b) have carried out an essentially parallel analysis using the same expansion and solution procedure. A closed-form analytic solution for this problem valid for *all* values of $\bar{\alpha}$ has been obtained by Zapryanov & Matakiev (1980).

3.1.1.3 Nonzero mean pressure gradient

If D is not zero, but very small (i.e. $D \ll \bar{\alpha}^{-1} \ll 1$), the solution can be obtained as a power series in D, with Lyne's solution the leading term (Smith 1975). For the large arteries, D is very large, and so are $\bar{\alpha}$ and R_s, e.g. in the canine aorta $D \approx 2000$, $\bar{\alpha} \approx 13$, and $R_s \approx 4200$. Because of the nonlinearity of the problem, the mean and oscillatory parts cannot be separated and one must simultaneously take D, $\bar{\alpha}$, and R_s to be large. Most relevant physiologically, in particular, is the ordering $1 \ll \bar{\alpha} \ll D < R_s$, with $D \ll \bar{\alpha}^3 R_s^{1/2}$, so that the amplitude of the oscillatory part of the pressure gradient is much larger than the mean; theoretically one would probably be most interested in the regimes in which transition occurs from outward secondary streaming in the core, characteristic of the steady pressure-gradient case, to inward streaming, characteristic of the purely oscillatory case. The only work done on this latter problem is that of Blennerhassett (1976), who studied cases for which $D \lesssim R_s$ in the limit $\bar{\alpha} \to \infty$. In this limit the Stokes layer, the thinnest layer on the wall, is the same as that calculated by Lyne (1971); hence one can focus on the steady component of the flow, which Blennerhassett shows can be determined from the solution of the steady equations (15) and (16), subject to boundary conditions (18), except that $\psi_r = 0$ is replaced by the slip velocity at the edge of the Stokes layer. In addition to some asymptotic results for this system, Blennerhassett obtains numerical solutions for general values of D and R_s. At not very large values of R_s and $D = O(R_s)$ the results are as expected, i.e. as D increases the flow pattern changes from one in which the secondary streaming is toward the inner wall and the peak axial velocity occurs near this wall (typical of purely oscillatory flow) to one in which the secondary streaming is toward the outer wall and the peak axial velocity occurs nearer that wall (typical of steady flow). Above some critical value of R_s, however, the situation becomes much more complicated, the volume flow-rate ratio Q_r (the mean flow rate in the curved tube divided by that in a straight tube at the same mean pressure gradient) becoming a multivalued function of D, implying that the flow at a given D depends on the nature of the approach to that state and the possibility of discontinuous jumps from one state of flow to another. For even larger values of R_s the situation appears to be still more complex, leading to the conclusion that the "problem of pulsatile flow in a curved tube is far from completely solved, even in the limit $\bar{\alpha} \to \infty$" (Pedley 1980).

Smith (1975) has reported on the little progress that is made if the limit $\bar{\alpha} \to \infty$ is not taken before any other limit, so that the Stokes layer is not necessarily the thinnest layer. In addition, he also treats asymptotically other limiting mean-pressure-gradient cases. (See also Blennerhassett 1976). Unfortunately, as noted by Pedley, for perhaps the most important application, the physiological one, not much headway has been made either asymptotically or numerically, in the former case because of the seeming intractability of the problem to such analysis.

3.1.2 GENERAL UNSTEADY FLOW   Solutions for the oscillating flows discussed above, even if available for values of the parameters $\bar{\alpha}$, D, and R_s appropriate to the aorta, would not be fully appropriate for the flow in this vessel because δ is not vanishingly small and the pressure gradient is not of the form assumed in (32). More relevant treatments for flow in the aorta, assuming δ finite and a more realistic pressure-gradient waveform, the flow being assumed to start from rest, are given in the form of small t-expansions by Farthing (1977) and Pedley (1980).

3.2 Developing Flow

For very much the same reasons as for steady flow, one would like to know for an unsteady flow in a curved pipe how the flow evolves from the entrance profile to a fully developed structure far downstream.

Singh's (1974) analysis of steady flow very near the entrance, discussed in Section 2.2, can be readily extended to unsteady flow; this has been done by Singh et al. (1978). Assuming that the total pressure across the cross section at the entrance is constant, the entrance profile is then

$$u = v = 0, \qquad w' = \frac{W_0(t)}{1 + \delta r \cos \alpha}. \tag{35}$$

(It is also assumed that $W_0 \geqslant 0$ so there is no backflow.) This extension, using the same inlet profile, has also been carried out by Pedley (1980), who shows in addition how Singh's analyses (Singh 1974, Singh et al. 1978) may be modified so as not to require vanishingly small δ nor constant curvature, modifications important in the aorta. Again the solutions are given as series expansions and are obtained by matching the boundary layer on the walls to the core flow. Results are given both for the case when the flow is quasi-steady ($s\dot{W}_0 / W_0^2 \ll 1$) and when it is not; we refer the reader to the original papers for a discussion of these results, and note here only the effect of the unsteadiness on the axial wall shear. The unsteadiness increases the skin friction during the accelerating phase ($\dot{W}_0 > 0$) through the accelerating pressure gradient, and decreases it during the decelerating phase ($\dot{W}_0 < 0$). Therefore the crossover of maximum shear (see Section 2.2) is inhibited in the acceleration phase (i.e. occurs at a larger value of s) and promoted in the deceleration phase. Since the effect of unsteadiness, or what Singh et al. call the entrance-velocity-fluctuation effect, on the skin friction increases as the flow develops further downstream, negative shear stresses, and therefore backflow near the wall, occur. During the accelerating phase the azimuthal skin friction falls below its steady value as s increases, perhaps ultimately becoming negative, but the range of validity of the series solution does not extend far enough to determine this.

The downstream extent of validity of this modified, unsteady, Singh series solution is the same as for the steady case (see Section 2.2).

Farthing (1977) has considered this problem when the curvature is slowly varying and has shown how, in an application to approximately quasi-static flow in the aorta, the small-s expansions of this section can be matched to the downstream diffusive flow, expressed as expansions in small t, referred to in Section 3.1.2.

For an experimental investigation of oscillatory flow development at the entry into a curved pipe, see Mullin & Greated (1980a).

3.3 Final Remarks

Numerical analyses of fully developed pulsatile flow are also presented in Rabadi et al. (1980), for small or large δ and a wide range of $\bar{\alpha}$ and D, and in Lin & Tarbell (1980). For experimental studies of pulsatile and oscillatory curved-tube or pipe flows, see Chandran et al. (1979), Chandran & Yearwood (1981), Lin & Tarbell (1980), Munson (1975), Mullin & Greated (1980a,b), and Talbot & Gong (1982).

4. FINITE BEND

Short segments of curved pipe, finite elbows, are ubiquitous and indispensible components in piping systems. The flow in a finite elbow is almost always a developing flow. Because of its importance in engineering applications, and its unique feature of changing the flow structure in the straight-pipe sections upstream and downstream of the elbow, we single it out from our previous discussion and use it as a vehicle to discuss the possible inlet conditions for a curved pipe in general.

Itō (1960) carried out an exhaustive set of measurements of the pressure losses in smooth-pipe bends for various elbow angles. Two long straight pipes were connected to the elbow. (Straight-pipe sections upstream and downstream of the elbow long enough for the establishment upstream and reestablishment downstream of Poiseuille flow eliminate the unknown influence of the inlet and exit conditions on the flow development in the elbow.) A typical plot of the pressure distribution in a 90° elbow system is given in Figure 6. Although the data in this figure are for a Reynolds number corresponding to turbulent flow, the general characteristics are equally relevant to laminar flow. The dramatic pressure variation near the inner and outer bends of the elbow is associated with the existence of the secondary flow. The most interesting feature of this figure is that the pressure drop starts to deviate from that of a Poiseuille flow 5–10 diameters upstream of the inlet of the elbow, the maximum pressure deviation occurring about 45° into the 90° elbow. At the inlet the pressure deviation is about 40–50% of its maximum value, at the exit about 20–40% of its maximum value. The pressure gradient gradually approaches that of a Poiseuille flow in the downstream straight section about 40–50 diameters from the elbow exit; for laminar flow the influence of the bend on the downstream flow almost certainly extends further. Thus, when a curved section of pipe is connected to straight sections very significant disturbances are communicated both upstream and downstream, from the curved into the straight sections and vice versa. Although the secondary flow may not reach its fully developed state in Itō's experiments because the elbow is short, the measurements indicate that a substantial secondary flow develops before the fluid enters the elbow, and that the secondary flow decays sharply before the fluid exits the curved section. The data also show that the flow development in the elbow is not a function of θ, the elbow angle, but depends on $\theta/\sqrt{\delta}$. Itō's measurements were for bends of circular cross section; for bends of rectangular cross section, see Ward-Smith (1968).

This discussion of short circular-arc bends has necessarily been brief. For a fuller discussion, see Ward-Smith (1980).

Having gained some appreciation of the "upstream effect" of a curved pipe, we now discuss briefly the inlet condition when such a pipe is

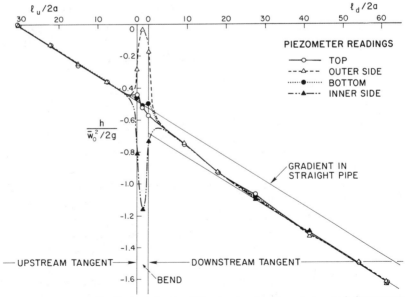

Figure 6 Pressure distribution along a 90° pipe bend and upstream and downstream connecting straight-pipe sections (Itō 1960). $R/a = 3.7$, $\mathrm{Re} = 2 \times 10^5$, h = head loss, l_u = length of upstream tangent, l_d = length of downstream tangent.

connected to a large vessel. Uniform axial velocity, the most common idealized inlet condition, is generated by the sudden contraction of the fluid passage. A strong favorable pressure gradient thins the boundary layer and results in an almost uniform axial-velocity profile. This has been measured by Agrawal et al. (1978). Apparently, the "upstream effect" is suppressed by the strong acceleration induced by the sudden contraction. On the other hand, the measured "crossover" point (see Section 2.2) agrees with Singh's solution with an assumed irrotational inlet flow condition. Strictly speaking, the inlet condition should be determined by simultaneously solving for the flows in the large vessel and the downstream curved pipe. Smith (1976b) studied the flow development in a straight pipe followed by a slightly coiled pipe. He treated the upstream effect by a coordinate transformation and neglected the dominant effect of geometric change, which is transmitted upstream via pressure disturbance. There is no other analytical (or numerical) treatment available.

Before closing this section, we briefly return to our discussion of the role δ plays in determining the general characteristics of curved-pipe flows. It has been demonstrated that the detailed geometric effects, represented by the terms multiplied by δ in Equations (8), contribute little to change the flow structure, at least for $\delta \leq 0.2$. This justifies the assertion that the limit $\delta \to 0$, $\delta^{1/2} \mathrm{Re}$ finite, of (8) indeed yields all the possible flow patterns for δ small but not necessarily infinitesimal. Functionally, then, the local shear depends only on Dean number, its coefficient varying weakly perhaps with δ. A good example of what can happen when δ is not small is a 180° elbow with $\delta = 1$. The axial flow may separate at the inner bend for very low Reynolds number, say $\mathrm{Re} \sim 10$, corresponding to $D = \sqrt{10}$, a small Dean number, but the flow pattern differs completely from small-Dean-number flow, suggesting that one must exercise great care when using the existing solutions for curved-pipe flows for small D when δ is not also small.

5. DIFFERENT GEOMETRIES, WALL CHARACTERISTICS, AND FLUID PROPERTIES

5.1 Different Cross Sections

Solutions have been obtained for steady fully developed flow in curved pipes of other than circular cross section. For low Dean number and a smooth, symmetric cross-sectional contour, such extensions would appear to present primarily mathematical difficulties. For an elliptic cross section and small curvature, Itō (1951) obtained, by successive approximation, the solution to second order in Dean number. Apparently unaware of this work, Cuming (1952) independently obtained the solution to first order, and Srivastava (1980) to second order. Srivastava's analysis is valid for larger curvature in that, following Topakoglu (1967) for a circular cross section, his solution is in the form of a double series expansion in Reynolds number and curvature ratio; he also gives explicitly an expression for the effect of curvature on flow resistance or the flux in the pipe. Numerical solutions for small but finite curvature ratio and moderate values of Dean number for an elliptic cross section have been obtained by Truesdell & Adler (1970). All of these solutions assume symmetric orientation of the elliptical cross section, i.e. that the axes of the ellipse are parallel and perpendicular to the symmetry plane of the pipe ($\alpha = 0$). For an elastico-viscous liquid Thomas & Walters (1965) obtain the solution, to first order in Dean number, for an elliptic cross section, the axes of which are arbitrarily oriented with respect to the symmetry plane.

Curved pipes of rectangular cross section are used extensively in industrial applications, in particular in heat- and mass-transfer apparatus, so considerably more work has been done on this cross section, both theoretically and experimentally. Itō (1951) and, for a square cross section, Cuming (1952) obtained solutions to first order in Dean number by successive approximation. Mori & Uchida (1967) obtained for a square cross section asymptotic solutions for large Dean number based on the boundary-layer approximation. Spanning the Dean-number range, from small to intermediate to fairly large values, and a wide range of cross-sectional aspect ratios, there are the numerical solutions of Cheng & Akiyama (1970), Cheng et al. (1976), and Schilling & Simon (1979). In their numerical solutions for a square cross section, Joseph et al. (1975) found that the expected secondary flow pattern with twin counterrotating vortices occurs at low Dean numbers, whereas at higher Dean numbers four secondary vortices, counterrotating in pairs, are present. The normal twin counterrotating secondary vortices are also present but are distorted to accommodate the additional pair of vortices, which are symmetrically located near the outer wall above and below the symmetry plane; though smaller than the normal vortices, they are equally strong in velocities. These four-vortex solutions for large Dean numbers were subsequently found for rectangular cross sections by Cheng & Akiyama and Cheng et al.

Dean & Hurst (1959) obtained approximate analytic solutions for flow in curved pipes of rectangular and circular cross sections assuming the secondary flow is a uniformly moving stream, from the inner to the outer bend.

To obtain analytic solutions for a general cross section at large Dean number, one can use boundary-layer ideas like those discussed in Section 2.1.1.4 and the situation is more complicated than that for low Dean number. Smith (1976a) has considered this problem and presents an asymptotic description for a general symmetric section. An explicit solution is presented for a triangular cross section. Solutions for flow in a triangular cross section have been obtained numerically for a wide range of Dean numbers by Collins & Dennis (1976a,b). For a rectangular cross section Smith puts forward analytical and numerical arguments that point fairly conclusively to the nonexistence of an attached laminar flow, near the inner bend at least. Humphrey et al. (1977) have numerically solved the Navier-Stokes equations for a strongly curved 90° bend of square cross section connected to upstream and downstream straight sections. They also assess the extent to which reduced, such as "parabo-lized," forms of the full equations can be utilized to describe the flow.

Nandakumar & Masliyah (1982) have studied general curved cross sections by solving the full Navier-Stokes equations numerically in a bipolar-toroidal coordinate system. They find that bifurcation of the two-vortex solution into two- and four-vortex solutions at large Dean numbers exists irrespective of the shape of the cross section. The proce-dure for obtaining these dual solutions and the explanation of their origin comes from the work of Benjamin (1978a,b).

5.2   Variable Curvature

Murata et al. (1976) have analyzed theoretically the steady fully devel-oped low-Dean-number flow through curved pipes of circular cross section and variable curvature, the center of curvature being assumed to lie along a specified two-dimensional curve. Pulsatile flow in such a pipe was considered by Inaba & Murata (1978). Pedley, in his account of the flow in curved pipes, extends existing theory, where possible, to account for the effect of nonuniform curvature.

5.3   Flexible Walls

Because sites of curvature, which occur frequently in the arterial system, are often also sites of atherosclerotic involvement, fluid-dynamic phe-nomena have been suggested as a causative factor in atherogenesis. A number of hypotheses for such a causal relationship have been put forth, beginning with that of Texon (1960), who implicated the "pulling" or suction effect on the intima of the low-pressure region near the inner bend, followed by Fry (1968), who suggested it was damage done to the endothelial cell lining in regions of high wall-shear stresses, then by Caro et al. (1971), who suggested a shear-dependent mass-transfer mechanism that implicated regions of low wall shear as being most dangerous. The resultant controversy, still unresolved (Nerem & Cornhill 1980), would obviously benefit from the availability of as complete a model of arterial flow as possible. To this end, Chandran et al. (1974) have presented an analytical model of fully developed oscillatory flow of a viscous incompressible fluid in a thin-walled elastic tube. The formulation and analysis are closely patterned after those of Morgan & Kiely (1954) and Womersley (1957) for oscillatory flow of a viscous fluid in a straight elastic tube; in fact, theirs is a perturbation of these analyses, assuming the curved tube to be of small curvature. The governing equations for the fluid and the thin elastic-tube wall (shell equations) are linearized, assuming (a) small radial wall displacements, (b) small a/λ, where λ is the wavelength of the pressure wave, (c) small w/c, where w is some characteristic axial velocity and c is the wave velocity, and finally (d) small δ. Only purely oscillatory motions are considered, the results showing that the secondary motions in the core are toward, and the peak axial velocity shifted to, the inner wall, both contrary to the case of steady flow and in agreement with the work of Lyne (1971) for rigid tubes. The wall shear is, again in contrast to steady flow, higher at the inner wall. Since arterial blood flows are subject to pulsatile pressure gradients with nonzero mean and many modes, the actual wall-shear distributions will depend on the relative importance of the steady- and oscillatory-flow components. In Chandran et al. (1979), these components are compared for a typical large curved artery and a typical extracorporeal device (a coiled-tube membrane oxygenator). Such comparisons, or indeed superpositions, of the steady and oscillatory components are often employed in arterial-flow analyses (McDonald 1974), but since they assume no interaction between these components they should be approached with caution (see Pedley 1980; also Section 3.1.1.3).

In addressing the question of the applicability of the unsteady, rigid-tube analyses of Section 3 to flow in the aorta, it should be pointed out that while wall distensibility is particularly important in propagation of the pulse wave and determining the local pressure gradient, it is less important in determining the fluid motion because λ is large compared with the length of the ascending aorta and w/c is small (Lighthill 1975).

5.4   Porous Tubes

Zapryanov & Matakiev (1980) have analyzed for a slightly curved porous tube the same unsteady problem considered for a nonporous tube by

Zalosh & Nelson (1973) and Mullin & Greated (1980b), namely the flow due to a purely oscillatory pressure gradient of small amplitude. They obtain a closed-form analytic solution valid for all values of the frequency parameter $\bar{\alpha}$.

5.5 Non-Newtonian Fluids

There have been extensions for a curved pipe of circular cross section of the original steady, fully developed, low-Dean-number analysis of Dean (1928) to non-Newtonian fluids. Among these are the work of Jones (1960) for a visco-inelastic, Reiner-Rivlin fluid; Kamel & Kaloni (1977) for a Cosserat fluid; Thomas & Walters (1963, 1965), the latter of these for an elliptic cross section, and Sharma & Prakash (1977) for a second-order viscoelastic fluid; and Jones (1967) for a third-order Coleman-Noll fluid. Questioning the adequacy of such second- or third-order fluid models at rates of deformation likely to occur in practical curved-pipe flows, Mashelkar & Devarajan (1976a) analyzed, for large Dean numbers, the flow of a power-law fluid using the momentum-integral approach of Itō (1969).

A number of papers have analyzed heat transfer in curved-pipe flows for non-Newtonian fluids. Raju & Rathna (1970) did so for a power-law fluid and found that dilatant fluids would be more efficient for the working of heat exchangers, while Srivastava & Sarin (1973a,b) did likewise for an elastico-viscous fluid for both circular and elliptic cross sections, for the latter case using the solution for the velocities obtained earlier by Thomas & Walters (1965). For unsteady flow, James (1975) has extended the analysis of Zalosh & Nelson (1973) for periodic pressure gradients to the case of an elastico-viscous liquid, in particular to study the effect of elasticity on the "negative centrifuging effect" discovered by Lyne (1971; see Section 3.1.1.2).

Measurements of friction factors for the flow of non-Newtonian fluids in curved pipes have been reported by Mishra & Gupta (1979), Singh & Mishra (1980), and Mashelkar & Devarajan (1976b), the last of these finding a "drag reduction" effect for viscoelastic fluids that could be correlated in terms of a Weissenberg number. Tsang & James (1980) noted that polymer additives caused a reduction of the secondary motion in curved tubes. Finally we might note a different application of such experiments in the work of Hayes & Hutton (1972), who used measurements of streamline patterns produced by flows in curved tubes and a third-order simple fluid model to determine the sum of the first and second normal-stress coefficients.

5.6   Multiphase Flow

Gould et al. (1974) have obtained correlations for predicting pressure distributions for two-phase flows. Morimoto & Ousaka (1974) did measurements on the effect of a 180° curved-pipe bend, through which an air-water mixture is flowing, on the flow in the upstream and downstream straight-pipe sections. Measurements of the same two-phase mixture flowing in bends of various curvatures and angles following straight-inlet conditions were carried out by Maddock et al. (1974), on the basis of which they suggested a flow structure for the developing region of a bend.

For more recent experimental work, plus references to other work in this area, see Whalley (1980). For the effect of curvature on boiling see Jensen & Bergles (1981, 1982).

6.   MIXING AND MASS TRANSPORT IN CURVED TUBES

6.1   Laminar Dispersion

Erdogan & Chatwin (1967; see also Erdogan 1968) analyzed the effect of curvature on laminar dispersion of solute injected into a curved circular tube through which a solvent is in steady laminar motion. Using the velocity distribution obtained by Dean (1927, 1928), they show that the classical Taylor-Aris theory for a straight pipe is valid for a uniformly curved tube, and derive an expression for the diffusivity that predicts that for all common liquids and most gases the diffusivity is reduced by the curvature. McConalogue (1970) has also considered this problem when molecular diffusion is negligible and convection is the dominant dispersion mechanism, employing as velocity distributions those obtained by him and Srivastava (McConalogue & Srivastava 1968) for higher Dean number. He also found the axial dispersion to be lower than that in a straight tube. Nunge et al. (1972) reconsidered the general dispersion problem analytically, using instead the velocity distributions obtained by Topakoglu (1967) so their solution was not limited to small values of the curvature. Their results, in contrast, predict that the dispersion coefficient may be significantly increased by curvature at low Reynolds numbers, particularly in liquid systems of biological interest. Experiments measuring laminar dispersion in curved tubes and the conditions under which the Taylor-Aris theory is applicable to such tubes have been reported by Caro (1966) and Nigam & Vasudeva (1976). In an experimental investigation of the use of the indicator dilution bolus-injection technique to measure flow parameters in curved tubes, Patel & Sirs (1977) found flow-rate estimations are more reliable than those of straight tubes due to secondary flows tending to disperse the indicator uniformly over the vessel cross section.

6.2 Mass Transfer

There are a number of interesting studies of the effect of tube curvature on mass transfer, covering a wide area of potential applications. Nunge & Adams (1973) describe, for desalination purposes, a new design for reverse osmosis in a curved-tube system, utilizing a split membrane over the circumference to take advantage of the scouring-silting nature of the secondary-flow pattern. Earlier, in an analytic study, Srinivasan & Tien (1971) found a significant increase in the mass-transfer rates in reverse osmosis due to the tube curvature. In an experimental study of condensation of water vapor from moist air in curved rectangular ducts, Smol'skiy & Chekhol'skiy (1978) found that heat- and mass-transfer rates and pressure drop were significantly higher than in straight channels. Since in membrane oxygenators the primary resistance to gas transfer comes from the blood boundary layer adjacent to the membrane, the induction of secondary flow, and therefore additional mixing, by curvature of the blood channel had been suggested almost from the beginning of the use of these devices (Weissman & Mockros 1968). Among the more recent work along these lines is that of Tanishita et al. (1978; see also Richardson et al. 1975), who report and study a new design of artificial membrane oxygenator consisting of curved tubes with periodically varying curvature wound in woven patterns. They find an augmentation in gas-transfer performance due to tube curvature similar to that found for helically coiled tubes but with lower flow resistance.

6.3 Flow of Fluid-Particle Mixtures

There have been a number of investigations of the motion of fluid-particle mixtures in curved tubes. Kaimal & Devanathan (1980) obtained trajectories of particles in such a flow for small δ as an asymptotic expansion in powers of Dean number. Yalamov & Muradyan (1980) obtained trajectories and capture coefficients of aerosol-particle mixtures in curved channels. In a theoretical investigation of diffusional deposition in a curved tube, Shaw & Rajendran (1977) showed that the secondary swirling motion in such a tube would increase diffusional deposition of airborne particles in the curved bronchial airways of the human respiratory tract. Yeung (1979) investigated analytically the motion of gas-particle mixtures in curved pipes to obtain information about impact velocities, and impingement angles and mass fluxes, quantities of particular interest in assessing pipe erosion in such flow systems. Jain & Prabha (1982) analyzed gas-particle flow in slightly curved pipes of circular and rectangular cross sections assuming, following Dean & Hurst (1959), a uniform outward streaming for the central portion of the cross section.

7. THERMAL EFFECTS

Extensive studies have been made of convective heat transfer in coiled pipes and finite elbows, because of their practical applications in heat exchangers. A variety of the flow models discussed earlier as well as full numerical approaches have been used to predict the heat-transfer rate. Representative papers are Akiyama & Cheng (1974), Berg & Bonilla (1950), Dravid et al. (1971), Hausen (1943), Janssen & Hoogendoorn (1978), Kalb & Seader (1972, 1974), Kubair & Kuloor (1966), Mori & Nakayama (1965, 1967), Oliver & Asghar (1976), Owhadi et al. (1968), Patankar et al. (1974), Schmidt (1967), Seban & McLaughlin (1963), Shchukin (1969), Simon et al. (1977), Singh & Bell (1974), Tarbell & Samuels (1973), and Zapryanov et al. (1980). The outstanding feature of convective heat transfer in curved pipes is that, due to the additional mixing by the secondary flow, the mean heat-transfer rate is larger than that in a straight pipe at the same flow rate. The local Nusselt number attains its maximum at the outer bend due to the local stagnation-like flow, and its minimum at the inner bend due to the reverse stagnation-point-like flow. Effects of natural convection have, however, not been considered in the literature cited above. Abul-Hamayel & Bell (1979) deduce from their experiments that the maximum Nusselt number does not always occur at the outer bend for a horizontal curved pipe. Yao & Berger (1978) obtained a series solution for a fully developed flow for small Dean number and the product $Re\,Ra$, where Ra is the Rayleigh number. For a horizontal curved pipe the centrifugal and buoyancy forces combine to yield two skewed vortices, which may be viewed as the superposition of two "horizontal" vortices induced by the centrifugal force and two "vertical" vortices induced by the buoyancy force. Consequently, the stagnation-point-like flow moves from $\alpha = 90°$ to somewhere between $0°$ and $90°$, where the maximum Nusselt number occurs. Prusa & Yao (1982) obtained a finite-difference solution for larger D and $Re\,Ra$. The location of the local maximum axial shear predicted by the series solution agrees well with that by the finite-difference method for the whole range of D and $Re\,Ra$. This may be because the location of the maximum axial shear (or Nusselt number) is determined by the ratio of

D to $Re\,Ra$, centrifugal force to buoyancy force, and not by their individual values.

Prusa & Yao compared the mass-flow rate in a heated curved pipe with that in a straight pipe under the same axial pressure gradient and found that large overheating can enhance the secondary flow, increase the flow resistance, and reduce the flow rate. Consequently, a heated curved pipe can be a poorer heat-transfer device than a straight pipe under the same axial pressure gradient. A flow-regime map is provided to determine when one should consider the effect of buoyancy force. The results of analyses that do not consider the effect of buoyancy can be assumed valid only when $Re\,Ra$ is very small compared with D.

8. EXPERIMENTAL STUDIES

The literature on experimental studies of curved-pipe and duct flows is extensive, and space limitations permit only a few representative investigations to be described, some of which already have been mentioned. References to additional work may be found in the literature cited.

The primary emphasis of the early experimental work on the flow in curved pipes of circular cross section was on measurements of the friction factor, or ratio, λ_c/λ_s, defined [see the paragraph above (21)] by White (1929) as the ratio of the resistances in a curved and straight pipe carrying the same flux. In Figure 7, friction ratios measured by White (1929), Adler (1934), and Itō (1959) are plotted as a function of the Dean number κ for different curvature ratios $\delta(=a/R)$, ranging from loosely coiled to more tightly coiled pipes. The departure of each curve for fixed δ from the envelope of experimental points, the κ at which this occurs increasing as δ increases, is usually ascribed to the onset of turbulence. For small Dean number and small δ, the series (21) obtained by Dean (1928) agrees very well with the experiments. For larger κ the envelope of experimental points, before the onset of turbulence, is well-fitted by the expression $\lambda_c/\lambda_s = 0.0969\kappa^{1/2} + 0.556$, proposed by Hasson (1955) to correlate White's data in the range $30 < \kappa < 2000$. This square-root variation wth κ is predicted by all the boundary-layer analyses (see Section 2.1.1.4 and Table 1), with coefficients of proportionality that differ only slightly from one another and Hasson's correlation. The finite-difference calculations (see Section 2.1.1.3) agree well with the experiments for low and intermediate κ for all the values of δ shown, and tend toward the boundary-layer results for larger κ. For example, Collins & Dennis (1975) fit their finite-difference solution to $\lambda_c/\lambda_s = 0.1028\kappa^{1/2} + 0.380$. Also shown in Figure 7 is Van Dyke's (1978) asymptotic series $(\kappa \to \infty)$ result $\lambda_c/\lambda_s \sim 0.47136\kappa^{1/4}$, obtained by his extension of Dean's series; this agrees well with the experimental data for $\delta \leqslant 1/250$ over much of the laminar range of κ, but diverges from the experimental data and the finite-difference and boundary-layer solutions for $\kappa \gtrsim 100$ when the coiling is tighter. Van Dyke suggested that this discrepancy was due to the inaccuracy or incorrectness of the then-available numerical and analytic solutions, but this contention has been weakened by later more refined numerical calculations (Dennis 1980). Neither can the subsequent discovery of the bifurcation, at large Dean numbers, of the two-vortex solution into two- and four-vortex solutions (Section 2.1.4) account for the discrepancy, since, as found by Dennis & Ng (1982), the friction ratios for the dual solutions are nearly the same. However, it is possible, as these latter authors suggest, that Van Dyke's asymptotic series solution is yet a third solution.

Adler also made extensive Pitot-tube measurements of axial-velocity distributions in fully developed steady laminar flow in curved circular pipes. His results exhibit clearly the now well-known skewing of the axial-velocity maximum toward the outer wall of the bend. Less complete data are available for the axial-velocity distributions in fully developed turbulent flow, but the results are similar to those of laminar flow.

More recent work has focused on developing flows in curved pipes and ducts, with inlet conditions being either an essentially inviscid, flat velocity profile such as is produced by a bell-mouth entry from a large chamber, or a fully developed straight-pipe velocity profile. The availability of laser velocimetry has made possible the measurement not only of the axial-velocity profiles, but also the profiles of the secondary motion, usually the component parallel to the plane of curvature but in a few instances the component perpendicular to this plane as well.

Agrawal et al. (1978) have measured axial- and secondary-velocity distributions in 180° curved circular pipes under steady laminar flow and

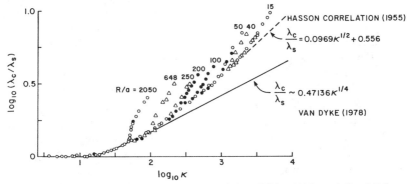

Figure 7 Friction ratios for various curvature ratios. ○, White (1929); ●, Adler (1934); △, Itō (1959) ($\kappa = 2D = (2a\overline{W}_0/\nu)(a/R)^{1/2}$). From Van Dyke (1978).

bell-mouth entry conditions for κ ranging from 138 to 679. Some of their principal findings are now described briefly.

The axial-velocity profiles very close to the entry of the curved section are found to have a potential-vortex structure, with the velocity maximum located in the neighborhood of the inner wall. With flow development, this inner-wall maximum is eroded by the action of the outwardly directed secondary flow in the core fluid, and the axial-velocity profiles are gradually transformed into the outwardly skewed, fully developed forms. At the lowest Dean number this transformation occurs in a relatively smooth fashion, but for κ > 500, doubly peaked axial-velocity profiles are found within the flow-development length. The value of κ at which doubly peaked axial velocity profiles first make their appearance is not accurately known, but is estimated to be in the neighborhood of 300.

The secondary motions associated with the "low" and "high" Dean-number regimes are found to be different. At low Dean numbers, the secondary motion is essentially the classical Dean-type circulation. At high Dean numbers, the occurrence of doubly peaked axial-velocity profiles appears also to coincide with the appearance of additional helical motions imbedded within the main Dean circulation. As the fully developed axial-velocity profiles are approached, these imbedded helical motions weaken and gradually disappear. Scarton et al. (1977) also report flow-visualization observations of additional inner-bend vortices in the entry-flow region, but their description of the motion is somewhat different from that of Agrawal et al. The existence of imbedded helical motions and of doubly peaked axial-velocity profiles has thus far not been predicted for pipes of circular cross section by theoretical or numerical analysis. However, doubly peaked profiles were found by Humphrey et al. (1977) in laminar flow at the exit of a 90° bend of square cross section under conditions of fully developed inlet conditions. Numerical calculations by these authors were in reasonable agreement with the measurements. Secondary velocities were not measured, but the calculated results suggest the presence of imbedded helical motions at the location where the axial profiles become double-peaked.

Although this review has dealt exclusively with laminar flows, it might be noted that measurements and calculations of turbulent flows in curved rectangular ducts exhibit many of the features observed in laminar flows. The interested reader is referred to Humphrey et al. (1981) and Taylor et al. (1982) for details and additional references.

Local wall-shear-rate measurements made by means of an electrochemical technique have been reported by Choi et al. (1979) and by Talbot & Wong (1982), for the same flow studied by Agrawal et al. Choi et al. found among other things that the wall regions adjacent to the locations of the imbedded helical secondary motions are regions of elevated shear rate. The measurements of Talbot & Wong indicate that the point on the inner bend where the circumferential boundary layers first collide is a point of minimum shear rate, as predicted by Stewartson et al. (1980), but that the structure of the flow downstream of the collision point differs from that predicted.

The axial distance required for the attainment of the fully developed curved-pipe velocity profiles has been investigated by Austin (1971) for the case of a parabolic inlet profile. He reports his correlation in terms of the angular distance θ_D required, and gives $\theta_D = 49(\kappa\delta)^{1/3}$ (degrees). The data of Agrawal et al. suggest that Austin's result may underestimate slightly the development length for the case of a flat entry profile, but it is of the right order of magnitude.

The behavior of laminar pulsatile curved-pipe entry flow is of particular interest because of its relevance to flows in large blood vessels, as for example, the aortic arch, and has been investigated experimentally by Talbot & Gong (1982). Here in addition to the Dean number as a governing parameter, one has, as mentioned earlier, the frequency parameter $\bar{\alpha} = a(\omega/v)^{1/2}$ and the secondary-flow Reynolds number $R_s = \delta\hat{W}_a^2/\omega v$, which is a measure of the amplitude of the mean pulsatile axial-flow component $\hat{W}_a$. In the case of fully developed pulsatile flow in a curved pipe, Smith (1975) has identified at least ten different flow regimes, depending on the relative magnitudes of κ, $\bar{\alpha}$, and R_s. From the experiments of Talbot & Gong, it appears that Smith's classifications may in general be applicable to entry flow as well, at least insofar as indicating when the flow may be expected to be quasi-steady or when significant modifications of the secondary motions from their steady-flow values may be expected due to interaction with the pulsatile component of the axial flow. Hot-film-anemometer measurements by Chandran & Yearwood (1981), while not as extensive, are in general agreement with the measurements of Talbot & Gong.

Finally, we note that an interesting "resonance" between the axial and secondary flow has been reported by Lin & Tarbell (1980) for fully developed pulsatile flow. The resonance is found to occur when the period of pulsation is equal to the average time required for fluid to travel around the secondary-flow circuit, and is characterized by a sharp increase in the friction factor.

ACKNOWLEDGMENTS

The authors wish to express appreciation to Dr. George K. Lea, Program Director of the Fluid Mechanics Program of the National Science Foundation, for generous research support in this area under several NSF grants in his program. L.-S. Y. would also like to acknowledge partial support from Mr. M. K. Ellingsworth of the Office of Naval Research.

Literature Cited

Abul-Hamayel, M. A., Bell, K. J. 1979. Heat transfer in helically-coiled tubes with laminar flow. *ASME Pap. No. 79-WA/HT-11*

Adler, M. 1934. Strömung in gekrümmten Rohren. *Z. Angew. Math. Mech.* 14:257–75

Agrawal, Y., Talbot, L., Gong, K. 1978. Laser anemometer study of flow development in curved circular pipes. *J. Fluid Mech.* 85:497–518

Akiyama, M., Cheng, K. C. 1971. Boundary vorticity method for laminar forced convection heat transfer in curved pipes. *Int. J. Heat Mass Transfer* 14:1659–75

Akiyama, M., Cheng, K. C. 1974. Graetz problem in curved pipes with uniform wall heat flux. *Appl. Sci. Res.* 29:401–18

Austin, L. 1971. *The development of viscous flow within helical coils.* PhD thesis. Univ. Utah, Salt Lake City

Austin, L. R., Seader, J. D. 1973. Fully developed viscous flow in coiled circular pipes. *AIChE J.* 19:85–94

Barua, S. N. 1963. On secondary flow in stationary curved pipes. *Q. J. Mech. Appl. Math.* 16:61–77

Benjamin, T. B. 1978a. Bifurcation phenomena in steady flows of a viscous fluid. I. Theory. *Proc. R. Soc. London Ser. A.* 359:1–26

Benjamin, T. B. 1978b. Bifurcation phenomena in steady flows of a viscous fluid. II. Experiments. *Proc. R. Soc. London Ser. A.* 359:27–43

Berg, R. R., Bonilla, C. F. 1950. Heating of fluids in coils. *Trans. NY Acad. Sci.* 13:12–18

Blennerhassett, P. 1976. *Secondary motion and diffusion in unsteady flow in a curved pipe.* PhD thesis. Imperial Coll., London

Caro, C. G. 1966. The dispersion of indicator flowing through simplified models of the circulation and its relevance to velocity profile in blood vessels. *J. Physiol.* 185:501–19

Caro, C. G., Fitz-Gerald, J. M., Schroter, R. C. 1971. Atheroma and arterial wall shear: observation, correlation and proposal of a shear dependent mass transfer mechanism for atherogenesis. *Proc. R. Soc. London Ser. B* 177:109–59

Chandran, K. B., Yearwood, T. L. 1981. Experimental study of physiological pulsatile flow in a curved tube. *J. Fluid Mech.* 111:59–85

Chandran, K. B., Hosey, R. R., Ghista, D. N., Vayo, V. W. 1979. Analysis of fully developed unsteady viscous flow in a curved elastic tube model to provide fluid mechanical data for some circulatory path-physiological situations and assist devices. *J. Biomech. Eng.* 101:114–23

Chandran, K. B., Swanson, W. M., Ghista, D. N. 1974. Oscillatory flow in thin-walled curved elastic tubes. *Ann. Biomed. Eng.* 2:392–412

Chandran, K. B., Yearwood, T. L., Wieting, D. M. 1979. An experimental study of pulsatile flow in a curved tube. *J. Biomech.* 12:793–805

Cheng, K. C., Akiyama, M. 1970. Laminar forced convection heat transfer in curved rectangular channels. *Int. J. Heat Mass Transfer* 13:471–90

Cheng, K. C., Lin, R-C., Ou, J-W. 1976. Fully developed laminar flow in curved rectangular channels. *J. Fluids Eng.* 98:41–48

Choi, U. S., Talbot, L., Cornet, I. 1979. Experimental study of wall shear rates in the entry region of a curved tube. *J. Fluid Mech.* 93:465–89

Coles, D. 1981. Prospects for useful research on coherent structure in turbulent shear flow. *Proc. Indian Acad. Sci. (Eng. Sci.)* 4:111–27

Collins, W. M., Dennis, S. C. R. 1975. The steady motion of a viscous fluid in a curved tube. *Q. J. Mech. Appl. Math.* 28:133–56

Collins, W. M., Dennis, S. C. R. 1976a. Steady flow in a curved tube of triangular cross-section. *Proc. R. Soc. London Ser. A* 352:189–211

Collins, W. M., Dennis, S. C. R. 1976b. Viscous eddies near a 90° and a 45° corner in flow through a curved tube of triangular cross-section. *J. Fluid Mech.* 76:417–32

Cuming, H. G. 1952. The secondary flow in curved pipes. *Aeronaut. Res. Counc. Rep. Mem. No. 2880*

Dean, W. R. 1927. Note on the motion of fluid in a curved pipe. *Philos. Mag.* 20:208–23

Dean, W. R. 1928. The streamline motion of fluid in a curved pipe. *Philos. Mag.* 30:673–93

Dean, W. R., Hurst, J. M. 1959. Note on the motion of fluid in a curved pipe. *Mathematika* 6:77–85

Dennis, S. C. R. 1980. Calculation of the steady flow through a curved tube using a new finite-difference method. *J. Fluid Mech.* 99:449–67

Dennis, S. C. R., Ng, M. 1980. Dual solutions for steady flow through a curved tube. *22nd Br. Theor. Mech. Colloq. Univ. Cambridge*

Dennis, S. C. R., Ng, M. 1982. Dual solutions for steady laminar flow through a curved tube. *Q. J. Mech. Appl. Math.* 35:305–24

Dravid, A. N., Smith, K. A., Merrill, E. W., Brian, P. L. T. 1971. Effect on secondary fluid motion of laminar flow heat transfer in helically coiled tubes. *AIChE J.* 17:1114–22

Erdogan, M. E. 1968. Laminar dispersion in a curved pipe. *Bull. Tech. Univ. Istanbul (Turkey)* 21:13–28

Erdogan, M. E., Chatwin, P. C. 1967. The effects of curvature and buoyancy on the laminar dispersion of solute in a horizontal tube. *J. Fluid Mech.* 29:465–84

Eustice, J. 1910. Flow of water in curved pipes. *Proc. R. Soc. London Ser. A* 84:107–18

Eustice, J. 1911. Experiments of streamline motion in curved pipes. *Proc. R. Soc. London Ser. A* 85:119–31

Fargie, D., Martin, B. W. 1971. Developing laminar flow in a pipe of circular cross-section. *Proc. R. Soc. London Ser. A* 321:461–76

Farthing, S. P. 1977. *Flow in the thoracic aorta and its relation to atherogenesis.* PhD thesis. Cambridge Univ.

Fry, D. L. 1968. Acute vascular endothelial changes associated with increased blood velocity gradients. *Circ. Res.* 22:165–97

Gould, T. L., Tek, M. R., Katz, D. L. 1974. Two-phase flow through vertical, inclined, or curved pipe. *J. Pet Technol.* 26:915–26

Greenspan, A. D. 1973. Secondary flow in a curved tube. *J. Fluid Mech.* 57:167–76

Grindley, J. H., Gibson, A. H. 1908. On the frictional resistance to the flow of air through a pipe. *Proc. R. Soc. London Ser. A* 80:114–39

Hasson, D. 1955. Streamline flow resistance in coils. *Res. Corresp.* 1:S1

Hausen, H. 1943. Darstellung des Warmeuberganges in Rohren durch verallgemeinerte Potenzbeziehungen. *Z. Ver. Dtsch. Ing., Beih. Verfahrenstech.* 4:91–98

Hawthorne, W. R. 1951. Secondary circulation in fluid flow. *Proc. R. Soc. London Ser. A* 206:374–87

Hayes, J. W., Hutton, J. F. 1972. Measurement of the normal stress coefficients of dilute polymer solutions from the flow in a curved tube. *Rheol. Acta* 11:89–92

Humphrey, J. A. C., Taylor, A. M. K., Whitelaw, J. H. 1977. Laminar flow in a square duct of strong curvature. *J. Fluid Mech.* 83:509–27

Humphrey, J. A. C., Whitelaw, J. H., Yee, G. 1981. Turbulent flow in a square duct with strong curvature. *J. Fluid Mech.* 103:443–63

Inaba, T., Murata, S. 1978. Pulsating laminar flow in a sinusoidally curved pipe. *Bull. JSME* 21:832–39

Itō, H. 1951. Theory on laminar flows through curved pipes of elliptic and rectangular cross-sections. *Rep. Inst. High Speed Mech., Tohoku Univ., Sendai, Jpn.*, Vol. 1, pp. 1–16

Itō, H. 1959. Friction factors for turbulent flow in curved pipes. *J. Basic Eng.* 81:123–34

Itō, H. 1960. Pressure losses in smooth pipe bends. *J. Basic Eng.* 82:131–43

Itō, H. 1969. Laminar flow in curved pipes. *Z. Angew. Math. Mech.* 49:653–63

Jain, R. K., Prabha, S. 1982. Gas-particulate flow through a curved pipe. *Z. Angew. Math. Mech.* 62:43–50

James, P. W. 1975. Unsteady elastico-viscous flow in a curved pipe. *Rheol. Acta* 14:679–87

Janssen, L. A. M., Hoogendoorn, C. J. 1978. Laminar convective heat transfer in helical coiled tubes. *Int. J. Heat Mass Transfer* 21:1179–1206

Jensen, M. K., Bergles, A. E. 1981. Critical heat flux in helically coiled tubes. *J. Heat Transfer* 103:661–66

Jensen, M. K., Bergles, A. E. 1982. Critical heat flux in helical coils with a circumferential heat flux tilt toward the outside surface. *Int. J. Heat Mass Transfer* 25:1383–95

Jones, D. T. 1967. *Some elastico-viscous flow problems.* PhD thesis. University Coll. of Wales, Aberystwyth

Jones, J. R. 1960. Flow of a non-Newtonian liquid in a curved pipe. *Q. J. Mech. Appl. Math.* 13:428–43

Joseph, B., Smith, E. P., Adler, R. J. 1975. Numerical treatment of laminar flow in helically coiled tubes of square cross-section. Part 1. Stationary helically coiled tubes. *AIChE J.* 21:965–74

Kaimal, M. R., Devanathan, R. 1980. Motion of a viscous fluid with suspended particles in a curved tube. *Int. J. Eng. Sci.* 18:847–54

Kalb, C. E., Seader, J. D. 1972. Heat and mass transfer phenomena for viscous flow in curved circular tubes. *Int. J. Heat Mass Transfer* 15:801–17. Errata in *Int. J. Heat Mass Transfer* 15:2680

Kalb, C. E., Seader, J. D. 1974. Fully developed viscous-flow heat transfer in curved circular tubes with uniform wall temperature. *AIChE J.* 20:340–46

Kamel, M. T., Kaloni, P. N. 1977. On the flow of a Cosserat fluid through a curved pipe. *Z. Angew. Math. Phys.* 28:551–76

Kubair, V., Kuloor, N. R. 1966. Heat transfer to Newtonian fluids in coiled pipes in laminar flow. *Int. J. Heat Mass Transfer* 9:63–75

Larrain, J., Bonilla, C. F. 1970. Theoretical analysis of pressure drop in the laminar flow of fluid in a coiled pipe. *Trans. Soc. Rheol.* 14:135–47

Lighthill, M. J. 1970. Turbulence. In *Osborne Reynolds and Engineering Science Today*, ed. D. M. McDowell, J. D. Jackson, pp. 83–146. New York: Barnes & Noble

Lighthill, M. J. 1975. *Mathematical Biofluiddynamics*. Philadelphia: Soc. Ind. Appl. Math. 281 pp.

Lin, J. J., Tarbell, J. M. 1980. An experimental and numerical study of periodic flow in a curved tube. *J. Fluid Mech.* 100:623–38

Lin, T.-S. 1972. *Laminar convective transport processes in strongly curved tubes*. PhD thesis. Clarkson Coll. Technol., Potsdam, N.Y. 193 pp.

Liu, N.-S. 1976. Finite-difference solution of the Navier-Stokes equations for incompressible three-dimensional internal flows. *Proc. 5th Int. Conf. Numer. Methods Fluid Dyn.*, pp. 330–35

Liu, N.-S. 1977. Developing flow in a curved pipe. *INSERM-Euromech 92* 71:53–64

Lyne, W. H. 1971. Unsteady viscous flow in a curved pipe. *J. Fluid Mech.* 45:13–31

Maddock, C., Lacey, P. M. C., Patrick, M. A. 1974. *The structure of two-phase flow in a curved pipe*. Presented at Symp. Multi-Phase Flow Syst., Glasgow, April 2–4. Preprints in *Inst. Chem. Eng. Symp. Ser.* No. 38: Multi-Phase Flow Systems, March 1974

Manlapaz, R. L., Churchill, S. W. 1980. Fully developed laminar flow in a helically coiled tube of finite pitch. *Chem. Eng. Commun.* 7:57–78

Mashelkar, R. A., Devarajan, G. V. 1976a. Secondary flows of non-Newtonian fluids: Part I—Laminar boundary layer flow of a generalized non-Newtonian fluid in a coiled tube. *Trans. Inst. Chem. Eng.* 54:100–7

Mashelkar, R. A., Devarajan, G. V. 1976b. Secondary flow of non-Newtonian fluids: Part II. Frictional losses in laminar flow of purely viscous and viscoelastic fluids through coiled tubes. *Trans. Inst. Chem. Eng.* 54:108–14

Masliyah, J. H. 1980. On laminar flow in curved semicircular ducts. *J. Fluid Mech.* 99:469–79

McConalogue, D. J. 1970. The effects of secondary flow on the laminar dispersion of an injected substance in a curved tube. *Proc. R. Soc. London Ser. A* 315:99–113

McConalogue, D. J., Srivastava, R. S. 1968. Motion of fluid in a curved tube. *Proc. R. Soc. London Ser. A* 307:37–53

McDonald, D. A. 1974. *Blood Flow in Arteries*. Baltimore: Williams & Wilkins. 496 pp. 2nd ed.

Mishra, P., Gupta, S. N. 1979. Momentum transfer in curved pipes. I. Newtonian fluids. *Ind. Eng. Chem. Process Des. Dev.* 18:130–37

Morgan, G. W., Kiely, J. P. 1954. Wave propagation in a viscous liquid contained in a flexible tube. *J. Acoust. Soc. Am.* 26:323–28

Mori, Y., Nakayama, W. 1965. Study on forced convective heat transfer in curved pipes (1st Report, Laminar region). *Int. J. Heat Mass Transfer* 8:67–82

Mori, Y., Nakayama, W. 1967. Study on forced convective heat transfer in curved pipes (3rd Report). *Int. J. Heat Mass Transfer* 10:681–95

Mori, Y., Uchida, Y. 1967. Study on forced convective heat transfer in curved square channel. *Trans. Jpn. Soc. Mech. Eng.* 33:1836–46

Morimoto, T., Ousaka, A. 1974. A study of air-water two phase flow in a curved pipe. The effects of a 180° bend on the flow in upstream and downstream side straight pipes in the case of horizontal lines. *Sci. Pap. Fac. Eng. Tokushima Univ. (Japan) No. 19*, pp. 57–73

Mullin, T., Greated, C. A. 1980a. Oscillatory flow in curved pipes. Part 1. The developing flow case. *J. Fluid Mech.* 98:383–95

Mullin, T., Greated, C. A. 1980b. Oscillatory flow in curved pipes. Part 2. The fully developed case. *J. Fluid Mech.* 98:397–416

Munson, B. R. 1975. Experimental results for oscillating flow in a curved pipe. *Phys. Fluids* 18:1607–9

Murata, S., Miyake, Y., Inaba, T. 1976. Laminar flow in a curved pipe with varying curvature. *J. Fluid Mech.* 73:735–52

Murata, S., Miyake, Y., Inaba, T., Ogawa, H. 1981. Laminar flow in a helically coiled pipe. *Bull. JSME* 24:355–62

Nandakumar, K., Masliyah, J. H. 1982. Bifurcation in steady laminar flow through curved tubes. *J. Fluid Mech.* 119:475–90

Nerem, R. M., Cornhill, J. F. 1980. The role of fluid mechanics in atherogenesis. *J. Biomech. Eng.* 102:181–89

Nigam, K. D. P., Vasudeva, K. 1976. Influence of curvature and pulsations on laminar dispersion. *Chem. Eng. Sci.* 31:835–37

Nunge, R. J., Adams, L. R. 1973. Reverse osmosis in laminar flow through curved tubes. *Desalination* 13:17–36

Nunge, R. J., Lin, T.-S. 1973. Laminar flow in strongly curved tubes. *AIChE J.* 19:1280–81

Nunge, R. J., Lin, T.-S., Gill, W. N. 1972. Laminar dispersion in curved tubes and channels. *J. Fluid Mech.* 51:363–83

Oliver, D. R., Asghar, S. M. 1976. Heat transfer to Newtonian and viscoelastic liquids during laminar flow in helical coils. *Trans. Inst. Chem. Eng.* 54:218–24

Olson, D. E. 1971. *Fluid mechanics relevant to respiration — flow within curved or elliptical tubes and bifurcating systems.* PhD thesis. Imperial Coll., London

Owhadi, A., Bell, K. J., Crain, B. Jr. 1968. Forced convection boiling inside helically coiled tubes. *Int. J. Heat Mass Transfer* 11:1779–93

Patankar, S. V., Pratap, V. S., Spalding, D. B. 1974. Prediction of laminar flow and heat transfer in helically coiled pipes. *J. Fluid Mech.* 62:539–51

Patel, I. C., Sirs, J. A. 1977. Indicator dilution measurements of flow parameters in curved tubes and branching networks. *Phys. Med. Biol.* 22:714–30

Pedley, T. J. 1980. *The Fluid Mechanics of Large Blood Vessels,* Chap. 4, pp. 160–234. Cambridge Univ. Press.

Prusa, J., Yao, L.-S. 1982. Numerical solution of fully-developed flow in heated curved tubes. *J. Fluid Mech.* 123:503–22

Rabadi, N. J., Simon, H. A., Chow, J. C. F. 1980. Numerical solution for fully developed, laminar pulsating flow in curved tubes. *Numer. Heat Transfer* 3:225–39

Raju, K. K., Rathna, S. L. 1970. Heat transfer for the flow of a power-law fluid in a curved pipe. *J. Indian Inst. Sci.* 52:34–47

Richardson, P. D., Tanishita, K., Galletti,

P. M. 1975. Mass transfer through tubes wound in serpentine shapes. *Lett. Heat Mass Transfer* 2:481–85

Riley, N., Dennis, S. C. R. 1976. *Flow in a curved pipe at high Dean numbers.* Presented at the Workshop on Viscous Interaction and Boundary-Layer Separation, Columbus, Ohio

Rushmore, W. L. 1975. *Theoretical investigation of curved pipe flows.* PhD thesis. State Univ. N.Y. at Buffalo. 160 pp.

Scarton, H. A., Shah, P. M., Tsapogas, M. J. 1977. Relationship of the spatial evolution of secondary flow in curved tubes to the aortic arch. In *Mechanics in Engineering,* pp. 111–31. Univ. Waterloo Press

Schilling, R., Simon, R. 1979. Berechnung der ausgebildeten Strömung in gekrümmten Kanälen mit rechteckigem Querschnitt. In *Recent Developments in Theoretical and Experimental Fluid Mechanics,* ed. U. Müller, K. G. Roesner, B. Schmidt, pp. 546–56. Berlin/Heidelberg: Springer-Verlag

Schlichting, H. 1979. *Boundary Layer Theory.* Transl. J. Kestin. New York: McGraw-Hill. 832 pp. 7th ed.

Schmidt, E. F. 1967. Wärmeübergang und Drukverlust in Rohrschlangen. *Chem. Ing. Tech.* 13:781–89

Seban, R. A., McLaughlin, E. F. 1963. Heat transfer in tube coils with laminar and turbulent flow. *Int. J. Heat Mass Transfer* 6:387–95

Sharma, H. G., Prakash, A. 1977. Flow of a second-order fluid in a curved pipe. *Indian J. Pure Appl. Math.* 8:546–57

Shaw, D. T., Rajendran, N. 1977. Diffusional deposition of airborne particles in curved bronchial airways. *J. Aerosol. Sci.* 8:191–97

Shchukin, V. K. 1969. Correlation of experimental data on heat transfer in curved pipes. *Therm. Eng.* 16:72–76

Simon, H. A., Chang, M. H., Chow, J. C. F. 1977. Heat transfer in curved tubes with pulsatile, fully developed, laminar flows. *J. Heat Transfer* 99:590–95

Singh, M. P. 1974. Entry flow in a curved pipe. *J. Fluid Mech.* 65:517–39

Singh, M. P., Sinha, P. C., Aggarwal, M. 1978. Flow in the entrance of the aorta. *J. Fluid Mech.* 87:97–120

Singh, R. P., Mishra, P. 1980. Friction factor for Newtonian and non-Newtonian fluid-flow in curved pipes. *J. Chem. Eng. Jpn.* 13:275–80

Singh, S. P. N., Bell, K. J. 1974. Laminar flow heat transfer in a helically-coiled tube. *Proc. 5th Int. Heat Transfer Conf., Tokyo,* 2:193–97

Smith, F. T. 1975. Pulsatile flow in curved pipes. *J. Fluid Mech.* 71:15–42

Smith, F. T. 1976a. Steady motion within a curved pipe. *Proc. R. Soc. London Ser. A* 347:345–70

Smith, F. T. 1976b. Fluid flow into a curved pipe. *Proc. R. Soc. London Ser. A* 351:71–87

Smol'skiy, B. M., Chekhol'skiy, A. S. 1978. Investigation of heat and mass transfer in condensation of water vapor from moist air in curved channels. *Heat Transfer — Sov. Res. (USA)* 10(4):162–69

Squire, H. B., Winter, K. G. 1949. The secondary flow in a cascade of aerofoils in a non-uniform stream. *R.A.E. Rep. Aero 2317*

Srinivasan, S., Tien, C. 1971. Reverse osmosis in a curved tubular membrane duct. *Desalination* 9:127–39

Srivastava, R. S. 1980. On the motion of a fluid in a curved pipe of elliptical cross-section. *Z. Angew. Math. Phys.* 31:297–303

Srivastava, R. S., Sarin, V. B. 1973a. Heat transfer effects for an elastico-viscous fluid in a curved pipe. *Indian J. Pure Appl. Math.* 4:437–48

Srivastava, R. S., Sarin, V. B. 1973b. Heat transfer effects for an elastico-viscous fluid in a curved pipe of elliptic cross-section. *Indian J. Pure Appl. Math.* 4:509–18

Stewartson, K., Simpson, C. J. 1982. On a singularity initiating a boundary-layer collision. *Q. J. Mech. Appl. Math.* 35:1–16

Stewartson, K., Cebeci, T., Chang, K. C. 1980. A boundary-layer collision in a curved duct. *Q. J. Mech. Appl. Math.* 33:59–75

Talbot, L., Gong, K. O. 1982. Pulsatile entrance flow in a curved pipe. *J. Fluid Mech.* In press

Talbot, L., Wong, S. J. 1982. A note on boundary layer collision in a curved pipe. *J. Fluid Mech.* 122:505–10

Tanishita, K., Nakano, K., Sakurai, Y., Hosokawa, T., Richardson, P. D., Galletti, P. M. 1978. Compact oxygenator design with curved tubes wound in weaving patterns. *Trans. Am. Soc. Artif. Intern. Organs.* 24:327–31

Tarbell, J. M., Samuels, M. R. 1973. Momentum and heat transfer in helical coils. *Chem. Eng. J.* 5:117–27

Taylor, A. K. M. P., Whitelaw, J. H., Yianneskis, M. 1982. Curved ducts with strong secondary motion: velocity measurements of developing laminar and turbulent flow. *J. Fluids Eng.* In press

Taylor, G. I. 1929. The criterion for turbulence in curved pipes. *Proc. R. Soc. London Ser. A* 124:243–49

Texon, M. 1960. The hemodynamic concept of atherosclerosis. *Bull. NY Acad. Med.* 36:263–74

Thomas, R. H., Walters, K. 1963. On the flow of an elastico-viscous liquid in a curved pipe under a pressure gradient. *J. Fluid Mech.* 16:228–42

Thomas, R. H., Walters, K. 1965. On the flow of an elastico-viscous fluid in a curved pipe of elliptic cross-section under a pressure gradient. *J. Fluid Mech.* 21:173–82

Thomson, J. 1876. On the origin of windings of rivers in alluvial plains, with remarks on the flow of water round bends in pipes. *Proc. R. Soc. London Ser. A* 25:5–8

Thomson, J. 1877. Experimental demonstration in respect to the origin of windings of rivers in alluvial plains, and to the mode of flow of water round bends of pipes. *Proc. R. Soc. London Ser. A.* 26:356–57

Topakoglu, H. C. 1967. Steady laminar flows of an incompressible viscous fluid in curved pipes. *J. Math. Mech.* 16:1321–38

Truesdell, L. C., Adler, R. J. 1970. Numerical treatment of fully developed laminar flow in helically coiled tubes. *AIChE J.* 16:1010–15

Tsang, H. Y., James, D. F. 1980. Reduction of secondary motion in curved tubes by polymer additives. *J. Rheol.* 24:955–56

Van Dyke, M. 1978. Extended Stokes series: laminar flow through a loosely coiled pipe. *J. Fluid Mech.* 86:129–45

Wang, C. Y. 1981. On the low-Reynolds-number flow in a helical pipe. *J. Fluid Mech.* 108:185–94

Ward-Smith, A. J. 1968. *Some aspects of fluid flow in ducts.* D. Phil. thesis. Univ. Oxford

Ward-Smith, A. J. 1980. *Internal Fluid Flow,* pp. 278–98. Oxford Univ. Press

Weissman, M. H., Mockros, L. F. 1968. Gas transfer to blood flowing in coiled circular tubes. *ASCE Proc., Eng. Mech. Div. J.* 94:857–72

Whalley, P. B. 1980. Air-water two-phase flow in a helically coiled tube. *Int. J. Multiphase Flow* 6:345–56

White, C. M. 1929. Streamline flow through curved pipes. *Proc. R. Soc. London Ser. A* 123:645–63

Williams, G. S., Hubbell, C. W., Fenkell, G. H. 1902. Experiments at Detroit, Mich., on the effect of curvature upon the flow of water in pipes. *Trans. ASCE* 47:1–196

Womersley, J. R. 1957. The mathematical analysis of the arterial circulation in a state of oscillatory motion. *Tech. Rep. WADC-TR 56-614*, Wright Air Dev. Cent.

Yalamov, Y. I., Muradyan, S. M. 1980. Trajectories and capture coefficient of aerosol-particles in curved channels. *Zh. Tekh. Fiz.* 50:400–2

Yao, L.-S., Berger, S. A. 1975. Entry flow in a curved pipe. *J. Fluid Mech.* 67:177–96

Yao, L.-S., Berger, S. A. 1978. Flow in heated curved pipes. *J. Fluid Mech.* 88:339–54

Yeung, W.-S. 1979. Erosion in a curved pipe. *Wear* 55:91–106

Yeung, W.-S. 1980. Laminar boundary-layer flow near the entry of a curved circular pipe. *J. Appl. Mech.* 47:697–702

Zalosh, R. G., Nelson, W. G. 1973. Pulsating flow in a curved tube. *J. Fluid Mech.* 59:693–705

Zapryanov, Z., Christov, Ch. 1977. Fully developed flow of viscous incompressible fluid in toroidal tube with circular cross-section. *Theor. Appl. Mech.* 8:11–17

Zapryanov, Z., Matakiev, V. 1980. An exact solution of the problem of unsteady fully-developed viscous flow in slightly curved porous tube. *Arch. Mech.* (*Poland*) 32:461–74

Zapryanov, Z., Christov, C., Toshev, E. 1980. Fully developed laminar flow and heat transfer in curved tubes. *Int. J. Heat Mass Transfer* 23:873–80

AUTHOR INDEX

(Names appearing in capital letters indicate authors of chapters in this volume.)

513

520   AUTHOR INDEX

SUBJECT INDEX

A

Accretion/erosion
two-phase flow and, 285
Acoustic-gravity waves, 321
Active transport
phloem flow and, 29
Adiabatic gaseous flames
instability region of, 191
Agglomeration
two-phase flow and, 285
Air
wave breaking and, 172
Airborne-powder avalanches,
47–48, 58–69
cloud formation mechanisms
and, 60–62
gravity current structure of,
62–64
impact force of, 69
large density-difference effects
and, 66–67
low-density clouds and,
64–66
sedimentation conditions
and, 68–69
snow-incorporation
mechanisms and, 67–68
Aircraft
autorotation and, 141–43
Airfoils
autorotation and, 125–27,
130–31
BoAR 80, 236–37
experimental testing of,
232–33
fluid mechanics and, 227–32
instability range of, 235
laminar separation bubbles
and, 228–32
Lissaman-Hibbs, 8025, 237
low-Reynolds-number,
223–39
scale effect and, 223
special-purpose, 236–38
steady flow characteristics of,
80
theoretical design of, 233–36
turbulators and, 232
Alfven waves, 333–34
Algebraic stress models
turbulent wall jets and,
453–57
Alternate-crests instability
wave breaking and, 157
Angular momentum
horizontal-axis wind turbines
and, 83–84
Anisotropic turbulence

homogeneous turbulence and,
209–12
Apoplasm
flow of water in, 35
Arnold diffusion
vortex motion and, 382
Atmospheres
magneto-atmospheric waves
and, 324–26
Autorotation, 123–45
applications of, 126–27
at arbitrary angle to the flow,
140–44
definition of, 124–27
parallel to the flow, 127–31
perpendicular to the flow,
132–39
vortex-induced vibration and,
144–45
Avalanche motion, 47–73
Avalanche wind, 72–73
Avalanches
airborne-powder, 47–48
cloud-formation
mechanisms and,
60–62
gravity current structure
of, 62–64
impact force of, 69
large density-difference
effects and, 66–67
low-density clouds and,
64–66
sedimentation conditions
and, 68–69
snow-incorporation
mechanisms and,
67–68
channelled, 49
classification of, 49–51
dense-snow, 47–48, 52–58
impact force of, 69
numerical models for,
57–58
properties of, 53–55
run-out conditions and, 58
velocity of, 55–56
dry-snow, 49
sequential photographs of,
52–53
ground, 49
impact force of, 69–73
loose-snow, 49
mixed, 49
open-slope, 49
release mechanisms and,
51–52
slab, 49
airborne-powder avalanche

formation and, 60
crown fracture of, 50
surface, 49
wet-snow, 49

B

Ballistics, 18–25
Batchelor-Prandtl theorem, 346
BBGKY hierarchy
distribution functions in
identical vortices and,
372
Benjamin-Feir instability
wave breaking and, 157
Bifurcation theory
vortex motion and, 352
Blast waves
apparatus for measuring
travel times of, 11
density profiles of, 13
intersection patterns of, 15
Schlieren photograph of, 18
BoAR 80 airfoil, 236–37
Boussinesq fluids
magneto-gravity waves and,
321
Boussinesq theory, 188
Boussinesq viscosity
turbulent wall jets and,
448–52
Bow waves
supersonic projectiles and, 19
Brunt-Väisälä frequency
magneto-atmospheric waves
and, 331
Burgers' equation
vortex motion and, 346–48
Butane-air flames
cellularly stable, 182

C

Carrier-Greenspan solution,
152
Cauchy stress tensor
decomposition of, 242
Cavitation
liquid-solid impact and, 102
Cellular flames
thermal-expansion-induced,
195–96
Channelled avalanches, 49
Cloud-in-cell technique
point vortices and, 374
Coiled pipe
flow in, 465–78
Cole-Hopf transformation
Burgers' equation and, 348
Collapsing breakers, 150

523

CUMULATIVE INDEXES

CONTRIBUTING AUTHORS, VOLUMES 11-15

529

CHAPTER TITLES, VOLUMES 11–15

Please list the volumes you wish to order by volume number. If you wish a standing order (the latest volume sent to you automatically each year), indicate volume number to begin order. Volumes not yet published will be shipped in month and year indicated. All prices subject to change without notice.

ANNUAL REVIEW SERIES

Annual Review of **ANTHROPOLOGY**		Prices Postpaid per volume USA/elsewhere	Regular Order Please send:	Standing Order Begin with:
			Vol. number	Vol. number
Vols. 1-10	(1972-1981)	$20.00/$21.00		
Vol. 11	(1982)	$22.00/$25.00		
Vol. 12	(avail. Oct. 1983)	$27.00/$30.00	Vol(s). _____	Vol. _____

Annual Review of **ASTRONOMY AND ASTROPHYSICS**

Vols. 1-19	(1963-1981)	$20.00/$21.00		
Vol. 20	(1982)	$22.00/$25.00		
Vol. 21	(avail. Sept. 1983)	$44.00/$47.00	Vol(s). _____	Vol. _____

Annual Review of **BIOCHEMISTRY**

$\begin{bmatrix} \text{Vols. 28-48} & \$18.00/\$18.50 \\ \text{Price effective through } 12/31/82 \end{bmatrix}$

Vols. 28-50	(1959-1981)	$21.00/$22.00		
Vol. 51	(1982)	$23.00/$26.00		
Vol. 52	(avail. July 1983)	$29.00/$32.00	Vol(s). _____	Vol. _____

Annual Review of **BIOPHYSICS AND BIOENGINEERING**

Vols. 1-10	(1972-1981)	$20.00/$21.00		
Vol. 11	(1982)	$22.00/$25.00		
Vol. 12	(avail. June 1983)	$47.00/$50.00	Vol(s). _____	Vol. _____

Annual Review of **EARTH AND PLANETARY SCIENCES**

Vols. 1-9	(1973-1981)	$20.00/$21.00		
Vol. 10	(1982)	$22.00/$25.00		
Vol. 11	(avail. May 1983)	$44.00/$47.00	Vol(s). _____	Vol. _____

Annual Review of **ECOLOGY AND SYSTEMATICS**

Vols. 1-12	(1970-1981)	$20.00/$21.00		
Vol. 13	(1982)	$22.00/$25.00		
Vol. 14	(avail. Nov. 1983)	$27.00/$30.00	Vol(s). _____	Vol. _____

1

Annual Review of **ENERGY**		Prices Postpaid per volume USA/elsewhere	Regular Order Please send:	Standing Order Begin with:
			Vol. number	Vol. number
Vols. 1-6	(1976-1981)	$20.00/$21.00		
Vol. 7	(1982)	$22.00/$25.00		
Vol. 8	(avail. Oct. 1983)	$56.00/$59.00	Vol(s). _____	Vol. _____

Annual Review of ENTOMOLOGY

Vols. 7-26	(1962-1981)	$20.00/$21.00		
Vol. 27	(1982)	$22.00/$25.00		
Vol. 28	(avail. Jan. 1983)	$27.00/$30.00	Vol(s). _____	Vol. _____

Annual Review of FLUID MECHANICS

Vols. 1-13	(1969-1981)	$20.00/$21.00		
Vol. 14	(1982)	$22.00/$25.00		
Vol. 15	(avail. Jan. 1983)	$28.00/$31.00	Vol(s). _____	Vol. _____

Annual Review of GENETICS

Vols. 1-15	(1967-1981)	$20.00/$21.00		
Vol. 16	(1982)	$22.00/$25.00		
Vol. 17	(avail. Dec. 1983)	$27.00/$30.00	Vol(s). _____	Vol. _____

Annual Review of IMMUNOLOGY — New Series 1983

Vol. 1	(avail. April 1983)	$27.00/$30.00	Vol(s). _____	Vol. _____

Annual Review of MATERIALS SCIENCE

Vols. 1-11	(1971-1981)	$20.00/$21.00		
Vol. 12	(1982)	$22.00/$25.00		
Vol. 13	(avail. Aug. 1983)	$64.00/$67.00	Vol(s). _____	Vol. _____

Annual Review of MEDICINE: Selected Topics in the Clinical Sciences

Vols. 1-3, 5-15	(1950-1952; 1954-1964)	$20.00/$21.00		
Vols. 17-32	(1966-1981)	$20.00/$21.00		
Vol. 33	(1982)	$22.00/$25.00		
Vol. 34	(avail. April 1983)	$27.00/$30.00	Vol(s). _____	Vol. _____

Annual Review of MICROBIOLOGY

Vols. 15-35	(1961-1981)	$20.00/$21.00		
Vol. 36	(1982)	$22.00/$25.00		
Vol. 37	(avail. Oct. 1983)	$27.00/$30.00	Vol(s). _____	Vol. _____

Annual Review of NEUROSCIENCE

Vols. 1-4	(1978-1981)	$20.00/$21.00		
Vol. 5	(1982)	$22.00/$25.00		
Vol. 6	(avail. March 1983)............	$27.00/$30.00	Vol(s). _____	Vol. _____

SEE ORDERING INFORMATION ON PAGE 4.